Ettore Majorana: Notes on Theoretical Physics

Fundamental Theories of Physics

An International Book Series on The Fundamental Theories of Physics: Their Clarification, Development and Application

Volume 133

Ettore Majorana: Notes on Theoretical Physics

Edited by

Salvatore Esposito
University of Naples "Federico II", Naples, Italy

Ettore Majorana Jr
Italian National Institute for Nuclear Physics, Site of Pisa, Italy

Alwyn van der Merwe
University of Denver, Denver, Colorado, U.S.A.

and

Erasmo Recami
University of Bergamo, Dalmine (BG), Italy

KLUWER ACADEMIC PUBLISHERS
DORDRECHT / BOSTON / LONDON

A C.I.P. Catalogue record for this book is available from the Library of Congress.

ISBN 1-4020-1649-2

Published by Kluwer Academic Publishers,
P.O. Box 17, 3300 AA Dordrecht, The Netherlands.

Sold and distributed in North, Central and South America
by Kluwer Academic Publishers,
101 Philip Drive, Norwell, MA 02061, U.S.A.

In all other countries, sold and distributed
by Kluwer Academic Publishers,
P.O. Box 322, 3300 AH Dordrecht, The Netherlands.

Printed on acid-free paper

Printed in the Netherlands.

... les savants croient qu'il y a une hiérarchie des faits et qu'on peut faire entre eux en choix judicieux; ils ont raison, puisque sans cela il n'y aurait pas de science et que la science existe. Il suffit d'ouvrir les yeux pour voir que les conquêtes de l'industrie qui ont enrichi tant d'hommes pratiques n'auraient jamais ou le jour si ces hommes pratiques avaient seuls existé, et s'ils n'avaient été devancés par des fous désinteressès qui sont morts pauvres, qui ne pensaient jamais à l'utile, et qui pourtant avaient un autre guide que leur caprice. C'est que, comme l'a dit Mach, ces fous ont économisé à leurs successeurs la peine de penser.

Henri Poincaré, *Science et Méthode*, 1927 [1]

[1] This paragraph has been extracted from a book belonging to the Ettore Majorana's library.

CONTENTS

Preface xiii

References xxv

Volumetto I 1

1	Electric potential	1
2	Retarded potential	4
3	Interaction energy of two electric or magnetic charge distributions	5
4	Skin effect in homogeneous cylindrical electric conductors	6
5	Thermodynamics of the thermoelectric cells	10
6	Energy of an isolated conductor	11
7	Attraction between masses which are far apart	12
8	Formulae	14
9	Electric lines	15
10	Density of a spherical mass distribution	18
11	Limit skin effect	18
12	Limit skin effect for simply-shaped conductors. Hints for arbitrary shapes	23
	12.1 Elliptic cross sections	23
	12.2 Effect of the irregularities of the boundary	24
13	Hysteresis in magnetic conductors in the limit skin effect regime	26
14	Field produced by a circular and homogeneous distribution of charges in its own plane	28
15	Field produced by a circular charge current in a plane	29
16	Weak skin effect in conductors, with an elliptic cross section, having the same magnetic permeability as the surrounding medium	30
17	Oscillating discharges in capacitors	31
18	Self-induction of a very long circular coil with many turns	32
19	Energy of a uniform circular distribution of electric or magnetic charges	33
20	Self-induction in a rectilinear coil with finite length	36
21	Mean distances of volume, surface, or line elements	38
22	Evaluation of some series	39
23	Self-induction of a finite length rectilinear coil with circular cross section and a finite (small) number of turns	41
24	Variation of the self-induction coefficient due to the skin effect	43
25	Mean error in estimating the probability of an event through a finite number of trials	45
26	Unbalance of a pure three-phase system	46
27	Table for the computation of $x!$	47
28	Influence of a magnetic field on the melting point	48
29	Specific heat of an oscillator	50
30	Do children of the same parents tend to be of the same sex?	52
31	Heat propagation from a certain cross section along an infinite-length bar endowed with another cross section acting as a heat well. A similarity with the crickets	53
32	Combinations	55
33	Energy and specific heat of a rotator	56

34 Gravitational attraction of an ellipsoid 58
35 Special cases: prolate ellipsoid and spheroid 63
36 Equilibrium of a rotating fluid 66
37 Definite integrals 69
38 Heat propagation in an isotropic and homogeneous medium 71
38.1 One-dimensional propagation 71
39 Conformal transformations 74
40 Wave mechanics of a mass point in a conservative field. Variational approach 77
41 Electromagnetic mass of the electron 78
42 Legendre polynomials 82
43 ∇^2 in spherical coordinates 83

Volumetto II 85

1 ∇^2 in cylindrical coordinates 85
2 Expansion of a harmonic function in a plane 85
3 Quantization of a linear harmonic oscillator 87
4 Diagonalization of a matrix 90
5 Wave quantization of a point particle that is attracted by a constant force towards a perfectly elastic wall 93
6 Relativistic hamiltonian for the motion of an electron 97
7 The Fermi function 101
8 The interatomic potential without statistics 105
9 Application of the Fermi potential 108
10 Statistical behavior of the fundamental terms in neutral atoms 111
11 Numbers to the fifth power 112
12 Diatomic molecule with identical nuclei 113
13 Numbers to the sixth power 114
14 Numbers to the seventh power 115
15 Second approximation for the potential inside the atom 116
16 Atomic polarizability 117
17 Fourier expansions and integrals 118
18 Blackbody 120
19 Radiation Theory (Part 1) 121
20 Moment of inertia of the Earth 127
21 Radiation Theory (Part 2) 131
22 About matrices 133
23 Radiation Theory (Part 3) 135
24 Perturbed keplerian motion in a plane 137
25 Radiation Theory (Part 4) 144
26 Definite integrals 146
27 Series expansions 148
28 Radiation Theory (Part 5): free electron scattering 150
29 De Broglie waves 154
30 $e^2 \simeq hc$? 155
31 The equation $y'' + Py = 0$ 156
32 Indeterminacy of vector and scalar potentials 160
33 On the spontaneous ionization of a hydrogen atom placed in a high potential region 161
34 Scattering of an α particle by a radioactive nucleus 177

35 Retarded potential 188
36 The equation $y'' = xy$ 190
37 Resonance degeneracy for many-electron atoms 190
38 Various formulae 192
38.1 Schwarz formula 192
38.2 Maximum value of random variables 193
38.3 Binomial coefficients 195
38.4 Expansion of $1/(1-x)^n$ 196
38.5 Relations between the binomial coefficients 197
38.6 Mean values of r^n between concentric spherical surfaces 198

Volumetto III 205

1 Evaluation of some series 205
2 The equation $\Box H = r$ 207
3 Equilibrium of a rotating heterogeneous liquid body (Clairaut problem) 212
4 Determination of a function from its moments 226
5 Probability curves 235
6 Evaluation of the integral $\int_0^{\pi/2} \frac{\sin kx}{\sin x}\, \mathrm{d}x$ 237
7 Infinite products 241
8 Bernoulli numbers and polynomials 242
9 Poisson brackets 243
10 Elementary physical quantities 248
11 "Chasing the dog" 250
12 Statistical potential in molecules 252
13 The group of proper unitary transformations in two variables 255
14 Exchange relations for infinitesimal transformations in the representations of continuous groups 265
15 Empirical relations for a two-electron atom 267
16 The group of rotations $O(3)$ 271
17 The Lorentz group 276
18 Dirac matrices and the Lorentz group 279
19 The spinning electron 287
20 Characters of $\mathcal{D}_j$ and reduction of $\mathcal{D}_j \times \mathcal{D}'_j$ 298
21 Intensity and selection rules for a central field 301
22 The anomalous Zeeman effect (according to the Dirac theory) 307
23 Complete sets of first-order differential equations 311

Volumetto IV 319

1 Connection between the susceptibility and the electric moment of an atom in its ground state 319
2 Ionization probability for a hydrogen atom in an electric field 322
3 Expansion of Legendre polynomials in the interval $-1 \leq x \leq 1$ 329
4 Multiplication rules for Legendre polynomials 329
5 Green functions for the differential equation $y'' + (2/x - 1)\, y + \phi(x) = 0$ 330

6 On the series expansion of the integral logarithm function 333
7 Fundamental characters of the group of permutations of f objects 335
8 Expansion of a plane wave in spherical harmonics 339
9 The Rutherford formula deduced from classical mechanics 342
10 The Rutherford formula deduced as a first approximation to the Born method 345
11 The Laplace equation 348
12 Polarization forces between hydrogen atoms 350
13 Integral representation of the Bessel functions 352
14 Cubic symmetry 355
15 Formulae 358
16 Plane waves in the Dirac theory 359
17 Improper operators 369
18 Integral representation of hydrogen eigenfunctions 372
19 Deflection of an alpha ray induced by a heavy nucleus (Classical Mechanics) 374
20 Scattering from a potential of the form $a/r - b/r^2$ 375
21 The set of orthogonal functions defined by the equation $y_a'' = (x-a)y_a$ 377
22 Fourier integral expansions 379
23 Circular integrals 381
24 Oscillation frequencies of ammonia 382
25 Spherical functions with spin one 385
26 Scattering of fast elections: relativistic Born method 396
27 Frequently used atomic quantities 402
28 Quasi-stationary states 403
29 Spherical functions with spin (II) 417

Volumetto V 419

1 Representations of the Lorentz group 419
2 Proton−neutron scattering 425
3 Zeros of half-order Bessel functions 427
4 Statistics and thermodynamics 428
4.1 Entropy of a system in equilibrium 428
4.2 Perfect gases 429
4.3 Monoatomic gas 430
4.4 Diatomic gas 430
4.5 Numerical expressions for the entropy of a gas 433
4.6 Free energy of diatomic gases 434
5 Frequently used polynomials 435
5.1 Legendre polynomials 435
6 Spinor transformations 435
7 Spherical functions with spin 1/2 441
8 Infinite-dimensional unitary representations of the Lorentz group 446
9 The equation $(\Box H + \lambda)A = p$ 450
10 Relevant formulas for the atomic eigenfunctions 454
11 Classical theory of multipole radiation 456
12 Hydrogen eigenfunctions 465

Selected facsimile pages 469

Index 481

PREFACE

HISTORICAL PRELUDE

Ettore Majorana's fame solidly rests on testimonies like the following, from the evocative pen of Giuseppe Cocconi. At the request of Edoardo Amaldi, he wrote from CERN (July 18, 1965):

"In January 1938, after having just graduated, I was invited, essentially by you, to come to the Institute of Physics at the University in Rome for six months as a teaching assistant, and once I was there I would have the good fortune of joining Fermi, Bernardini (who had been given a chair at Camerino a few months earlier) and Ageno (he, too, a new graduate), in the research of the products of disintegration of μ "mesons" (at that time called mesotrons or yukons), which are produced by cosmic rays [...]

"It was actually while I was staying with Fermi in the small laboratory on the second floor, absorbed in our work, with Fermi working with a piece of Wilson's chamber (which would help to reveal mesons at the end of their range) on a lathe and me constructing a jalopy for the illumination of the chamber, using the flash produced by the explosion of an aluminum ribbon shortcircuited on a battery, that Ettore Majorana came in search of Fermi. I was introduced to him and we exchanged few words. A dark face. And that was it. An easily forgettable experience if, after a few weeks while I was still with Fermi in that same workshop, news of Ettore Majorana's disappearance in Naples had not arrived. I remember that Fermi busied himself with telephoning around until, after some days, he had the impression that Ettore would never be found.

"It was then that Fermi, trying to make me understand the significance of this loss, expressed himself in quite a peculiar way; he who was so objectively harsh when judging people. And so, at this point, I would like to repeat his words, just as I can still hear them ringing in my memory: 'Because, you see, in the world there are various categories of scientists: people of a secondary or tertiary standing, who do their best but do not go very far. There are also those of high standing, who come to discoveries of great importance, fundamental for the development of science' (and here I had the impression that he placed himself in that category). 'But then there are geniuses like Galileo and Newton. Well, Ettore was one of them. Majorana had what no one else in the world had [...]'"

And, with first-hand knowledge, Bruno Pontecorvo, adds: "Some time after his entry into Fermi's group, Majorana already possessed such an erudition and had reached such a high level of comprehension of physics that he was able to speak on the same level with Fermi about scientific problems. Fermi himself held him to be the greatest theoretical physicist of our time. He often was astounded [...]. I remember exactly these words that Fermi spoke: 'If a problem has already been proposed, no one in the world can resolve it better than Majorana.' " (See also (Pontecorvo, 1972).)

Ettore Majorana disappeared rather mysteriously on March 26, 1938, and was never seen again (Recami, 1991). The myth of his "disappearance" has contributed to nothing more than the notoriety he was entitled to, for being a true genius and a genius well ahead of his time.

In this volume we are finally publishing his notebooks or *Volumetti*, which comprise his study notes written in Rome between 1927, when he abandoned his studies in engineering to take up physics, and 1931. Those manuscripts are a paragon not only of order, based on argument and even supplied with an index, but also of conciseness, essentiality and originality; so much so that the notebooks can be regarded as an excellent modern text of theoretical physics, even after more than seventy years, and a "gold-mine" of seminal new theoretical, physical, and mathematical ideas and hints, quite stimulating and useful for modern research.

Let us recall that Majorana, after having switched to physics at the beginning of 1928, graduated with Fermi on July 6, 1929, and went on to collaborate with the famous group created by Enrico Fermi and Franco Rasetti (at the start with O. M. Corbino's important help); a theoretical subdivision of which was formed mainly (in the order of their entrance into the Institute) by Ettore Majorana, Gian Carlo Wick, Giulio Racah, Giovanni Gentile Jr., Ugo Fano, Bruno Ferretti, and Piero Caldirola. The members of the experimental subgroup were: Emilio Segré, Edoardo Amaldi, Bruno Pontecorvo, Eugenio Fubini, Mario Ageno, Giuseppe Cocconi, along with the chemist Oscar D'Agostino. Afterwards, Majorana qualified for university teaching of theoretical physics ("Libera Docenza") on November 12, 1932; spent about six months in Leipzig with W. Heisenberg during 1933; and then, for some unknown reasons, stopped participating in the activities of Fermi's group. He even ceased publishing the results of his research, except for his paper "Teoria simmetrica dell'elettrone e del positrone," which (ready since 1933) Majorana was persuaded by his colleagues to remove from a drawer and publish just prior to the 1937 Italian national competition for three full-professorships.

With respect to the last point, let us recall that in 1937 there were numerous Italian competitors for these posts, and many of them were of exceptional caliber; above all: Ettore Majorana, Giulio Racah, Gian Carlo Wick, and Giovanni Gentile Jr. (the son of the famous philosopher bearing the same name, and the inventor of "parastatistics" in quantum mechanics). The judging committee was chaired by E. Fermi and had as members E. Persico, G. Polvani, A. Carrelli, and O. Lazzarino. On the recommendation of the judging committee, the Italian Minister of National Education installed Majorana as professor of theoretical physics at Naples University because of his "great and well-deserved fame," independently of the competition itself; actually, "the Commission hesitated to apply the normal university competition procedures to him." The attached report on the scientific activities of Ettore Majorana, sent to the minister by the committee, stated:

"Without listing his works, all of which are highly notable both for their originality of the methods utilized as well as for the importance of the achieved results, we limit ourselves to the following:

"In modern nuclear theories, the contribution made by this researcher to the introduction of the forces called "Majorana forces" is universally recognized as the one, among the most fundamental, that permits us to theoretically comprehend the reasons for nuclear stability. The work of Majorana today serves as a basis for the most important research in this field.

"In atomic physics, the merit of having resolved some of the most intricate questions on the structure of spectra through simple and elegant considerations of symmetry is due to Majorana.

"Lastly, he devised a brilliant method that permits us to treat the positive and negative electron in a symmetrical way, finally eliminating the necessity to rely on the extremely artificial and unsatisfactory hypothesis of an infinitely large electrical charge diffused in space, a question that had been tackled in vain by many other scholars."

One of the most important works of Ettore Majorana, the one that introduces his "infinite-components equation" was not mentioned, since it had not yet been understood. It is interesting to note, however, that the proper light was shed on his theory of electron and anti-electron symmetry (today climaxing in its application to neutrinos and anti-neutrinos) and on his resulting ability to eliminate the hypothesis known as the "Dirac sea," a hypothesis that was defined as "extremely artificial and unsatisfactory," despite the fact that in general it had been uncritically accepted.

The details of Majorana and Fermi's first meeting were narrated by E. Segré (Segré, 1971): "The first important work written by Fermi in Rome ['Su alcune proprietà statistiche dell'atomo' (On certain statistical properties of the atom)] is today known as the Thomas-Fermi method.... When Fermi found that he needed the solution to a non-linear differential equation characterized by unusual boundary conditions in order to proceed, in a week of assiduous work with his usual energy, he calculated the solution with a little hand calculator. Majorana, who had entered the Institute just a short time earlier and who was always very skeptical, decided that Fermi's numeric solution probably was wrong and that it would have been better to verify it. He went home, transformed Fermi's original equation into a Riccati equation, and resolved it without the aid of any calculator, utilizing his extraordinary aptitude for numeric calculation. When he returned to the Institute and skeptically compared the little piece of paper on which he had written his results to Fermi's notebook, and found that their results coincided exactly, he could not hide his amazement." We have indulged in the foregoing anecdote since the pages on which Majorana solved Fermi's differential equation have in the end been found, and it has been shown recently (Esposito, 2002) that he actually followed two independent (and quite original) paths to the same mathematical result, one of them leading to an Abel, rather than a Riccati, equation.

ETTORE MAJORANA'S PUBLISHED PAPERS

Majorana published few scientific articles: nine, actually, besides his sociology paper entitled "Il valore delle leggi statistiche nella fisica e nelle scienze sociali" (The value of statistical laws in physics and the social sciences), which was however published not by Majorana but (posthumously) by G. Gentile Jr., in *Scientia* [**36** (1942) 55-56]. We already know that Majorana switched from engineering to physics in 1928 (the year in which he published his first article, written in collaboration with his friend Gentile) and then went on to publish his works in theoretical physics only for a very few years, practically only until 1933. Nevertheless, even his *published* works are a mine of ideas and techniques of theoretical physics that still remains partially unexplored. Let us list his nine published articles:

(1) "Sullo sdoppiamento dei termini Roentgen ottici a causa dell'elettrone rotante e sulla intensità delle righe del Cesio," in collaboration with Giovanni Gentile Jr., *Rendiconti Accademia Lincei* **8** (1928) 229-233.

(2) "Sulla formazione dello ione molecolare di He," *Nuovo Cimento* **8** (1931) 22-28.

(3) "I presunti termini anomali dell'Elio," *Nuovo Cimento* **8** (1931) 78-83.

(4) "Reazione pseudopolare fra atomi di Idrogeno," *Rendiconti Accademia Lincei* **13** (1931) 58-61.

(5) "Teoria dei tripletti *P'* incompleti," *Nuovo Cimento* **8** (1931) 107-113.

(6) "Atomi orientati in campo magnetico variabile," *Nuovo Cimento* **9** (1932) 43-50.

(7) "Teoria relativistica di particelle con momento intrinseco arbitrario," *Nuovo Cimento* **9** (1932) 335-344.

(8) "Über die Kerntheorie," *Zeitschrift für Physik* **82** (1933) 137-145; "Sulla teoria dei nuclei," *La Ricerca Scientifica* **4**(1) (1933) 559-565.

(9) "Teoria simmetrica dell'elettrone e del positrone," *Nuovo Cimento* **14** (1937) 171-184.

The first papers, written between 1928 and 1931, concern atomic and molecular physics: mainly questions of atomic spectroscopy or chemical bonds (within quantum mechanics, of course). As E. Amaldi has written (Amaldi, 1966 and 1986), an in-depth examination of these works leaves one struck by their superb quality: They reveal both a deep knowledge of the experimental data, even in the minutest detail, and an uncommon ease, without equal at that time, in the use of the symmetry properties of the quantum states in order to qualitatively simplify problems and choose the most suitable method for their quantitative resolution. Among the first papers, "Atomi orientati in campo magnetico variabile" (Atoms oriented in a variable magnetic field) deserves special mention. It is in this article, famous among atomic physicists, that the effect now known as the *Majorana-Brossel effect* is introduced. In it, Majorana predicts and calculates the modification of the spectral line shape due to an oscillating magnetic field. This work has also remained a classic in the treatment of non-adiabatic spin-flip. Its results —once generalized, as suggested by Majorana himself, by Rabi in 1937 and by Bloch and Rabi in 1945— established the theoretical basis for the experimental method used to reverse the spin also of neutrons by a radio-frequency field, a method that is still practiced today, for example, in all polarized-neutron spectrometers. The Majorana paper introduces moreover the so-called *Majorana sphere* (to represent spinors by a set of points on the surface of a sphere), as noted not long ago by R. Penrose and others (Penrose, 1987, 1993 and 1996).

Majorana's last three articles are all of such importance that none of them can be set aside without comment.

The article "Teoria relativistica di particelle con momento intrinseco arbitrario" (Relativistic theory of particles with arbitrary spin) is a typical example of a work that is so far ahead of its time that it became understood and evaluated in depth only many years later. Around 1932 it was commonly thought that one could write relativistic quantum equations only in the case of particles with zero or half spin. Convinced of the contrary, Majorana —as we know from his manuscripts— began constructing suitable quantum-relativistic equations (Mignani et al., 1974) for higher spin values (one, three-halves, etc.); and he even devised a method for writing the equation for a generic spin-value. But still he published nothing, until he discovered that one could write a single equation to cover an infinite series of cases, that is, an entire infinite family of particles of arbitrary spin (even if at that time the known particles could be counted on one hand). In order to implement his programme with these "infinite components" equations, Majorana invented a technique for the representation of a group several years before Eugene Wigner did. And, what is more, Majorana obtained the infinite-dimensional unitary representations of the Lorentz group that will be re-discovered by Wigner in his 1939 and 1948 works. The entire theory was re-invented by Soviet mathematicians (in particular Gelfand and collaborators) in a series of articles from 1948 to 1958 and finally applied by physicists years later. Sadly, Majorana's initial article remained in the shadows for a good 34 years until D. Fradkin, informed by E. Amaldi, released [*Am. J. Phys.* **34** (1966) 314] what Majorana many years earlier had accomplished.

As soon as the news of the Joliot-Curie experiments reached Rome at the beginning of 1932, Majorana understood that they had discovered the "neutral proton" without having realized it. Thus, even before the official announcement of the discovery of the neutron, made soon afterwards by Chadwick, Majorana was able to explain the structure and stability of atomic nuclei with the help of protons and neutrons, antedating in this way also the pioneering work of D. Ivanenko, as both

Segré and Amaldi have recounted. Majorana's colleagues remember that even before Easter he had concluded that protons and neutrons (indistinguishable with respect to the nuclear interaction) were bound by the "exchange forces" originating from the exchange of their spatial positions alone (and not also of their spins, as Heisenberg would propose), so as to produce the alpha particle (and not the deuteron) saturated with respect to the binding energy. Only after Heisenberg had published his own article on the same problem was Fermi able to persuade Majorana to meet his famous colleague in Leipzig; and finally Heisenberg was able to convince Majorana to publish his results in the paper "Über die Kerntheorie." Majorana's paper on the stability of nuclei was immediately recognized by the scientific community –a rare event, as we know, from his writings– thanks to that timely "propaganda" made by Heisenberg himself. We seize the present opportunity to quote two brief passages from Majorana's letters from Leipzig. On February 14, 1933, he writes his mother (the italics are ours): "The environment of the physics institute is very nice. I have good relations with Heisenberg, with Hund, and with everyone else. *I am writing some articles in German. The first one is already ready....*" The work that is already ready is, naturally, the cited one on nuclear forces, which, however, remained *the only paper* in German. Again, in a letter dated February 18, he tells his father (we italicize): "*I will publish in German, after having extended it, also my latest article which appeared in Nuovo Cimento.*" Actually, Majorana published nothing more, either in Germany or after his return to Italy, except for the article (in 1937) of which we are about to speak. It is therefore of importance to know that Majorana was engaged in writing other papers: in particular, he was expanding his article about the infinite-components equations.

As we said, from the existing manuscripts it appears that Majorana was also formulating the essential lines of his symmetric theory of electrons and anti-electrons during the years 1932-1933, even though he published this theory only years later, when participating in the aforementioned competition for a professorship, under the title "Teoria simmetrica dell'elettrone e del positrone" (Symmetrical theory of the electron and positron), a publication that was initially noted almost exclusively for having introduced the Majorana representation of the Dirac matrices in real form. A consequence of this theory is that a neutral fermion has to be identical with its anti-particle, and Majorana suggested that neutrinos could be particles of this type. As with Majorana's other writings, this article also started to gain prominence only decades later, beginning in 1957; and nowadays expressions like Majorana spinors, Majorana mass, and Majorana neutrinos are fashionable.

As already mentioned, Majorana's publications (still little known, despite it all) is a potential gold-mine for physics. Recently, for example, C. Becchi pointed out how, in the first pages of the present paper, a clear formulation of the quantum action principle appears, the same principle that in later years, through Schwinger's and Symanzik's works, for example, has brought about quite important advances in quantum field theory.

ETTORE MAJORANA'S UNPUBLISHED PAPERS

Majorana also left us several unpublished scientific manuscripts, all of which have been catalogued (Baldo et al., 1987), (Recami, 1999) and kept at Domus Galilaeana. Our analysis of these manuscripts has allowed us to ascertain that all the existing material seems to have been written by 1933; even the rough copy of his last article, which Majorana proceeded to publish in 1937 —as already mentioned— seems to have been ready by 1933, the year in which the discovery of the positron was confirmed. Indeed, we are unaware of what he did in the following years from 1934 to 1938, except for a series of 34 letters written by Majorana between March 17, 1931, and November 16, 1937, in reply to his uncle Quirino —a renowned experimental physicist and at a time president of the Italian Physical Society— who had been pressing Majorana for theoretical explanations of his own experiments. By contrast, his sister Maria recalled that, even in those years, Majorana —who had reduced his visits to Fermi's Institute, starting from the beginning of 1934 (that is, after his return from Leipzig)— continued to study and work at home many hours during the day and at night. Did he continue to dedicate himself to physics? From a letter of his to Quirino, dated January 16, 1936, we find a first answer, because we get to learn that Majorana had been occupied "since some time, with quantum electrodynamics"; knowing Majorana's modesty and love for understatements, this no doubt means that by 1935 Majorana had profoundly dedicated himself to original research in the field of quantum electrodynamics.

Do any other unpublished scientific manuscripts of Majorana exist? The question, raised by his letters from Leipzig to his family, becomes of greater importance when one reads also his letters addressed to the National Research Council of Italy (CNR) during that period. In the first one (dated January 21, 1933), Majorana asserts: "At the moment, I am occupied with the elaboration of a theory for the description of arbitrary-spin particles that I began in Italy and of which I gave a summary notice in *Nuovo Cimento*...." In the second one (dated March 3, 1933) he even declares, referring to the same work: "I have sent an article on nuclear theory to Zeitschrift für Physik. I have the manuscript of a new theory on elementary particles ready, and will send it to the same journal in a few days." Considering that the article described here as a "summary notice" of a new theory was already of a very high level, one can imagine how interesting it would be to discover a copy of its final version, which went unpublished. [Is it still, perhaps, in the *Zeitschrift für Physik* archives? Our own search ended in failure.] One must moreover not forget that the above-cited letter to Quirino Majorana, dated January 16, 1936, revealed that his nephew continued to work on theoretical physics even subsequently, occupying himself in depth, at least, with quantum electrodynamics.

Some of Majorana's other ideas, when they did not remain concealed in his own mind, have survived in the memories of his colleagues. One such reminiscence we owe to Gian Carlo Wick. Writing from Pisa on October 16, 1978, he recalls: "...The scientific contact [between Ettore and me], mentioned by Segré, happened in Rome on the occasion of the 'A. Volta Congress' (long before Majorana's sojourn in Leipzig). The conversation took place in Heitler's company at a restaurant, and therefore without a blackboard...; but even in the absence of details, what Majorana described in words was a 'relativistic theory of charged particles of zero spin based on the idea of field quantization' (second quantization). When much later I saw Pauli and Weisskopf's article [*Helv. Phys. Acta* **7** (1934) 709], I remained absolutely convinced that what Majorana had discussed was the same thing...."

THIS VOLUME

In the present book, we reproduce and translate, for the first time, five neatly organized notebooks, known, in Italian, as "Volumetti" (booklets). Written in Rome by Ettore Majorana between 1927 and 1932, the original manuscripts are kept at the Domus Galilaeana in Pisa. Each of them is composed of about 100–150 sequentially numbered pages of approximate size 11 cm × 18 cm. Every notebook is prefaced by a table of contents, which evidently was gradually made out by the author when a particular line of thought was finished; and a date, penned on its first blank (i.e., on the initial blank page of each notebook) records when it was completed —except for the last, and smallest, booklet, which is undated, probably because it remained unfinished.

Numbered blank pages appear in the original manuscript in some cases between the end of a Section and the beginning of the next one; we have deleted these blanks in this volume.

Most likely, Majorana used to approach the issues treated in his notebooks following well-defined schemes arising from his studies. Each notebook was written during a period of about one year, starting from the years during which Ettore Majorana was completing his studies at the University of Rome. Thus the contents of these notebooks range from typical topics covered in academic courses to topics at the frontiers of research. Despite this unevenness in the level of sophistication (which becomes apparent on inspection of different notebooks or even a single notebook), the style in which any particular topic is treated is never obvious. As an example, we refer here to Majorana's study of the shift in the melting point of a substance when it is placed in a magnetic field or, more interestingly, his examination of heat propagation using the "cricket simile." Also remarkable is his treatment of contemporary physics topics in an original and lucid manner, such as Fermi's explanation of the electromagnetic mass of the electron, the Dirac equation with its applications, and the Lorentz group, revealing in some cases the literature preferred by him. As far as frontier research arguments are concerned, we here quote only two illuminating examples: the study of quasi-stationary states, anticipating Fano's theory by about 20 years, and Fermi's theory of atoms, reporting analytic solutions of the Thomas-Fermi equation with appropriate boundary conditions in terms of simple quadratures, which to our knowledge is still lacking.

In the translation of the notebooks we have attempted to adhere to the original Italian version as much as possible, adopting personal interpretations and notations only in a very few cases where the meaning of some paragraphs or the followed procedures were not clear enough. Nevertheless, for compactness' sake, we have replaced Planck's constant h, used throughout the original text, by the more current $2\pi\hbar$, except where results of the old quantum theory are involved. All changes from the original, introduced in the English version, are pointed out in footnotes. Additional footnotes have been introduced, as well, where the interpretation of some procedures or the meaning of particular parts require further elaboration. Footnotes which are not present in the original manuscript are denoted by the symbol @.

The major effort we have made to carefully check and type all equations and tables was motivated by our desire to facilitate the reading of Majorana's notebooks as much as possible, with a hope of rendering their intellectual treasure accessible for the first time to the widest audience.

Figures appearing in the notebooks have been reproduced without the use of photographic or scanning devices but are otherwise true in form to the original drawings. The same holds for tables, which in almost all cases have been reproduced independently of the source; i.e., we have performed our own calculations, following the methods used in the text. Several tables exhibit gaps, revealing that in these cases the author for some reason did not perform the corresponding calculations: In such instances, we have completed the tables whenever possible, filling the gaps with the appropriate expressions. Other minor changes, mainly related to typos in the original manuscript, are pointed out in footnotes.

For a better understanding of the style adopted by Ettore Majorana in composing his notebooks, and also for giving an idea of the method of translation and editing followed in this volume, we have reproduced, by scanning, a whole section (Sec. 3.3) from the original manuscript; which can illustrate some of peculiarities of Ettore Majorana's *Volumetti*. These selected pages are reported at the end of the book.

A short bibliography follows this Preface. Far from being exhaustive, it provides only some references about the topics touched upon in this introduction.

ACKNOWLEDGMENTS

This work has been partially funded by a grant provided by the Italian Embassy in Washington (and in this connection we are most grateful to former Ambassador Ferdinando Salleo and the Scientific Attaché Prof. Alexander Tenenbaum for their support), and by COFIN funds (coordinated by P.Tucci) of the Italian MURST. For his kind helpfulness, we are very indebted to Carlo Segnini, the present curator of the Domus Galilaeana at Pisa (as well as to previous curators and directors). Special thanks are moreover due to Francesco Bassani, the president of the Italian Physical Society, for a useful discussion; to many colleagues (in particular Dharam Ahluwalia, Roberto Battiston and Enrico Giannetto) for kind cooperation over the years; to Maria Alessandra Papa for her collaboration in the translation and to Cristiano Palomba for a preliminary reading of this book. Finally, we should express our appreciation to Jackie Gratrix for her help in preparing the camera-ready version of this book.

The Editors

REFERENCES

E. Amaldi, *La Vita e l'Opera di E. Majorana* (Accademia dei Lincei, Rome, 1966); "Ettore Majorana: Man and scientist," in *Strong and Weak Interactions. Present problems*, A. Zichichi, ed. (Academic, New York, 1966).

M. Baldo, R. Mignani, and E. Recami, "Catalogo dei manoscritti scientifici inediti di E. Majorana," in *Ettore Majorana: Lezioni all'Università di Napoli*, pp.175-197.

S. Esposito, "Majorana solution of the Thomas-Fermi equation," *Am. J. Phys.* **70** (2002) 852-856; "Majorana transformation for differential equations," *Int. J. Theor. Phys.* **41** (2002) 2417-2426.

R. Mignani, E. Recami and M. Baldo: "About a Dirac–like equation for the photon, according to E.Majorana," *Lett. Nuovo Cimento* **11** (1974) 568.

R. Penrose, "Newton, quantum theory and reality," in *300 Years of Gravitation*, S. W. Hawking and W. Israel, eds. (University Press, Cambridge, 1987). J. Zimba and R. Penrose, *Stud. Hist. Phil. Sci.* **24** (1993) 697. R. Penrose, *Ombre della Mente (Shadows of the Mind)* (Rizzoli, 1996), pp.338–343 and 371–375.

B. Pontecorvo, *Fermi e la fisica moderna* (Editori Riuniti, Rome, 1972); in *Proceedings of the International Conference on the History of Particle Physics, Paris, July 1982*, *Physique* **43** (1982).

B. Preziosi, ed. *Ettore Majorana: Lezioni all'Università di Napoli* (Bibliopolis, Napoli, 1987).

E. Recami, *Il caso Majorana: Epistolario, Documenti, Testimonianze*, 2nd edn (Oscar Mondadori, Milan, 1991), p.230; 4th edn (Di Renzo, Rome, 2002), p.273.

E. Recami, "Ettore Majorana: L'opera edita ed inedita," *Quaderni di Storia della Fisica (of the Giornale di Fisica)* (SIF, Bologna) **5** (1999) 19-68.

E. Segré, *Enrico Fermi, Fisico* (Zanichelli, Bologna, 1971).

1

VOLUMETTO I: 8 MARCH 1927

1. ELECTRIC POTENTIAL

$$\mathbf{E} = -\boldsymbol{\nabla} V, \qquad \boldsymbol{\nabla}^2 V = -4\pi\rho.$$

The electric potential at a point O of space $\mathcal{S}$ surrounded by the surface σ is given by

$$V_O = \int_\sigma k\,V\,\mathrm{d}\sigma + \int_S \rho\,(1/r - U)\,\mathrm{d}S, \tag{1.1}$$

where r is the distance between O and P, k is the effective surface charge[1] density generated by a unit charge at P responsible for the electric effects outside the surface, and U is the potential from such a distribution. Thus

$$V_O = \int_\sigma k\,V\,\mathrm{d}\sigma + \frac{1}{4\pi}\int_S U\,\boldsymbol{\nabla}^2 V\,\mathrm{d}S - \frac{1}{4\pi}\int_S \frac{\boldsymbol{\nabla}^2 V}{r}\,\mathrm{d}S. \tag{1.2}$$

In the region $\mathcal{S}$, we have

$$U\,\boldsymbol{\nabla}^2 V = \boldsymbol{\nabla}\cdot\left(U\,\boldsymbol{\nabla} V - V\,\boldsymbol{\nabla} U\right), \tag{1.3}$$

so that[2]

$$\begin{aligned} V_O = & \int_\sigma k\,V\,\mathrm{d}\sigma + \frac{1}{4\pi}\int_\sigma U\,\frac{\partial V}{\partial n}\,\mathrm{d}\sigma \\ & - \frac{1}{4\pi}\int_\sigma V\left(\frac{\partial U}{\partial n}\right)_{\mathrm{i}}\mathrm{d}\sigma - \frac{1}{4\pi}\int_S \frac{\boldsymbol{\nabla}^2 V}{r}\,\mathrm{d}S. \end{aligned} \tag{1.4}$$

[0]@ The label @ distinguishes editorial comments from the author's footnotes.

[1]@ In the original manuscript, the author often used the word "mass" in the place of "charge"; here, for clarity, we always use the second term.

[2]@ Here, the author is using the indices "i" and "e" to denote the regions internal and external to any given surface, respectively. The index n labels the component of any given vector along the external normal $\boldsymbol{n}$ to this surface.

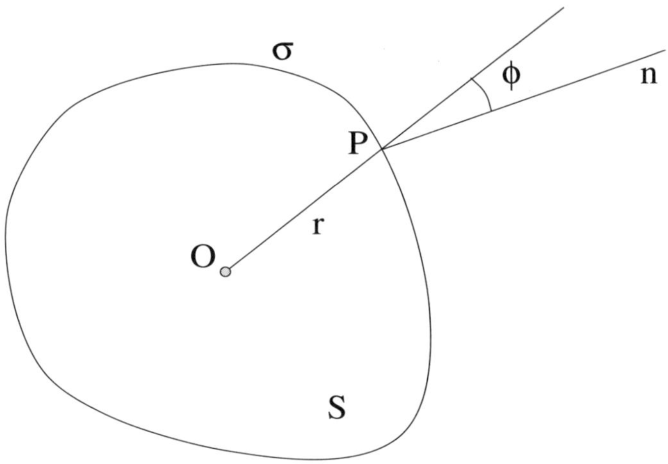

Fig. 1.1. Definition of some quantities used in the text.

On the surface we have instead

$$U = 1/r \,, \tag{1.5}$$

$$\left(\frac{\partial U}{\partial n}\right)_{\rm i} = -E_{n{\rm i}} = -E_{ne} + 4\pi k = -\frac{1}{r^2}\cos\phi + 4\pi k \,, \tag{1.6}$$

and by substitution we get

$$V_O = \frac{1}{4\pi}\int_\sigma \left(V\,\cos\phi + r\,\frac{\partial V}{\partial n}\right)\frac{{\rm d}\sigma}{r^2} - \frac{1}{4\pi}\int_S \frac{\nabla^2 V}{r}\,{\rm d}S. \tag{1.7}$$

This formula holds for any arbitrary functions V, since we can always find a charge distribution generating the potential V in the region $\mathcal{S}$. If there is no charge in S, then

$$V_O = \frac{1}{4\pi}\int_\sigma \left(V\,\cos\phi + r\,\frac{\partial V}{\partial n}\right)\frac{{\rm d}\sigma}{r^2}. \tag{1.8}$$

Let us now prove Eq. (1.7) directly. We set

$$V_O' = \frac{1}{4\pi}\int_\sigma \left(V\,\cos\phi + r\,\frac{\partial V}{\partial n}\right)\frac{{\rm d}\sigma}{r^2} - \frac{1}{4\pi}\int_S \frac{\nabla^2 V}{r}\,{\rm d}S. \tag{1.9}$$

Let us consider an infinitesimal *homothety* with center at O that transforms the surface σ into σ' and the space $\mathcal{S}$ into $\mathcal{S}'$. The integration regions transform accordingly. It is simple to evaluate the variations of the integrals by using the homothety relations. Actually, if $1+{\rm d}\alpha$ is the homothety ratio, the following relations, connecting each given quantity to the corresponding one, hold:

$$\begin{aligned}
\delta V &= \mathrm{d}\alpha\,\mathbf{OP}\cdot\boldsymbol{\nabla} V, && (1.10)\\
\delta\cos\phi &= 0, && (1.11)\\
\delta r &= \mathrm{d}\alpha\; r, && (1.12)\\
\delta\frac{\mathrm{d}\sigma}{r^2} &= 0, && (1.13)\\
\delta\boldsymbol{\nabla}^2 V &= \mathrm{d}\alpha\,\mathbf{OP}\cdot\boldsymbol{\nabla}\,\boldsymbol{\nabla}^2 V, && (1.14)\\
\delta\frac{\partial V}{\partial n} &= \frac{\partial}{\partial n}\mathbf{OP}\cdot\boldsymbol{\nabla} V\,\mathrm{d}\alpha - \frac{\partial V}{\partial n}\,\mathrm{d}\alpha, && (1.15)\\
\delta\frac{\mathrm{d}S}{r} &= 2\,\mathrm{d}\alpha\,\frac{\mathrm{d}S}{r}. && (1.16)
\end{aligned}$$

From these equations, we get

$$\begin{aligned}
\delta V_0' &= \frac{\mathrm{d}\alpha}{4\pi}\int_\sigma\left(\mathbf{OP}\cdot\boldsymbol{\nabla} V\,\cos\phi + r\,\frac{\partial}{\partial n}\,\mathbf{OP}\cdot\boldsymbol{\nabla} V\right)\frac{\mathrm{d}\sigma}{r^2}\\
&\quad - \frac{\mathrm{d}\alpha}{4\pi}\int_S\left(\frac{\mathbf{OP}\cdot\boldsymbol{\nabla}\,\boldsymbol{\nabla}^2 V}{r} + 2\,\frac{\boldsymbol{\nabla}^2 V}{r}\right)\mathrm{d}S. \qquad (1.17)
\end{aligned}$$

The surface integral can be viewed as the outward flux, through the surface σ, of the vector:

$$\mathbf{M} = \mathbf{OP}\cdot\boldsymbol{\nabla} V\,\frac{\mathbf{OP}}{r^3} + \frac{1}{r}\boldsymbol{\nabla}\left(\mathbf{OP}\cdot\boldsymbol{\nabla} V\right). \qquad (1.18)$$

This vector is infinite at O; however it is only a first order infinity, so the surface integral can be transformed into the volume integral

$$\int_S \boldsymbol{\nabla}\cdot\mathbf{M}\,\mathrm{d}S.$$

Moreover, it is easy to show that

$$\boldsymbol{\nabla}\cdot\mathbf{M} = \frac{\mathbf{OP}\cdot\boldsymbol{\nabla}\,\boldsymbol{\nabla}^2 V}{r} + 2\,\frac{\boldsymbol{\nabla}^2 V}{r}, \qquad (1.19)$$

so that we have $\delta V_0' = 0$. Now, if the surface σ becomes infinitesimal around O, then the volume integral in Eq. (1.9) vanishes, and the surface integral tends to $4\pi V_O$. We thus get

$$V_O' = V_O, \qquad \text{q.e.d.} \qquad (1.20)$$

2. RETARDED POTENTIAL

Let H be a function of space and time that obeys the differential equation

$$\nabla^2 H = \frac{1}{c^2}\frac{\partial^2 H}{\partial t^2}. \tag{1.21}$$

Let O denote a point in space, r the distance of another point P from O, and m a function of P and of t; we then set

$$\overline{m\,(P,t)} = m\,(P, t - r/c)\,. \tag{1.22}$$

If we consider the function

$$H_1(P,t) = H\,(P, t - r/c)\,, \tag{1.23}$$

it is easy to find the differential equation satisfied by it:

$$\nabla^2 H_1 = -\frac{2}{c}\frac{\partial^2 H_1}{\partial r \partial t} - \frac{2}{rc}\frac{\partial H_1}{\partial t}. \tag{1.24}$$

If O belongs to the region $\mathcal{S}$ confined by the surface σ, then, by using Eq. (1.7) and noting that $H_{1O} = H_O$ at O, we find

$$\begin{aligned} H_O &= \frac{1}{4\pi}\int_\sigma \left(H_1 \cos\phi + r\frac{\partial H_1}{\partial n}\right)\frac{\mathrm{d}\sigma}{r^2} \\ &+ \frac{1}{4\pi}\int_S \left(\frac{2}{rc}\frac{\partial^2 H_1}{\partial r\partial t} + \frac{2}{r^2 c}\frac{\partial H_1}{\partial t}\right)\mathrm{d}S. \end{aligned} \tag{1.25}$$

Let us decompose the region $\mathcal{S}$ into cones having their vertices at O. The volume element of a cone with aperture angle $\mathrm{d}\omega$ between two spheres centered at O and having radius r and $r + \mathrm{d}r$, respectively, is $\mathrm{d}\omega\, r^2 \mathrm{d}r$. The integral over the volume of the cone is then

$$\mathrm{d}\omega \int_0^r \frac{2}{c}\left(\frac{\partial H_1}{\partial t} + r\frac{\partial^2 H_1}{\partial r \partial t}\right)\mathrm{d}r = \mathrm{d}\omega\,\frac{2r}{c}\frac{\partial H_1}{\partial t}, \tag{1.26}$$

where on the r.h.s. the term $\partial H_1/\partial r$ has to be evaluated at the base of the cone on the surface σ. If the area of this base is $\mathrm{d}\sigma$ and ϕ is the angle between the cone axis and the outward normal direction, then we have

$$\mathrm{d}\omega\,\frac{2r}{c}\frac{\partial H_1}{\partial t} = \frac{2}{rc}\frac{\partial H_1}{\partial t}\cos\phi\,\mathrm{d}\sigma, \tag{1.27}$$

and the integral over the entire region $\mathcal{S}$ turns into the surface integral

$$\int_\sigma \frac{2}{rc}\frac{\partial H_1}{\partial t}\cos\phi\,\mathrm{d}\sigma. \tag{1.28}$$

Inserting it into 1.25, one obtains

$$H_O = \frac{1}{4\pi} \int_\sigma \left(H_1 \cos\phi + r \frac{\partial H_1}{\partial n} + \frac{2r}{c} \frac{\partial H_1}{\partial t} \cos\phi \right) \frac{d\sigma}{r^2}; \tag{1.29}$$

and, noting that

$$H_1 = \overline{H}, \tag{1.30}$$

$$\frac{\partial H_1}{\partial n} = \overline{\frac{\partial H}{\partial n}} - r/c \cos\phi \, \overline{\frac{\partial H}{\partial t}}, \tag{1.31}$$

$$\frac{\partial H_1}{\partial t} = \overline{\frac{\partial H}{\partial t}}, \tag{1.32}$$

we obtain

$$H_O = \frac{1}{4\pi} \int_\sigma \left(\overline{H} \cos\phi + r \overline{\frac{\partial H}{\partial n}} + \frac{r \cos\phi}{c} \overline{\frac{\partial H}{\partial t}} \right) \frac{d\sigma}{r^2}. \tag{1.33}$$

If we define

$$\overline{\overline{m(P,t)}} = m(P, t + r/c), \tag{1.34}$$

and

$$H_2(P,t) = H\left(P, t + \frac{r}{c}\right), \tag{1.35}$$

the differential equation satisfied by H_2 becomes

$$\nabla^2 H_2 = \frac{2}{c} \frac{\partial^2 H_2}{\partial r \partial t} + \frac{2}{rc} \frac{\partial H_2}{\partial t}. \tag{1.36}$$

Similarly, we now find

$$H_O = \frac{1}{4\pi} \int_\sigma \left(\overline{\overline{H}} \cos\phi + r \overline{\frac{\partial H}{\partial n}} - \frac{r \cos\phi}{c} \overline{\frac{\partial H}{\partial t}} \right) \frac{d\sigma}{r^2}. \tag{1.37}$$

3. INTERACTION ENERGY OF TWO ELECTRIC OR MAGNETIC CHARGE DISTRIBUTIONS

Let us consider two electric or magnetic charge distributions located in different regions of space. Let σ be a surface (which may be simply-connected or not) bounding the space region $\mathcal{S}$ which contains all the charges of the first distribution and no charges of the second distribution. Let V be the potential of the field $\mathbf{E}$ produced by the first distribution of charges $m_1, m_2, \ldots, m_n$ located at points $P_1, P_2, \ldots, P_n$. Let V' be the potential of the field $\mathbf{E}'$ produced by the second distribution. With obvious notations, we have

$$U = \sum_{i=1}^{n} m_i V_i', \tag{1.38}$$

and, by applying Eq. (1.8),

$$U = \frac{1}{4\pi} \int_\sigma \left(V' \sum_{i=1}^{n} \frac{m_i}{r_i^2} \cos\phi_i + E_n' \sum_{i=1}^{n} \frac{m_i}{r_i} \right) \mathrm{d}\sigma, \tag{1.39}$$

where E_n' is the component of $\mathbf{E}'$ along the *inward* direction perpendicular to σ. Now we have

$$\sum_{i=1}^{n} \frac{m_i}{r_i^2} \cos\phi_i = E_n, \tag{1.40}$$

$$\sum_{i=1}^{n} \frac{m_i}{r_i} = V, \tag{1.41}$$

where E_n is the component of $\mathbf{E}$ along the *outward* normal to σ. By substitution, we find the relevant formula

$$U = \frac{1}{4\pi} \int_\sigma (E_n V' + E_n' V)\, \mathrm{d}\sigma. \tag{1.42}$$

4. SKIN EFFECT IN HOMOGENEOUS CYLINDRICAL ELECTRIC CONDUCTORS

Let us consider a cylindrical conductor whose cross section (assumed to be circular) is small with respect to the length of the conductor. The potential can then be considered uniform on any given cross section, and the current density as depending only on the distance a from the axis. Let us denote by $I = I_1 + iI_2$ the complex quantity[3] representing

[3]@ The author is considering a conductor in which there flows an alternating current of frequency ω.

the current intensity flowing through a circle that is coaxial with the cross section of the conductor and has radius a. Let us indicate by $D = D_1 + iD_2$ the current density at a distance a from the axis, with μ the magnetic permeability of the conductor, A the radius of the cross section and ρ the electric resistivity. Then the "counter-electromotive forces"[4] per unit length along a current line at a distance a from the axis, due to the Joule effect[5] and to the variation of the induction in the conductor, are

$$D\,\rho \tag{1.43}$$

and

$$2\,\mu\,\omega\,i\int_a^A \frac{I}{x}\,\mathrm{d}x, \tag{1.44}$$

respectively. Since all other electromotive forces are equal for the different current lines, we can conclude that

$$D\,\rho + 2\,\mu\,\omega\,i\int_a^A (I/x)\,\mathrm{d}x \;=\; \text{constant.} \tag{1.45}$$

By differentiating, we then find

$$\rho\,\mathrm{d}D \;=\; 2\,\mu\,\omega\,i\,\frac{I}{a}\,\mathrm{d}a. \tag{1.46}$$

We can consider I and D as functions of s, the area of the circle having radius a, since

$$D \;=\; \frac{\mathrm{d}I}{\mathrm{d}s}, \tag{1.47}$$

$$2\,\frac{\mathrm{d}a}{a} \;=\; \frac{\mathrm{d}s}{s}, \tag{1.48}$$

which yields

$$\rho\,\mathrm{d}\frac{\mathrm{d}I}{\mathrm{d}s} \;=\; \mu\,\omega\,i\,\frac{I}{s}\,\mathrm{d}s, \tag{1.49}$$

or equivalently

$$\frac{\mathrm{d}^2 I}{\mathrm{d}s^2} \;=\; \frac{\mu\,\omega\,i}{\rho}\,\frac{I}{s}. \tag{1.50}$$

By setting

$$p \;=\; \frac{\mu\,\omega}{\rho}\,s, \tag{1.51}$$

[4]@ That is, the forces blocking the flow of the current.

[5]@ In the original manuscript, the Joule effect is called an "Ohmic effect." However, we prefer to use the widely-known terminology of Joule effect.

we get

$$\frac{\mathrm{d}^2 I}{\mathrm{d}p^2} = i\,\frac{I}{p}. \tag{1.52}$$

This equation clearly shows that p is independent of the fundamental measure units of the electromagnetic system. Therefore, to make the computation easier, given that p is proportional to s, we choose the length measure unit in such a way that $p = s$, without altering p. Since $I = 0$ when $p = 0$, we can easily integrate Eq. (1.52) using a series expansion. Remembering that $I = I_1 + iI_2$, we find

$$I_1 = m\left(p - \frac{1}{2!^2 \cdot 3}p^3 + \frac{1}{4!^2 \cdot 5}p^5 - \frac{1}{6!^2 \cdot 7}p^7 + \ldots\right), \tag{1.53}$$

$$I_2 = m\left(\frac{1}{2}p^2 - \frac{1}{3!^2 \cdot 4}p^4 + \frac{1}{5!^2 \cdot 6}p^6 - \frac{1}{7!^2 \cdot 8}p^8 + \ldots\right), \tag{1.54}$$

wherein m is a constant factor that may be chosen to be real after an appropriate shift of the origin of time. Given our convention $p = s$, by differentiation with respect to p, we get

$$D_1 = m\left(1 - \frac{1}{2!^2}p^2 + \frac{1}{4!^2}p^4 - \frac{1}{6!^2}p^6 + \ldots\right), \tag{1.55}$$

$$D_2 = m\left(p - \frac{1}{3!^2}p^3 + \frac{1}{5!^2}p^5 - \frac{1}{7!^2}p^7 + \ldots\right). \tag{1.56}$$

The mean heat per unit time dissipated along a length ℓ of the conductor due to Joule effect is

$$Q_1 = \frac{1}{2}m^2\rho\,\ell \int_0^p \left[\left(1 - \frac{1}{2!^2}p^2 + \frac{1}{4!^2}p^4 - \frac{1}{6!^2}p^6 + \ldots\right)^2 + \left(p - \frac{1}{3!^2}p^3 + \frac{1}{5!^2}p^5 - \frac{1}{7!^2}p^7 + \ldots\right)^2\right] \mathrm{d}p. \tag{1.57}$$

Instead, the heat that would be produced if the current was uniformly distributed is given by

$$Q = \frac{1}{2}m^2\rho\,\frac{1}{p}\left[\left(p - \frac{1}{2!^2 \cdot 3}p^3 + \frac{1}{4!^2 \cdot 5}p^5 - \frac{1}{6!^2 \cdot 7}p^7 + \ldots\right)^2 + \left(\frac{1}{2}p^2 - \frac{1}{3!^2 \cdot 4}p^4 + \frac{1}{5!^2 \cdot 6}p^6 - \frac{1}{7!^2 \cdot 8}p^8 + \ldots\right)^2\right]. \tag{1.58}$$

Table 1.1. Some values for the skin effect (see the text).

p	R_1/R
1	1.0782
2	1.2646
3	1.4789
4	1.6779
6	2.0067
10	2.5069
24	3.7274
60	5.7357
100	7.3277

Denoting by R_1 the apparent resistance of the conductor in the AC regime and by R the resistance in the DC regime, we find

$$\frac{R_1}{R} = \frac{Q_1}{Q} = \frac{\displaystyle\int_0^p \left[\left(1 - \frac{1}{2!^2}p^2 + \ldots\right)^2 + \left(p - \frac{1}{3!^2}p^3 + \ldots\right)^2\right] \mathrm{d}p}{\displaystyle\frac{1}{p}\left[\left(p - \frac{1}{2!^2{\cdot}3}p^3 + \ldots\right)^2 + \left(\frac{1}{2}p^2 - \frac{1}{3!^2{\cdot}4}p^4 + \ldots\right)^2\right]}. \tag{1.59}$$

Both the numerator and the denominator in its r.h.s. may be expanded in power series of p. By performing this expansion and dividing by p, we find

$$\frac{R_1}{R} = \frac{1 + \frac{1}{3!}p^2 + \frac{1}{2!^2{\cdot}5!}p^4 + \frac{1}{3!^2{\cdot}7!}p^6 + \frac{1}{4!^2{\cdot}9!}p^8 + \ldots}{1 + \frac{1}{2!{\cdot}3!}p^2 + \frac{1}{2!{\cdot}3!{\cdot}5!}p^4 + \frac{1}{3!{\cdot}4!{\cdot}7!}p^6 + \frac{1}{4!{\cdot}5!{\cdot}9!}p^8 + \ldots}. \tag{1.60}$$

By relaxing the constraints on the measure units, in this expression we have $p = \mu\omega s/\rho = \mu\ell\omega/R$ or, using the Ohm as the resistance measure unit and m as the measure unit of length,

$$p = \frac{\mu\,\omega\,\ell}{10^7\,R} = \frac{2\pi f\,\mu\,\ell}{10^7\,R}.$$

In Table 1.1 we report some values of R_1/R for different values[6] of p. For small values of p ($p << 1$), we can use the expression

$$\frac{R_1}{R} = 1 + \frac{1}{12}p^2 - \frac{1}{180}p^4, \tag{1.61}$$

[6]@ Notice that the author used Eq. (1.60) to obtain the values in this Table up to $p = 6$ and then invoked the expansion (1.62) to complete the Table.

while, for large values of p,

$$\frac{R_1}{R} = \sqrt{\frac{1}{2}p} + \frac{1}{4} + \frac{3}{64}\left(\sqrt{\frac{1}{2}p}\right)^{-1} \tag{1.62}$$

or, in the simplest case,

$$\frac{R_1}{R} = \sqrt{\frac{1}{2}p} + \frac{1}{4}, \tag{1.63}$$

where the last but one equation yields practically exact results for $p > 10$ (relative error less than 0.0001).

5. THERMODYNAMICS OF THERMOELECTRIC CELLS

Let us suppose that a unitary quantity of electricity[7] can be related to some amount of entropy S which depends on the nature and on the temperature T of the conductor. If a quantity of electricity q flows in the conductor, its entropy varies from qS to $q(S + \mathrm{d}S)$, where $\mathrm{d}S$ can be either infinitesimal or finite, depending on the equal or different nature of the conductor's ends. If we neglect the Joule effect, which can be taken into account separately, the motion of the charges inside the conductor should be considered reversible, and therefore the entropy increase $q\mathrm{d}S$ can be related to the absorption of a quantity $qT\mathrm{d}S$ of heat that takes place where the nature of the conductor changes or, in a homogeneous conductor, where the temperature changes (Thomson effect). Thus, if q electric charges flow through a closed circuit, the total absorbed heat is

$$q \int T\,\mathrm{d}S,$$

where the integral over the whole circuit is, in general, different from zero only if the temperature is not equal in all the elements of the circuit and if there are at least two different elements. Thus, if E is the mechanical equivalent of heat, energy conservation requires an electromotive force e to appear in the circuit:

$$e = E \int T\,\mathrm{d}S. \tag{1.64}$$

[7]@ That is, for instance, a quantity of electric charge (flowing in a conductor) corresponding to the chosen measure unit.

From this, the fundamental laws of the electric cell follow.

6. ENERGY OF AN ISOLATED CONDUCTOR

Let σ be a conducting charged surface with unitary electric charge, k the surface charge density, ϵ the energy of the system, and V the potential of the conductor. Now suppose that the surface σ deforms into the surface σ_1, with surface charge density k_1, energy ϵ_1, and potential V_1. Let ϵ_{m} be the mutual energy of the two distributions and $\epsilon(k - k_1)$ the total energy of the first distribution and of the second one with reversed sign. Clearly, we have

$$\epsilon(k - k_1) = \epsilon + \epsilon_1 - \epsilon_{\mathrm{m}}. \tag{1.65}$$

Let us assume that σ_1 is completely external to σ. The potential of the field produced by the distribution k_1 will be equal to V_1 in all points of σ, so that we'll have $\epsilon_{\mathrm{m}} = V_1$, and, since $\epsilon_1 = V_1/2$, $\epsilon_{\mathrm{m}} = 2\epsilon_1$. By substitution, we obtain

$$\epsilon - \epsilon_1 = \epsilon(k - k_1). \tag{1.66}$$

Assuming that σ_1 is very close to σ, the field produced by the difference of the two distributions is zero inside σ, finite between σ, and σ_1, and infinitesimal outside σ_1. Thus, the energy per unit volume of such a field is zero inside σ, finite between σ and σ_1, and a *second-order* infinitesimal outside σ_1. Since the distance between σ and σ_1 is a first-order infinitesimal, if we neglect infinitesimals of order higher than the first, we should only consider the volume energy contained between σ and σ_1. But in this region the field produced by the second distribution is zero. Consequently we can say that, for an infinitesimal variation of σ, as long as the resulting surface is completely external to σ, the electrostatic energy decreases by an amount that is equal to the energy that was originally contained between σ and the new surface. This can also be stated in a different way. Let $\mathrm{d}\boldsymbol{\sigma}$ be an element of σ; the volume element between σ, σ_1, and the normals to the boundary of $\mathrm{d}\boldsymbol{\sigma}$ is $\mathrm{d}\boldsymbol{\sigma}\cdot\mathrm{d}\boldsymbol{\alpha}$, quantity $\mathrm{d}\boldsymbol{\alpha}$ being the distance between the two surfaces σ and σ_1. Neglecting infinitesimal quantities, the magnitude of the field inside such an element is $4\pi k = F$, so that the energy contained in the element is $F(k/2)\mathrm{d}\sigma\mathrm{d}\alpha$. On the other hand, $k\mathrm{d}\sigma$ is the charge $\mathrm{d}m$ distributed on $\mathrm{d}\sigma$, and thus $F(k/2)\mathrm{d}\sigma\mathrm{d}\alpha = (\mathrm{d}m/2)F\mathrm{d}\alpha$. By integrating over all the space between

σ and σ_1, we find

$$\epsilon - \epsilon_1 = -\delta\epsilon = \frac{1}{2}\int \mathbf{F}\cdot\delta\boldsymbol{\alpha}\,dm, \tag{1.67}$$

and

$$V - V_1 = -\delta V = \int \mathbf{F}\cdot\delta\boldsymbol{\alpha}\,dm. \tag{1.68}$$

It is very easy to see that this equation holds even when σ_1 is not entirely external to σ, as long as F is the force external to σ and the sign of $d\alpha$ is positive or negative, according to whether σ_1 is locally external or internal to σ.

7. ATTRACTION BETWEEN MASSES WHICH ARE FAR APART

Let us consider a system of gravitating masses $m_1, m_2, \ldots, m_n$ located at points $P_1, P_2, \ldots, P_n$, respectively. Let O be the center of mass of the system and m its total mass. Let us fix a Cartesian reference frame with origin in 0. The potential at point P, defined by the coordinates x, y, z is

$$\begin{aligned} V &= \sum_{i=1}^{n} m_i \left[(x-x_i)^2 + (y-y_i)^2 + (z-z_i)^2\right]^{-1/2} \\ &= \sum_{i=1}^{n} m_i \left[x^2 + y^2 + z^2 - 2(xx_i + yy_i + zz_i) + x_i^2 + y_i^2 + z_i^2\right]^{-1/2}. \end{aligned}$$

Denoting by r the distance between P and O, and with α, β, γ the direction cosines of the straight line OP, we will have

$$\begin{aligned} V &= \sum_{i=1}^{n} m_i \left[r^2 - 2r(\alpha x_i + \beta y_i + \gamma z_i) + x_i^2 + y_i^2 + z_i^2\right]^{-1/2} \\ &= \frac{1}{r}\sum_{i=1}^{n} m_i \left[1 - (2/r)(\alpha x_i + \beta y_i + \gamma z_i) + (x_i^2 + y_i^2 + z_i^2)/r^2\right]^{-1/2}. \end{aligned}$$

If r is infinitely large, then the quantity inside the square brackets differs from unity by an infinitesimal of the same order as $1/r$. On performing an expansion of this quantity in powers of such infinitesimal up to the fourth order and neglecting in the sum third-order terms (because of the

$1/r$ factor), we get

$$\begin{aligned} V &= \frac{1}{r}\sum_{i=1}^{n} m_i + \frac{1}{r^2}\sum_{i=1}^{n} m_i(\alpha x_i + \beta y_i + \gamma z_i) \\ &+ \frac{1}{r^3}\sum_{i=1}^{n} m_i\left[\frac{3}{2}(\alpha x_i + \beta y_i + \gamma z_i)^2 - \frac{1}{2}\left(x_i^2 + y_i^2 + z_i^2\right)\right]. \end{aligned}$$

On noting that $\sum_i m_i = m$ and $\sum_i m_i x_i = \sum_i m_i y_i = \sum_i m_i z_i = 0$ and transforming the last term in the previous equation, we find

$$\begin{aligned} V &= \frac{m}{r} + \frac{1}{r^3}\sum_{i=1}^{n} m_i\left[\left(x_i^2 + y_i^2 + z_i^2\right) - \frac{3}{2}\left(\alpha^2(y_i^2 + z_i^2)\right.\right. \\ &\left.\left. + \beta^2(x_i^2 + z_i^2) + \gamma^2(x_i^2 + y_i^2) - 2\alpha\beta x_i y_i - 2\alpha\gamma x_i z_i - 2\beta\gamma y_i z_i\right)\right]. \end{aligned}$$

Thus, introducing the polar moment of inertia $\mathcal{I}_p$ with respect to the center of mass of the given system and the moment of inertia $\mathcal{I}$ of the same system with respect to the direction OP, we find

$$V = \frac{m}{r} + \frac{1}{r^3}\left(\mathcal{I}_p - \frac{3}{2}\mathcal{I}\right) + \mathrm{O}\left(\frac{1}{r^4}\right). \tag{1.69}$$

We can then say that the potential at large distances generated by a system of Newtonian masses is determined, up to fourth-order infinitesimals, by the mass and by the moments of inertial (the "inertia central core") of the system. Since, as Eq. (1.69) shows, up to third-order terms we have $V/m = 1/r$, then in the second term of (1.69) we may replace $1/r$ with the approximate value V/m. On solving the equation with respect to $1/r$, we get, up to fourth-order terms,

$$\frac{1}{r} = \frac{V}{m} - \frac{V^3}{m^4}\left(\mathcal{I}_p - \frac{3}{2}\mathcal{I}\right), \tag{1.70}$$

or, taking the reciprocal of both sides, to *second* order we find

$$r = \frac{m}{V} + \frac{V}{m^2}\left(\mathcal{I}_p - \frac{3}{2}\mathcal{I}\right). \tag{1.71}$$

Since it is always possible to find an equivalent body ("omeoid") having the same mass and the same "central core of inertia" as the given system, we can conclude that the equipotential surfaces of the field produced at large distances by any mass distribution are, up to second-order terms, ellipses that have a common focal point and whose axes coincide with the principal axes of inertia of the mass distribution; at first order, the equipotential surfaces are spheres centered at the center of mass.

8. FORMULAE

In what follows, we denote by $\mathcal{S}$ a region of space enclosed by a surface σ:

(1) $$\nabla \cdot (m\,\mathbf{F}) = m\,\nabla \cdot \mathbf{F} + \nabla m \cdot \mathbf{F},$$

(2) $$\nabla \cdot \mathbf{E} \times \mathbf{F} = \nabla \times \mathbf{E} \times \mathbf{F} - \mathbf{E} \times \nabla \times \mathbf{F},$$

(3) $$\nabla\,(m\,n) = m\,\nabla n + n\,\nabla m,$$

(4) $$\nabla^2\,(m\,n) = m\,\nabla^2 n + 2\,\nabla m \cdot \nabla n + n\,\nabla^2 m,$$

(5) $$\nabla \times \nabla \times \mathbf{E} = -\,\nabla^2 \mathbf{E} + \nabla\,\nabla \cdot \mathbf{E},$$

(6) [8] $$\int_\sigma E_n\,\mathrm{d}\sigma = \int_S \nabla \cdot \mathbf{E}\,\mathrm{d}S,$$

(7) $$\int_\sigma m\,E_n\,\mathrm{d}\sigma = \int_S (m\,\nabla \cdot \mathbf{E} + \nabla m \cdot \mathbf{E})\,\mathrm{d}S,$$

(8) $$\int_\sigma \mathbf{n} \times \mathbf{F}\,\mathrm{d}\sigma = \int_S \nabla \times \mathbf{F}\,\mathrm{d}S,$$

(9) $$\int_\sigma p\,\mathbf{n}\,\mathrm{d}\sigma = \int_S \nabla p\,\mathrm{d}S,$$

(10)
$$\int_\sigma q\,\mathbf{n}\,\mathrm{d}\sigma = \int_S \left(\frac{\partial q\,\mathbf{i}}{\partial x} + \frac{\partial q\,\mathbf{j}}{\partial y} + \frac{\partial q\,\mathbf{k}}{\partial z}\right)\mathrm{d}S,$$ if q is a *homography*,[9]

(11) $$\nabla \times m\mathbf{F} = m\,\nabla \times \mathbf{F} + \nabla m \times \mathbf{F},$$

(12)

---

[8]@ E_n denotes the component of the vector $\mathbf{E}$ along the outward normal $\boldsymbol{n}$ to the surface σ.

[9]@ $\mathbf{i}, \mathbf{j}, \mathbf{k}$ are the unit vectors along the coordinate axes x, y, z, respectively.

$$\int_\sigma \mathbf{OP}\times q\,\mathbf{n}\,\mathrm{d}\sigma = \int_S \Big[\mathbf{OP}\times\left(\frac{\partial q\,\mathbf{i}}{\partial x}+\frac{\partial q\,\mathbf{j}}{\partial y}+\frac{\partial q\,\mathbf{k}}{\partial z}\right) + \mathbf{i}\times q\,\mathbf{i}+\mathbf{j}\times q\,\mathbf{j}+\mathbf{k}\times q\,\mathbf{k}\Big]\,\mathrm{d}S,$$

(13) $$\int_\sigma \mathbf{E}\,E_n\,\mathrm{d}\sigma = \int_S \left(\mathbf{E}\,\boldsymbol{\nabla}\cdot\mathbf{E} - \mathbf{E}\times\boldsymbol{\nabla}\times\mathbf{E}+\frac{1}{2}\boldsymbol{\nabla}E^2\right)\mathrm{d}S,$$

(14) $$\int_\sigma \left(\mathbf{E}\,E_n - \frac{1}{2}E^2\,\mathbf{n}\right)\mathrm{d}\sigma = \int_S (\mathbf{E}\,\boldsymbol{\nabla}\cdot\mathbf{E} - \mathbf{E}\times\boldsymbol{\nabla}\times\mathbf{E})\,\mathrm{d}S.$$

(15) Let it be $U_1 = U_1(x_1, x_2, x_3)$ and[10]

$$x_1 = x_1(x,y,z),\quad x_2 = x_2(x,y,z),\quad x_3 = x_3(x,y,z). \tag{1.72}$$

By setting $U(x,y,z) = U_1(x_1,x_2,x_3)$, we deduce

$$\begin{aligned}\boldsymbol{\nabla}^2 U &= \frac{\partial^2 U_1}{\partial x_1^2}\,|\boldsymbol{\nabla}x_1|^2 + \ldots + \frac{\partial^2 U_1}{\partial x_1\partial x_2}\cdot 2\,\boldsymbol{\nabla}x_1\cdot\boldsymbol{\nabla}x_2+\ldots\\ &\quad+\frac{\partial U_1}{\partial x_1}\boldsymbol{\nabla}^2 x_1 + \frac{\partial U_1}{\partial x_2}\boldsymbol{\nabla}^2 x_2 + \frac{\partial U_1}{\partial x_3}\boldsymbol{\nabla}^2 x_3.\end{aligned} \tag{1.73}$$

Similar formulae hold for transformations involving spaces with an arbitrary number of dimensions and also for transformations among spaces with different dimensionalities.

9. ELECTRIC LINES

Let r,L,C, and g be the resistance, self-inductance, capacitance and dispersion per unit length of an electric line. Let us assume that they are constant. If an (alternating) current of frequency $\omega/2\pi$ is flowing along the line, the general expressions for the (complex) current intensity[11] i and for the potential V at a distance x from a point O of the line (chosen as the origin), are

$$V = A\cosh px + B\sinh px, \tag{1.74}$$

$$i = -A\,q\sinh px - B\,q\cosh px. \tag{1.75}$$

[10]@ In the original manuscript, the author used both notations x_1, x_2, x_3 and x_1, y_1, z_1.

[11]@ Here, $\sqrt{-1}$ will be denoted by j.

A and B are arbitrary constants, whereas

$$p = \sqrt{r + L\omega j}\,\sqrt{g + C\omega j}, \quad q = \sqrt{g + C\omega j}/\sqrt{r + L\omega j}.$$

Let ℓ be the length of the line and $V_0 = V(x = 0)$, $V_1 = V(x = \ell)$, $i_0 = i(x = 0)$, and $i_1 = i(x = \ell)$. Let us also suppose that V_0 is known and that the line is closed on a (complex) resistance R.

On putting $x = 0$ in Eq. (1.74), we find

$$V_0 = A, \tag{1.76}$$

while for $x = \ell$, with the value just found for A, we get

$$V_1 = V_0 \cosh p\ell + B \sinh p\ell, \tag{1.77}$$

$$i = -V_0\, q \sinh p\ell - B\, q \cosh p\ell. \tag{1.78}$$

Since under our conditions it is $V_1 = R i_1$, we have

$$V_0 \cosh p\ell + B \sinh p\ell + V_0\, R\, q \sinh p\ell + B\, R\, q \cosh p\ell = 0, \tag{1.79}$$

that is,

$$B = -V_0 \frac{\cosh p\ell + Rq \sinh p\ell}{\sinh p\ell + Rq \cosh p\ell}. \tag{1.80}$$

On substituting the last expression into Eqs. (1.74) and (1.75) and letting x take appropriate values, we easily find the following expressions:

$$i_0 = V_0\, q \frac{\cosh p\ell + Rq \sinh p\ell}{\sinh p\ell + Rq \cosh p\ell}, \tag{1.81}$$

$$V_1 = \frac{V_0\, R\, q}{\sinh p\ell + Rq \cosh p\ell}, \tag{1.82}$$

$$i_1 = \frac{V_0\, q}{\sinh p\ell + Rq \cosh p\ell}. \tag{1.83}$$

Some particular cases follow:

- For $R = \infty$:

$$i_0 = V_0\, q \frac{\sinh p\ell}{\cosh p\ell}, \tag{1.84}$$

$$V_1 = \frac{V_0}{\cosh p\ell}, \tag{1.85}$$

$$i_1 = 0. \tag{1.86}$$

- For $R = 0$:

$$i_0 = V_0\, q \frac{\cosh p\ell}{\sinh p\ell}, \tag{1.87}$$

$$V_1 = 0, \tag{1.88}$$

$$i_1 = \frac{V_0\, q}{\sinh p\ell}. \tag{1.89}$$

- For[12] $r = g = 0$:

$$
\begin{aligned}
i_0 &= V_0 \sqrt{C/L}\, \frac{\cos\sqrt{LC}\omega\ell + jR\sqrt{C/L}\sin\sqrt{LC}\omega\ell}{R\sqrt{C/L}\cos\sqrt{LC}\omega\ell + j\sin\sqrt{LC}\omega\ell}, && (1.90)\\
V_1 &= \frac{V_0 R\sqrt{C/L}}{R\sqrt{C/L}\cos\sqrt{LC}\omega\ell + j\sin\sqrt{LC}\omega\ell}, && (1.91)\\
i_1 &= \frac{V_0 \sqrt{C/L}}{R\sqrt{C/L}\cos\sqrt{LC}\omega\ell + j\sin\sqrt{LC}\omega\ell}. && (1.92)
\end{aligned}
$$

- For $r = g = 0$ and $R = \infty$:

$$
\begin{aligned}
i_0 &= V_0 \sqrt{C/L}\, j\, \frac{\sin\sqrt{LC}\omega\ell}{\cos\sqrt{LC}\omega\ell} = \sqrt{C/L}\, j\, \tan\sqrt{LC}\omega\ell, && (1.93)\\
V_1 &= \frac{V_0}{\cos\sqrt{LC}\omega\ell}, && (1.94)\\
i_1 &= 0. && (1.95)
\end{aligned}
$$

- For $r = g = R = 0$:

$$
\begin{aligned}
i_0 &= -V_0 \sqrt{\frac{C}{L}}\, j\, \frac{1}{\tan\sqrt{LC}\omega\ell}, && (1.96)\\
V_1 &= 0, && (1.97)\\
i_1 &= -V_0 \sqrt{\frac{C}{L}}\, j\, \frac{1}{\sin\sqrt{LC}\omega\ell}. && (1.98)
\end{aligned}
$$

[ ] [13]

---

[12]@ In the original manuscript, the expression for i_1 is lacking.

[13]@ At this point, a crossed out page occurs, whose content is the following:
"If Z is the impedance of the line considered and Y the transfer admittance, denoting by V_0, and i_0 the input potential and current, respectively, and by V_1, and i_1 the output potential and current, we have

$$
\begin{aligned}
V_1 &= V_0 \cosh\sqrt{YZ} + i_0 \frac{Z}{Y} \sinh\sqrt{YZ}, && (1.99)\\
i_i &= i_0 \cosh\sqrt{YZ} + V_0 \frac{Y}{Z} \sinh\sqrt{YZ}. && (1.100)
\end{aligned}
$$

After performing a series expansion, the first terms read

$$
V_1 = V_0 \left(1 + \frac{YZ}{2}\right) + i_0 Z \left(1 + \frac{YZ}{6}\right), \qquad (1.101)
$$

10. DENSITY OF A SPHERICAL MASS DISTRIBUTION

Let us consider a mass distribution (obeying Newton's law) on a spherical surface whose density K is not constant. Denoting by V_0 and K_0 the potential and the mass density, respectively, at point P_0, with V the potential in an arbitrary point P at a distance d from P_0 and with r the radius of the sphere, the following relation holds:

$$K_0 = \frac{1}{4\pi}\left(\frac{V_0}{r} + \int_\sigma \frac{V_0 - V}{\pi d^2}\, \mathrm{d}\sigma\right), \tag{1.107}$$

where the integral is evaluated over the entire spherical surface.

11. LIMIT SKIN EFFECT

Let us consider a conductor having uniform cross section (but arbitrary shape) in which an AC current is flowing. As the frequency of the current increases, the current flows inside an increasingly thin surface layer of the conductor. In the limiting case, we may think of the current as a purely surface phenomenon and consider a linear current density, defined as the current intensity flowing through a length of the conductor cross section edge. In the extreme situation, for a given total current intensity, the surface current density inside the conductor is zero and, clearly, the magnetic field vanishes as well. Now, the magnetic field inside the conductor is due to the current flowing on the surface and to the mag-

---

$$i_i = i_0\left(1 + \frac{YZ}{2}\right) + V_0\, Y\left(1 + \frac{YZ}{6}\right). \tag{1.102}$$

The T−network method should give

$$V_1 = V_0\left(1 + \frac{YZ}{2}\right) + i_0\, Z\left(1 + \frac{YZ}{4}\right). \tag{1.103}$$

$$i_i = i_0\left(1 + \frac{YZ}{2}\right) + V_0\, Y, \tag{1.104}$$

while with the Π−network method we obtain

$$V_1 = V_0\left(1 + \frac{YZ}{2}\right) + i_0\, Z, \tag{1.105}$$

$$i_i = i_0\left(1 + \frac{YZ}{2}\right) + V_0\, Y\left(1 + \frac{YZ}{4}\right). \tag{1.106}$$

netization of the conductor (if this is a magnetic one) in the thin surface layer where the current is flowing. The second contribution to the magnetic field tends to zero, because the volume of the considered surface layer tends to zero, while the magnetization magnitude doesn't increase indefinitely. It then follows that inside the conductor the field produced by the current is also zero in the limiting case. Let us decompose the current that flows through every element of the boundary of the conductor into two components, one with zero phase and the other having phase equal to $\pi/2$. Inside the conductor both the field due to the first component and the field due the second component must vanish. We can replace the elementary currents of equal phase with elementary DC currents having the same effective intensity; the field due to the former is the same as the effective field due to the latter. It is well known that the magnetic field produced by a set of parallel and rectilinear DC currents is orthogonal and numerically equal to the electric field generated by a set of charges distributed along straight lines coinciding with the current axis and having linear charge densities numerically equal to the current intensities. In our case, if we replace the elementary currents flowing through the edge of the cross section of the conductor with such charge densities, we have a surface charge density distribution on the whole conductor that is numerically equal to the linear current density. On the other hand, such a distribution must produce no field inside the conductor, so it must be the same as the distribution of an isolated charged conductor. This distribution is perfectly determined up to a constant factor, and since its surface densities are proportional to the linear (phase 0 or phase $\pi/2$) current densities, we can conclude that:

(1) The elementary currents that flow on the surface of the conductor all have the same phase.
(2) The linear densities of these currents are proportional to the surface density (which has to be computed on the surface elements where the currents are flowing) of a certain charge distribution as in an isolated charged conductor.

Let us now study the dependence of the surface current density upon the depth of the thin conductor layer. Since the thickness of this layer is infinitesimal, we can assume the surface of the conductor to be a plane in a region having a large extension with respect to its thickness, so that both the current density and the field are functions of the depth of the layer only. Let us fix a right-handed Cartesian reference frame with origin on the surface, its x axis along the direction of the current and its z axis along the inward direction normal to the surface. Clearly, neglecting small quantities, the magnetic field is aligned with the y axis, even though its orientation does not necessarily coincide with it. If u is the (complex) current density, H the (complex) magnetic field, and ρ the electric resistivity, neglecting the displacement current which has no relevant role, Maxwell's equations take the form

$$\frac{\partial H}{\partial z} = -4\pi\, u, \tag{1.108}$$

$$\frac{\partial u}{\partial z} = -\frac{\mu\,\omega\, j}{\rho}\, H. \tag{1.109}$$

If the permeability is constant, we obtain the equation

$$\frac{\partial^2 u}{\partial z^2} = \frac{4\pi\,\mu\,\omega\, j}{\rho}\, u, \tag{1.110}$$

whose general solution is

$$u = a\,\mathrm{e}^{\sqrt{2\pi\mu\omega/\rho}\,(1+j)\,z} + b\,\mathrm{e}^{-\sqrt{2\pi\mu\omega/\rho}\,(1+j)\,z}. \tag{1.111}$$

The first term of the expression above must have a zero coefficient, since it diverges for large z; thus we'll have

$$u = u_0\,\mathrm{e}^{-\sqrt{2\pi\mu\omega/\rho}\,(1+j)\,z}. \tag{1.112}$$

This equation represents a damped wave travelling from outside to inside. The damping factor is equal to the dephasing constant, as for heat propagation, and it is $\sqrt{2\pi\mu\omega/\rho} = 2\pi\sqrt{\mu f/\rho}$. The wavelength and the propagation speed are, respectively,

$$\lambda = \sqrt{\frac{2\pi\rho}{\mu\omega}} = \sqrt{\frac{\rho}{\mu f}}, \tag{1.113}$$

$$v = f\,\lambda = \sqrt{\frac{\rho f}{\mu}}, \tag{1.114}$$

while the linear current density is

$$d = \int_0^\infty u\,\mathrm{d}r = \sqrt{\frac{\rho}{4\pi\,\mu\,\omega}}\,\frac{u_0}{\sqrt{\delta}} \tag{1.115}$$

$$= \frac{1}{2\pi}\sqrt{\frac{\rho}{2\mu\, f}}\,\frac{u_0}{\sqrt{\delta}} = \frac{\lambda}{2\pi\sqrt{2}}\,\frac{u_0}{\sqrt{\delta}}. \tag{1.116}$$

It follows that there is a 45° phase difference between the whole current in the conductor and the current flowing in the outer surface layer. In mechanical measure units, the heat per unit time and unit conductor surface produced by the Joule effect is

$$\begin{aligned} q &= \int_0^\infty \rho\,|u|^2\,\mathrm{d}z \;-\; |u_0|^2 \int_0^\infty \rho\,\mathrm{e}^{-4\pi\sqrt{\mu f/\rho}z}\,\mathrm{d}z \\ &= \rho\,|u_0|^2\,\frac{1}{4\pi}\sqrt{\frac{\rho}{\mu\,f}} \;=\; |u_0|^2\,\frac{\rho\,\lambda}{4\pi}. \end{aligned} \tag{1.117}$$

We shall call equivalent a layer of thickness s such that the same quantity of heat is produced when a current flows in it with uniform density at any depth. Then, we would have

$$\rho\,\frac{|d|^2}{s} \;=\; \rho\,|u_0|^2\,\frac{\lambda}{4\pi}; \tag{1.118}$$

and, using Eq. (1.116), we deduce that

$$s \;=\; \frac{\lambda}{2\pi} \;=\; \frac{1}{2\pi}\sqrt{\frac{\rho}{\mu\,f}}. \tag{1.119}$$

As far as the Ohmic resistance is concerned, the current flows inside the equivalent layer with a density that is independent of its depth but which varies along the boundary of the conductor. Thus, it is not correct to compute the resistance per unit length by dividing the resistivity by the area of the cross section of the whole equivalent layer. This is correct only if the cross section is circular, while in all other cases this method underestimates the real values.

Let us now consider a conductor with a circular cross section of radius r. The equivalent cross section has the shape of a ring, with outer radius r and thickness s; its area is then $2\pi rs - \pi s^2$. Note that s is infinitesimal and that it was determined in first approximation, i.e., up to second-order terms. Consequently, in order to prove the consistency of Eq. (1.119) under inclusion of the term πs^2 in the expression for the area above, we should reason as follows. Let us denote by A the area of the equivalent cross section. Then, by using Eq. (1.63), we find approximately

$$\frac{\pi r^2}{A} \;=\; \sqrt{\frac{1}{2}p} + \frac{1}{4} \;=\; \sqrt{\frac{\mu\,\omega}{2\rho}\,\pi r^2} + \frac{1}{4} \;=\; \pi r\sqrt{\frac{\mu\,f}{\rho}} + \frac{1}{4}. \tag{1.120}$$

On multiplying by A (which is a first order infinitesimal) and dividing by the r.h.s. term (which is a first-order infinite), we find, up to third-order terms,

$$A \;=\; \frac{\pi r^2}{\pi r\sqrt{\dfrac{\mu f}{\rho}} + \dfrac{1}{4}} \;=\; r\sqrt{\frac{\rho}{\mu\,f}}\,\frac{1}{1 + \dfrac{1}{4\pi r}\sqrt{\dfrac{\rho}{\mu f}}}$$

$$= r\sqrt{\frac{\rho}{\mu f}} - \frac{1}{4\pi}\frac{\rho}{\mu f}, \tag{1.121}$$

and, from Eq. (1.119), we finally get

$$A = 2\pi r s - \pi s^2 = 2\pi \left(r - \frac{s}{2} \right) s, \tag{1.122}$$

as we anticipated.

Let us now turn to cross sections of arbitrary shape. Our method will be that of considering, as far as the resistance is concerned, circular equivalent cross sections for infinitely large skin effect. We must note, though, that, as the frequency goes to infinity, such equivalence holds, in general, only to first order. Thus, by computing the radius of the equivalent circle and the cross section of the equivalent layer using Eq. (1.122), we will be making a second-order error and not one of order three, as we may think from Eq. (1.122). Despite the fact that the error that we make in estimating A by Eq. (1.122) is of the same order as $-\pi s^2$, nevertheless it is better to take into account such a term, as this yields a better approximation.

Apart from a constant factor, the heat produced per unit time and unit length of the conductor by the linear current density d and by each element $\mathrm{d}\ell$ of its boundary is $d^2 \mathrm{d}\ell$, so that the total heat per unit length and time is

$$Q = c \int d^2 \mathrm{d}\ell, \tag{1.123}$$

while the total current intensity is

$$i = \int d\, \mathrm{d}\ell. \tag{1.124}$$

On replacing the given cross section with the equivalent circle of length p, we find

$$Q = c\,\frac{1}{p} \left(\int d\, \mathrm{d}\ell \right)^2, \tag{1.125}$$

from which one gets

$$p = \left(\int d\, \mathrm{d}\ell \right)^2 \Big/ \int d^2 \mathrm{d}\ell. \tag{1.126}$$

In all cases, we have

$$p \leq \ell. \tag{1.127}$$

Table 1.2. Some values for the radius r of the equivalent circle for the limit skin effect in a conductor with elliptic cross section (with semi-axes a and b). In the last two columns, the radii r_A and r_p of the circles having the same area and the same length, respectively, of the elliptic cross section are also reported.

a	b	r	r_A	r_p
1	0.9	0.949	0.949	0.951
1	0.8	0.897	0.894	0.903
1	0.7	0.843	0.837	0.857
1	0.6	0.787	0.775	0.813
1	0.5	0.728	0.707	0.771
1	0.4	0.666	0.632	0.733
1	0.3	0.598	0.548	0.698
1	0.2	0.520	0.447	0.669
1	0.1	0.425	0.318	0.647
1	0	0	0	0.637

12. LIMIT SKIN EFFECT FOR SIMPLY-SHAPED CONDUCTORS. HINTS FOR ARBITRARY SHAPES

12.1 Elliptic Cross Sections

As is well known, the quantity d introduced in the previous Section is proportional to the projection of the position vector on the direction normal to the conductor surface. Let us consider a conductor with elliptic cross section, whose semi-axes are a and b; then, at an arbitrary point $(a\cos t, b\sin t)$, we have,

$$d = \frac{c}{\sqrt{a^2\sin^2 t + b^2\cos^2 t}}, \tag{1.128}$$

$$\mathrm{d}\ell = \sqrt{a^2\sin^2 t + b^2\cos^2 t}\,\mathrm{d}t, \tag{1.129}$$

where c is a constant. If we denote by r the radius of the equivalent circle, we get, from Eq. (1.126),

$$p = 2\pi r = 4\pi^2 \left(\int_0^{2\pi} \frac{\mathrm{d}t}{\sqrt{a^2\sin^2 t + b^2\cos^2 t}} \right)^{-1}, \tag{1.130}$$

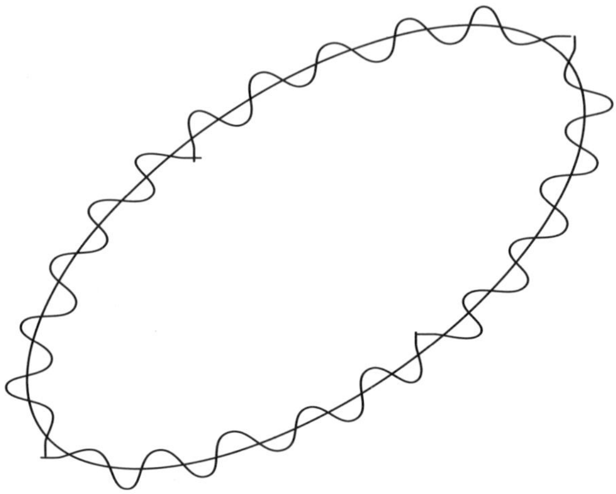

Fig. 1.2. A regular section of a conductor on which a wave-like edge has been superimposed.

or, restricting the integration to 1/4 of the ellipse:

$$r = \frac{\pi}{2}\left(\int_0^{2\pi} \frac{\mathrm{d}t}{\sqrt{a^2 \sin^2 t + b^2 \cos^2 t}}\right)^{-1}. \qquad (1.131)$$

In Table 1.2 we report some numerical values for r. From this Table we see that the equivalent circle is always better approximated by the circle having the same area as the cross section, rather than by the one with the same length; although the ratio between the radii of the equivalent circle and the circle having the same area tends to infinity for infinite eccentricity. Nevertheless, for $b/a = 0.1$, this ratio doesn't reach the value 1.35. The suggestion of some authors of replacing an irregular cross section with the circle having the same length, rather than with the circle of the same area, seems then erroneous, even as an approximation. This conclusion is confirmed by the following discussion.

12.2 Effect of the Irregularities of the Boundary

Let us assume that a cross section with a regular boundary (i.e., one with a radius of curvature that is never too small with respect to the dimensions of the cross section) is replaced with another cross section, nearly overlapping the first but with a wave-like edge. It is clear that the area enclosed by these two differently shaped boundaries will be nearly the same, whereas their length might be significantly different. The

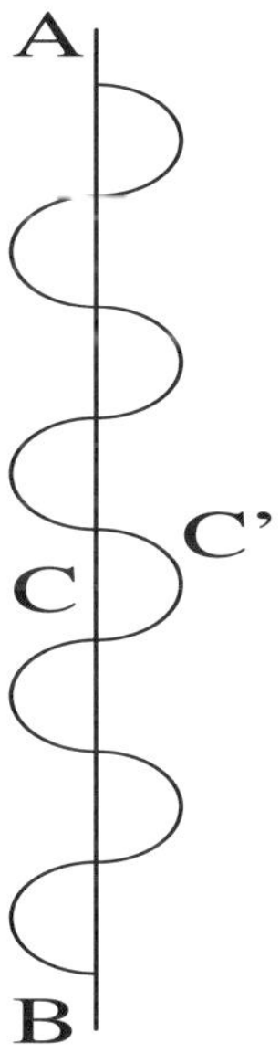

Fig. 1.3. A small part of the section of Figure 1.2 (see text).

question to be addressed is how these differences are reflected in the apparent resistance in the regime of an infinite skin effect. To facilitate the calculations, we shall assume that the wave pattern superimposed on the regular edge is infinitely small. Consider a small part of the boundary of the regular cross section, containing however many oscillations of the edge wave pattern. Let the conductor be charged with a charge q per unit length; if d now is the charge density, the length of the equivalent circumference will be (cfr. Eq. (1.126))

$$p = q^2 \left(\int d^2 \, \mathrm{d}\ell \right)^{-1} . \tag{1.132}$$

Apart from a factor 2π, the term $d^2 \mathrm{d}\ell$ has the same numerical value as the electrostatic stress on the surface element[14] $\mathrm{d}\boldsymbol{\ell} \cdot \hat{\boldsymbol{u}}$. It is thus clear that the charge distribution does not vary much if we take the irregular edge instead of the regular one, as long as we consider parts of it that are long enough to contain many oscillations.

It follows (see Fig. 1.3) that the same electrostatic stress acts on the plane[15] $\hat{\boldsymbol{s}} \otimes \hat{\boldsymbol{u}}$ relative to the regular cross section as well as on $\hat{\boldsymbol{s}}' \otimes \hat{\boldsymbol{u}}$

[14]@ The unit vector $\hat{\boldsymbol{u}}$ represents a generic direction at the considered point of the edge.

[15]@ Using an analogy with the electrostatic stress, the author needs a it surface on which this acts. It is defined as that surface that contains the edge of the cross section (which is a

relative to the wavy one. However, in the first case the resulting stress derives from the composition of nearly parallel elementary stresses, whereas in the second case the elementary stresses are pointing in different directions. It then follows that in the latter case the arithmetic sum of the elementary stresses is larger. Consequently, from Eq. (1.132) we deduce that p, relative to the irregular cross section, is larger than the corresponding one for the regular section, whilst the radius of the equivalent circle is smaller. By combining this result with those we derived above for elliptic cross sections, we can intuitively conclude that, for a rather long cross section with small irregularities (such as that of a railroad), the radius of the equivalent circle is only slightly larger than that of the circle having the same area as the cross section, while it is sensibly smaller than the radius of the circle having the same length.

13. HYSTERESIS IN MAGNETIC CONDUCTORS IN THE LIMIT SKIN EFFECT REGIME

The results in Sec. 1.11 for the limit skin effect were obtained by assuming a constant magnetic permeability and neglecting hysteresis. To some extent the effects of hysteresis can be roughly taken into account by introducing a phase delay α in the induction field with respect to the magnetic field. Using the notation of the symbolic method, the quantity μ, which is the ratio between different AC quantities having different phase values, is imaginary and has argument $-\alpha$, which we shall regard as constant in order to facilitate the calculations. Let us write μ as

$$\mu = \mu_0 \, \mathrm{e}^{-i\alpha}. \qquad (1.133)$$

All the formulae in Sec. 1.11 in symbolic notation will still hold as long as μ is treated as a complex number. If we then insert Eq. (1.133) into Eqs. (1.112) and (1.115), we get

$$u = u_0 \exp\{-2\pi\sqrt{\mu_0 f/\rho}\,[\cos(45^\circ - \alpha/2) + j\sin(45^\circ - \alpha/2)]\, z\} \qquad (1.134)$$

one-dimensional line, which thus does not suffice to define the surface) and an auxiliary unit vector. In the same manner, a surface is also defined in the case of a wavy-edge cross section. The unit vectors $\hat{s}$ and $\hat{s}'$ are the curvilinear abscissas on the "regular" and the "irregular" edge, respectively.

and

$$d = \frac{u_0}{2\pi}\sqrt{\frac{\rho}{2\mu_0 f}}\left[\cos\left(45^\circ - \frac{\alpha}{2}\right) - j\,\sin\left(45^\circ - \frac{\alpha}{2}\right)\right]. \tag{1.135}$$

It follows that the delay between the current and the electric field on the surface of the conductor is $45^\circ - \alpha/2$. Equations (1.113), (1.114), and (1.119) then become

$$\lambda = \sqrt{\frac{\rho}{2\mu_0 f}}\,\frac{1}{\sin(45^\circ - \alpha/2)}, \tag{1.136}$$

$$v = \sqrt{\frac{\rho f}{2\mu_0}}\,\frac{1}{\sin(45^\circ - \alpha/2)}, \tag{1.137}$$

$$q_1 = |u_0|^2\,\frac{1}{4\pi}\sqrt{\frac{\rho}{2\mu_0 f}}\,\frac{\rho}{\cos(45^\circ - \alpha/2)}, \tag{1.138}$$

$$s_1 = \frac{\cos(45^\circ - \alpha/2)}{\pi}\sqrt{\frac{\rho}{2\mu_0 f}}. \tag{1.139}$$

From Eqs. (1.136) and (1.139), it follows that the hysteresis increases the wavelength and decreases the losses due to Joule effect. However, the quantity q_1 in Eq. (1.138) is the heat due only to the Joule effect, and in Eq. (1.139) s_1 is not the true thickness of the equivalent layer, but only the one related to the Joule effect. We shall then denote by q the total energy loss and by s the thickness of the actual equivalent layer, also accounting for hysteresis. The energy flowing in a unit time through the surface of the conductor that is transformed into heat is, by the Poynting theorem,

$$q = \frac{E\,H}{4\pi}\,\cos\phi, \tag{1.140}$$

where E is the effective electric field at the surface of the conductor, H the effective magnetic field, and ϕ the phase difference between the electric and magnetic fields. In our case we'll have

$$E = |u_0|\,\rho, \tag{1.141}$$

$$H = 4\pi\,|d| = 2\,u_0\sqrt{\frac{\rho}{2\mu_0 f}}, \tag{1.142}$$

$$\phi = 45^\circ - \frac{\alpha}{2}, \tag{1.143}$$

and thus

$$q = \frac{\rho\,|u_0|^2}{2\pi}\sqrt{\frac{\rho}{2\mu_0 f}}\,\cos\left(45^\circ - \frac{\alpha}{2}\right), \tag{1.144}$$

$$s = \frac{1}{2\pi\,\cos(45^\circ - \alpha/2)}\sqrt{\frac{\rho}{2\mu_0 f}}. \tag{1.145}$$

Denoting by q_2 the heat related to hysteresis only, we get

$$q_2 = q - q_1 = \frac{\rho\,|u_0|^2}{4\pi}\sqrt{\frac{\rho}{2\mu_0 f}}\,\frac{\sin\alpha}{\cos(45^\circ - \alpha/2)}, \tag{1.146}$$

$$\frac{q_2}{q_1} = \sin\alpha. \tag{1.147}$$

Introducing then the heat q_0 that would be produced for the same current if there were no hysteresis or for the same value of $|u_0|$ [which is the same, because of Eq. (1.135)], we find

$$q_0 = \rho\,|u_0|^2\,\frac{1}{4\pi}\sqrt{\frac{\rho}{\mu_0 f}}, \tag{1.148}$$

$$\frac{q}{q_0} = \cos\frac{\alpha}{2} + \sin\frac{\alpha}{2}, \tag{1.149}$$

$$\frac{q_0 - q_1}{q_2} = \frac{\sin\alpha/2 + \cos\alpha/2 - 1}{\sin\alpha}. \tag{1.150}$$

From the last equation, we note that for a small hysteresis one half of the loss due to hysteresis is compensated by the decreased loss due to the Joule effect; in the strong hysteresis regime this balancing is relatively smaller. From Eq. (1.147) we can also see that the ratio between the loss due to hysteresis and that due to Joule effect is independent of the frequency.

Finally, we observe that if on a straight line we consider the points O, Q_0, Q_1, and Q, with $\overline{OQ_0} = q_0$, $\overline{OQ_1} = q_1$, $\overline{OQ} = q$, we obtain a harmonic group.[16]

14. FIELD PRODUCED BY A CIRCULAR AND HOMOGENEOUS DISTRIBUTION OF CHARGES IN ITS OWN PLANE[17]

Let r be the radius of a circular charge distribution whose linear density is K. If x is the distance from the center, the field inside the circle is

[16]@ Such geometric properties seem to derive from Eqs. (1.147), (1.149), and (1.150).

[17]@ In the original manuscript, the heading refers to Newtonian masses rather than to charges, while the material presented both in this and in the following Section leads one to consider charges and the electric or magnetic field produced by them. However, the results are quite general, due to the similarity between the Newton and Coulomb laws.

given by[18]

$$\mathbf{E} = \frac{2\pi K}{r}\left(\frac{1}{2}\cdot\frac{x}{r} + \frac{1}{2}\cdot\frac{9}{8}\cdot\frac{x^3}{r^3} + \frac{1}{2}\cdot\frac{9}{8}\cdot\frac{25}{24}\cdot\frac{x^5}{r^5} + \dots\right)\hat{\boldsymbol{r}}\,, \tag{1.151}$$

whereas outside the circle the field is

$$\mathbf{E} = \frac{2\pi K}{r}\left(\frac{r^2}{x^2} + \frac{3}{4}\cdot\frac{r^4}{x^4} + \frac{3}{4}\cdot\frac{15}{16}\cdot\frac{r^6}{x^6} + \frac{3}{4}\cdot\frac{15}{16}\cdot\frac{35}{36}\cdot\frac{r^8}{x^8} + \dots\right)\hat{\boldsymbol{r}}\,. \tag{1.152}$$

In both the series, which always converge, the coefficients of the terms $(x/r)^{\pm n}$ approach the asymptotic value $2/\pi$ as $n \to \infty$.

15. FIELD PRODUCED BY A CIRCULAR CHARGE CURRENT IN A PLANE

Let i be the current intensity and r the radius of the circle. If x is the distance from the center of the circle, the field inside it is[19]:

$$\mathbf{H} = \frac{2\pi i}{r}\left(1 + \frac{3}{4}\cdot\frac{x^2}{r^2} + \frac{3}{4}\cdot\frac{15}{16}\cdot\frac{x^4}{r^4} + \frac{3}{4}\cdot\frac{15}{16}\cdot\frac{35}{36}\cdot\frac{x^6}{r^6} + \dots\right)\hat{\boldsymbol{n}}\,, \tag{1.153}$$

while outside it reads:

$$\mathbf{H} = -\frac{2\pi i}{r}\left(\frac{1}{2}\cdot\frac{r^3}{x^3} + \frac{1}{2}\cdot\frac{9}{8}\cdot\frac{r^5}{x^5} + \frac{1}{2}\cdot\frac{9}{8}\cdot\frac{25}{24}\cdot\frac{r^7}{x^7} + \frac{1}{2}\cdot\frac{9}{8}\cdot\frac{25}{24}\cdot\frac{49}{48}\cdot\frac{r^9}{x^9} + \dots\right)\hat{\boldsymbol{n}}\,. \tag{1.154}$$

The above equations can easily be derived from those of the previous paragraph.

[18]@ In the following two equations, $\hat{\boldsymbol{r}}$ denotes the radial unit vector. In the original manuscript the author did not use a vector notation for the field **E**, and thus its direction (given by $\hat{\boldsymbol{r}}$) is not explicitly specified. Also in other places of this book the author regarded as understood the vector directions.

[19]@ In the following two equations, $\hat{\boldsymbol{n}}$ denotes the unit vector normal to the plane considered here. In the original manuscript the author did not write **H** as a vector, and thus its direction (given by $\hat{\boldsymbol{n}}$) is left implicit.

16. WEAK SKIN EFFECT IN CONDUCTORS, WITH AN ELLIPTIC CROSS SECTION, HAVING THE SAME MAGNETIC PERMEABILITY AS THE SURROUNDING MEDIUM

For the weak skin effect, the apparent resistance in an AC conductor may be cast in the form

$$R_a = R_c \left(1 + c p^2\right), \tag{1.155}$$

where R_c is the DC resistance and $p = \mu\omega/\rho$, with μ the magnetic permeability of the conductor and ρ its resistance per unit length. The coefficient c depends on the shape of the cross section and on the permeability of the conductor, as well as on that of the medium; for conductors with circular cross section, c is always $1/12$. When both the medium and the conductor have the same permeability, c becomes a shape coefficient. In general, c can be computed assuming that the difference in the electromotive force between two current lines due to flux variations inside the conductor is, to a first approximation, the same as that one would observe with a uniform distribution of current. If we have an elliptical cross section defined by the equation $x^2/a^2 + y^2/b^2 = 1$, and if the conductor and the medium have the same permeability, then the difference in the electromotive force between the central current line and the line that intersects the cross section in the point (x, y) is

$$E = 2\pi\,\mu\,\omega\,u \left(\frac{b}{a+b}\,x^2 + \frac{a}{a+b}\,y^2\right), \tag{1.156}$$

where u is the current density. The electromotive force is at 90° with respect to the current. It is then straightforward to compute c; we find

$$c = \frac{3a^2 - 2ab + 3b^2}{12(a+b)^2}, \tag{1.157}$$

or, on setting $k = b/a$,

$$c = \frac{3 - 2k + 3k^2}{12(1+k)^2}. \tag{1.158}$$

In Table 1.3 we show the value of the coefficient c for some vales of k.

Table 1.3. Some values of the shape coefficient c for conductors with elliptic cross sections, with axis ratio k. In the last column we also report the difference between the coefficient c for a given k and for the previous k value.

k	c	$(c(k_i) - c(k_{i-1})) \times 10^4$
1.00	0.0833	
0.90	0.0838	5
0.80	0.0854	16
0.70	0.0885	31
0.60	0.09375	52
0.50	0.1019	82
0.40	0.1139	120
0.30	0.1317	178
0.20	0.1574	257
0.10	0.1949	375
0.00	0.2500	551

17. OSCILLATING DISCHARGES IN CAPACITORS

If a capacitor with capacitance C charged with a charge Q is closed in a circuit with a resistance R and self-induction coefficient L, then an oscillating discharge is produced, provided that the resistance is not too high. Let T be the period of the oscillation, t the time it takes the current to reach the maximum intensity $i_{\max}$, and k the ratio of the current intensity at time t and at time $t-T$. On setting $R_1 = R/\sqrt{4L/C}$, the following relations are seen to hold:

(a) For $R_1 < 1$:

$$T = \frac{2\pi\sqrt{LC}}{\sqrt{1-R_1^2}}, \tag{1.159}$$

$$t = \frac{\sqrt{LC}}{\sqrt{1-R_1^2}} \arccos R_1, \tag{1.160}$$

$$\frac{t}{T} = \frac{\arccos R_1}{2\pi}, \tag{1.161}$$

$$\begin{aligned} i_{\max} &= \frac{Q}{\sqrt{LC}} \exp\left\{-\frac{Rt}{2L}\right\} = \frac{Q}{\sqrt{LC}} \exp\left\{-R_1 \frac{\arccos R_1}{\sqrt{1-R_1^2}}\right\} \qquad (1.162) \\ &= \frac{Q}{\sqrt{LC}} \exp\left\{-\frac{RT}{2L}\frac{\arccos R_1}{2\pi}\right\} = \frac{Q}{\sqrt{LC}} \exp\left\{-\frac{tR_1}{\sqrt{LC}}\right\}, \end{aligned}$$

$$k = \exp\left\{-\frac{TR_1}{\sqrt{LC}}\right\} = \exp\left\{-\frac{R_1 2\pi}{\sqrt{1-R_1^2}}\right\}. \tag{1.163}$$

(b) For $R_1 > 1$:

$$t = \sqrt{LC}\,\frac{\log\left(R_1 + \sqrt{R_1^2-1}\right)}{\sqrt{R_1^2-1}}, \tag{1.164}$$

$$\begin{aligned} i_{\max} &= \frac{Q}{\sqrt{LC}}\left(R_1 + \sqrt{R_1^2-1}\right)^{-\frac{R_1}{\sqrt{R_1^2-1}}} \\ &= \frac{Q}{\sqrt{LC}}\exp\left\{\frac{t-R_1}{\sqrt{LC}}\right\}. \end{aligned} \tag{1.165}$$

(c) For very large R_1:

$$t = \sqrt{LC}\,\frac{\log 2R_1}{R_1} = \frac{2L}{r}\log 2R_1, \tag{1.166}$$

$$i_{\max} = \frac{Q}{\sqrt{LC}}\left(\frac{1}{2R_1} - \frac{\log 2R_1 - 1/2}{4R_1^3}\right). \tag{1.167}$$

In Table 1.4 we list the values assumed by these quantities as R_1 varies.

18. SELF-INDUCTION OF A VERY LONG CIRCULAR COIL WITH MANY TURNS

If a current i circulates in the coil, on equating the electromagnetic energy of the system to $(1/2)Li^2$, we obtain

$$L = 4\pi^2 n^2 \ell\left(\frac{1}{2}r_1^2 + \frac{1}{3}r_1 r_2 + \frac{1}{6}r_2^2\right), \tag{1.168}$$

with n denoting the total number of turns per cm, ℓ the length of the coil, r_1 its inner radius, and r_2 its outer radius. The previous equation may also be written as

$$L = 4\pi n^2 \ell S, \tag{1.169}$$

with

$$S = \frac{3S_1 + 2\sqrt{S_1 S_2} + S_2}{8}, \tag{1.170}$$

Table 1.4. Oscillating discharges in capacitors: numeric values for some quantities defined in the text.

$R/\sqrt{\frac{4L}{C}}$	$\frac{T}{2\pi\sqrt{LC}}$	$\frac{t}{2\pi\sqrt{LC}}$	$\frac{t}{T}$	$i_{\max}/\frac{Q}{\sqrt{LC}}$	k
0	1.000	0.2500	0.2500	1.000	1.000
0.1	1.005	0.2352	0.2341	0.863	0.532
0.2	1.021	0.2224	0.2180	0.756	0.277
0.3	1.048	0.2112	0.2015	0.672	0.139
0.4	1.091	0.2013	0.1845	0.603	0.064
0.5	1.155	0.1925	0.1667	0.546	0.027
0.6	1.250	0.1845	0.1476	0.499	0.0090
0.7	1.400	0.1773	0.1266	0.459	0.0021
0.8	1.667	0.1707	0.1024	0.424	0.00023
0.9	2.294	0.1647	0.0718	0.394	0.000002
1	∞	0.1592	0.0000	0.368	0.000
2		0.1210		0.218	
10		0.0479		0.049	
100		0.0084		0.005	

where S_1 and S_2 are the inner and outer cross sections of the coil, respectively. If the relative difference between S_2 and S_1 is not very large, we can approximately express S as

$$S = \frac{1}{3}(2S_1 + S_2). \tag{1.171}$$

Clearly, Eq. (1.169) also holds for non-circular cross sections, as long as the various layers of turns are uniformly stacked and have "homothetic" cross sections.[20]

19. ENERGY OF A UNIFORM CIRCULAR DISTRIBUTION OF ELECTRIC OR MAGNETIC CHARGES

Let R be now the radius of a circular plate α on which the charges are uniformly distributed, ρ the charge density and Q the total charge. The

[20]@ Namely, whose surfaces are connected by an homothety.

potential at the center of the circle is

$$V_0 = 2\pi R\rho = \frac{2}{R}Q. \tag{1.172}$$

Having fixed a coordinate system $Oxyz$ with its origin in the center of the circle and with its z-axis perpendicular to it, the following relation holds for the components E_x, E_y, E_z of the field at any point outside the charge distribution:

$$\frac{\partial E_x}{\partial x} + \frac{\partial E_y}{\partial y} + \frac{\partial E_z}{\partial z} = 0. \tag{1.173}$$

This equation is no longer valid at the position of the charges, on the circle, since $\partial E_z/\partial z$ is infinite there. In spite of this, Eq. (1.173) can still be used for these points, if we replace the value of $\partial E_z/\partial z$ at any one of them with its limiting value, got by considering an infinitesimal region near, but external, to the plane xy. Now, in general, we have

$$E_z = \rho\,\omega, \tag{1.174}$$

where ω is the solid angle subtended by an infinitesimal area surrounding an arbitrary point of the circle α. Thus,

$$\frac{\partial E_z}{\partial z} = \rho\,\frac{\partial \omega}{\partial z}. \tag{1.175}$$

Note, from this expression, that $\partial E_r/\partial r$ takes the same numerical value (apart from the sign) as the component H_z of the magnetic field produced by a current of intensity ρ flowing along the circumference of α. The expression for such a quantity is known in terms of a series expansion (see Sec. 1.15). On substituting it into Eq. (1.175), we can derive, for the points inside α, that

$$\frac{\partial E_z}{\partial z} = -\frac{2\pi\rho}{R}\left(1 + \frac{3}{4}\cdot\frac{r^2}{R^2} + \frac{3}{4}\cdot\frac{15}{16}\cdot\frac{r^4}{R^4} + \frac{3}{4}\cdot\frac{15}{16}\cdot\frac{35}{36}\cdot\frac{r^6}{R^6} + \ldots\right), \tag{1.176}$$

quantity r denoting the distance from the center.

The component E_r of the field in the xy plane is radial and in this plane depends only on r. We have

$$E_x = \frac{x}{r}\,E_r, \tag{1.177}$$

$$E_y = \frac{y}{r}\,E_r. \tag{1.178}$$

Taking their derivatives and substituting them into the Laplace equation,[21] we find

$$\frac{E_r}{r} + \frac{\partial E_r}{\partial r} = \frac{2\pi\rho}{R}\left(1 + \frac{3}{4}\cdot\frac{r^2}{R^2} + \frac{3}{4}\cdot\frac{15}{16}\cdot\frac{r^4}{R^4} + \frac{3}{4}\cdot\frac{15}{16}\cdot\frac{35}{36}\cdot\frac{r^6}{R^6} + \dots\right). \tag{1.179}$$

This equation allows us to express E_r as a series expansion in r; one gets

$$E_r = \frac{\pi\rho}{R}\left(r + \frac{1}{2}\cdot\frac{3}{4}\cdot\frac{r^3}{R^2} + \frac{1}{3}\cdot\frac{3}{4}\cdot\frac{15}{16}\cdot\frac{r^5}{R^4} + \frac{1}{4}\cdot\frac{3}{4}\cdot\frac{15}{16}\cdot\frac{35}{36}\cdot\frac{r^7}{R^6} + \dots\right). \tag{1.180}$$

The potential at a distance r from the center will be

$$\begin{aligned} V &= V_0 - \int_0^r E_r \,\mathrm{d}r \\ &= \frac{Q}{R}\left(2 - \frac{1}{2}\cdot\frac{r^2}{R^2} - \frac{1}{2}\cdot\frac{1}{2^2}\cdot\frac{3}{4}\cdot\frac{r^4}{R^4} - \frac{1}{2}\cdot\frac{1}{3^2}\cdot\frac{3}{4}\cdot\frac{15}{16}\cdot\frac{r^6}{R^6}\right. \\ &\qquad \left. - \frac{1}{2}\cdot\frac{1}{4^2}\cdot\frac{3}{4}\cdot\frac{15}{16}\cdot\frac{35}{36}\cdot\frac{r^8}{R^8} + \dots\right) \\ &= \frac{Q}{R}\left[2 - \frac{1}{2}\cdot\left(\frac{r^2}{R^2} + \frac{1}{4}\cdot\frac{3}{4}\cdot\frac{r^4}{R^4} + \frac{1}{9}\cdot\frac{3}{4}\cdot\frac{15}{16}\cdot\frac{r^6}{R^6}\right.\right. \\ &\qquad \left.\left. + \frac{1}{16}\cdot\frac{3}{4}\cdot\frac{15}{16}\cdot\frac{35}{36}\cdot\frac{r^8}{R^8} + \dots\right)\right] \qquad (1.181) \\ &= \frac{Q}{R}\frac{2}{\pi}\int_0^\pi \frac{1 + (r/R)\cos\alpha}{\sqrt{1 + 2(r/R)\cos\alpha + (r^2/R^2)}}\,\mathrm{d}\alpha. \qquad (1.182) \end{aligned}$$

The potential V_R on the boundary is

$$V_R = \frac{Q}{R}\left[2 - \frac{1}{2}\left(4 - \frac{8}{\pi}\right)\right] = \frac{Q}{R}\frac{4}{\pi} = 1.2732\,\frac{Q}{R}. \tag{1.183}$$

The mean potential V_{m} will be

$$\begin{aligned} V_{\mathrm{m}} &= \int_0^R V\, r\,\mathrm{d}r \Big/ \int_0^R r\,\mathrm{d}r \\ &= \frac{Q}{R}\left[2 - \frac{1}{2}\cdot\left(\frac{1}{1\cdot 2} + \frac{1}{4\cdot 3}\cdot\frac{3}{4} + \frac{1}{9\cdot 4}\cdot\frac{3}{4}\cdot\frac{15}{16} + \dots\right.\right. \\ &\qquad \left.\left. + \frac{1}{n^2(n-1)}\cdot\frac{3}{4}\cdot\frac{15}{16}\cdot\frac{35}{36}\cdot\frac{63}{64}\cdots\frac{4(n-1)^2 - 1}{4(n-1)^2}\cdots + \dots\right)\right] \end{aligned}$$

[21]@ Or, more precisely, into the first one of Maxwell's equations.

$$
\begin{aligned}
&= \frac{Q}{R}\left[2 - \frac{1}{2}\left(4 - \frac{32}{3\pi}\right)\right] = \frac{Q}{R}\frac{16}{3\pi} = \frac{16}{3}R\rho \\
&= \frac{4}{3}V_R = 1.69765\,\frac{Q}{R}.
\end{aligned}
\tag{1.184}
$$

The energy of the distribution then becomes

$$
E = \frac{1}{2}Q\,V_{\mathrm{m}} = \frac{Q^2}{R}\frac{8}{3\pi} = 0.84883\,\frac{Q^2}{R}. \tag{1.185}
$$

Just to make a comparison, it is interesting to note that the potential due to a charge Q distributed on a conducting circular plate of radius R is $(\pi/2)Q/R$ in the absence of other conductors. It follows that the ratio of the energy associated with the uniform charge distribution to the energy corresponding to the minimum energy configuration is

$$
\frac{V_{\mathrm{m}}}{\pi Q/2R} = \frac{16/3\pi}{\pi/2} = \frac{32}{3\pi^2} = 1.08076. \tag{1.186}
$$

20. SELF-INDUCTION IN A RECTILINEAR COIL WITH FINITE LENGTH

Were the coil of infinite length, then, regardless of the winding width and of the shape of its cross section, the field at any point P would be directed along the coil and would have the magnitude $4\pi ni$, where i is the current intensity and n is the number of turns per unit length. If, by contrast, the coil has a finite length, the resulting field gets one more contribution. In this case we should in fact also add the field generated by two surface distributions of magnetic charges σ_1 and σ_2, located at the beginning and at the end of the coil, with surface density ni and $-ni$, respectively. Let us assume that a unitary current intensity is flowing in the coil; then the densities of the distributions σ_1 and σ_2 can be simply written n and $-n$. If we now fix a reference frame with its x axis along the direction of the coil, the components of the resulting field at an arbitrary point will be

$$
\begin{aligned}
H_x &= 4\pi\,n + H'_x, \\
H_y &= H'_y, \\
H_z &= H'_z,
\end{aligned}
\tag{1.187}
$$

H'_x, H'_y, H'_z being the contributions to the total field from the σ_1 and σ_2 distributions. The total energy of the system is

$$\epsilon = \frac{1}{8\pi} \int (H_x^2 + H_y^2 + H_z^2)\,\mathrm{d}V, \tag{1.188}$$

where the integral extends over all of space. Since $i = 1$, it is also true that

$$\epsilon = \frac{1}{2}\,L, \tag{1.189}$$

which yields

$$\begin{aligned} L &= \frac{1}{4\pi} \int (H_x^2 + H_y^2 + H_z^2)\,\mathrm{d}V \\ &= \int 4\pi\, n^2\,\mathrm{d}V \;+\; 1\overline{8}\pi \int (H_x'^2 + H_y'^2 + H_z'^2)\,\mathrm{d}V \;+\; \int 2n\, H'_x\,\mathrm{d}V. \end{aligned} \tag{1.190}$$

The first of the three terms is the self-induction L_1 of the coil when its transverse dimensions are negligible compared to its length or, more precisely, if the flux in each turn of the coil is the same as for a coil of infinite length. The second term is twice the proper energy ϵ' of the joint σ_1 and σ_2 distributions.

As for the third term, we note that the coefficient n must be taken as zero not only around the coil but also next to its ends. Then, the only place where the integrand does not vanish is on the cylindrical region that has such end surfaces as bases. On carrying out first the integration with respect to x, and letting ℓ denote the length of the coil and S its normal section, one finds

$$\int 2n\, H'_x\,\mathrm{d}V = \int_S \mathrm{d}y\,\mathrm{d}z \int_a^{a+\ell} 2n\, H'_x\,\mathrm{d}x, \tag{1.191}$$

a being the abscissa of the lower side of the coil. Now, the integral $\int_a^{a+\ell} H'_x \mathrm{d}x$ is the difference in the magnetic potential due to the σ_1 and σ_2 distributions between two corresponding points in the outermost cross sections of the coil. If E is the magnetic potential due to σ_1 and σ_2 at a point on the "upper" side, then, by symmetry, the potential at the corresponding point on the "lower" side must be $-E$. It is then possible to rewrite the foregoing equation as

$$\int 2n\, H'_x\,\mathrm{d}V = -\int_S 4n\, E\,\mathrm{d}y\,\mathrm{d}z. \tag{1.192}$$

Since $\int_S nE\mathrm{d}x\mathrm{d}z = \epsilon'$, by substitution into Eq. (1.190), we have $L = L_1 + 2\epsilon' - 4\epsilon' = L_1 - 2\epsilon'$, and, on setting

$$L = K\,L_1, \tag{1.193}$$

we finally obtain

$$K = 1 - \frac{2\epsilon'}{L_1}. \tag{1.194}$$

21. MEAN DISTANCES OF VOLUME, SURFACE, OR LINE ELEMENTS[22]

(1) Harmonic mean of the distances among the volume elements of a sphere of radius R:

$$d_{\rm m} = \frac{5}{6}R = 0.8333\,R. \tag{1.195}$$

(2) Harmonic mean of the distances among the surface elements of a circle of radius R:

$$d_{\rm m} = \frac{3\pi}{16}R = 0.58905\,R. \tag{1.196}$$

(3) Geometric mean of the distances among the surface elements of a circle of radius R:

$$d_{\rm m} = R\,{\rm e}^{-1/4} = 0.7788\,R. \tag{1.197}$$

(4) Geometric mean of the distances among the elements of a rectilinear segment of length a:

$$d_{\rm m} = R\,{\rm e}^{-3/2} = 0.2231\,a. \tag{1.198}$$

(5) Arithmetic mean of the distances among the surface elements of a circle of radius R:

$$d_{\rm m} = \frac{128}{45\pi}R = 0.9054\,R. \tag{1.199}$$

(6) Root mean square of the distances among surface elements of a circle of radius R:

$$d_{\rm m} = R. \tag{1.200}$$

[22]See Sec. 2.38.6.

(7) The n-th root of the mean value of the n-th power of the distances among the surface elements of a circle of radius R for even and odd n, respectively, are

$$d_{\rm m} = 2R\sqrt[n]{\frac{16}{(n+2)(n+4)}\,\frac{1{\cdot}3{\cdot}5{\cdot}\cdots{\cdot}(n+1)}{2{\cdot}4{\cdot}6{\cdot}\cdots{\cdot}(n+2)}}, \qquad (1.201)$$

$$d_{\rm m} = 2R\sqrt[n]{\frac{32}{\pi(n+2)(n+4)}\,\frac{2{\cdot}4{\cdot}6{\cdot}\cdots{\cdot}(n+1)}{3{\cdot}5{\cdot}7{\cdot}\cdots{\cdot}(n+2)}}. \qquad (1.202)$$

22. EVALUATION OF SOME SERIES[23]

(1)
$$\left(1-\frac{2}{\pi}\right)+\frac{1}{2}\left(\frac{3}{4}-\frac{2}{\pi}\right)+\frac{1}{3}\left(\frac{3}{4}\frac{15}{16}-\frac{2}{\pi}\right)+\ldots = \frac{4\log 2-2}{\pi/2}. \qquad (1.203)$$

(2)
$$1+\frac{1}{4}\cdot\frac{3}{4}+\frac{1}{9}\cdot\frac{3{\cdot}15}{4{\cdot}16}+\frac{1}{16}\cdot\frac{3{\cdot}15{\cdot}35}{4{\cdot}16{\cdot}36}+\frac{1}{25}\cdot\frac{3{\cdot}15{\cdot}35{\cdot}63}{4{\cdot}16{\cdot}36{\cdot}64}+\ldots = 4-\frac{8}{\pi}. \qquad (1.204)$$

(3)
$$\frac{1}{2}\cdot\frac{1}{1^2}+\frac{1}{3}\cdot\frac{1}{2^2}\cdot\frac{3}{4}+\frac{1}{4}\cdot\frac{1}{3^2}\cdot\frac{3{\cdot}15}{4{\cdot}16}+\frac{1}{5}\cdot\frac{1}{4^2}\cdot\frac{3{\cdot}15{\cdot}35}{4{\cdot}16{\cdot}36}+\ldots = 4-\frac{32}{3\pi}. \qquad (1.205)$$

(4)
$$1+\frac{1}{4}+\frac{1}{9}+\ldots = \frac{\pi^2}{6}. \qquad (1.206)$$

(5)[24]
$$1+\frac{1}{8}+\frac{1}{27}+\frac{1}{64}\ldots = 1.2021. \qquad (1.207)$$

[23]See Secs. 2.28 and 3.1.

[24]@ In the original manuscript, the sum of this series, which is the Riemann $\zeta(s)$ function for $s=3$, is not reported.

(6)
$$1+\frac{1}{16}+\frac{1}{81}+\frac{1}{256}+\ldots = \frac{\pi^4}{90}. \tag{1.208}$$

(7)
$$1+x+x^2+\ldots = \frac{1}{1-x}. \tag{1.209}$$

(8)
$$x+2x^2+3x^3+\ldots = \frac{x}{(1-x)^2}. \tag{1.210}$$

(9)
$$x+4x^2+9x^3+16x^4+\ldots = \frac{x(1+x)}{(1-x)^3}. \tag{1.211}$$

(10)
$$x+8x^2+27x^3+64x^4+\ldots = \frac{x(1+4x+x^2)}{(1-x)^4}. \tag{1.212}$$

(11)
$$\sin x+\frac{1}{3}\sin 3x+\frac{1}{5}\sin 5x+\ldots = \frac{\pi}{4}, \quad 0<x<\pi. \tag{1.213}$$

(12)
$$\sin x+\frac{1}{2}\sin 2x+\frac{1}{3}\sin 3x+\ldots = \frac{\pi+x}{2}, \quad 0<x<2\pi. \tag{1.214}$$

(13)
$$\cos x+\cos 2x+\cos 3x+\ldots+\cos nx$$
$$= \frac{\sin\left(n+1/2\right)x}{2\sin x/2}-\frac{1}{2}. \tag{1.215}$$

(14)
$$\frac{\sin^2 x}{1}+\frac{\sin^2 2x}{4}+\frac{\sin^2 3x}{9}+\ldots = x\,\frac{\pi-x}{2}, \quad 0<x<\pi. \tag{1.216}$$

(15)
$$\sin^2 x+\frac{\sin^2 3x}{9}+\frac{\sin^2 5x}{25}+\frac{\sin^2 7x}{49}+\ldots = \frac{\pi}{4}\,x,$$
$$0<x<\pi/2. \tag{1.217}$$

(16)
$$\frac{\cos x}{1}+\frac{\cos 2x}{4}+\frac{\cos 3x}{9}+\ldots = \frac{1}{4}\,x^2-\frac{\pi}{2}\,x+\frac{1}{6}\,\pi^2,$$
$$0<x<2\pi. \tag{1.218}$$

23. SELF-INDUCTION OF A FINITE LENGTH RECTILINEAR COIL WITH CIRCULAR CROSS SECTION AND A FINITE (SMALL) NUMBER OF TURNS[25]

Let us consider a coil with N turns, length ℓ, and diameter d; its self-induction coefficient can be cast in the form

$$L = K\,\pi^2\,d^2\,\frac{N^2}{\ell}, \tag{1.219}$$

where K is a numerical coefficient lower than 1 and approaching 1 as the ratio d/ℓ decreases. This coefficient can be computed from an expression given in Sec. 1.20. If $d/\ell \leq 1$, we can use the series expansion

$$\begin{aligned} K \;=\; & 1 - \frac{4}{3\pi}\,\frac{d}{\ell} + \frac{1}{8}\left(\frac{d}{\ell}\right)^2 - \frac{1}{64}\left(\frac{d}{\ell}\right)^4 + \frac{5}{1024}\left(\frac{d}{\ell}\right)^6 \\ & - \frac{35}{16384}\left(\frac{d}{\ell}\right)^8 + \frac{147}{131072}\left(\frac{d}{\ell}\right)^{10} + \ldots \\ & \pm \frac{1}{(n+1)(2n-1)}\left(\frac{1{\cdot}3{\cdot}5\cdots(2n-1)}{2{\cdot}4{\cdot}6\cdots 2n}\right)^2\left(\frac{d}{\ell}\right)^{2n} \mp \ldots . \end{aligned} \tag{1.220}$$

If, instead, we set $p = d^2/(\ell^2 + d^2)$, the following series expansion holds in all cases:

$$\begin{aligned} K \;=\; & 1 - \frac{4}{3\pi}\,\frac{d}{\ell} + \frac{1}{8}\,p + \frac{7}{64}\,p^2 + \frac{101}{1024}\,p^3 + \frac{1485}{16384}\,p^4 \\ & + \frac{11059}{131072}\,p^5 + \frac{83139}{1048576}\,p^6 + \ldots . \end{aligned} \tag{1.221}$$

Denoting by $b_n p^n$ the generic p^n term in the above series and by $a_n (d/\ell)^{2n}$ the term corresponding to $(d/\ell)^{2n}$ in the series (1.220), we get the relation

$$b_n \;=\; a_1 - n\,a_2 + \frac{n(n-1)}{2}\,a_3 - \frac{n(n-1)(n-2)}{3!}\,a_4 + \ldots \pm n\,a_{n-1} \mp a_n . \tag{1.222}$$

In the limit $n \to \infty$, we have

$$\lim_{n\to\infty} \frac{b_n \;-\; 4/3\pi\sqrt{\pi n}}{b_n} \;=\; 0. \tag{1.223}$$

[25]This is the continuation of Sec. 1.20

Table 1.5. Some values of the correction factor K vs the ratio d/ℓ

d/ℓ	K
0.1	0.9588
0.2	0.9201
0.3	0.8838
0.4	0.8499
0.5	0.8181
0.6	0.7885
0.7	0.7609
0.8	0.7351
0.9	0.7110
1	0.6884

The first terms of the expansion (correct to seven decimal digits) are

$$\begin{aligned} K &= 1 - 0.4244132\, d/\ell + 0.125\, p + 0.109375\, p^2 + 0.0986328\, p^3 \\ &\quad + 0.0906372\, p^4 + 0.0843735\, p^5 + 0.0792875\, p^6 + \ldots . \end{aligned} \tag{1.224}$$

If d/ℓ is very large, the following approximate formula may be used:

$$K = \frac{2}{\pi\, d/\ell} \left[\log\left(4\,\frac{d}{\ell}\right) - \frac{1}{2} \right] = \frac{2}{\pi\, d/\ell} \left[\log\left(\pi\,\frac{d}{\ell}\right) - 0.258 \right]. \tag{1.225}$$

This result is easily derived from the well-known expression for the self-induction coefficient of a circular coil. In Table 1.5 we report some values[26] of K corresponding[27] to $d/\ell \leq 10$.

If d/ℓ is somewhat larger than one, the series expansion in Eq. (1.224) converges very slowly and the computation of the coefficients becomes very cumbersome. In this case it is then convenient to use the following

[26]The following approximate formulae for K may be useful as d/ℓ increases:

$$\begin{aligned} K &= 1 - \frac{4}{3\pi}\frac{d}{\ell} + \frac{p}{8-7p}, \\ K &= 1 - \frac{4}{3\pi}\frac{d}{\ell} + \frac{48p - 29p^2}{384 - 568p + 194p^2}. \end{aligned}$$

[27]@ In the original manuscript, this Table consists of 40 entries, ranging from $d/\ell = 0.1$ to $d/\ell = 10$, but only the first 10 corresponding values of K are reported. By deviating from the method we have usually adopted, we here prefer not to include the remaining values in the Table, since it is not clear which formula the author would have used to compute K for d/ℓ larger than one. We also note that the reported values have probably been obtained from Eq. (1.220) with $n = 10$.

expansion[28]:

$$K = \sqrt{1+\frac{d^2}{4\ell^2}} - \frac{d}{\ell}\left(\frac{4}{3\pi}+\frac{1}{2{\cdot}4}c_1 - \frac{1{\cdot}3}{2{\cdot}4{\cdot}6}c_2 + \ldots \pm \frac{1{\cdot}3{\cdot}(2n-1)}{2{\cdot}4{\cdots}(2n+2)}c_n \mp \ldots\right), \tag{1.226}$$

in which

$$c_n = \frac{1}{(2n)!}\,\frac{\mathrm{d}^{2n}\sqrt{1+x^2}}{\mathrm{d}x^{2n}}\bigg|_{x=\frac{2\ell}{d}}. \tag{1.227}$$

Computation of the first terms yields

$$\begin{aligned} K &= \sqrt{1+(d/2\ell)^2} - \frac{d}{\ell}\left(\frac{4}{3\pi}+\frac{1}{16}\frac{1}{\left[1+(2\ell/d)^2\right]^{3/2}}\right.\\ &+\frac{1}{128}\frac{1-4\,(2\ell/d)^2}{\left[1+(d/2\ell)^2\right]^{7/2}}+\frac{5}{2048}\frac{1-12\,(2\ell/d)^2+8\,(2\ell/d)^4}{\left[1+(d/2\ell)^2\right]^{11/2}}\\ &\left.+\frac{7}{32768}\frac{5-120\,(2\ell/d)^2+240\,(2\ell/d)^4-64\,(2\ell/d)^6}{\left[1+(d/2\ell)^2\right]^{15/2}}+\ldots\right). \end{aligned}$$

24. VARIATION OF THE SELF-INDUCTION COEFFICIENT DUE TO THE SKIN EFFECT

The self-induction of an electric conductor with a circular cross section can be divided into two parts: one, due to the flux circulating outside the conductor, is generally more important and is independent of the frequency; the other is due to the induction lines (that get closed inside the conductor), depends on the skin effect, and thus, for any conductor, is frequency-dependent. Denoting by ℓ this latter part of the self induction coefficient per unit length, we have, if the skin effect is negligible,

$$\ell = \mu/2. \tag{1.228}$$

[28]@ Notice that the author is again using a Taylor series expansion, but of a particular kind, as it can be deduced from the expression for c_n.

In general, if E is the complex electric field at the surface of the conductor (i.e., the field due both to the potential drop and to the external flux variations), R_1 the AC resistance per unit length of the conductor, ω the angular frequency, and $i = a + bj$ the overall current,[29] we get

$$E = (a + b\,j)\,(R_1 + \ell\,\omega\,j)\,. \tag{1.229}$$

On setting $p = \mu\omega S/\rho$ (S is the cross section of the conductor and ρ its resistivity) or, in practical units,

$$p = \frac{\mu\,\omega}{10^{10}\,R}, \tag{1.230}$$

R being the DC resistance measured in ohm per km of conductor, we have (see Sec. 1.4)

$$\begin{aligned} a &= m\left(p - \frac{1}{2!^2{\cdot}3}p^3 + \frac{1}{4!^2{\cdot}5}p^5 - \frac{1}{6!^2{\cdot}7}p^7 + \dots\right), & (1.231)\\ b &= m\left(\frac{1}{2}p^2 - \frac{1}{3!^2{\cdot}4}p^4 + \frac{1}{5!^2{\cdot}6}p^6 - \frac{1}{7!^2{\cdot}8}p^8 + \dots\right). & (1.232)\end{aligned}$$

The field E can be computed by multiplying ρ by the surface current density; with the units used in Sec. 1.4, it is $\rho = \mu\omega$, and thus one obtains

$$\begin{aligned} E &= m\,\mu\,\omega\left(1 - \frac{p^2}{2!^2} + \frac{p^4}{4!^2} - \frac{p^6}{6!^2} + \dots\right)\\ &\quad + m\,\mu\,\omega\,j\left(p - \frac{p^3}{3!^2} + \frac{p^5}{5!^2} - \frac{p^7}{7!^2} + \dots\right). \end{aligned} \tag{1.233}$$

From Sec. 1.4, we derive the expression for R_1, which can then be introduced into Eq. (1.229). Thus, having set $E = u + vj$, we get

$$\ell\,\omega = \frac{a\,v - b\,u}{a^2 + b^2}, \tag{1.234}$$

from which we deduce

$$\ell = \frac{\mu}{2}\,\frac{1 + \dfrac{p^2}{2!^2{\cdot}3!} + \dfrac{p^4}{3!^2{\cdot}5!} + \dfrac{p^6}{4!^2{\cdot}7!} + \dfrac{p^8}{5!^2{\cdot}9!} + \dots}{1 + \dfrac{p^2}{2!3!} + \dfrac{p^4}{2!3!5!} + \dfrac{p^6}{3!4!7!} + \dfrac{p^8}{4!5!9!} + \dots}. \tag{1.235}$$

The values of R_1/R and ℓ/μ as functions of p shown[30] in Table 1.6 have been derived by using this equation and Eq. (1.60).

---

[29]@ Notice that the author is using the (electro-technical) notation j for the imaginary unit.

[30]@ In the original manuscript, the results of this Table corresponding to values from $p = 4.5$ to 100 were lacking. Moreover, a few values of R_1/R differ slightly from the ones reported here, which have just been obtained following the text.

Table 1.6. Influence of the skin effect on the effective resistance and self-induction coefficient of an electric conductor.

p	R_1/R	ℓ/μ	p	R_1/R	ℓ/μ
0.1	1.0008	0.4998	2.5	1.372	0.4100
0.2	1.0033	0.4992	3	1.479	0.3857
0.3	1.0075	0.4981	3.5	1.581	0.3633
0.4	1.0132	0.4967	4	1.678	0.3432
0.5	1.0205	0.4949	4.5	1.768	0.3253
0.6	1.0293	0.4927	5	1.853	0.3096
0.7	1.0396	0.4901	6	2.007	0.2836
0.8	1.0512	0.4873	7	2.146	0.2630
0.9	1.0641	0.4841	8	2.274	0.2464
1.0	1.0782	0.4806	9	2.394	0.2326
1.1	1.0934	0.4768	10	2.507	0.2210
1.2	1.1096	0.4728	15	3.005	0.1814
1.3	1.1267	0.4686	20	3.427	0.1582
1.4	1.1447	0.4642	25	3.799	0.1430
1.5	1.1634	0.4597	30	4.135	0.1327
1.6	1.1827	0.4550	40	4.732	0.1203
1.7	1.2026	0.4501	50	5.256	0.1135
1.8	1.2229	0.4452	60	5.730	0.1096
1.9	1.2436	0.4403	80	6.537	0.1055
2.0	1.2646	0.4352	100	7.167	0.1036

25. MEAN ERROR IN ESTIMATING THE PROBABILITY OF AN EVENT THROUGH A FINITE NUMBER OF TRIALS

Let p be the probability of a given event, and suppose that in a series of n trials the event happens m times. On estimating p by the ratio m/n, we make an error e defined by the relation

$$p - m/n + e. \tag{1.236}$$

We now look for an estimate of the mean square value of e. Let X be a quantity such that, at each trial, it takes the value $1-p$ or $-p$, depending on whether the event has occurred or not. The mean value of X is zero, while its mean square value is $p(1-p)^2+p^2(1-p)=p(1-p)$. Considering n trials, the mean square value of the variable $\sum_i X_i$ is

$\sqrt{np(1-p)}$; but if the event has occurred m times we'll have:

$$\sum_i X_i = m\,(1-p) - (n-m)\,p = m - n\,p = -\,n\,e. \tag{1.237}$$

We thus deduce that the mean value of e is $\sqrt{p(1-p)/n}$ or, using the usual notations,

$$p = \frac{m}{n} \pm \sqrt{\frac{p(1-p)}{n}}. \tag{1.238}$$

For the case in which p is unknown and only its approximate value m/n is given, if we can assume that the difference between these quantities is so small that the substitution of the latter for the former into the expression for the mean error doesn't change it significantly, we can approximately write

$$p = \frac{m}{n} \pm \frac{1}{n}\sqrt{\frac{m(n-m)}{n}}. \tag{1.239}$$

If n is much larger than m, a further simplification is possible:

$$p = \frac{m}{n} \pm \frac{\sqrt{m}}{n}. \tag{1.240}$$

On multiplying the previous relations by n, they become

$$np = m \pm \sqrt{\frac{m(n-m)}{n}}, \tag{1.241}$$

$$np = m \pm \sqrt{m} \quad \left(\text{for small } \frac{m}{n} \text{ values}\right). \tag{1.242}$$

26. UNBALANCE OF A PURE THREE-PHASE SYSTEM

Let V_1, V_2 and V_3 be the values of three AC "intensive quantities" forming a pure, direct and unbalanced three-phase system. This system may be seen as the sum of two balanced systems: The first one is direct and has the magnitude A, the second one is inverted and has the magnitude B. If the unbalance is not too big, A and B may be computed using the approximate relations

$$A = (1/3)\,(V_1 + V_2 + V_3), \tag{1.243}$$

$$B = \sqrt{(2/3)\,[(V_1 - A)^2 + (V_2 - A)^2 + (V_3 - A)^2]}. \tag{1.244}$$

Table 1.7. The function $x!$ for $0 \leq x \leq 1$.

x	$x!$
0	1.0000
0.05	0.9735
0.1	0.9514
0.15	0.9330
0.2	0.9182
0.25	0.9064
0.3	0.8975
0.35	0.8911
0.4	0.8873
0.45	0.8857
0.5	0.8862
0.55	0.8889
0.6	0.8935
0.65	0.9001
0.7	0.9086
0.75	0.9191
0.8	0.9314
0.85	0.9456
0.9	0.9618
0.95	0.9799
1	1.0000

27. TABLE FOR THE COMPUTATION[31] OF $x!$

The difference $\log n! - n(\log n - 1) - (1/2)\log n$ tends to a finite value as $n \to \infty$. This means that, for very large n, we can set[32]

$$n! = \sqrt{C\,n}\;(n/e)^n . \tag{1.246}$$

Let us now determine C. Let x be the probability that in $2n$ trials an event with probability $1/2$ occurs t times. If $2n$ is very large, we can

[31]@ It is not clear how the author obtained the values in Table 1.7, since in this Section he considered only the limit for large x of the function $x!$. Probably, half of this Table was derived from the formula

$$x!\,(1-x)! = \frac{\pi\,x\,(1-x)}{\sin\pi\,x}, \tag{1.245}$$

which appears near this Table in the original manuscript.

[32]@ Here e is the Napier constant.

represent $x = x(t)$ by an error function. The latter can be obtained from the following constraints: The area beneath the error curve must be equal to 1, the mean value of t must be n, and the mean-square value of the deviation of t from n must be $n/2$ (see Sec. 1.25). We then find

$$x = \frac{1}{\sqrt{\pi n}} \exp\left\{-\frac{(t-n)^2}{n}\right\}. \tag{1.247}$$

The maximum value of x is $x_0 = 1/\sqrt{\pi n}$, which can also be derived directly from the combinatorial theory:

$$x_0 = \frac{1}{2^{2n}} \binom{2n}{n} = \frac{(2n)!}{2^{2n}\,(n!)^2}. \tag{1.248}$$

On substituting Eq. (1.246) into this expression, and comparing the result with Eq. (1.247), we find $C = 2\pi$; so that, in the considered limit, it holds

$$n! = \sqrt{2\pi n}\left(\frac{n}{e}\right)^n. \tag{1.249}$$

Notice that, for large n, we also have

$$\frac{2^{2n}\,(n!)^2}{(2n)!} = \sqrt{\pi n}. \tag{1.250}$$

28. INFLUENCE OF A MAGNETIC FIELD ON THE MELTING POINT

Let us consider the system shown in Fig. 1.4 and suppose it is in equilibrium. If, using any procedures, we move a unit volume of solid from vessel 2 to vessel 1 and place it in a thin layer at the boundary between solid and liquid, we have to do the work

$$L_1 = h\,(\gamma_1 - \gamma_2) \tag{1.251}$$

against gravity, where γ_1 and γ_2 are the specific weights of the solid and the liquid, respectively. If we assume, for the moment, that the solid is a magnetic material while the liquid is not, it is easy to compute how much work the magnetic field has to perform on the solid unit volume during the process described above:

$$L_2 = \frac{H^2}{8\pi}\,\frac{\mu_1 - 1}{\mu_1}, \tag{1.252}$$

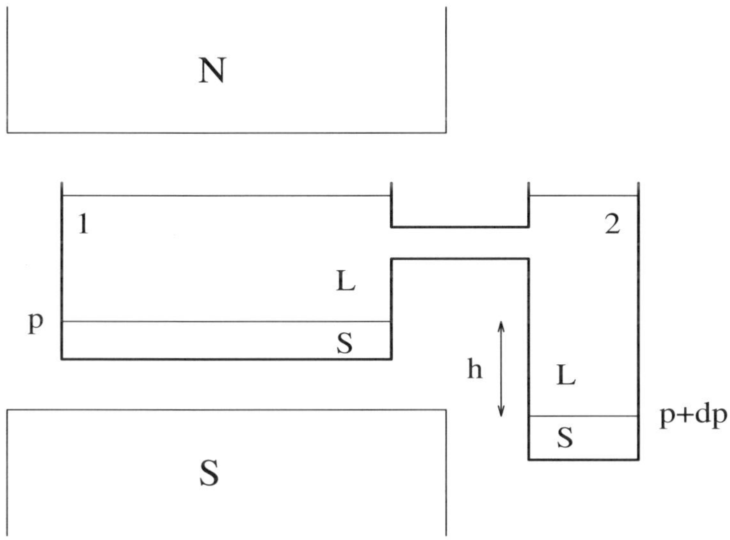

Fig. 1.4. Solid-liquid phase transition in presence of a magnetic field.

μ_1 being the magnetic permeability of the solid. In order to exclude the possibility of perpetual motion, it must hold

$$L_1 = L_2, \tag{1.253}$$

wherefrom we obtain

$$h = \frac{H^2}{8\pi} \frac{\mu_1 - 1}{\mu_1} \frac{1}{\gamma_1 - \gamma_2}. \tag{1.254}$$

Since we have assumed the liquid to be non-magnetic, the pressure distribution inside it is hydrostatic, and thus we find (cfr. Fig. 1.4)

$$\Delta p = h\,\gamma_2 = \frac{H^2}{8\pi} \frac{\mu_1 - 1}{\mu_1} \frac{\gamma_2}{\gamma_1 - \gamma_2}, \tag{1.255}$$

or, introducing the specific volumes,

$$\Delta p = \frac{H^2}{8\pi} \frac{\mu_1 - 1}{\mu_1} \frac{V_1}{V_2 - V_1}. \tag{1.256}$$

Denoting by T the melting temperature in the absence of a magnetic field, at pressure p, and by $T + \Delta T$ the temperature in the presence of a magnetic field at the same pressure, we have that $T + \delta T$ is the fusion temperature, under ordinary conditions, corresponding to a pressure $p + \Delta p$. From the Clapeyron equation, one gets

$$\Delta T = \left(\frac{T}{\rho}\right) (V_2 - V_1) \Delta p, \tag{1.257}$$

and, substitution into Eq. (1.256), we find

$$\Delta T = \frac{TH^2}{8\pi} \frac{\mu_1 - 1}{\mu_1} \frac{V_1}{\rho}. \tag{1.258}$$

The generalization of Eqs. (1.255) and (1.258) to the case where the liquid has an arbitrary magnetic permeability μ_2 is obviously the following:

$$\Delta p = \frac{H^2}{8\pi} \left(\frac{\mu_1 - 1}{\mu_1} \frac{\gamma_2}{\gamma_1 - \gamma_2} + \frac{\mu_2 - 1}{\mu_2} \frac{\gamma_1}{\gamma_2 - \gamma_1} \right), \tag{1.259}$$

$$\Delta T = T \frac{H^2}{8\pi} \left(\frac{\mu_1 - 1}{\mu_1} \frac{V_1}{\rho} - \frac{\mu_2 - 1}{\mu_2} \frac{V_2}{\rho} \right). \tag{1.260}$$

If $\mu_1 = \mu_2 = \mu$, we obtain

$$\Delta p = - \frac{H^2}{8\pi} \frac{\mu - 1}{\mu}, \tag{1.261}$$

$$\Delta T = \frac{TH^2}{8\pi} \frac{\mu - 1}{\mu} \frac{V_1 - V_2}{\rho}. \tag{1.262}$$

In this case the boundary surface between solid and liquid would be at the same level in both vessels of Fig. 1.4 but, due to the magnetization of the liquid, the pressure distribution would not be hydrostatic, thus resulting in $\Delta p \neq 0$.

Similar relations hold if the magnetic field is replaced with an electric field or if the different phases are liquid-vapor or solid-vapor instead of solid-liquid.

29. SPECIFIC HEAT OF AN OSCILLATOR

The mean energy of an oscillator with frequency ν at temperature T is

$$\epsilon = \frac{h\nu}{e^{h\nu/kT} - 1}, \tag{1.263}$$

where h is the action quantum[33] and $k = R/N$ the Boltzmann constant. On taking the derivative with respect to the temperature and setting

[33]@ In this Section we follow the author in using Planck's constant h instead of replacing it with the reduced Planck constant $\hbar$.

Table 1.8. Specific heat and mean energy of an oscillator (see the text for notations).

$\frac{1}{p} = \frac{T}{T_0}$	$p = \frac{T_0}{T}$	$\frac{c}{k}$	$\frac{\epsilon}{kT}$	$\frac{\epsilon}{kT_0}$	$\frac{kT-\epsilon}{kT_0}$
0	∞	0	0.0000	0.0000	0.0000
0.2	5	0.1707	0.0338	0.0068	0.1932
0.4	2.5	0.6089	0.2236	0.0894	0.3106
0.6	1.67	0.7967	0.3873	0.2319	0.3669
0.8	1.75	0.8794	0.5019	0.4016	0.3984
1	1	0.9207	0.5820	0.5820	0.4180
1.2	0.83	0.9445	0.6417	0.7732	0.4316
1.4	0.71	0.9590	0.6867	0.9671	0.4413
1.6	0.625	0.9681	0.7198	1.1517	0.4483
1.8	0.556	0.9746	0.7476	1.3447	0.4539
2	0.500	0.9794	0.7707	1.5415	0.4585
2.5	0.400	0.9868	0.8133	2.0332	0.4668
3	0.333	0.9908	0.8427	2.5307	0.4723
4	0.250	0.9948	0.8802	3.5208	0.4792
5	0.200	0.9967	0.9033	4.5167	0.4833
10	0.100	0.9992	0.9508	9.5083	0.4917
∞	0	1	1	∞	0.5000

$p = h\nu/kT = T_0/T$ for brevity, we obtain the following expression for the specific heat:

$$c = \frac{\mathrm{d}\epsilon}{\mathrm{d}T} = \frac{kp^2\,\mathrm{e}^p}{(\mathrm{e}^p - 1)^2} = k\left(\frac{p}{\mathrm{e}^{p/2} - \mathrm{e}^{-p/2}}\right)^2. \tag{1.264}$$

The ratio c/k is always less than 1 and is given in Table 1.8 as a function[34] of p. For very large T, Eq. (1.263) becomes

$$\epsilon = kT - \frac{1}{2}h\nu = k\left(T - \frac{1}{2}T_0\right),$$

$T_0 = h\nu/k$ denoting the temperature at which the mean energy of the oscillator, computed in the framework of classical mechanics, is the same as that of the lowest quantum energy level.

[34]@ In the original manuscript, this Table was almost entirely empty. Apart from the values in the first two columns (which are input values), the author wrote down only the first and the last value in the third column and the first value in the fourth column.

30. DO CHILDREN OF THE SAME PARENTS TEND TO BE OF THE SAME SEX?

The *a priori* probability that in a certain region a newborn child is male can be cast in the form

$$W = 1/2 + \alpha, \tag{1.265}$$

where α is, in general, positive. On the other hand, the probability for a couple of parents to have a male child may not be the same as W, and we shall write it as

$$W_1 = W + \beta = 1/2 + \alpha + \beta. \tag{1.266}$$

The mean value of β is zero, whereas its mean-square value measures the tendency to generate children of the same sex. Indicating with $\overline{\beta}$ this mean-square value, Eq. (1.266) can then be recast as follows, on adopting the usual notations:

$$W_1 = W \pm \overline{\beta} = 1/2 + \alpha \pm \overline{\beta}. \tag{1.267}$$

In order to statistically estimate $\overline{\beta}$ by samples, the easiest way is the following. Let us consider a couple of parents who have had n children, ℓ of which are male and m female. The most probable value[35] of $(\ell - m)^2$ is (see Sec. 1.25)

$$\text{prob. value of } (\ell - m)^2 = n + 4(\alpha + \beta)^2 (n^2 - n). \tag{1.268}$$

If we write the previous expression for a large number of families and then sum term by term, the sum on the l.h.s. can be replaced with the sum of the actual values of $(\ell - m)^2$, and by doing this we make a relative error that tends to zero. We then obtain

$$\sum (\ell - m)^2 = \sum n + 4 \sum (\alpha + \beta)^2 (n^2 - n) \tag{1.269}$$

$$= \sum n + 4\alpha^2 \sum (n^2 - n) + 4 \sum \beta^2 (n^2 - n) + 8\alpha \sum \beta (n^2 - n). \tag{1.270}$$

Since we have implicitly assumed that the mean value of β is zero independently of n, the mean value of the last term on the r.h.s. of Eq. (1.270) is also zero. We can thus neglect it and write

$$\sum (\ell - m)^2 = \sum n + 4\alpha^2 \sum (n^2 - n) + 4 \sum \beta^2 (n^2 - n), \tag{1.271}$$

[35]@ That is, n times the probability for the event to occur.

or, since we assume that β does not depend on n,

$$\sum (\ell - m)^2 = \sum n + 4(\alpha^2 + \beta^2) \sum (n^2 - n). \tag{1.272}$$

The quantity α can be estimated through a larger statistical sample, so that $\overline{\beta}$ is determined by Eq. (1.272):

$$\overline{\beta} = \sqrt{\frac{\sum(\ell - m)^2 - \sum n}{4\sum(n^2 - n)} - \alpha^2}. \tag{1.273}$$

If α is unknown, it may be approximated by the relation

$$\alpha = \frac{\sum \ell}{\sum n} - \frac{1}{2}; \tag{1.274}$$

and, by substitution in Eq. (1.273), we get

$$\overline{\beta}^2 = \frac{\sum(\ell - m)^2 - \sum n}{4\sum(n^2 - n)} - \left(\frac{\sum \ell}{\sum n} - \frac{1}{2}\right)^2. \tag{1.275}$$

Taking into account that α in Eq. (1.274) is affected by an error which is of the order of $1/2\sqrt{\sum n}$ (see Sec. 1.25), this expression should be replaced with

$$\overline{\beta}^2 = \frac{\sum(\ell - m)^2 - \sum n}{4\sum(n^2 - n)} - \left(\frac{\sum \ell}{\sum n} - \frac{1}{2}\right)^2 + \frac{1}{4\sum n}. \tag{1.276}$$

31. HEAT PROPAGATION FROM A CERTAIN CROSS SECTION ALONG AN INFINITE LENGTH BAR ENDOWED WITH ANOTHER CROSS SECTION ACTING AS A HEAT WELL. A SIMILARITY WITH THE CRICKETS

Suppose that N individuals are initially all located at point O of a straight line x and that each one of them jumps at time intervals $\mathrm{d}t$ a length $\mathrm{d}x$ to the left or to the right of his current position with equal probability, and also suppose that the ratio $\mathrm{d}x^2/\mathrm{d}t = \mu^2$ is finite. Furthermore, let us assume that at a distance ℓ from O there is located a deadly trap. The problem is to determine the number linear density

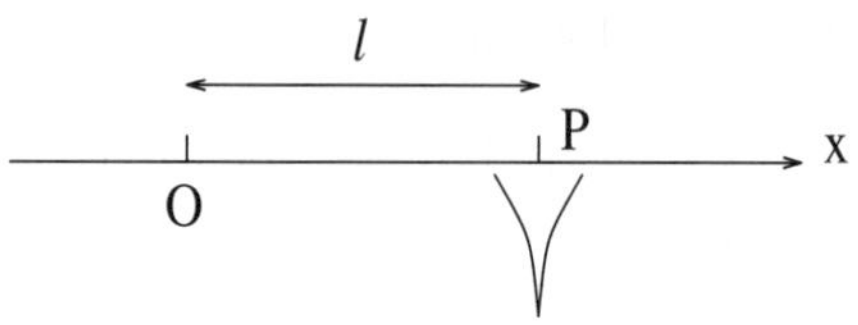

Fig. 1.5. Heat propagation in one dimension (see text).

$U(x,t)$ of survivors at time t and at a point x. Note that, if there were no traps, the linear density would be

$$U_0(x,t) = \frac{N}{\mu\sqrt{2\pi t}}\, \mathrm{e}^{-x^2/2\mu^2 t}. \tag{1.277}$$

Moreover, notice that we can still view the individuals falling into the trap as alive and kicking after their death, as long as, starting from the moment they fall, another, negative signed individual is associated with each of them. Then, in order to derive the actual density U, all we have to do is to subtract from U_0 the density U_1 of the negative individuals. The last quantity can be easily obtained by observing that, for $x > \ell$,

$$U_1(x,t) = U_0(x,t), \tag{1.278}$$

while, for symmetry reasons, for $x < \ell$ one has

$$U_1(x,t) = U_1(2\ell - x, t) = U_0(2\ell - x, t). \tag{1.279}$$

Thus

$$\begin{aligned} U(x,t) &= U_0(x,t) - U_0(2\ell - x, t) \\ &= \frac{N}{\mu\sqrt{2\pi t}} \left[\mathrm{e}^{-x^2/2\mu^2 t} - \mathrm{e}^{-(2\ell - x)^2/2\mu^2 t} \right]; \end{aligned} \tag{1.280}$$

and for large t this may be written as

$$U(x,t) = \frac{2N\,\ell(\ell - x)}{\mu^3\, t\sqrt{2\pi t}}\, \mathrm{e}^{-(\ell - x)^2/2\mu^2 t}\, \mathrm{e}^{-\ell^2/2\mu^2 t}. \tag{1.281}$$

For a given t, we then deduce

$$U_{\max} = U(\ell - \mu\sqrt{t}, t) = \frac{2N\,\ell\, \mathrm{e}^{-1/2}}{\mu^2\, t\sqrt{2\pi}}\, \mathrm{e}^{-\ell^2/2\mu^2 t}, \tag{1.282}$$

while the number of survivors (for large t) becomes

$$N_a = \frac{2N\,\ell}{\mu\sqrt{2\pi t}}\, \mathrm{e}^{-\ell^2/2\mu^2 t}. \tag{1.283}$$

If we now compute the moment of the position with respect to P,

$$\int (\ell - x)\, \mathrm{d}N_a = N\,\ell, \tag{1.284}$$

we find that the "center of gravity" of the distribution of the living individuals and of the dead ones (these are supposed to be concentrated in the trap at P), does not move from O, as was clear *a priori*. The survival probability curve has, at first, an inflection point between O and P, but this moves towards the trap and then disappears at $t = \ell^2/3\mu^2$. At this time, which is when we have the highest mortality, $N/12$ individuals have died.

The function U_0 obeys the differential equation

$$\frac{\partial U_0}{\partial t} = \frac{\mu^2}{2}\frac{\partial^2 U_0}{\partial x^2}, \tag{1.285}$$

and thus is suited to represent the way a quantity of heat $Q = N$ propagates from a localized cross section of an infinitely long bar, if it is given that

$$\mu^2 = 2c/\gamma\delta, \tag{1.286}$$

with c denoting the heat transmission coefficient, γ the specific heat, and δ the density. Note that μ^2 given by Eq. (1.286) represents the mean square value of the heat displacement per unit time in any one direction. The square of the total displacement in space per unit time will be $3\mu^2 = 6c/\gamma\delta$.

32. COMBINATIONS [36]

The sum of the probabilities that an event having probability 1/2 will take place n times in n trials or in $n+1$ trials or in $n+2$... or in $2n$ trials is equal to 1; expressed as a formula, it writes

$$\sum_{r=0}^{n} \frac{1}{2^{n+r}} \binom{n+r}{n} = 1. \tag{1.287}$$

Indeed, it holds

$$\sum_{r=0}^{n+1} \frac{1}{2^{n+1+r}} \binom{n+1+r}{n+1}$$

[36]See Sec. 2.38.5.

$$
\begin{aligned}
&= \frac{1}{2}\sum_{r=0}^{n+1}\frac{1}{2^{n+r}}\binom{n+r}{n} + \frac{1}{2}\sum_{r=1}^{n+1}\frac{1}{2^{n+r}}\binom{n+r}{n+1} \\
&= \frac{1}{2}\sum_{r=0}^{n}\frac{1}{2^{n+r}}\binom{n+r}{n} + \frac{1}{2^{2n+2}}\binom{2n+1}{n} \\
&\quad + \frac{1}{2}\sum_{r=1}^{n+2}\frac{1}{2^{n+r}}\binom{n+r}{n+1} - \frac{1}{2^{2n+3}}\binom{2n+2}{n+1};
\end{aligned}
$$

and, since

$$
\begin{aligned}
\frac{1}{2^{2n+2}}\binom{2n+1}{n} &= \frac{1}{2^{2n+3}}\binom{2n+2}{n+1}, \\
\sum_{r=1}^{n+2}\frac{1}{2^{n+r}}\binom{n+r}{n+1} &= \sum_{r=0}^{n+1}\frac{1}{2^{n+1+r}}\binom{n+1+r}{n+1},
\end{aligned}
$$

one obtains

$$
\sum_{r=0}^{n+1}\frac{1}{2^{n+1+r}}\binom{n+1+r}{n+1} = \sum_{r=0}^{n}\frac{1}{2^{n+r}}\binom{n+r}{n}. \qquad (1.288)
$$

Thus, if Eq. (1.287) holds for $n = k$, it also holds for $n = k+1$; and, since it holds for $n = 1$, it will hold for any n.

In the same way, one can prove the relation

$$
\sum_{r=0}^{\infty}\frac{1}{2^{n+r}}\binom{n+r}{n} = 2. \qquad (1.289)
$$

33. ENERGY AND SPECIFIC HEAT OF A ROTATOR

Let $\mathcal{I}$ be the moment of inertia of a rotator. Sommerfeld's constraints yield[37]

$$
\mathcal{I}\,\omega = \frac{nh}{2\pi}, \qquad (n = 0, 1, \ldots), \qquad (1.290)
$$

and thus[38]

$$
\epsilon = \frac{1}{2}\mathcal{I}\,\omega^2 = \frac{n^2h^2}{8\pi^2\mathcal{I}} = \frac{nh\nu}{2}, \qquad (1.291)
$$

[37]@ In this Section we adhere to the author's use of h, rather than rewriting it in terms of $2\pi\hbar$.

[38]@ Here ϵ and ν are the energy and frequency of the rotator, while ω is its angular frequency.

$$\nu = \frac{\omega}{2\pi} = \frac{nh}{4\pi^2\mathcal{I}}. \tag{1.292}$$

According to Boltzmann's law, the mean energy at a temperature T is

$$\bar{\epsilon} = \frac{\sum_{n=0}^{\infty} \frac{n^2h^2}{8\pi^2\mathcal{I}} \exp\left\{-\frac{n^2h^2}{8\pi^2\mathcal{I}kT}\right\}}{\sum_{n=0}^{\infty} \exp\left\{-\frac{n^2h^2}{8\pi^2\mathcal{I}kT}\right\}} = \frac{\sum_{n=0}^{\infty} \frac{h\nu_0}{2}\, n^2 \exp\left\{-\frac{h\nu_0}{2kT}n^2\right\}}{\sum_{n=0}^{\infty} \exp\left\{-\frac{h\nu_0}{2kT}n^2\right\}}, \tag{1.293}$$

$\nu_0 = h/4\pi^2\mathcal{I}$ denoting the fundamental frequency. On setting

$$p = \frac{1}{2}\frac{h\nu_0}{kT} = \frac{h^2}{8\pi^2\mathcal{I}kT},$$

we get

$$\bar{\epsilon} = kT \sum_{n=0}^{\infty} pn^2\, \mathrm{e}^{-pn^2} \Big/ \sum_{n=0}^{\infty} \mathrm{e}^{-pn^2}. \tag{1.294}$$

For $p \to 0$, which means $T \to \infty$, we obviously find

$$\lim_{p\to 0} \bar{\epsilon} = \frac{1}{2}kT. \tag{1.295}$$

The specific heat c is obtained by taking the derivative of Eq. (1.294) with respect to T and noting that $\mathrm{d}p/\mathrm{d}T = -p/T$:

$$c = \frac{\mathrm{d}\bar{\epsilon}}{\mathrm{d}T} = k\,p^2 \left[\frac{\sum_{n=0}^{\infty} n^4\, \mathrm{e}^{-pn^2}}{\sum_{n=0}^{\infty} \mathrm{e}^{-pn^2}} - \left(\frac{\sum_{n=0}^{\infty} n^2\, \mathrm{e}^{-pn^2}}{\sum_{n=0}^{\infty} \mathrm{e}^{-pn^2}} \right)^2 \right]. \tag{1.296}$$

If T_0 is the temperature corresponding to the classical mean energy given by the energy of the lowest non-vanishing quantum state, one gets

$$T_0 = \frac{1}{2}\frac{h\nu_0}{k}, \tag{1.297}$$

$$p = T_0/T. \tag{1.298}$$

In Table 1.9 the specific heat and the mean energy are shown for different temperatures.[39]

[39]@ In the original manuscript, the values in the third and fourth column were missing.

Table 1.9. Specific heat and mean energy of a rotator vs. its temperature.

$\frac{1}{p} = \frac{T}{T_0}$	$p = \frac{T_0}{T}$	$\bar{\epsilon} \Big/ \frac{1}{2} kT$	$c \Big/ \frac{1}{2} k$
0.2	5.00	1.3375	1.1118
0.4	2.50	1.5548	1.1235
0.6	1.67	1.7763	1.0867
0.8	1.25	2.0182	0.9955
1.0	1.00	2.2896	0.8525
1.2	0.83	2.5927	0.6829
1.4	0.71	2.9244	0.5169
1.6	0.62	3.2784	0.3740
1.8	0.56	3.6486	0.2613
2.0	0.50	4.0297	0.1776
3.0	0.33	6.0022	0.0200
4.0	0.25	8.0001	0.0018

34. GRAVITATIONAL ATTRACTION OF AN ELLIPSOID

Let us consider a mass distribution on the ellipsoidal surface

$$\frac{x^2}{a^2} + \frac{y^2}{b^2} + \frac{z^2}{c^2} = 1, \tag{1.299}$$

such that the surface density σ in every point is proportional to the *projection* of the position vector (starting from the ellipsoid center) on the normal to the surface:

$$\sigma = \rho \Big/ \sqrt{x^2/a^4 + y^2/b^4 + z^2/c^4}\,. \tag{1.300}$$

The total mass m can be computed easily. Indeed, our distribution can be viewed as the limit for $\alpha \to 0$ of a spatially uniform distribution having volume density ρ/α and filling the space between the ellipsoid of semi-axes a, b, c and the ellipsoid of semi-axes $a(1+\alpha), b(1+\alpha), c(1+\alpha)$. We then have

$$m = \lim_{\alpha \to 0} \frac{\rho}{\alpha} \frac{4}{3} \pi\, a\, b\, c \left[(1+\alpha)^3 - 1 \right] = 4\pi\, a\, b\, c\, \rho. \tag{1.301}$$

As is well known, the ellipsoid in Eq. (1.299) is an equipotential surface, so the field inside it is zero whereas the field outside it, but near it, is normal to the surface and has the magnitude

$$F = 4\pi\, \sigma\, K = 4\pi\, K\, \rho \left[x^2/a^4 + y^2/b^4 + z^2/c^4 \right]^{-1/2}, \tag{1.302}$$

where K is the coefficient in Newton's law.[40] On introducing the total mass, we have

$$F = \frac{mK}{abc}\left[x^2/a^2 + y^2/b^2 + z^2/c^2\right]^{-1/2}. \quad (1.303)$$

In particular, at the ends of the symmetry axes the force will be

$$F_a = \frac{mK}{b\,c}, \quad F_b = \frac{mK}{c\,a}, \quad F_c = \frac{mK}{a\,b}, \quad (1.304)$$

respectively.

Let us now construct the equipotential surface that is infinitesimally close to our ellipsoid. In order to do this, consider a point P on the outward normal direction originating from another point $P_0(x_0, y_0, z_0)$, such that the distance between P_0 and P is

$$\mathrm{d}s = -\frac{\mathrm{d}U}{F} = -\,\mathrm{d}U\,\frac{a\,b\,c}{mK}\sqrt{\frac{x^2}{a^2} + \frac{y^2}{b^2} + \frac{z^2}{c^2}}.$$

The coordinates of point P will be

$$\begin{aligned} x &= x_0 + (-\,\mathrm{d}U)\,\frac{a\,b\,c}{mK}\,\frac{x_0}{a^2}, \\ y &= y_0 + (-\,\mathrm{d}U)\,\frac{a\,b\,c}{mK}\,\frac{y_0}{b^2}, \\ z &= z_0 + (-\,\mathrm{d}U)\,\frac{a\,b\,c}{mK}\,\frac{z_0}{c^2}, \end{aligned} \quad (1.305)$$

so that, by neglecting higher-order infinitesimals and setting

$$\mathrm{d}t = -\,2\frac{a\,b\,c}{mK}\,\mathrm{d}U, \quad (1.306)$$

and

$$\begin{aligned} x &= x_0 + \frac{1}{2}\,\frac{x_0}{a^2}\,\mathrm{d}t, \\ y &= y_0 + \frac{1}{2}\,\frac{y_0}{b^2}\,\mathrm{d}t, \\ z &= z_0 + \frac{1}{2}\,\frac{z_0}{c^2}\,\mathrm{d}t; \end{aligned} \quad (1.307)$$

---

[40]@ Note that F is the gravitational force field related to the gravitational potential U (see below). Equation (1.302) then is a relation analogous to Coulomb's theorem for the electrostatic field in the proximity of a conductor.

one can write

$$\begin{aligned} \frac{x}{\sqrt{a^2+\mathrm{d}t}} &= \frac{x_0}{a}, \\ \frac{y}{\sqrt{b^2+\mathrm{d}t}} &= \frac{y_0}{b}, \\ \frac{z}{\sqrt{c^2+\mathrm{d}t}} &= \frac{z_0}{c}. \end{aligned} \tag{1.308}$$

On squaring and summing the above expressions, we obtain the equation that describes the equipotential surface under consideration:

$$\frac{x^2}{a^2+\mathrm{d}t} + \frac{y^2}{b^2+\mathrm{d}t} + \frac{z^2}{c^2+\mathrm{d}t} = 1. \tag{1.309}$$

Neglecting infinitesimals of order higher than the first, this surface is itself an ellipsoid, and the same considerations developed above apply to it as well. Thus, by introducing another error of order greater than the first, the surface

$$\frac{x^2}{a^2+2\mathrm{d}t} + \frac{y^2}{b^2+2\mathrm{d}t} + \frac{z^2}{c^2+2\mathrm{d}t} = 1 \tag{1.310}$$

is again an equipotential surface. In general, up to n infinitesimal errors of order higher than the first, the surface

$$\frac{x^2}{a^2+n\mathrm{d}t} + \frac{y^2}{b^2+n\mathrm{d}t} + \frac{z^2}{c^2+n\mathrm{d}t} = 1 \tag{1.311}$$

is equipotential. This means that the ratio of the error to ndt remains infinitesimal for any ndt. If $n \to \infty$ in such a way that $ndt = t$ remains finite, the surface

$$\frac{x^2}{a^2+t} + \frac{y^2}{b^2+t} + \frac{z^2}{c^2+t} = 1 \tag{1.312}$$

will be exactly equipotential. This one, then, is the general expression for the equipotential surfaces external to the ellipsoidal mass distribution; t can take any positive value.[41]

From Eq. (1.308) we can derive the general expression for the lines of force:

$$\begin{aligned} x &= \alpha\sqrt{a^2+t}, \\ y &= \beta\sqrt{b^2+t}, \\ z &= \gamma\sqrt{c^2+t}, \end{aligned} \tag{1.313}$$

[41]One can formally prove this by showing that it is possible to construct a function $U = U(t)$ that obeys Laplace's equation $\nabla^2 U = 0$ and is zero for $t \to \infty$.

with $\alpha^2 + \beta^2 + \gamma^2 = 1$. The constants α, β, γ are evidently the direction cosines of the asymptotes of the lines of force, which are straight lines passing through the ellipsoid center.

In order to obtain the potential $U = U(t)$ on the ellipsoid in Eq. (1.312), notice that the potential difference between two infinitesimally close equipotential surfaces can be deduced from Eq. (1.306). On integrating it between $t = \infty$ and $t = t_0$, one finds

$$U(t_0) = \frac{mK}{2} \int_{t_0}^{\infty} \frac{\mathrm{d}t}{\sqrt{(a^2+t)(b^2+t)(c^2+t)}}. \tag{1.314}$$

In particular, on the ellipsoid in Eq. (1.299) the potential will be

$$U(0) = \frac{mK}{2} \int_{0}^{\infty} \frac{\mathrm{d}t}{\sqrt{(a^2+t)(b^2+t)(c^2+t)}}. \tag{1.315}$$

As far as the effects outside the ellipsoid are concerned, we may replace the original distribution of total mass m with a similar distribution of the same mass placed on a confocal ellipsoid. Thus, we may generalize Eq. (1.303) to points outside an arbitrary ellipsoid:

$$\begin{aligned} F &= \frac{mK}{\sqrt{(a^2+t)(b^2+t)(c^2+t)}} \\ &\times \frac{1}{\sqrt{x^2/(a^2+t)^2 + y^2/(b^2+t)^2 + z^2/(c^2+t)^2}}. \end{aligned} \tag{1.316}$$

From Eq. (1.315), we can immediately derive the "gravitational capacitance" of the ellipsoid:

$$C = 2 \left(\int_{0}^{\infty} \frac{\mathrm{d}t}{\sqrt{(a^2+t)(b^2+t)(c^2+t)}} \right)^{-1}. \tag{1.317}$$

Now let us consider an ellipsoidal space region

$$\frac{x^2}{a^2} + \frac{y^2}{b^2} + \frac{z^2}{c^2} < 1$$

filled with matter of uniform volume density ρ. The force in the internal region is a linear function of the coordinates, and its components along the x, y, and z axes are, respectively,

$$-L\,x, \; -M\,y, \; -N\,z; \quad L + M + N = 4\pi\,K\,\rho. \tag{1.318}$$

In particular, at the edge of the semi-axis of length a, the force is directed along the inward normal direction, and the absolute value of its magnitude is La. In order to compute L, let us decompose our (filled) ellipsoid into elementary ellipsoidal mass distributions over an infinite set of ellipsoidal homothetic surfaces obeying the equation

$$\frac{x^2}{p^2a^2} + \frac{y^2}{p^2b^2} + \frac{z^2}{p^2c^2} = 1, \tag{1.319}$$

with $0 < p < 1$. The mass between two ellipsoids having semi-axes of lengths pa, pb, pc and $(p + \mathrm{d}p)a, (p + \mathrm{d}p)b, (p + \mathrm{d}p)c$, respectively, is

$$\mathrm{d}m = 4\pi\, a\, b\, c\, p^2\, \rho\, \mathrm{d}p. \tag{1.320}$$

This mass acts on the unit mass placed at $(a, 0, 0)$ with a force that is directed along the x axis and whose magnitude, apart from a sign, is

$$\begin{aligned} \mathrm{d}F &= \frac{K\,\mathrm{d}m}{\sqrt{[a^2 + p^2(b^2 - a^2)]\,[a^2 + p^2(c^2 - a^2)]}} \\ &= \frac{4\pi\, a\, b\, c\, \rho\, K\, p^2\, \mathrm{d}p}{\sqrt{[a^2 + p^2(b^2 - a^2)]\,[a^2 + p^2(c^2 - a^2)]}}. \end{aligned} \tag{1.321}$$

On setting

$$p = \frac{a}{\sqrt{a^2 + t}}, \tag{1.322}$$

we find

$$t = (a/p)^2\,(1 - p^2), \tag{1.323}$$

$$\mathrm{d}F = \frac{4\pi\, a^2\, b\, c\, K\, \rho\, \mathrm{d}t}{2\,(a^2 + t)\,\sqrt{(a^2 + t)(b^2 + t)(c^2 + t)}}, \tag{1.324}$$

and

$$\mathrm{d}F = -\,4\pi\, a^2\, b\, c\, K\, \rho\, \mathrm{d}t\, \frac{\partial}{\partial a^2}\, \frac{1}{\sqrt{(a^2 + t)(b^2 + t)(c^2 + t)}}. \tag{1.325}$$

The expression for the resulting force due to the completely filled ellipsoid is obtained by varying p between 0 and 1, or t between 0 and ∞, and by summing the contributions

$$L\, a = -\,4\pi\, a^2\, b\, c\, K\, \rho\, \frac{\partial}{\partial a^2} \int_0^\infty \frac{\mathrm{d}t}{\sqrt{(a^2 + t)(b^2 + t)(c^2 + t)}}, \tag{1.326}$$

that is,

$$L = -\,4\pi\, a\, b\, c\, K\, \rho\, \frac{\partial}{\partial a^2} \int_0^\infty \frac{\mathrm{d}t}{\sqrt{(a^2 + t)(b^2 + t)(c^2 + t)}}; \tag{1.327}$$

while similar relations hold for M and N. Thus, the force inside and on the surface of the ellipsoid results to be completely determined:

$$\mathbf{F} = -Lx\,\mathbf{i} - My\,\mathbf{j} - Nz\,\mathbf{k}\,. \tag{1.328}$$

In order to determine the force at points outside the surface, we observe that, m being the total mass of the ellipsoid, Eq. (1.327) can be cast in the form

$$L = -3K\,m\,\frac{\partial}{\partial a^2}\int_0^\infty \frac{\mathrm{d}t}{\sqrt{(a^2+t)(b^2+t)(c^2+t)}}, \tag{1.329}$$

and similarly for M and N. Moreover, we can immediately verify, by decomposing the (filled) ellipsoid into shells, that a homogeneously filled ellipsoid is equivalent, as far as the effects outside it are concerned, to any other filled confocal ellipsoid endowed with the same total mass. Thus, given an external point $P(x, y, z)$ and having determined t in such a way that

$$\frac{x^2}{a^2+t} + \frac{y^2}{b^2+t} + \frac{z^2}{c^2+t} = 1, \tag{1.330}$$

the force acting on the unit mass at P will be

$$\mathbf{F} = -L(t)\,\mathbf{i}\,x - M(t)\,\mathbf{j}\,y - N(t)\,\mathbf{k}\,z, \tag{1.331}$$

with

$$L(t) = -4\pi\,a\,b\,c\,K\,\rho\,\frac{\partial}{\partial a^2}\int_t^\infty \frac{\mathrm{d}t}{\sqrt{(a^2+t)(b^2+t)(c^2+t)}}. \tag{1.332}$$

In particular, for $t = 0$, i.e., on the surface of the ellipsoid, we find again Eq. (1.327) for L, and similar equations for M and N.

35. SPECIAL CASES: PROLATE ELLIPSOID AND SPHEROID

I. Let us suppose that a and b are much smaller than c, namely, $a, b \ll c$. By introducing

$$t_1 = \frac{1}{2}\left(t + \sqrt{(a^2+t)(b^2+t)} - a\,b\right) \tag{1.333}$$

into the expression

$$\int_0^\infty \frac{\mathrm{d}t}{\sqrt{(a^2+t)(b^2+t)(c^2+t)}}, \tag{1.334}$$

the latter becomes

$$\int_0^\infty \frac{\mathrm{d}t_1}{\left[(1/4)(a+b)^2+t_1\right]\,\sqrt{(c^2+t_1)}}\sqrt{\frac{c^2+t_1}{c_2+t}}. \tag{1.335}$$

The difference between t and t_1 is of the order a^2 or b^2 and, since c is much greater than both a and b, the factor $\sqrt{(c^2+t_1)/(c^2+t)}$ in the above integrand is always very close to 1. Since all the other factors do not change sign in the limit of very large c, we can write

$$\begin{aligned}\int_0^\infty \frac{\mathrm{d}t}{\sqrt{(a^2+t)(b^2+t)(c^2+t)}} &= \int_0^\infty \frac{\mathrm{d}t_1}{\left[(1/4)(a+b)^2+t_1\right]\,\sqrt{(c^2+t_1)}} \\ &= \frac{2}{\sqrt{c^2-(1/4)(a+b)^2}}\log\frac{c+\sqrt{c^2-(1/4)(a+b)^2}}{(1/2)(a+b)},\end{aligned} \tag{1.336}$$

and, since the previous relation holds anyway only as a first approximation,

$$\int_0^\infty \frac{\mathrm{d}t}{\sqrt{(a^2+t)(b^2+t)(c^2+t)}} = \frac{2}{c}\log\frac{4c}{a+b}. \tag{1.337}$$

The potential of the filled ellipsoid of mass m will be

$$U_0 = \frac{m\,K}{c}\log\frac{4c}{a+b}, \tag{1.338}$$

while the "capacitance" of the ellipsoid is

$$C = \frac{c}{\log\left[4c/(a+b)\right]}. \tag{1.339}$$

At distances small with respect to the semi-axis length c, the constants L, M, N describing the attraction inside the ellipsoid, and the functions $L(t)$,$M(t)$,$N(t)$ describing the force outside the ellipsoid, will be given in first approximation by

$$\begin{aligned} L &= 4\pi\,K\,\rho\,\frac{b}{a+b}, \\ M &= 4\pi\,K\,\rho\,\frac{a}{a+b}, \\ N &= 4\pi\,K\,\rho\,\frac{a\,b}{c^2}\left(\log\frac{4c}{a+b}-1\right), \end{aligned} \tag{1.340}$$

and by

$$\begin{aligned} L(t) &= 4\pi\, K\, \rho\, \frac{a}{\sqrt{a^2+t}}\, \frac{b}{\sqrt{a^2+t}\, +\, \sqrt{b^2+t}}, \\ M(t) &= 4\pi\, K\, \rho\, \frac{b}{\sqrt{b^2+t}}\, \frac{a}{\sqrt{a^2+t}\, +\, \sqrt{b^2+t}}, \\ N(t) &= 4\pi\, K\, \rho\, \frac{a\, b}{c^2} \left(\log \frac{4c}{\sqrt{a^2+t}\, +\, \sqrt{b^2+t}} - 1 \right), \end{aligned} \tag{1.341}$$

respectively.

II. Let us now suppose $a = b$, while c is arbitrary. We shall then have

$$\int_0^\infty \frac{\mathrm{d}t}{\sqrt{(a^2+t)(b^2+t)(c^2+t)}} = \int_0^\infty \frac{\mathrm{d}t}{(a^2+t)\, \sqrt{(c^2+t)}}$$

$$= \begin{cases} \dfrac{2}{\sqrt{c^2-a^2}} \log \dfrac{c+\sqrt{c^2-a^2}}{a}, & c > a, \\[2ex] \dfrac{2}{\sqrt{a^2-c^2}} \arccos \dfrac{c}{a}, & c < a. \end{cases} \tag{1.342}$$

On introducing the eccentricity e of the meridian ellipse, one gets

$$\int_0^\infty \frac{\mathrm{d}t}{\sqrt{(a^2+t)(b^2+t)(c^2+t)}} = \begin{cases} \dfrac{1}{c\, e} \log \dfrac{1+e}{1-e}, & c > a, \\[2ex] \dfrac{2}{a\, e} \arcsin e, & c < a. \end{cases} \tag{1.343}$$

The capacitance and the potential of the filled ellipsoid of mass m are

$$U_0 = \begin{cases} \dfrac{m\, K}{2\, c\, e} \log \dfrac{1+e}{1-e}, & c > a, \\[2ex] \dfrac{m\, K}{a\, e} \arcsin e, & c < a, \end{cases} \tag{1.344}$$

and

$$C = \begin{cases} \dfrac{2ce}{\log{(1+e)/(1-e)}}, & c > a, \\[2ex] a\, \dfrac{e}{\arcsin e}, & c < a, \end{cases} \tag{1.345}$$

respectively. For spheroids, then, the constants L, M, N become

$$\begin{aligned} L &= M = \frac{2\pi K \rho}{e^2}\left(1 - \frac{1-e^2}{2e}\log\frac{1+e}{1-e}\right), \\ N &= 4\pi K \rho\,\frac{1-e^2}{e^2}\left(\frac{1}{2e}\log\frac{1+e}{1-e} - 1\right), \end{aligned} \tag{1.346}$$

for $c > a$, and

$$\begin{aligned} L &= M = 2\pi K \rho\,\frac{\sqrt{1-e^2}}{e^2}\left(\frac{\arcsin e}{e} - \sqrt{1-e^2}\right), \\ N &= \frac{4\pi K \rho}{e^2}\left(1 - \frac{\arcsin e}{e}\sqrt{1-e^2}\right), \end{aligned} \tag{1.347}$$

for $c < a$. The functions $L(t), M(t), N(t)$ for a point P outside the ellipsoid can be computed by replacing ρ and e, of the previous expressions, with the corresponding values for the homothetic ellipsoid passing through P and having the same mass as the given ellipsoid.

36. EQUILIBRIUM OF A ROTATING FLUID

The equilibrium configuration of a rotating fluid may be an ellipsoid of revolution. In order for the fluid ellipsoid, with surface

$$\frac{x^2+y^2}{a^2} + \frac{z^2}{c^2} = 1 \tag{1.348}$$

and rotating with angular velocity ω around the z axis, to be in equilibrium, it is necessary that the sum of the attraction potential and the centripetal potential be constant throughout the entire surface. This means that in all the points of the surface the following must hold:

$$\frac{1}{2}\omega^2\left(x^2+y^2\right) - \frac{1}{2}L\left(x^2+y^2\right) - \frac{1}{2}N z^2 = \text{constant}, \tag{1.349}$$

or also

$$\left(L - \omega^2\right)\left(x^2+y^2\right) + N z^2 = \text{constant}. \tag{1.350}$$

As a consequence, the following equation must be true:

$$\frac{L-\omega^2}{N} = \frac{c^2}{a^2} = 1 - e^2, \tag{1.351}$$

e denoting the eccentricity of the meridian section. On using L and N from Eqs. (1.347), we find

$$\epsilon = \frac{(3-2e^2)\sqrt{1-e^2}\arcsin e - 3e + 3e^3}{e^3}, \tag{1.352}$$

where we have set

$$\epsilon = \frac{\omega^2}{2\pi K \rho} = \frac{2\omega^2}{4\pi K \rho}. \tag{1.353}$$

In what follows, we shall put

$$\eta = \frac{3}{2}\epsilon = \frac{\omega^2}{(4/3)\pi K \rho}, \tag{1.354}$$

$$s = 1 - \sqrt{1-e^2}. \tag{1.355}$$

Notice that s is the flattening of the ellipsoid, while ϵ measures the ratio between the repulsive drag force field and the attractive gravitational field inside the fluid ellipsoid. In particular, $\eta = (3/2)\epsilon$ is the *ratio* between the "centrifugal" force acting on a mass m at a distance r from the rotation axis and the attraction force that would be exerted on the same mass located on the surface of a sphere with radius r and density ρ. In general, the known quantity is ϵ (or η). Then Eq. (1.352) shows that to a given value of e it corresponds only one value of ϵ while, by contrast, for each value of ϵ (smaller than a given limit value) there exist two values of e. This means that two equilibrium configurations are possible: One is obtained for weak flattening and is stable, whereas the other corresponds to strong flattening and is probably unstable. For increasing ϵ, the two solutions get closer, and it exists a value of ϵ such that they coincide. Beyond this limit, i.e., above some angular velocity value for a given density, equilibrium is no longer possible. For weak flattening, we have

$$s = \frac{1}{2}e^2 = \frac{15}{8}\epsilon = \frac{5}{4}\eta. \tag{1.356}$$

In Table 1.10, we report[42] the values of ϵ, η, and $1000/\rho T^2$ as functions of the flattening, ρ being the density relative to water and T the revolution period in hours. In our computation, K has been set equal to $1/(1.5 \times 10^7)$ (c.g.s. units), so that $1000/\rho T^2 = (432/\pi)\epsilon = 137.51\epsilon$.

[42]@ In the original manuscript, only the values of s (first column) were reported. We calculated the corresponding values for the remaining columns from Eqs. (1.352), (1.354), and (1.353), respectively.

Table 1.10. Equilibrium configurations for a rotating fluid (see text).

s	ϵ	η	$\frac{1000}{\rho T^2}$	s	ϵ	η	$\frac{1000}{\rho T^2}$
0.01	0.005322	0.07983	0.7318	0.37	0.1713	0.2570	23.56
0.02	0.01062	0.01593	1.450	0.38	0.1748	0.2622	24.04
0.03	0.01589	0.02384	2.185	0.39	0.1782	0.2674	24.51
0.04	0.02114	0.03171	2.907	0.40	0.1816	0.2724	24.97
0.05	0.02636	0.03954	3.625	0.41	0.1848	0.2772	25.41
0.06	0.03155	0.04733	4.339	0.42	0.1880	0.2819	25.85
0.07	0.03672	0.05508	5.049	0.43	0.1910	0.2865	26.26
0.08	0.04185	0.06278	5.755	0.44	0.1939	0.2909	26.67
0.09	0.04696	0.07043	6.457	0.45	0.1968	0.2952	27.06
0.10	0.05203	0.07804	7.154	0.46	0.1995	0.2992	27.43
0.11	0.05706	0.08569	7.847	0.47	0.2021	0.3031	27.79
0.12	0.06207	0.09310	8.535	0.48	0.2046	0.3069	28.13
0.13	0.06703	0.1005	9.218	0.49	0.2067	0.3104	28.46
0.14	0.07196	0.1079	9.896	0.50	0.2092	0.3138	28.77
0.15	0.07685	0.1153	10.57	0.51	0.2113	0.3170	29.06
0.16	0.08170	0.1223	11.24	0.52	0.2133	0.3199	29.33
0.17	0.08651	0.1298	11.90	0.53	0.2151	0.3227	29.58
0.18	0.09128	0.1369	12.55	0.54	0.2168	0.3252	29.81
0.19	0.09600	0.1440	13.20	0.55	0.2184	0.3275	30.03
0.20	0.1007	0.1510	13.84	0.56	0.2197	0.3296	30.22
0.21	0.1053	0.1580	14.48	0.57	0.2210	0.3315	30.39
0.22	0.1099	0.1648	15.11	0.58	0.2220	0.3330	30.53
0.23	0.1144	0.1716	15.73	0.59	0.2229	0.3344	30.65
0.24	0.1189	0.1783	16.35	0.60	0.2236	0.3354	30.75
0.25	0.1233	0.1850	16.96	0.61	0.2242	0.3362	30.83
0.26	0.1277	0.1915	17.56	0.62	0.2245	0.3367	30.87
0.27	0.1320	0.1980	18.15	0.63	0.2247	0.3370	30.89
0.28	0.1362	0.2043	18.73	0.64	0.2246	0.3369	30.89
0.29	0.1404	0.2106	19.31	0.65	0.2243	0.3365	30.85
0.30	0.1445	0.2168	13.87	0.70	0.2196	0.3294	30.20
0.31	0.1486	0.2228	20.43	0.75	0.2084	0.3126	28.66
0.32	0.1525	0.2288	20.98	0.80	0.1895	0.2842	26.06
0.33	0.1564	0.2347	21.51	0.85	0.1613	0.2419	22.18
0.34	0.1603	0.2404	22.04	0.90	0.1220	0.1830	16.77
0.35	0.1640	0.2461	22.56	0.95	0.06919	0.1038	9.514
0.36	0.1677	0.2516	23.06	1.00	0.0000	0.0000	0.0000

Notice that the limiting value of ϵ is obtained[43] for $s = 0.6$ and is $\epsilon_{\max} = 0.224$. For the computation of ϵ, the following expansion holds:

$$\begin{aligned}\epsilon &= (1-s)\left[\frac{8}{15}s + \frac{44}{105}s^2 + \frac{4}{15}s^3 + \frac{32{\cdot}17}{5{\cdot}7{\cdot}9{\cdot}11}s^4 + \frac{800}{7{\cdot}9{\cdot}11{\cdot}13}s^5\right.\\ &\quad\left. + \frac{736}{3{\cdot}5{\cdot}7{\cdot}11{\cdot}13}s^6 + \ldots + k_n s^n + \ldots\right], \qquad (1.357)\end{aligned}$$

with

$$k_n = \frac{n(3n+5)\,n!}{1{\cdot}3{\cdot}5{\cdot}7{\cdots}(2n+3)}. \qquad (1.358)$$

37. DEFINITE INTEGRALS

We have[44]:

(1) [45]

$$\begin{aligned}\int_0^\infty \frac{1}{r}\sin nr\,\mathrm{e}^{-k^2r^2}\,\mathrm{d}r &= \sqrt{\pi}\int_0^{n/2k}\mathrm{e}^{-x^2}\,\mathrm{d}x\\ &= \frac{\pi}{2}\,\theta\left(\frac{n}{2k}\right). \qquad (1.359)\end{aligned}$$

(2) [46]

$$\int_{-\infty}^{+\infty}\cos nr\,\mathrm{e}^{-k^2r^2}\,\mathrm{d}r = \frac{\sqrt{\pi}}{k}\,\mathrm{e}^{-n^2/4k^2}. \qquad (1.360)$$

(3)

$$\int_0^\pi x\,\sin x\,\mathrm{d}x = \pi. \qquad (1.361)$$

(4)

$$\int_0^\pi x^2\sin x\,\mathrm{d}x = \pi^2 - 2{\cdot}2. \qquad (1.362)$$

(5)

$$\int_0^\pi x^3\sin x\,\mathrm{d}x = \pi^3 - 6\pi. \qquad (1.363)$$

(6)

$$\int_0^\pi x^4\sin x\,\mathrm{d}x = \pi^4 - 12\pi^2 + 2{\cdot}24. \qquad (1.364)$$

[43]@ More precisely, the maximum is reached at $s = 0.632$, corresponding to $\epsilon_{\max} = 0.22467$.

[44]See Sec. 2.26.

[45]@ Notice that $\theta(x)$ is the error function.

[46]Notice that the quantity k on the r.h.s. is positive.

(7) For integer $n \geq 0$ [and, of course, $(-1)^0 = 1$], one has

$$\int_0^\pi x^{2n+1} \sin x \,\mathrm{d}x = (-1)^n (2n+1)! \left(\pi - \frac{\pi^3}{3!} + \ldots \pm \frac{\pi^{2n+1}}{(2n+1)!} \right). \tag{1.365}$$

(8) For integer $n \geq 0$ [and $(-1)^0 = 1$], one gets

$$\int_0^\pi x^{2n} \sin x \,\mathrm{d}x = (-1)^n (2n)! \left(1{\cdot}1 - \frac{\pi^2}{2!} + \frac{\pi^4}{4!} - \ldots \pm \frac{\pi^{2n}}{(2n)!} \right). \tag{1.366}$$

(9) By using the series expansions for $\sin\pi$ and $\cos\pi$, Eqs. (1.365) and (1.366) can be combined into the single expression:

$$\int_0^\pi x^n \sin x \,\mathrm{d}x = n!\pi^n \left(\frac{\pi^2}{(n+2)!} - \frac{\pi^4}{(n+4)!} + \frac{\pi^6}{(n+6)!} - \ldots \right), \tag{1.367}$$

which probably holds for $n > -1$, even for non-integer values. For very large n, we obtain, as a first approximation:

$$\int_0^\pi x^n \sin x \,\mathrm{d}x = \frac{\pi^{n+2}}{(n+1)(n+2)}. \tag{1.368}$$

(10)

$$\int_{-\infty}^{+\infty} \mathrm{e}^{-x^2} \cos nx \,\mathrm{d}x = \mathrm{e}^{-n^2/4} \sqrt{\pi}, \tag{1.369}$$

$$\int_0^{+\infty} \mathrm{e}^{-kx^2} \cos nx \,\mathrm{d}x = \mathrm{e}^{-n^2/4k} \sqrt{\frac{\pi}{k}}. \tag{1.370}$$

(11)

$$\begin{aligned} \int_{-\infty}^{+\infty} \frac{x^3 \,\mathrm{d}x}{\mathrm{e}^x - 1} &= \sum_{k=1}^{\infty} \int_{-\infty}^{+\infty} x^3 \,\mathrm{e}^{-kx} \,\mathrm{d}x \\ &= 6 \left(1 + \frac{1}{2^4} + \frac{1}{3^4} + \ldots \right) = \frac{\pi^4}{15}; \end{aligned} \tag{1.371}$$

see Eq. (1.208).

(12)

$$\int_{-\infty}^{+\infty} \frac{\sin^2 kx}{x^2} \,\mathrm{d}x = k\,\pi. \tag{1.372}$$

38. HEAT PROPAGATION IN AN ISOTROPIC AND HOMOGENEOUS MEDIUM

Let us consider an isotropic and homogeneous medium with a transmission coefficient c, a specific heat γ, and a density δ. The mean square value of the heat displacement in a given direction in a unit time interval is (see Eq. (1.286))

$$\mu^2 = \frac{2c}{\gamma\,\delta}, \tag{1.373}$$

and the differential equation for the temperature T can be written as

$$\frac{\partial T}{\partial \tau} = \frac{1}{2}\,\mu^2\,\nabla^2\,T. \tag{1.374}$$

To determining the temperature distribution, which depends on the specific problem at hand, we can either use the method of sources or derive particular solutions giving the temperature as the product of a time-dependent function and a space-dependent function. In the following, we are going to study quantitatively the heat propagation along one, two and possibly three dimensions.[47]

38.1 One-dimensional Propagation

1.38.1.1. Method of sources. The heat quantity $\mathrm{d}Q$ initially concentrated on a cross section at position x_0 of an infinite-length bar with unitary cross section, propagates and distributes itself in such a way that, from Eq. (1.373), the heat volume density at a point x and at time τ is

$$\rho(x,\tau) = \frac{\mathrm{d}Q}{\mu\,\sqrt{2\pi\tau}}\,\exp\left\{-\frac{(x-x_0)^2}{2\mu^2\tau}\right\}. \tag{1.375}$$

If T_0 is the initial temperature at point x_0, the quantity of heat existing between the cross sections at x_0 and at $x_0 + \mathrm{d}x_0$ can be written as $T_0\mathrm{d}x_0\gamma\delta$. On substituting this expression for $\mathrm{d}Q$ into Eq. (1.375) and then integrating, we can derive the heat density. By dividing by $\gamma\delta$, we can compute the temperature at any point and at any time:

$$T(x,\tau) = \int_{-\infty}^{+\infty} \frac{T_0}{\mu\,\sqrt{2\pi\tau}}\,\exp\left\{-\frac{(x-x_0)^2}{2\mu^2\tau}\right\}\mathrm{d}x_0. \tag{1.376}$$

[47]@ Actually, in the original manuscript, only the one-dimensional case was written down.

Notice that, weren't the bar of infinite length, we would have to consider suitable boundary conditions. However, there are cases where the problem can easily be reduced to that of a bar having infinite length. Let us work out an example:

Consider a finite bar bound by the cross sections S_1 and S_2 at x_1 and x_2, respectively, with $x_1 < x_2$. Let $T_0(x_0)$ be the initial temperature at point x_0, with $x_1 < x_0 < x_2$, and assume that the temperatures T_1 and T_2 of the two ends are constant. The problem is to determine the temperature $T(x,\tau)$ at an arbitrary time τ and at an arbitrary point x between x_1 and x_2. To this end, we shall use the linearity of the heat propagation equations and decompose the temperature distribution, at an arbitrary time, into the sum of two terms, one of which is the distribution that describes only the effects of the boundary conditions and is constant in time. In other words, it holds

$$T(x,\tau) \;=\; T_1 + \frac{x-x_1}{x_2-x_1}\,(T_2 - T_1) + T'(x,\tau), \tag{1.377}$$

$$T_0(x_0) \;=\; T_1 + \frac{x-x_1}{x_2-x_1}\,(T_2 - T_1) + T_0'(x_0). \tag{1.378}$$

Given the initial conditions, the problem is thus reduced to that of finding the temperature distribution along the points of a bar whose ends are at zero temperature. To determining $T'(x,\tau)$, consider an infinite length bar and a point x_0 on it that is initially at the temperature $T_0'(x_0)$. The quantity T_0', whatever its analytic expression may be, is for the moment defined only for x_0 between x_1 and x_2. If $x_1 < x_0 < x_2$ and n is an even integer, let us set

$$T_0'\left(x_0 + n(x_2 - x_1)\right) \;=\; T_0'(x_0), \tag{1.379}$$

while, for odd n,

$$T_0'\left(x_0 + n(x_2 - x_1)\right) \;=\; -\,T_0'\left(x_1 + x_2 - x_0\right). \tag{1.380}$$

The initial temperature is thus defined on every cross section of the bar, except for a discrete number of cross sections — which is irrelevant to the solution of the problem. Notice that the initial temperature takes on opposite values at points that are symmetric with respect to the cross sections S_1 or S_2, so that these cross sections are always at zero temperature. It then follows that the temperature distribution in the infinite-length bar is, at points between x_1 and x_2, exactly the quantity $T'(x,\tau)$ we are looking for. Now, from Eq. (1.376), we deduce

$$T'(x,\tau) \;=\; \int_{-\infty}^{+\infty} \frac{T_0'}{\mu\sqrt{2\pi\tau}} \exp\left\{-\frac{(x-x_0)^2}{2\mu^2\tau}\right\} \mathrm{d}x_0, \tag{1.381}$$

and the problem is thus solved.

1.38.1.2. Particular solutions. By setting $T(x,\tau) = X(x)Y(\tau)$, we obtain

$$\frac{1}{Y}\frac{\mathrm{d}Y}{\mathrm{d}\tau} = \frac{\mu^2}{2}\frac{1}{X}\frac{\mathrm{d}^2Y}{\mathrm{d}x^2} = \lambda. \tag{1.382}$$

We can infer some particular solutions; for instance:

$$T = A\,\mathrm{e}^{-c_1\tau}\sin\left(\frac{\sqrt{2c_1}}{\mu}\,x - c_2\right), \tag{1.383}$$

which, for $\lambda < 0$, can be cast also in the form

$$T = A\,\mathrm{e}^{-\omega^2\mu^2\tau/2}\sin(\omega\,x - c)\,. \tag{1.384}$$

Another solution is

$$T = A\,\mathrm{e}^{-\sqrt{\omega}x/\mu}\sin\left(\omega\,\tau - \frac{\sqrt{\omega}}{\mu}\,x + c\right). \tag{1.385}$$

The solutions in Eqs. (1.384) and (1.385) are special cases of the following, more general solution, from which they can be obtained by setting $\alpha = 0$ and $\alpha = \beta$, respectively:

$$T = A\,\mathrm{e}^{(\alpha^2-\beta^2)\tau/2}\,\mathrm{e}^{-\alpha x/\mu}\sin\left(\alpha\,\beta\,\tau - \frac{\beta}{\mu}\,x + c\right). \tag{1.386}$$

Equation (1.386) represents a surface in the space x, τ, T. Its intersections with planes parallel to the T-axis are, in general, damped sinusoidal curves. By contrast, for planes parallel to the straight-line

$$\alpha\,\beta\,\tau - \frac{\beta}{\mu}\,x = 0,$$

or to

$$\frac{\tau}{2}(\alpha^2 - \beta^2)\,\tau - \frac{\alpha}{\mu}\,x = 0,$$

the intersections are exponential or (non-damped) sinusoidal curves, respectively. The geometric peculiarity of the solutions in Eq. (1.384) and (1.385) lies in the fact that the surfaces represented by those equations have sinusoidal intersections with planes parallel to the $\tau = 0$ plane or with planes parallel to the $x = 0$ plane, respectively.

The problem of the cooling of a finite length bar whose ends are at zero temperature, which has been solved above using the method of sources, can also be solved by using solutions of the kind (1.384). In fact, we only need to decompose T_0' into eigenfunctions of the type in Eq. (1.384), relative to the interval $[x_1, x_2]$, where one sets $\tau = 0$; that is, into sinusoidal curves that vanish for $x = x_1$ and $x = x_2$ and have periods $2(x_2 - x_1)/1$, $2(x_2 - x_1)/2$, $2(x_2 - x_1)/3$, $2(x_2 - x_1)/4$, etc. If we write the expansion of T_0' as

$$T_0' = A_1 \sin\pi \frac{x - x_1}{x_2 - x_1} + A_2 \sin 2\pi \frac{x - x_1}{x_2 - x_1} + \ldots, \tag{1.387}$$

then we obviously have

$$\begin{aligned} T'(x, \tau) &= A_1 \exp\left\{-\frac{\pi^2 \mu^2 \tau}{2(x_2 - x_1)^2}\right\} \sin\pi \frac{x - x_1}{x_2 - x_1} \\ &+ A_2 \exp\left\{-\frac{4\pi^2 \mu^2 \tau}{2(x_2 - x_1)^2}\right\} \sin 2\pi \frac{x - x_1}{x_2 - x_1} + \ldots . \end{aligned} \tag{1.388}$$

39. CONFORMAL TRANSFORMATIONS

Let

$$\begin{aligned} x_1' &= x_1'(x_1, \ldots, x_r, \ldots, x_n), \\ &\ldots \\ x_r' &= x_r'(x_1, \ldots, x_r, \ldots, x_n), \\ &\ldots \\ x_n' &= x_n'(x_1, \ldots, x_r, \ldots, x_n) \end{aligned} \tag{1.389}$$

represent a transformation such that

$$\sum_i \mathrm{d}x_i'^2 \Big/ \sum_i \mathrm{d}x_i^2 = f(x_1, \ldots, x_n). \tag{1.390}$$

Such a condition can be analytically verified by requiring that, at the same point, the gradients of the quantities x' take on the same absolute value, and that the scalar products between any pair of such gradients are

$$\nabla x_i' \cdot \nabla x_j' = m = \begin{cases} f(x_1, \ldots, x_n), & \text{for } i = j, \\ 0, & \text{for } i \neq j. \end{cases} \tag{1.391}$$

Let us put

$$\frac{\partial^2 x_i'}{\partial x_r \partial x_s} = k(i, r, s).$$

From the theorem of the mixed partial derivatives[48], from the condition that all the derivatives of the absolute value of the gradients of x' with respect to the same variable take on the same value, and from the condition that all the derivatives of the scalar products of such quantities vanish, we get the following equations (where we are assuming the x axes to be parallel to the gradients of x'):

$$\begin{aligned} k(i,r,s) &= k(i,s,r), \\ k(i_1,i_1,s) &= k(i_2,i_2,s), \\ k(i,r,s) &= -\,k(r,i,s), \quad \text{for } i \neq r. \end{aligned} \tag{1.392}$$

Now, let us put

$$p_r = \frac{\partial^2 x'_r}{\partial x_r^2}.$$

It is then easy to show that all the derivatives of x' can be expressed in terms of p_r. Indeed, from Eq. (1.392) one deduces that:

(a) for $i \neq r \neq s$,

$$\begin{aligned} k(i,r,s) &= k(i,s,r) = -\,k(s,i,r) = -\,k(s,r,i) \\ &= k(r,s,i) = k(r,i,s) = -\,k(i,r,s) = 0; \end{aligned} \tag{1.393}$$

(b) for $r = i$,

$$k(i,i,s) = k(s,s,s) = \frac{\partial^2 x'_s}{\partial x_s^2} = p_s; \tag{1.394}$$

(c) for $s = i$,

$$k(i,r,i) = k(i,i,r) = p_r; \tag{1.395}$$

(d) for $r = s \neq i$,

$$k(i,r,r) = -\,k(r,i,r) = -\,p_i. \tag{1.396}$$

We can also check that, whatever the n quantities $p_1,p_2,\ldots,p_n$ are, the quantities $k(i,r,s)$ given by Eqs. (1.393), (1.394), (1.395), (1.396) satisfy Eq. (1.392).

---

[48]@ That is, the Schwartz theorem.

If we consider matrices whose r, s element is given by the derivative $k(i, r, s)$, we can build a set of n matrices conjugate to the n quantities x'. They have the following form[49]:

$$\left[\frac{\partial^2 x'_1}{\partial x_r \partial x_s}\right] = \begin{pmatrix} p_1 & p_2 & \dots & p_{n-1} & p_n \\ p_2 & -p_1 & \dots & 0 & 0 \\ \dots & & & & \\ p_{n-1} & 0 & \dots & -p_1 & 0 \\ p_n & 0 & \dots & 0 & -p_1 \end{pmatrix}, \quad \dots,$$

$$\left[\frac{\partial^2 x'_i}{\partial x_r \partial x_s}\right] = \begin{pmatrix} -p_i & 0 & 0 & \dots & p_1 & \dots & 0 & 0 \\ 0 & -p_i & 0 & \dots & p_2 & \dots & 0 & 0 \\ 0 & 0 & -p_i & & p_3 & & 0 & 0 \\ \dots & & & & & & & \\ p_1 & p_2 & p_3 & \dots & p_i & \dots & p_k & 0 \\ \dots & & & & & & & \\ 0 & 0 & 0 & \dots & p_k & \dots & 0 & 0 \\ \dots & & & & & & & \\ 0 & 0 & 0 & \dots & p_n & \dots & 0 & -p_i \end{pmatrix}, \quad \dots .$$

We deduce that

$$\nabla^2 x'_i = -(n-2)\, p_i. \tag{1.397}$$

It follows that the quantities x' are not harmonic functions, unless $n = 2$ or unless all the p vanish, so that Eq. (1.389) simply represents a similitude. If $n = 2$, then the quantities x' are always harmonic functions. In this case, if $U'(x', y')$ is a harmonic function, by setting $U(x, y) = U'(x', y')$, we deduce from Eqs. (1.391) and (1.397) and (15) of Sec. 1.8 that $U(x, y)$ is a harmonic function as well. The transformation of the xy plane into the $x'y'$ plane is a conformal transformation of one into another plane. Such a transformation preserves the form of the infinitesimal geometric figures but can either invert the rotation direction or not. To obtain a conformal transformation we need only to set

$$x' + i\, y' = f(x + iy) \tag{1.398}$$

or

$$x' - i\, y' = f(x + iy), \tag{1.399}$$

where $f(x + iy)$ is an arbitrary analytic function. In the first case, the rotation direction is preserved, while in the second case it is inverted. The analytical considerations developed above may have an interesting,

[49]@ The generic i matrix has to be understood as the one whose ith row or column is given by $(p_1, p_2, p_3, \dots, p_n)$, while all the other diagonal elements are equal to $-p_i$.

brief confirmation as follows. Let us consider a conformal transformation (1.389) changing a point of the n-dimensional space $\mathcal{S}$ into a point of another space $\mathcal{S}'$. From the constraints (1.390) we have that corresponding infinitesimal geometric figures are connected by a similitude, and the "similitude ratio" $k = \sqrt{f(x_1, \ldots, x_n)}$ depends in general on the considered point. Let U be a x-dependent function and let us require that

$$U'(x'_1, ;x'_n) = U(x_1, \ldots, x_n). \tag{1.400}$$

The flux of the gradient of U through a surface element $\mathrm{d}S$ can be written as

$$\mathrm{d}\phi = |\nabla U| \, \mathrm{d}S \cos\alpha, \tag{1.401}$$

while the flux of $\nabla U'$ through the corresponding element is

$$\mathrm{d}\phi' = \left|\nabla U'\right| \, \mathrm{d}S' \cos\alpha'. \tag{1.402}$$

Since we have a conformal transformation,

$$\left|\nabla U'\right| = \frac{1}{k} \left|\nabla U\right|, \tag{1.403}$$

$$\mathrm{d}s' = k^{n-1} \, \mathrm{d}S, \tag{1.404}$$

and thus

$$\mathrm{d}\phi' = k^{n-2} \, \mathrm{d}\phi. \tag{1.405}$$

It follows that, if $n = 2$ and the flux of U through a closed surface vanishes, then the flux of U' through any closed surface vanishes as well. In other words, the conformal transformation of a plane into another plane preserves the harmonic behavior of the harmonic functions. If n is different from 2, such a property is not in general preserved; unless k is a constant, in which case the conformal transformation reduces to a simple similitude.

40. WAVE MECHANICS OF A MASS POINT IN A CONSERVATIVE FIELD. VARIATIONAL APPROACH

Let E be an eigenvalue of the equation[50]

$$\nabla^2 \psi + \frac{2m}{\hbar^2} (E - U) \psi = 0. \tag{1.406}$$

[50]@ In the original manuscript, the old notation $h/2\pi$ is used, while we here denote the same quantity by $\hbar$.

Let us then consider a variation δU of the potential U; for Eq. (1.406) to have a finite and single-valued solution, E has to undergo a variation δE. Let us write the solution of the novel equation as $\psi_i = \psi\,(1+\alpha)$. On substituting into Eq. (1.406), we obtain the equation satisfied by α:

$$\psi\,\nabla^2\alpha + 2\,\nabla\psi\cdot\nabla\alpha + \frac{2m}{\hbar^2}\,(\delta E - \delta U)\,\psi = 0, \tag{1.407}$$

or, multiplying[51] by ψ:

$$\nabla\cdot\left(\psi^2\,\nabla\alpha\right) + \frac{2m}{\hbar^2}\,(\delta E - \delta U)\,\psi^2 = 0. \tag{1.408}$$

Since the quantity $\psi^2\nabla\alpha$ is, in general, an infinitesimal term of order higher than two, on integrating over the whole space $\mathcal{S}$ one gets

$$\int (\delta E - \delta U)\,\psi^2\,\mathrm{d}S = 0, \tag{1.409}$$

that is,

$$\delta E = \int \psi^2\,\delta U\,\mathrm{d}S \Big/ \int \psi^2\,\mathrm{d}S\,. \tag{1.410}$$

41. ELECTROMAGNETIC MASS OF THE ELECTRON[52]

Let us assume that the charge e is distributed over a sphere of radius a. By taking into account the Lorentz contraction along the direction of motion, we obtain the following equations for the electrostatic and for the magnetic energy, respectively:

$$E_{\mathrm{s}} = \frac{e^2}{6a}\left(\frac{2}{\sqrt{1-v^2/c^2}} + \sqrt{1-v^2/c^2}\right), \tag{1.411}$$

$$E_{\mathrm{m}} = \frac{e^2}{3ac^2}\,\frac{v^2}{\sqrt{1-v^2/c^2}}, \tag{1.412}$$

[51]@ Notice that, if ψ is complex, one has to multiply by ψ^*, and in what follows ψ^2 must be replaced with $|\psi|^2$.

[52]@ It is interesting to read about this subject, for instance, E. Fermi, "Correzione di una contraddizione tra la teoria elettrodinamica e quella relativistica delle masse elettromagnetiche" (Correcting a contradiction between the electrodynamic and the relativistic theory about the electromagnetic masses), *Nota Interna della Scuola Normale Superiore di Pisa* (1924). Among the subsequent, related papers by Fermi group's members, cfr., for example, P.Caldirola, *Nuovo Cimento A* **4**, 497 (1979) and refs. therein.

and

$$E_{em} = E_{\rm s} + E_{\rm m} = \frac{e^2}{6a}\left(2\,\frac{1+v^2/c^2}{\sqrt{1-v^2/c^2}} + \sqrt{1-v^2/c^2}\right). \quad (1.413)$$

For an electron at rest one finds

$$E_{\rm s} = e^2/2a, \quad (1.414)$$

$$E_{\rm m} = 0, \quad (1.415)$$

$$E_{em} = e^2/2a. \quad (1.416)$$

The electromagnetic moment will be

$$Q = \frac{2}{v}\,E_{\rm m} = \frac{2e^2}{3ac^2}\,\frac{v}{\sqrt{1-v^2/c^2}}, \quad (1.417)$$

and, assuming no inertial component of the electron mass,[53] it will be

$$m = \frac{2}{3}\,\frac{e^2}{ac^2}\,\frac{1}{\sqrt{1-v^2/c^2}} = \frac{m_0}{\sqrt{1-v^2/c^2}}, \quad (1.418)$$

where m_0 is the rest mass, which is then given by

$$m_0 = \frac{2}{3}\,\frac{e^2}{ac^2} = \frac{4}{3}\,\frac{E_{em}}{c^2}. \quad (1.419)$$

Under the assumption made above of no inertial mass contribution, the mass-energy equivalence (apart from a factor c^2) leads us to conclude that the proper energy of an electron is[54]

$$E_0' = \frac{1}{3}\,E_{\rm em} = \frac{e^2}{6a}, \quad (1.420)$$

or, for a moving electron,

$$\begin{aligned} E' = mc^2 - E_{\rm em} &= \frac{2}{3}\,\frac{e^2}{a}\,\frac{1}{\sqrt{1-v^2/c^2}} \\ &\quad - \frac{e^2}{6a}\left(2\,\frac{1+v^2/c^2}{\sqrt{1-v^2/c^2}} + \sqrt{1-v^2/c^2}\right) \\ &= \frac{e^2}{6a}\sqrt{1-v^2/c^2} = E_0'\sqrt{1-v^2/c^2}, \end{aligned} \quad (1.421)$$

[53]@ That is, that the electron mass is entirely electromagnetic in nature.

[54]@ The author considered the total energy E of an electron to have two components: an electromagnetic one, $E_{\rm em}$, and a self-energy term, E', so that $E = E_{\rm em} + E'$. The energy E' is obtained by requiring that $E = mc^2$.

i.e., the proper energy is proportional to a volume term.[55] There is no discrepancy between this result and the assumption that there is no inertial component. Indeed, the electron does experience an (attractive) strain force, which has to balance the repulsive electromagnetic force, originated by the electron charge, that would lead to a spreading-out of the charge itself. This implies that the proper energy of the electron "core"[56] slows down the forward (with respect to the motion direction) part of the charge, while the backward part of the charge gets accelerated. It follows that there is an energy flux through the core in the direction opposite to the motion one. The momentum associated with this flux *must* be equal and opposite to the momentum associated with the energy E' (which moves with the electron velocity). This can be directly checked by considering a specific (but arbitrary) "bounding model": for example, by assuming that charges are maintained on a spherical surface, at a fixed distance from the center, by strings. In such a case, the energy flux through any cross section is obtained by computing the string tension, and multiplying it by the component of the electron velocity along the string direction. The ratio between such an energy flux and c^2 yields the momentum per unit length; so that, by taking the vectorial sum of the momenta relative to each string (and noting that the elementary momentum in each string lies along the string direction), we get the energy flux one gets the energy flux through the whole core.

Furthermore, Eq. (1.421) shows another peculiar feature: The proper energy of the core decreases with the electron speed and vanishes for speeds close to that of light. Then the problem arises of the energy balance for the core. We easily recognize that the energy decreases with the speed is due to the electron's contraction along the direction of motion. In fact, let us divide the electron into two parts, by means of a yz plane normal to the motion direction and passing through the center of the electron (at which center we put the origin of our reference frame). The charge de distributed over each one of the two spherical surface elements, which get projected into the same element $dy\,dz$ on

[55]@ Cf. the next footnote.

[56]@ In the author's model, the electron has two essential components: the external surface (on which its charge is distributed); and an internal part (which in this translation is called the electron "core"), whose proper energy is proportional to the volume of the core. Actually, in the original manuscript, the electron proper energy is regarded also as the energy of the core. Let us recall, incidentally, that "mechanical" models for the electron were in fashion at that time, for instance along the Abraham–Lorentz's line (the famous Dirac's papers, e.g., appeared in 1938, ten years later).

the chosen plane, will be

$$\mathrm{d}e \;=\; \mathrm{d}y\,\mathrm{d}z\,\frac{a}{\sqrt{a^2-y^2-z^2}}\,\frac{e}{4\pi a^2} \;=\; \frac{e\,\mathrm{d}y\,\mathrm{d}z}{4\pi a\sqrt{a^2-y^2-z^2}}. \tag{1.422}$$

Now, we can imagine that the core is made of longitudinal and transverse strings, each linking two elementary charges symmetric with respect the yz plane. The longitudinal strings compensate the mechanical actions of the electric field components along the direction of motion. The transverse strings compensate the mechanical actions of the transverse component of the electric field as well as of the magnetic field (whose components are by themselves normal to the motion direction) of the moving charges. The tension in the string tying up the two above-considered elements is

$$\mathrm{d}t \;=\; \frac{1}{2}\,\frac{e}{a^2}\,\frac{\sqrt{a^2-y^2-z^2}}{a}\,\mathrm{d}e \;=\; \frac{e^2}{8\pi a^4}\,\mathrm{d}y\,\mathrm{d}z, \tag{1.423}$$

which does not depend on the velocity. The string length is

$$\ell \;=\; 2\sqrt{a^2-y^2-z^2}\,\sqrt{1-v^2/c^2}. \tag{1.424}$$

If the electron velocity increases in magnitude by $\mathrm{d}v$ without changing its direction, the transverse strings do not change their length and thus the associated energy does not change either. Instead, the longitudinal strings get shorter and the energy variation $\mathrm{d}\alpha$ of each of them is obtained by multiplying the string tension by the length variation:

$$\begin{aligned}\mathrm{d}\,(\mathrm{d}\alpha) \;&=\; \mathrm{d}t\,\mathrm{d}\ell \;=\; -\,\frac{2e^2}{8\pi a^4}\,\mathrm{d}y\,\mathrm{d}z\,\sqrt{a^2-y^2-z^2}\;z\frac{(v/c^2)\,\mathrm{d}v}{\sqrt{1-v^2/c^2}} \\ &=\; -\,\frac{e^2\,v\sqrt{a^2-y^2-z^2}}{4\pi a^4\,c^2\sqrt{1-v^2/c^2}}\,\mathrm{d}y\,\mathrm{d}z\,\mathrm{d}v.\end{aligned} \tag{1.425}$$

On integrating with respect to y and z, and considering that

$$\int_{yz\ \mathrm{plane}} \mathrm{d}\alpha \;=\; E', \tag{1.426}$$

we obtain

$$\mathrm{d}E' \;=\; -\,\frac{e^2\,v}{6a\,c^2\sqrt{1-v^2/c^2}}\,\mathrm{d}v. \tag{1.427}$$

The same result is found by differentiating Eq. (1.421), which proves that the proper energy decrease with the speed is actually due to the

Lorentz contraction. The fraction of the total energy contained in the core is

$$\frac{E'}{mc^2} = \frac{1}{4}\left(1 - \frac{v^2}{c^2}\right),$$

which is equal to 1/4 for small velocities. For larger velocities, this ratio decreases, due to both the decrease of the proper energy and the increase of the electromagnetic energy; while at the speed of light the entire electron energy tends to be electromagnetic in nature.

42. LEGENDRE POLYNOMIALS

The Legendre polynomials are defined by the equation

$$P_n = \frac{1}{2^n n!} \frac{\mathrm{d}^n (x^2 - 1)^n}{\mathrm{d}x^n} \tag{1.428}$$

and satisfy the relation

$$\int_{-1}^{1} P_m P_n \,\mathrm{d}x = \begin{cases} 0\,, & \text{for } m \neq n, \\ \dfrac{2}{2n+1}\,, & \text{for } m = n. \end{cases} \tag{1.429}$$

Moreover,

$$P_n(1) = 1\,, \qquad P_n(-1) = (-1)^n. \tag{1.430}$$

The first polynomials read as follows:

$$\begin{aligned} P_0 &= 1, & \int_{-1}^{1} P_n^2 \,\mathrm{d}x &= 2 \\ P_1 &= x, & & \frac{2}{3} \\ P_2 &= \frac{3}{2}x^2 - \frac{1}{2}, & & \frac{2}{5} \\ P_3 &= \frac{5}{2}x^3 - \frac{3}{2}x, & & \frac{2}{7} \end{aligned}$$

$$\begin{aligned}
P_4 &= \frac{35}{8}x^4 - \frac{15}{4}x^2 + \frac{3}{8}, & \frac{2}{9}\\
P_5 &= \frac{63}{8}x^5 - \frac{35}{4}x^3 + \frac{15}{8}x, & \frac{2}{11}\\
P_6 &= \frac{231}{16}x^6 - \frac{315}{16}x^4 + \frac{105}{16}x^2 - \frac{5}{16}, & \frac{2}{13}\\
P_7 &= \frac{429}{16}x^7 - \frac{693}{16}x^5 + \frac{315}{16}x^3 - \frac{35}{16}x. & \frac{2}{15}
\end{aligned}$$

43. ∇^2 IN SPHERICAL COORDINATES

Let V be a function of the variables x, y, z which, in terms of r, θ, ϕ, write

$$\begin{aligned}
r &= \sqrt{x^2 + y^2 + z^2},\\
\theta &= \arccos z/r,\\
\phi &= \arctan y/r.
\end{aligned} \tag{1.431}$$

In view of the relations

$$|\nabla r| = 1, \quad |\nabla \theta| = \frac{1}{r}, \quad |\nabla \phi| = \frac{1}{r\sin\theta}, \tag{1.432}$$

$$\nabla r \times \nabla \theta = \nabla \theta \times \nabla \phi = \nabla \phi \times \nabla r = 0, \tag{1.433}$$

$$\nabla^2 r = \frac{2}{r}, \quad \nabla^2 \theta = \frac{\cot\theta}{r^2}, \quad \nabla^2 \phi = 0, \tag{1.434}$$

we deduce, from (15) of Sec. 1.8, that

$$\nabla^2 V = \frac{\partial^2 V}{\partial r^2} + \frac{2}{r}\frac{\partial V}{\partial r} + \frac{1}{r^2}\left(\frac{1}{\sin^2\theta}\frac{\partial^2 V}{\partial \phi^2} + \frac{\partial^2 V}{\partial \theta^2} + \frac{\partial V}{\partial \theta}\cot\theta\right). \tag{1.435}$$

2

VOLUMETTO II: 23 APRIL 1928

1. ∇^2 IN CYLINDRICAL COORDINATES

Let x and y be the following functions of r and ϕ:

$$r = \sqrt{x^2 + y^2}, \tag{2.1}$$

$$\phi = \arctan \frac{y}{x}. \tag{2.2}$$

Since:

$$|\nabla r| = 1, \quad |\nabla \phi| = \frac{1}{r}, \tag{2.3}$$

$$\nabla r \times \nabla \phi = 0, \tag{2.4}$$

$$\nabla^2 r = \frac{1}{r}, \quad \nabla^2 \phi = 0, \tag{2.5}$$

then, from (15) in Sec. 1.8,

$$\nabla^2 V = \frac{\partial^2 V}{\partial z^2} + \frac{1}{r}\frac{\partial V}{\partial r} + \frac{1}{r^2}\frac{\partial^2 V}{\partial \phi^2}. \tag{2.6}$$

2. EXPANSION OF A HARMONIC FUNCTION IN A PLANE

Let us consider the expansion of a harmonic function V near the point $O(0,0)$. We have

$$V = P_0 + P_1 + P_2 + P_3 + \ldots + P_n + \ldots, \tag{2.7}$$

where P_n is a homogeneous integer polynomial of order n in the variables x and y. Since $\nabla^2 V = 0$, we obtain

$$\nabla^2 P_n = 0, \tag{2.8}$$

which means that P_n is a harmonic function. Let us note that P_n has $n+1$ coefficients while $\nabla^2 P_n$, being a polynomial of order $n-2$, has $n-1$ coefficients, which are related to the P_n coefficients by bilinear relations. Moreover, all these coefficients must be equal to zero. It follows that only two of the P_n coefficients may be chosen arbitrarily with the exception of P_0, which is an arbitrary constant. The most general expression for P_n is then

$$P_n = a_n P'_n + b_n P''_n, \tag{2.9}$$

with P'_n and P''_n two known homogenous and integer polynomials of order n. We can now write

$$P_n = r^n \mu_n(\phi), \quad P'_n = r^n \mu'_n(\phi), \quad P''_n = r^n \mu''_n(\phi) \tag{2.10}$$

and thus

$$\mu_n(\phi) = a_n \mu'_n(\phi) + b_n \mu''_n(\phi). \tag{2.11}$$

We can now choose P'_n and P''_n in such a way that μ'_n and μ''_n are orthogonal in the interval $(0, 2\pi)$. It will suffice to prove that, for $m \neq n$, we always have

$$\int_0^{2\pi} \mu_m \mu_n \, \mathrm{d}\phi = 0, \quad m \neq n. \tag{2.12}$$

In order to do this, let us consider the outgoing flux from a circle of radius r of the vector

$$r^m \mu_m \nabla r^n \mu_n - r^n \mu_n \nabla r^m \mu_m. \tag{2.13}$$

From Eq. (2.10), it follows that

$$\begin{aligned} \frac{\partial}{\partial r} r^m \mu_m &= m r^{m-1} \mu_m, \\ \frac{\partial}{\partial r} r^n \mu_n &= n r^{n-1} \mu_n. \end{aligned} \tag{2.14}$$

Thus the outgoing flux is given by the expression

$$F = (n-m) r^{n+m-1} \int_0^{2\pi} \mu_m \mu_n \, \mathrm{d}\phi. \tag{2.15}$$

On the other hand, the divergence of the vector in Eq. (2.13) is

$$\nabla \cdot (r^m \mu_m \nabla r^n \mu_n - r^n \mu_n \nabla r^m \mu_m)$$

$$= r^m \mu_m \nabla^2 r^n \mu_n - r^n \mu_n \nabla^2 r^m \mu_m = 0, \tag{2.16}$$

from which Eq. (2.12) follows.

3. QUANTIZATION OF A LINEAR HARMONIC OSCILLATOR

Let us consider an oscillator of mass m acted on by the force $-K\,q$ acts; its Hamiltonian can be written as

$$H(q,p) = \frac{1}{2}\,K\,q^2 + \frac{p^2}{2m} = E, \tag{2.17}$$

and the oscillator frequency will be

$$\nu = \frac{1}{2\pi}\sqrt{\frac{K}{m}}. \tag{2.18}$$

The matrices representing p and q, and the ones derived from them, obey the sum and product rules of matrices and satisfy the following conditions:

(a) differentiation with respect to time: $\dot{M}_{rs} = (i/\hbar)\,(E_r - E_s)\,M_{rs}$;

(b) $p\,q - q\,p = \hbar/i$;

(c) H should be diagonal: $H = \mathrm{diag}(E_1, E_2, \ldots)$;

(d) the matrices must be Hermitian .

From these conditions, it follows that the Hamilton equations hold:

$$\dot{q} = \frac{\partial H}{\partial p}, \tag{2.19}$$

$$\dot{p} = -\frac{\partial H}{\partial q}. \tag{2.20}$$

If we combine condition (a) with the others, it becomes

$$\dot{M} = \frac{i}{\hbar}\,(H\,M - M\,H)\,. \tag{2.21}$$

In our case, Eqs. (2.19) and (2.20) read

$$m\,\dot{q} = p, \tag{2.22}$$

$$\dot{p} = -\,K\,q. \tag{2.23}$$

From Eq. (2.22), it follows that the elements of p and q are connected with each other by the relation

$$p_{rs} = q_{rs}\,\frac{im}{\hbar}\,(E_r - E_s)\,. \tag{2.24}$$

Moreover, since

$$m\ddot{q} + K q = 0, \tag{2.25}$$

we obtain

$$\left[K - \frac{m}{\hbar^2}(E_r - E_s)^2\right] q_{rs} = 0. \tag{2.26}$$

Thus, q_{rs} is different from zero only if

$$K = \frac{m}{\hbar^2}(E_r - E_s)^2, \tag{2.27}$$

which is equivalent to

$$E_r - E_s = \pm h\nu. \tag{2.28}$$

Let us now evaluate the (r, r) element of the matrix $pq - qp$. Due to the constraint (b), we have

$$\frac{\hbar}{i} = \sum_\alpha (p_{r\alpha} q_{\alpha r} - q_{r\alpha} p_{\alpha r}), \tag{2.29}$$

or, from Eq. (2.24):

$$\frac{\hbar}{i} = \sum_\alpha (E_r - E_\alpha)\frac{im}{\hbar}(q_{r\alpha} q_{\alpha r} + q_{\alpha r} q_{r\alpha}). \tag{2.30}$$

This is equivalent to

$$\sum_\alpha (E_r - E_\alpha)|q_{rs}|^2 = \frac{\hbar^2}{2m}. \tag{2.31}$$

All the terms in this sum add up to zero, apart from the two for which, respectively,

$$E_\alpha = E_r + h\nu, \tag{2.32}$$
$$E_\alpha = E_r - h\nu. \tag{2.33}$$

It follows that, if an eigenvalue E_r exists, it must also exist one of the two eigenvalues $E_r - h\nu$ and $E_r + h\nu$. Due to the form of H, there exists at least an eigenvalue E_0 such that the eigenvalue $E_0 - h\nu$ does not exist. However, the eigenvalue $E_1 = E_0 + h\nu$ does exist. Thus, on setting $r = 0$ in Eq. (2.31), we obtain

$$h\nu|q_{01}|^2 = \frac{\hbar^2}{2m}, \tag{2.34}$$

that is,

$$|q_{01}|^2 = \frac{\hbar}{4\pi m\nu}. \tag{2.35}$$

On setting $r = 1$ in Eq. (2.31), we find that also the eigenvalue $E_2 = E_1 + h\nu = E_0 + 2h\nu$ must exist, and we can deduce that

$$|q_{12}|^2 = \frac{2\hbar}{4\pi m\nu}. \qquad (2.36)$$

By iteration of Eq. (2.31), we can show (by induction) that the eigenvalue $E_0 + nh\nu$ exists, n being an arbitrary integer, and we also find that

$$|q_{n-1,n}|^2 = \frac{n\hbar}{4\pi m\nu}. \qquad (2.37)$$

It follows that the matrix q and, from Eq. (2.24), the matrix p have both vanishing elements except for the ones adjacent to the diagonal. Note that both q and p matrices are known, apart from a not essential (phase) factor, since their arguments are obviously equal and opposite to those of the complex conjugates of q_{rs} and q_{sr}, p_{rs} and p_{sr}.

It is now possible to determine the matrix H, to verify that condition (c) is met and to determine E_0 at the same time. From Eq. (2.17) we have

$$H_{rs} = \frac{1}{2} K \sum_\alpha q_{r\alpha}\, q_{\alpha s} + \frac{1}{2m} \sum_\alpha p_{r\alpha}\, p_{\alpha s}, \qquad (2.38)$$

which, by Eq. (2.24), becomes

$$H_{rs} = \sum_\alpha \left[\frac{1}{2} K + \frac{m}{2\hbar^2}\,(E_r - E_\alpha)\,(E_s - E_\alpha)\right] q_{r\alpha}\, q_{\alpha s}. \qquad (2.39)$$

All the terms in the sum are zero, apart from those for which α differs from r and from s by one. In order for α to take one of these values, one of the following cases must apply:

$$\begin{aligned} &(\mathrm{I}) && r = s + 2, \\ &(\mathrm{II}) && r = s = n, \\ &(\mathrm{III}) && r = s - 2. \end{aligned} \qquad (2.40)$$

In case I, the sum reduces to a single term, which can be obtained by setting $\alpha = s + 1$. We then find:

$$\begin{aligned} H_{s+2,s} &= \left[\frac{1}{2} K + \frac{m}{2\hbar^2}\,(E_{s+2} - E_{s+1})(E_s - E_{s+1})\right] q_{s+2,s+1}\, q_{s+1,s} \\ &= \left(\frac{1}{2} K - 2\pi^2 \nu^2 m\right) q_{s+2,s+1}\, q_{s+1s} = 0. \end{aligned} \qquad (2.41)$$

We can reason in a similar manner in case III, and thus the corresponding matrix is diagonal. In case II, two terms are non-vanishing in the sum, and they may be obtained by setting $\alpha = r-1 = s-1 = n-1$ or $\alpha = r+1 = s+1 = n+1$. We then get

$$H_n = E_n = \left[\frac{1}{2}K + \frac{m}{2\hbar^2}(E_n - E_{n-1})^2\right] q_{n,n-1}\, q_{n-1,n} + \left[\frac{1}{2}K + \frac{m}{2\hbar^2}(E_n - E_{n+1})^2\right] q_{n,n+1}\, q_{n+1,n}. \tag{2.42}$$

From Eqs. (2.36) and (2.37) we find

$$(E_n - E_{n-1})^2 = (E_n - E_{n+1})^2 = h^2\nu^2, \tag{2.43}$$

$$q_{n-1,n}\, q_{n,n-1} = |q_{n-1,n}|^2 = \frac{n\hbar}{4\pi m\nu}, \tag{2.44}$$

$$q_{n,n+1}\, q_{n+1,n} = |q_{n,n+1}|^2 = \frac{(n+1)\hbar}{4\pi m\nu}, \tag{2.45}$$

from which

$$E_n = \left(\frac{1}{2}K + 2\pi^2 m\nu^2\right)\frac{n\hbar}{4\pi m\nu} + \left(\frac{1}{2}K + 2\pi^2 m\nu^2\right)\frac{(n+1)\hbar}{4\pi m\nu}; \tag{2.46}$$

and thus, since

$$\frac{1}{2}K + 2\pi^2 m\nu = K,$$

one gets:

$$E_n = \frac{K}{8\pi^2 m\nu}(2n+1)\,h = \frac{\nu}{2}(2n+1)\,h = h\nu\left(n+\frac{1}{2}\right). \tag{2.47}$$

In particular:

$$E_0 = \frac{1}{2}h\nu. \tag{2.48}$$

4. DIAGONALIZATION OF A MATRIX

Let H be any Hermitian matrix and S a matrix such that

$$S\,S^{-1} = 1, \tag{2.49}$$

quantity S^{-1} being defined by the relation

$$S^{-1}_{rs} = S^*_{sr}. \tag{2.50}$$

Condition (2.49) can also be written as

$$\sum_i S_{ri}\, S^{-1}_{is} \;=\; \delta_{rs} \quad \text{or} \quad \sum_i S_{ri}\, S^*_{si} \;=\; \delta_{rs}, \tag{2.51}$$

where[1]

$$\delta_{rs} \;=\; \begin{cases} 1, & r = s, \\ 0, & r \neq s. \end{cases}$$

The matrix SHS^{-1} corresponds to the matrix H to which the transformation S is applied. We then have

$$\begin{aligned} \left(S\,H\,S^{-1}\right)_{rs} \;&=\; \sum_i S_{ri}\left(H\,S^{-1}\right)_{is} \;=\; \sum_i S_{ri}\sum_k H_{ik}\, S^{-1}_{ks} \\ &=\; \sum_i\sum_k H_{ik}\, S_{ri}\, S^*_{sk}; \end{aligned} \tag{2.52}$$

and, since we require the matrix $W = SHS^{-1}$ to be diagonal, we obtain

$$\sum_i\sum_k H_{ik}\, S_{ri}\, S^*_{sk} \;=\; E_r\, \delta_{rs}. \tag{2.53}$$

On replacing the indices r and s with m and l and multiplying the resulting expression by S_{ln}, we find

$$\sum_i\sum_k H_{ik}\, S_{mi}\, S^*_{lk}\, S_{ln} \;=\; E_m\, S_{ln}\, \delta_{ml}; \tag{2.54}$$

and, by summing over the index l,

$$\sum_i\sum_k\sum_l H_{ik}\, S_{mi}\, S^*_{lk}\, S_{ln} \;=\; E_m\, S_{mn}. \tag{2.55}$$

The l.h.s. term may also be written as

$$\begin{aligned} &\sum_i\sum_k\sum_l H_{ik}\, S_{mi}\, S^*_{lk}\, S_{ln} \;=\; \sum_i\sum_k H_{ik}\, S_{mi}\sum_l S^*_{lk}\, S_{ln} \\ &= \sum_i\sum_k H_{ik}\, S_{mi}\sum_l S_{kl}\, S^*_{nl} \;=\; \sum_i\sum_k H_{ik}\, S_{mi}\, \delta_{kn} \\ &= \sum_i S_{mi}\, H_{in}, \end{aligned}$$

so that Eq. (2.55) becomes

$$\sum_i S_{mi}\, H_{in} \;=\; E_m\, \delta_{mn}, \tag{2.56}$$

[1]@ In the original manuscript, the definition of the Kronecker symbol was given after Eq. (2.53).

that is

$$\sum_i S_{mi}\,(H_{in} - \delta_{in}\,E_m) = 0. \tag{2.57}$$

We can also derive Eq. (2.56), and then (2.57), by exploiting the associative property of the matrix product. In order to demonstrate this property it suffices to show that, given any three matrices a, b, and c,

$$(a\,b)\,c = a\,b\,c. \tag{2.58}$$

Indeed,

$$\begin{aligned}((a\,b)\,c)_{rs} &= \sum_i (a\,b)_{ri}\,c_{is} = \sum_i\sum_k a_{rk}\,b_{ki}\,c_{is}\,0 \\ &= \sum_k a_{rk}\sum_i b_{ki}\,c_{is} = \sum_k a_{rk}\,(b\,c)_{ks} \\ &= (a\,(b\,c))_{rs} = (a\,b\,c)_{rs}; \quad \text{q.e.d.}\end{aligned}$$

From the relation

$$S\,H\,S^{-1} = W, \tag{2.59}$$

we then obtain

$$W\,S = \left(S\,H\,S^{-1}\right)\,S = S\,H\,S^{-1}\,S = S\,H, \tag{2.60}$$

from which Eq. (2.56) immediately follows.

If the matrix is finite and has N rows or columns, by varying the index n in Eq. (2.57) from 1 to N, we obtain N homogeneous linear equations among the N elements of the n-th row of S. Now, since the elements S_{mn} cannot be all zero, due to Eq. (2.51), the determinant of the coefficients of such homogenous equations must vanish. The following must then hold:

$$\det\begin{pmatrix} H_{11}-E_m & H_{12} & H_{13} & \dots & H_{1N} \\ H_{21} & H_{22}-E_m & H_{23} & \dots & H_{2N} \\ H_{31} & H_{32} & H_{33}-E_m & \dots & H_{3N} \\ \dots & & & & \\ H_{N1} & H_{N2} & H_{N3} & \dots & H_{NN}-E_m \end{pmatrix} = 0. \tag{2.61}$$

It follows that the diagonal elements of the matrix W can be nothing but the roots of Eq. (2.61). If these are all distinct, it is possible to construct W in such a way that the diagonal elements are all equal to E. Matrix S may then be constructed from Eq. (2.57). Indeed, the elements of the nth row of S may be determined, up to a constant factor, by solving the system of linear equations mentioned above. The constant factor may then be derived from the normalization condition given by Eq. (2.51). It is thus possible to find all the rows of S, each of which is associated with one of the roots of Eq. (2.61). We can then demonstrate that the orthogonality condition satisfied by the rows of S, which is demanded by Eq. (2.51), is automatically obtained if H is Hermitian. If there are q coinciding roots, there exists an infinite number of ways that we can use to construct q rows of S that are associated with those q coinciding roots.[2]

5. WAVE QUANTIZATION OF A POINT PARTICLE THAT IS ATTRACTED BY A CONSTANT FORCE TOWARDS A PERFECTLY ELASTIC WALL

Let us consider the one-dimensional motion along the direction perpendicular to an elastic surface. If x is the distance from the wall and g the acceleration of the particle, its potential energy is given by

$$U = \begin{cases} m\,g\,x, & x > 0, \\ \infty, & x < 0, \end{cases} \tag{2.62}$$

and the Schrödinger equation reads

$$\begin{cases} \dfrac{\mathrm{d}^2\psi}{\mathrm{d}x^2} + \dfrac{2m}{\hbar^2}\,(E - m\,g\,x)\,\psi = 0, & x > 0, \\ \\ \psi = 0, & x \leq 0. \end{cases} \tag{2.63}$$

Suppose that ψ is a solution of the above equation with eigenvalue E. On setting

$$x_1 = (m\,g\,x - E)\,\sqrt[3]{2/(m\hbar^2 g^2)}, \tag{2.64}$$

Eq. (2.63) becomes, viewing ψ as a function of x_1:

$$\begin{cases} \dfrac{\mathrm{d}^2\psi}{\mathrm{d}x_1^2} = x_1\,\psi, \\ \\ \psi(x_1 = \alpha) = 0, \end{cases} \tag{2.65}$$

---

[2]@ In the original manuscript, after this section, the author planned writing a section on Laguerre polynomials. However, such a new section was never written.

where[3]

$$\alpha = -E\sqrt[3]{2/(m\hbar^2 g^2)}.$$

The function ψ must also be regular and finite for $\alpha < x_1$; as we shall see, this condition completely determines ψ as a function of x_1 up to a constant factor. If $F(x)$ is such a function, the second one of conditions (2.65) can be expressed by saying that α is a zero of function ψ. One can then obtain all the possible energy levels from the relation

$$E = -\alpha\sqrt[3]{m\hbar^2 g^2/2}. \tag{2.66}$$

We can now also evaluate the periodicity modulus for the action, which we shall denote by S. This will be useful in order to compare the results from wave mechanics with those that we can deduce from the Sommerfeld conditions. We have

$$S = 2\int_0^{E/mg} m\sqrt{\frac{2}{m}(E - mgx)}\,\mathrm{d}x = \frac{4}{3g}\sqrt{\frac{2}{m}}\,E^{3/2} \tag{2.67}$$

or, on writing E as in Eq. (2.66),

$$\frac{S}{h} = \frac{2}{3\pi}(-\alpha)^{3/2}, \tag{2.68}$$

whereas the Sommerfeld conditions would yield

$$S/h = n, \tag{2.69}$$

with n a non-negative integer.

Let us now try to construct the function $F(x) = y$. Two particular solutions of Eq. (2.65) are the following (see Sec. 2.31):

$$\begin{aligned} M &= 1 + \frac{1}{2{\cdot}3}x^3 + \frac{1}{2{\cdot}3{\cdot}5{\cdot}6}x^6 + \frac{1}{2{\cdot}3{\cdot}5{\cdot}6{\cdot}8{\cdot}9}x^9 + \ldots, \\ N &= x + \frac{1}{3{\cdot}4}x^4 + \frac{1}{3{\cdot}4{\cdot}6{\cdot}7}x^7 + \frac{1}{3{\cdot}4{\cdot}6{\cdot}7{\cdot}9{\cdot}10}x^{10} + \ldots. \end{aligned} \tag{2.70}$$

The general solution is a combination of M and N, and, since M and N go to infinity as $x \to \infty$, up to a constant factor it must be true that

$$y = M - \lambda N, \tag{2.71}$$

[3]@ The second equation (2.65) corresponds to $\psi(x = 0) = 0$; obviously, the wavefunction vanishes also for $x < 0$ and $x_1 < \alpha$.

with

$$\lambda = \lim_{x\to\infty} \frac{M}{N}. \tag{2.72}$$

We shall prove soon that λ is finite. We now show that $\log y$ does indeed go to zero sufficiently fast as $x \to \infty$. Let us cast an arbitrary solution of Eq. (2.65) in the form

$$y = e^u. \tag{2.73}$$

We then have

$$u'' + u'^2 = x \tag{2.74}$$

and, for very large x,

$$u' = \pm\sqrt{x}. \tag{2.75}$$

The upper sign corresponds to solutions tending to infinity for large x, while the lower sign corresponds to solutions tending to zero for large x. Now it is simple to show that the expansion of u in decreasing powers of $\sqrt{x}$ is identical for all the ys that tend to infinity, apart from an additive constant. It then follows that the ratio among two solutions that tend to infinity is a non-zero constant. But if y has the form (2.71), we also have that

$$\lim_{x\to\infty} \frac{y}{M} = \frac{M - \lambda N}{M} = 0, \tag{2.76}$$

and thus y tends to zero sufficiently fast. In order to determine λ, let us start by setting

$$\phi(0) = 1; \quad \phi(3) = \frac{1}{2\cdot 3}; \quad \phi(3n) = \frac{1}{2\cdot 3\cdot 5\cdot 6\cdots(3n-1)\cdot(3n)}. \tag{2.77}$$

We thus have

$$M = \phi(0) + \phi(3)\,x^3 + \phi(6)\,x^6 + \phi(9)\,x^9 \ldots. \tag{2.78}$$

We can define $\phi(x)$ for any $x > -2$, by making use of the functional equation

$$\phi(x+3) = \frac{\phi(x)}{(x+2)(x+3)} \tag{2.79}$$

and by assuming that $\phi(x)$ is evaluated in the limit of very large x by linear logarithmic interpolation between $\phi(3n)$ and $\phi(3n+3)$, with $3n < x < 3n+3$. In this limit we evidently have

$$\frac{x^{2/3}\phi(x+1)}{\phi(x)} = 1 \tag{2.80}$$

or, more generally

$$\lim \frac{x^{2\alpha/3}\phi(x+\alpha)}{\phi(x)} = 1, \tag{2.81}$$

from which it is easy to deduce that

$$\lim_{x\to\infty} \frac{\phi(0)+\phi(3)x^3+\phi(6)x^6+\phi(9)x^9+\ldots}{\phi(\alpha)x^\alpha+\phi(\alpha+3)x^{\alpha+3}+\phi(\alpha+6)x^{\alpha+6}+\ldots} = 1 \tag{2.82}$$

(for $\alpha > -2$). In particular,

$$\begin{aligned}&\lim_{x\to\infty} \frac{\phi(0)+\phi(3)x^3+\phi(6)x^6+\phi(9)x^9+\ldots}{\phi(1)x+\phi(4)x^4+\phi(7)x^7+\phi(10)x^{10}+\ldots}\\ &= \lim_{x\to\infty} \frac{M}{\phi(1)\,N} = 1,\end{aligned} \tag{2.83}$$

from which $\lambda = \phi(1)$. Under the form of an infinite product, we have [see item (3) in Sec. 3.7] that

$$\lambda^3 = [\phi(1)]^3 = \frac{1}{2}\cdot\frac{4^2\cdot 7}{5^3}\cdot\frac{7^2\cdot 10}{8^3}\cdot\frac{10^2\cdot 13}{11^3}\cdot\frac{13^2\cdot 16}{14^3}\cdots, \tag{2.84}$$

from which we derive

$$\lambda = \phi(1) = \sqrt[3]{3}\,\frac{\Gamma(2/3)}{\Gamma(1/3)} = \frac{\sqrt[3]{3}}{2}\,\frac{(2/3)!}{(1/3)!} = 0.729. \tag{2.85}$$

We finally have

$$F(x) = \phi(0) - \phi(1)\,x + \phi(3)\,x^3 - \phi(4)\,x^4 + \phi(6)\,x^6 - \phi(7)\,x^7 + \ldots. \tag{2.86}$$

Note that, for $x > 0$, we always have $F(x) > 0$, $F'(x) < 0$, $F''(x) > 0$, while, for $x < 0$, function F has an infinite number of zeros. From the differential equation for F, it is easy to show that, if α_n and α_{n+1} are two consecutive zeros with $\alpha_n > \alpha_{n+1}$, then the following relation holds

$$\alpha_n - \alpha_{n+1} = \frac{\pi}{\sqrt{\xi}}, \quad \text{with} \quad \alpha_{n+1} < -\xi < \alpha_n. \tag{2.87}$$

It follows that

$$\frac{S_{n+1}-S_n}{h} = \sqrt{\frac{\xi_1}{\xi}}, \quad \text{with} \quad \alpha_{n+1} < -\xi_1 < \alpha_n. \tag{2.88}$$

For large quantum numbers, we have

$$(S_{n+1}-S_n)\,/h = 1, \tag{2.89}$$

which is consistent with Sommerfeld's constraints. The first zero of F is at[4]

$$-\alpha_1 \simeq 2.33, \tag{2.90}$$

[4]@ A more precise calculation yields $-\alpha_1 \simeq 2.33811$.

which corresponds to[5]

$$S_1 \simeq 7.49\,h. \tag{2.91}$$

6. RELATIVISTIC HAMILTONIAN FOR THE MOTION OF AN ELECTRON

Let ϕ be the scalar potential and V_x, V_y, V_z the components of the vector potential. Let us set

$$\begin{gathered} C_0 = \phi, \quad C_1 = -i\,V_x, \quad C_2 = -i\,V_y, \quad C_3 = -i\,V_z, \\ x_0 = i\,c\,t, \quad x_1 = x, \quad x_2 = y, \quad x_3 = z, \\ \mathrm{d}s^2 = \textstyle\sum_i \mathrm{d}x_i^2, \end{gathered}$$

and consider the spacetime action[6]

$$\frac{cP}{i} = \int mc^2\,\mathrm{d}s + e\int C_i\,\mathrm{d}x_i. \tag{2.92}$$

We have

$$\delta\left(\frac{cP}{i}\right) = \int mc^2\,\dot{x}_i\,\mathrm{d}\delta x_i + e\int C_i\,\mathrm{d}\delta x_i + e\int \frac{\partial C_i}{\partial x_j}\,\delta x_j\,\mathrm{d}x_i, \tag{2.93}$$

that is,

$$\begin{aligned} \delta\left(\frac{cP}{i}\right) &= \Big[\Big(mc^2\,\dot{x}_i + e\,C_i\Big)\,\delta x_i\Big]_a^b - \int mc^2\,\ddot{x}_i\,\delta x_i\,\mathrm{d}s \\ &\quad - e\int \frac{\partial C_i}{\partial x_j}\,\delta x_i\,\dot{x}_j\,\mathrm{d}s + e\int \frac{\partial C_i}{\partial x_j}\,\delta x_j\,\dot{x}_i\,\mathrm{d}s. \end{aligned} \tag{2.94}$$

The constraint that P be stationary reads

$$mc^2\,\ddot{x}_i = e\left(\frac{\partial C_j}{\partial x_i} - \frac{\partial C_i}{\partial x_j}\right)\dot{x}_j, \tag{2.95}$$

---

[5]@ In the original manuscript, the approximate value $S_1 \simeq 0.76\,h$ was given.

[6]@ Throughout this section, the author adopted the summing convention over repeated indices.

and this is equivalent to the following four equations:

$$\frac{\mathrm{d}}{\mathrm{d}t}\frac{mc^2}{\sqrt{1-v^2/c^2}} = -\frac{e}{c}\left(\frac{\partial C_x}{\partial t}\frac{\partial x}{\partial t}+\frac{\partial C_y}{\partial t}\frac{\partial y}{\partial t}+\frac{\partial C_z}{\partial t}\frac{\partial z}{\partial t}\right)$$

$$-e\left(\frac{\partial \phi}{\partial x}\frac{\partial x}{\partial t}+\frac{\partial \phi}{\partial y}\frac{\partial y}{\partial t}+\frac{\partial \phi}{\partial z}\frac{\partial z}{\partial t}\right),$$

$$\frac{\mathrm{d}}{\mathrm{d}t}\frac{m\,\mathrm{d}x/\mathrm{d}t}{\sqrt{1-v^2/c^2}} = -e\frac{\partial \phi}{\partial x}-\frac{e}{c}\frac{\partial C_x}{\partial t}+\frac{e}{c}\frac{\mathrm{d}y}{\mathrm{d}t}\left(\frac{\partial C_y}{\partial x}-\frac{\partial C_x}{\partial y}\right)$$

$$-\frac{e}{c}\frac{\mathrm{d}z}{\mathrm{d}t}\left(\frac{\partial C_x}{\partial z}-\frac{\partial C_z}{\partial x}\right),$$

$$\frac{\mathrm{d}}{\mathrm{d}t}\frac{m\,\mathrm{d}y/\mathrm{d}t}{\sqrt{1-v^2/c^2}} = -e\frac{\partial \phi}{\partial y}-\frac{e}{c}\frac{\partial C_y}{\partial t}+\frac{e}{c}\frac{\mathrm{d}z}{\mathrm{d}t}\left(\frac{\partial C_z}{\partial y}-\frac{\partial C_y}{\partial z}\right)$$

$$-\frac{e}{c}\frac{\mathrm{d}x}{\mathrm{d}t}\left(\frac{\partial C_y}{\partial x}-\frac{\partial C_x}{\partial y}\right),$$

$$\frac{\mathrm{d}}{\mathrm{d}t}\frac{m\,\mathrm{d}z/\mathrm{d}t}{\sqrt{1-v^2/c^2}} = -e\frac{\partial \phi}{\partial z}-\frac{e}{c}\frac{\partial C_z}{\partial t}+\frac{e}{c}\frac{\mathrm{d}x}{\mathrm{d}t}\left(\frac{\partial C_x}{\partial z}-\frac{\partial C_z}{\partial x}\right)$$

$$-\frac{e}{c}\frac{\mathrm{d}y}{\mathrm{d}t}\left(\frac{\partial C_z}{\partial y}-\frac{\partial C_y}{\partial z}\right),$$

which then are the equations of motion for the electron. Now, given an arbitrary surface (that may even degenerate into a single point), the action P along a line that ends at some fixed point and starts from a point lying on the surface, such that δP is stationary at the lower extreme, is a function of position. Since the variation of P is stationary for fixed end points, on keeping the lower end fixed and changing the upper end by an infinitesimal vector $(\mathrm{d}x_0, \mathrm{d}x_1, \mathrm{d}x_2, \mathrm{d}x_3)$, the variation of P reduces to (see Eq. (2.94)):

$$\mathrm{d}\,(cP/i) = \left(mc^2\,\dot{x}_i + e\,C_i\right)\mathrm{d}x_i, \tag{2.96}$$

or

$$\mathrm{d}P = -\frac{mc^2}{\sqrt{1-v^2/c^2}}\mathrm{d}t - e\,\phi\,\mathrm{d}t + \left(\frac{m}{\sqrt{1-v^2/c^2}}\frac{\mathrm{d}x}{\mathrm{d}t}-\frac{e}{c}C_x\right)\mathrm{d}x$$

$$+\left(\frac{m}{\sqrt{1-v^2/c^2}}\frac{\mathrm{d}y}{\mathrm{d}t}-\frac{e}{c}C_y\right)\mathrm{d}y$$
$$+\left(\frac{m}{\sqrt{1-v^2/c^2}}\frac{\mathrm{d}z}{\mathrm{d}t}-\frac{e}{c}C_z\right)\mathrm{d}z. \tag{2.97}$$

Let us define the momenta conjugateto t, x, y, z as

$$\begin{aligned}
p_t &= -\frac{mc^2}{\sqrt{1-v^2/c^2}}-e\,\phi = -W,\\
p_x &= \frac{m}{\sqrt{1-v^2/c^2}}\frac{\mathrm{d}x}{\mathrm{d}t}+\frac{e}{c}C_x,\\
p_y &= \frac{m}{\sqrt{1-v^2/c^2}}\frac{\mathrm{d}y}{\mathrm{d}t}+\frac{e}{c}C_y,\\
p_z &= \frac{m}{\sqrt{1-v^2/c^2}}\frac{\mathrm{d}z}{\mathrm{d}t}+\frac{e}{c}C_z.
\end{aligned} \tag{2.98}$$

From Eq. (2.97) it then follows that

$$\frac{\partial P}{\partial t} = -W = p_t, \quad \frac{\partial P}{\partial x} = p_x, \quad \frac{\partial P}{\partial y} = p_y, \quad \frac{\partial P}{\partial z} = p_z. \tag{2.99}$$

The four momenta (2.98) are connected by the relation

$$-\left(\frac{W}{c}-\frac{e}{c}\phi\right)^2+\left(p_x-\frac{e}{c}C_x\right)^2$$
$$+\left(p_y-\frac{e}{c}C_y\right)^2+\left(p_z-\frac{e}{c}C_z\right)^2+m^2c^2 = 0. \tag{2.100}$$

From Eq. (2.100) we can deduce the Hamiltonian of the system, apart from a factor $1/2m$. Indeed, if τ is the proper time, denoting by M the l.h.s. of Eq. (2.100) divided by $1/2m$, we get

$$\begin{aligned}
\frac{\partial M}{\partial p_t} &= -\frac{\partial M}{\partial W} = \frac{1}{\sqrt{1-v^2/c^2}} = \frac{\mathrm{d}t}{\mathrm{d}\tau} = \dot{t},\\
\frac{\partial M}{\partial p_x} &= \frac{1}{\sqrt{1-v^2/c^2}}\frac{\mathrm{d}x}{\mathrm{d}t} = \frac{\mathrm{d}x}{\mathrm{d}\tau} = \dot{x},
\end{aligned}$$

$$
\begin{aligned}
\frac{\partial M}{\partial p_y} &= \dot{y}, \quad \frac{\partial M}{\partial p_z} = \dot{z}, \\
\frac{\partial M}{\partial t} &= \frac{e}{\sqrt{1-v^2/c^2}} \frac{\partial \phi}{\partial t} - \frac{e}{c} \frac{1}{\sqrt{1-v^2/c^2}} \frac{\mathrm{d}C_x}{\mathrm{d}t} \frac{\mathrm{d}x}{\mathrm{d}t} \\
& \quad - \frac{e}{c} \frac{1}{\sqrt{1-v^2/c^2}} \frac{\mathrm{d}C_y}{\mathrm{d}t} \frac{\mathrm{d}y}{\mathrm{d}t} - \frac{e}{c} \frac{1}{\sqrt{1-v^2/c^2}} \frac{\mathrm{d}C_z}{\mathrm{d}t} \frac{\mathrm{d}z}{\mathrm{d}t} \\
&= \frac{e}{c} \frac{1}{\sqrt{1-v^2/c^2}} \left[\frac{\mathrm{d}\phi}{\mathrm{d}t} - \frac{1}{c} \left(\frac{\mathrm{d}C_x}{\mathrm{d}t} \frac{\mathrm{d}x}{\mathrm{d}t} + \frac{\mathrm{d}C_y}{\mathrm{d}t} \frac{\mathrm{d}y}{\mathrm{d}t} + \frac{\mathrm{d}C_z}{\mathrm{d}t} \frac{\mathrm{d}z}{\mathrm{d}t} \right) \right. \\
& \quad \left. - \frac{\partial \phi}{\partial x} \frac{\mathrm{d}x}{\mathrm{d}t} - \frac{\partial \phi}{\partial y} \frac{\mathrm{d}y}{\mathrm{d}t} - \frac{\partial \phi}{\partial z} \frac{\mathrm{d}z}{\mathrm{d}t} \right] \\
&= e \frac{\mathrm{d}\phi}{\mathrm{d}\tau} + \frac{\mathrm{d}}{\mathrm{d}\tau} \frac{mc^2}{\sqrt{1-v^2/c^2}} = \frac{\mathrm{d}W}{\mathrm{d}\tau} = \dot{W} = -\dot{p}_t, \\
\frac{\partial M}{\partial x} &= \frac{e}{\sqrt{1-v^2/c^2}} \frac{\partial \phi}{\partial x} \\
& \quad - \frac{e}{c} \frac{1}{\sqrt{1-v^2/c^2}} \left(\frac{\partial C_x}{\partial x} \frac{\mathrm{d}x}{\mathrm{d}t} + \frac{\partial C_y}{\partial x} \frac{\mathrm{d}y}{\mathrm{d}t} + \frac{\partial C_z}{\partial x} \frac{\mathrm{d}z}{\mathrm{d}t} \right) \\
&= \frac{e}{\sqrt{1-v^2/c^2}} \frac{\partial \phi}{\partial x} + \frac{e}{c} \frac{1}{\sqrt{1-v^2/c^2}} \frac{\partial C_x}{\partial t} \\
& \quad - \frac{e}{c} \frac{1}{\sqrt{1-v^2/c^2}} \frac{\mathrm{d}y}{\mathrm{d}t} \left(\frac{\partial C_y}{\partial x} - \frac{\partial C_x}{\partial y} \right) \\
& \quad + \frac{e}{c} \frac{1}{\sqrt{1-v^2/c^2}} \frac{\mathrm{d}z}{\mathrm{d}t} \left(\frac{\partial C_x}{\partial z} - \frac{\partial C_z}{\partial x} \right) - \frac{e}{c} \frac{1}{\sqrt{1-v^2/c^2}} \frac{\mathrm{d}C_x}{\mathrm{d}t} \\
&= - \frac{e}{\sqrt{1-v^2/c^2}} \frac{\mathrm{d}}{\mathrm{d}t} \left(\frac{m}{\sqrt{1-v^2/c^2}} \frac{\mathrm{d}x}{\mathrm{d}t} + \frac{e}{c} C_x \right) \\
&= - \frac{\mathrm{d}p_x}{\mathrm{d}\tau} = -\dot{p}_x, \\
\frac{\partial M}{\partial y} &= -\dot{p}_y, \quad \frac{\partial M}{\partial z} = -\dot{p}_z.
\end{aligned}
$$

In the expressions above, the quantity e is the electric charge of the particle (including its sign); considering electrons with negative charge and denoting now by e the absolute value of the charge, Eqs. (2.98) and

(2.100) become

$$
\begin{aligned}
p_t &= -\frac{mc^2}{\sqrt{1-v^2/c^2}} + e\,\phi = -W, \\
p_x &= \frac{m}{\sqrt{1-v^2/c^2}}\frac{\mathrm{d}x}{\mathrm{d}t} - \frac{e}{c}C_x, \\
p_y &= \frac{m}{\sqrt{1-v^2/c^2}}\frac{\mathrm{d}y}{\mathrm{d}t} - \frac{e}{c}C_y, \\
p_z &= \frac{m}{\sqrt{1-v^2/c^2}}\frac{\mathrm{d}z}{\mathrm{d}t} - \frac{e}{c}C_z;
\end{aligned}
\tag{2.101}
$$

$$
\begin{aligned}
&-\left(\frac{W}{c} + \frac{e}{c}\phi\right)^2 + \left(p_x + \frac{e}{c}C_x\right)^2 + \left(p_y + \frac{e}{c}C_y\right)^2 \\
&+ \left(p_z + \frac{e}{c}C_z\right)^2 + m^2c^2 = 0.
\end{aligned}
\tag{2.102}
$$

7. THE FERMI FUNCTION[7]

The Fermi function [*i.e., the Thomas-Fermi function*] is the solution of the following differential equation under the given boundary conditions:

$$
\phi'' = \frac{\phi^{3/2}}{\sqrt{x}}, \quad \phi(0) = 1, \quad \phi(\infty) = 0. \tag{2.103}
$$

The function is tabulated[8] in Table 2.1.

[7]@ The subject of this section is more commonly known as the Thomas-Fermi function. The author refers here to E.Fermi, *Z. Phys.* **48** (1928) 73.

[8]@ How the author obtained the *numerical* values of the now so-called Thomas-Fermi function is not clear; but they are very accurate.

Table 2.1. The Thomas-Fermi function $\phi(x)$, and its derivative, for several points.

x	$\phi(x)$	$-\phi'(x)$
0	1	1.58
0.05	0.936	1.15
0.1	0.882	0.995
0.2	0.793	0.79
0.3	0.721	0.66
0.4	0.660	0.56
0.5	0.607	0.49
0.6	0.561	0.43
0.7	0.521	0.38
0.8	0.486	0.34
0.9	0.453	0.31
1	0.424	0.29

x	$\phi(x)$
1.1	0.398
1.2	0.374
1.3	0.353
1.4	0.333
1.5	0.315
2	0.243
2.5	0.193
3	0.157
3.5	0.129
4	0.108
5	0.079
6	0.059
7	0.046
8	0.036

x	$\phi(x)$
9	0.029
10	0.024
12	0.017
14	0.012
16	0.009
18	0.007
20	0.0056
25	0.0034
30	0.0022
40	0.0011
50	0.0006
60	0.0004
80	0.0002
100	0.0001

Setting:[9]

$$t = 1 - \frac{1}{12}\sqrt{x^3\,\phi}, \tag{2.104}$$

$$\phi = \exp\left\{\int u(t)\,\mathrm{d}t\right\}, \tag{2.105}$$

we find

$$\begin{aligned} \frac{du}{\mathrm{d}t} &= \frac{16}{3(1-t)} + \left(8 + \frac{1}{3(1-t)}\right) u \\ &\quad + \left(\frac{7}{3} - 4t\right) u^2 - \frac{2}{3}\, t(1-t)\, u^3, \end{aligned} \tag{2.106}$$

$$u(0) = \infty, \tag{2.107}$$

$$\phi = \exp\left\{\int_1^t u(t)\,\mathrm{d}t\right\}. \tag{2.108}$$

[9]@ The author looked for a parametric solution of the Thomas-Fermi equation in the form

$$x = x(t), \quad \phi = \phi(t),$$

where t is a parameter. At this point, he performed the change of variables represented by Eqs. (2.104) and (2.105). Schematically, the method was then the following: Consider x and ϕ as functions of t, given (implicitly) by Eqs. (2.104) and (2.105), respectively. Afterwards, evaluate from them their first and second derivatives with respect to t, and substitute the results into the Thomas-Fermi equation (2.103); note that this equation contains derivatives of ϕ with respect to x. The outcome is a first-order (Abel) differential equation for the unknown function $u(t)$, namely Eq. (2.106). Finally, the boundary conditions (2.103) are taken into account in Eqs. (2.107) and (2.108).

Setting instead[10]

$$t = 144^{-1/6}\, x^{1/2}\, \phi^{1/6}, \tag{2.109}$$

$$u = -\sqrt[3]{16/3}\,\phi^{-4/3}\,\phi', \tag{2.110}$$

the following relation is seen to hold

$$\frac{du}{dt} = 8\,\frac{tu^2-1}{1-t^2u}. \tag{2.111}$$

For $x=0$, we have $t=0$ and

$$u(0) = -\sqrt[3]{16/3}\,\phi_0'.$$

For $x=\infty$, from the asymptotic expansion of ϕ we find $u=1$, $t=1$. The branch of u corresponding to ϕ is defined at the points between $t=0$ and $t=1$. This branch can be obtained by a series expansion (which is always convergent) starting from $t=1$. More precisely, setting $t_1 = 1-t$, we have[11]

$$u = a_0 + a_1\,t_1 + a_2\,t_1^2 + a_3\,t_1^3 + \ldots, \tag{2.112}$$

---

[10]@ In the remaining part of this section, only the substitutions in Eqs. (2.109) and (2.110) are considered. Notice that the method used here by the author is quite different from the previous one, although it looks similar. The author considered the parametric description of x and ϕ:

$$x = x(t), \quad \phi = \phi(x(t))$$

(note that now ϕ is a function of t only through its dependence on x). He then translated the problem in terms of t and $u(t)$, by use of Eqs. (2.109) and (2.110). The procedure he then adopted is the following: The derivative with respect to t of Eq. (2.110) is taken [considering $\phi=\phi(x(t))$] and the Thomas-Fermi equation in (2.103) is substituted in it. Then x (and its t-derivative) from Eq. (2.109) is got and substituted into the relation just obtained. The result is a first-order differential equation for $u(t)$ (with the boundary conditions reported in the text), which is solved by a series expansion (see below). Once $u(t)$ was obtained, the author did not deduce the expression for the Thomas-Fermi function ϕ from Eqs. (2.109) and (2.110), but he looked again for a parametric solution, of the type

$$x = x(t), \quad \phi = \phi(t),$$

setting

$$\phi(t) = \exp\left\{\int w(t)\mathrm{d}t\right\},$$

where $w(t)$ is a function that can be expressed in terms of $u(t)$ on substituting the above expression for $\phi(t)$ into Eqs. (2.109) and (2.110). The final result is expressed by Eqs. (2.116) and (2.117), where the boundary conditions are taken into account as well.

[11]@ In the original manuscript, the exponents of the variable t_1 were forgotten.

where $a_0 = 1$, $a_1 = 9 - \sqrt{73}$, while the remaining coefficients can be evaluated from the following linear "iterating" relation[12]

$$\sum_{n=0}^{m} \frac{1}{2} \left[(m-n+1)\, a_{m-n+1} \left(\delta_n - (a_n - 2\, a_{n-1} + a_{n-2}) \right) \right.$$
$$+ (n+1)\, a_{n+1} \left(\delta_{m-n} - (a_{m-n} - 2\, a_{m-n-1} + a_{m-n-2}) \right)$$
$$\left. +16\, a_{m-n}\, a_n - 8\, (a_{m-n}\, a_{n-1} + a_n\, a_{m-n} - 1) \right] = 0. \qquad (2.114)$$

The first coefficients are as follows[13]

$$\begin{array}{llll} a_0 \simeq 1.000000, & a_1 \simeq 0.455996, \\ a_2 \simeq 0.304455, & a_3 \simeq 0.222180, \\ a_4 \simeq 0.168213, & a_5 \simeq 0.129804, \\ a_6 \simeq 0.101300, & a_7 \simeq 0.0796351, \\ a_8 \simeq 0.0629230, & a_9 \simeq 0.0499053, \\ a_{10} \simeq 0.0396962. & \end{array}$$

Setting $t_1 = 1$ in the expansion (2.112), from Eq. (2.110) we obtain

$$-\phi_0' = \left(\frac{3}{16} \right)^{1/3} (1 + a_1 + a_2 + \ldots). \qquad (2.115)$$

The series in the r.h.s. has all positive terms and has a geometric convergence; the ratio between one term and the previous one tends to approximately 4/5. Once u is known, we can obtain the parametric equation for the function ϕ with simple quadratures. We find

$$\phi = \exp \left\{ - \int_0^t \frac{6ut}{1 - t^2 u} \mathrm{d}t \right\}, \qquad (2.116)$$

$$x = t^2 \left(\frac{144}{\phi} \right)^{1/3}. \qquad (2.117)$$

---

[12]@ The author solved Eq. (2.111) by series; he substituted (2.112) into (2.111), thus obtaining an "iterating" relation for the coefficients a_1, a_2, a_3, ... (the first coefficient a_0 is given by the boundary condition for $x = 0$). Using this procedure, one gets the iterative relation reported in Eq. (2.114). Strangely enough, this is different from the one reported in the original manuscript, which reads

$$a_1(a_n - 2a_{n-1} + a_{n-2}) + 2a_2(a_{n-1} - 2a_{n-2} + a_{n-3})$$
$$+3a_3(a_{n-2} - 2a_{n-3} + a_{n-4}) + \ldots + na_n(a_1 - 2a_0)$$
$$+8(a_0 a_n + a_1 a_{n-1} + \ldots + a_n a_0)$$
$$-8(a_0 a_{n-1} + a_1 a_{n-2} + \ldots + a_{n-1} a_0) = 0. \qquad (2.113)$$

However, the remaining discussion and the results presented in the manuscript are all correct, thus indicating the presence only of an oversight [probably due to the particular hurry]. Notice also that, as stated by the author, the equations determining the coefficients a_2, a_3, ... are linear; while the equation for a_1 is quadratic, and we have to choose the smaller solution, that is, $a_1 = 9 - \sqrt{73}$.

[13]@ In the original manuscript, only the first five coefficients were given.

The other coefficients of the expansion are the following:[14]

$$\begin{array}{lllllll} a_{11} & \simeq & 0.0396962, & \qquad & a_{12} & \simeq & 0.0252838, \\ a_{13} & \simeq & 0.0202322, & & a_{14} & \simeq & 0.0162136, \\ a_{15} & \simeq & 0.0130101, & & a_{16} & \simeq & 0.0104518, \\ a_{17} & \simeq & 0.00840558, & & a_{18} & \simeq & 0.00676660, \\ a_{19} & \simeq & 0.00545216, & & a_{20} & \simeq & 0.00439678. \end{array}$$

8. THE INTERATOMIC POTENTIAL WITHOUT STATISTICS

To first order, the problem of the electron distribution in heavy atoms may be solved as follows: Neglect the inversions of the periodic system and suppose all the electron orbits are circular. From Pauli's principle we have two electrons in a circular orbit of quantum number 1, eight electrons in an orbit of quantum number 2, and in general $2n^2$ electrons in an orbit of quantum number n. If Z denotes the atomic number, then

$$Z = \sum_1^n 2n^2 \tag{2.118}$$

and, for very large Z,

$$Z = \frac{2}{3} n^3. \tag{2.119}$$

The pth electron will be in an orbit having a quantum number

$$Q = \sqrt[3]{3p/2}; \tag{2.120}$$

and, since it feels an effective charge $Z-p$, its distance from the nucleus will be

$$r = \frac{\hbar^2 \, (3p/2)^{2/3}}{m\, e^2\, (Z-p)}. \tag{2.121}$$

In other words, if we set

$$x_1 = \frac{r}{\mu_1} = \frac{1}{2}\frac{r}{\mu}\left(\frac{\pi}{2}\right)^{2/3} \simeq \frac{x}{1.480}, \tag{2.122}$$

[14]@ In the original manuscript, the numerical values of all these coefficients were missing.

with[15]

$$\mu_1 = \frac{\hbar^2\,(3/2)^{2/3}}{m\,e^2\,Z^{1/3}} \simeq 6.93\times 10^{-9}\,Z^{-1/3} \tag{2.123}$$

and thus

$$\mu = \frac{(3\pi)^{2/3}\,\hbar^2}{2^{7/3}\,m\,e^2\,Z^{1/3}} \simeq 4.7\times 10^{-9}\,Z^{-1/3}, \tag{2.124}$$

and, furthermore, write the interatomic potential as

$$V_1 = \frac{Ze}{r}\,\phi_1, \tag{2.125}$$

we then obtain

$$x_1 = \frac{(1-\phi_1+x\phi_1')^{2/3}}{\phi_1 - x\phi_1'}. \tag{2.126}$$

Indeed,

$$\frac{Z-p}{Z} = \phi_1 - x\,\phi_1'. \tag{2.127}$$

If we set[16]

$$t = \phi_1 - x\,\phi_1' \tag{2.128}$$

and thus

$$x_1 = \frac{(1-t)^{2/3}}{t}, \tag{2.129}$$

then, knowing that $\phi_1(\infty) = 0$, we'll have[17]

$$\phi_1 = -\,x_1 \int_\infty^{x_1} \frac{t}{x^2}\,\mathrm{d}x \tag{2.130}$$

and, after integration,

$$\phi_1 = \frac{9}{4t}\left[1 - (1-t)^{2/3}\right] - \frac{3}{2} + \frac{t}{4}, \tag{2.131}$$

which, with

$$x = 2\left(\frac{2}{\pi}\right)^{2/3}\frac{(1-t)^{2/3}}{t}, \tag{2.132}$$

---

[15]@ The numerical value reported in the original manuscript is slightly different: 6.96×10^{-9}. As already stated, we usually rewrite all the equations in terms of the reduced Planck constant $\hbar$ instead of using h.

[16]@ Here the author applied the same method used in the previous section (a change of variable) to find the Thomas-Fermi–like function ϕ_1.

[17]@ Indeed, notice that

$$-\frac{t}{x^2} = -\frac{\phi_1}{x^2} + \frac{\phi_1'}{x} = \frac{\mathrm{d}}{\mathrm{d}x}\frac{\phi_1}{x}.$$

Table 2.2. Comparison between the functions ϕ and ϕ_1 for several points.

x	ϕ	ϕ_1
0	1	1
0.1	0.883	0.883
0.2	0.793	0.793
0.3	0.722	0.721
0.4	0.660	0.660
0.5	0.607	0.608
0.6	0.562	0.564
0.7	0.521	0.525
0.8	0.486	0.491
0.9	0.453	0.462
1	0.424	0.435
2	0.243	0.276

yields the parametric equation for ϕ_1 in the measurement units introduced by Fermi.

It is interesting to make a comparison with Fermi's ϕ (see[18] Table 2.2). From this, we conclude that our approximate method yields a value for the electron density near the nucleus that is roughly a sixth of the actual value, and, for the potential generated by the negative charges in the vicinity of the nucleus, a value that is smaller than the actual one by roughly 4%. The approximate potential derived above can thus be used for the K and L series and, with a small error, also for the M series. It fails though to give correct results for the outermost regions of the atom. The reason for this is twofold: We have neglected the inversions of the periodic system and we have approximated the elliptical with circular orbits. The elliptical orbits are particularly relevant for strongly non-Coulombian fields, such as the ones that are present in the outermost regions, since, for any given total quantum number, they are closer to the nucleus than the circular orbits. The expansions of ϕ and ϕ_1 are

$$\phi = 1 - 1.58\,x + \frac{4}{3}\,x^{3/2} + \dots, \tag{2.133}$$

$$\phi_1 = 1 - 1.52\,x + 1.11\,x^{3/2} + \dots. \tag{2.134}$$

[18]@ In the original manuscript, the numerical values corresponding to $x = 0.2, 0.4, 0.5, 0.7, 0.8, 0.9$ were missing. Note the presence of some slightly different numerical values for ϕ as compared to the ones in Table 2.1.

9. APPLICATION OF THE FERMI POTENTIAL

Let us take the unit charge to be the nuclear charge and the unit length to be $1/\mu$, with, as usual,

$$\mu = \frac{(3\pi)^{2/3}\,\hbar^2}{2^{7/3}\,m\,e^2\,Z^{1/3}} = 4.7\times 10^{-9}\,Z^{-1/3}. \tag{2.135}$$

In a heavy atom, the potential and the field at a distance x are

$$V = \frac{\phi}{x}, \tag{2.136}$$

$$E = \frac{\phi - x\phi'}{x^2}. \tag{2.137}$$

This means that at a distance exceeding x there exists a negative charge $\phi - x\phi'$.

Let us now see how we can evaluate the potential and kinetic energy of the atom by means of statistical arguments. Let us first understand the relation between the initial slope of ϕ and the total energy. This will be useful for verifying the results of the direct computation of the potential and of the kinetic energy. From the expression for μ, we can derive that the energy of the atom is proportional to the atomic number to the power 7/3:

$$\epsilon = K\,Z^{7/3}. \tag{2.138}$$

If, instead of the atomic number Z, we consider the number $\alpha = \log Z$, we find that the differential of the energy is

$$\mathrm{d}\epsilon = \frac{7}{3}\,\epsilon\,\mathrm{d}\alpha. \tag{2.139}$$

We can think of this as the result of an addition of a positive charge $Ze\,\mathrm{d}\alpha$ to the nucleus, increasing the number of electrons of the quantity $Z\,\mathrm{d}\alpha$. Since $Ze = 1$ in our practical measure units, it will suffice to add to the nucleus a charge $\mathrm{d}\alpha$, and to add to the atom as many electrons as is necessary in order to have an equal negative charge. Assuming –as is likely– that the quantum numbers of the initial electrons are not altered by this procedure[19] (the adiabatic principle does not ensure this) and that the variation in energy due to the introduction of new electrons in

[19]Even if this hypothesis was not correct, the conclusions we have drawn from it would still hold, because, in any case, the energy variations would be of the second order.

the outmost regions is of the second order, conservation of energy will take the form

$$d\epsilon = V_0' \, d\alpha, \tag{2.140}$$

in which V_0' is the potential on the nucleus due to the electronic charges only. The linear density of the negative charges at a distance x is $x\phi''$, so that

$$V_0' = \int_0^\infty \frac{1}{x} \, x \, \phi'' \, dx = \phi_0'. \tag{2.141}$$

From Eqs. (2.139), (2.140), and (2.141) we infer

$$\epsilon = \frac{3}{7} \, \phi_0'. \tag{2.142}$$

Computing the potential energy is easy. While moving the electrons to infinity with a flux that is constant at every point and proportional to the initial density, the potential at a distance x varies linearly from the value ϕ/x to $1/x$. The potential energy thus is

$$U = -\int_0^\infty \frac{1+\phi}{2x} \, x\phi'' \, dx = \phi_0' + \frac{1}{2} \int_0^\infty \phi'^2 \, dx. \tag{2.143}$$

If instead we want to consider the sum of the potential energies from each electron, we'll have

$$U_1 = -\int_0^\infty \frac{\phi}{x} \, x\phi'' \, dx = \phi_0' + \int_0^\infty \phi'^2 \, dx. \tag{2.144}$$

Computation of the kinetic energy is not as easy. This is due to the fact that, although the pressure in a perfect gas has no meaning, nevertheless it is possible to define a fictitious stress *homography* which is formally equivalent to that of ordinary fluids. Since such a homography should have a symmetry axis along the radial direction, we shall consider only a radial pressure p' and a transverse pressure p''. The static equations then reduce to

$$4\pi \left[x^2 \, \frac{dp'}{dx} + 2x \, (p' - p'') \right] = -\frac{\phi\phi''}{x} + \phi'\phi''. \tag{2.145}$$

Denoting by T' the kinetic energy (per unit volume) due to the radial component of the velocity and T'' that due to the transverse component, the following relations hold:

$$p' = 2T', \tag{2.146}$$
$$p'' = T''. \tag{2.147}$$

On multiplying the two sides of Eq. (2.145) by $x/2$, we find

$$4\pi\left(x^3\,\frac{\mathrm{d}T'}{\mathrm{d}x} + 2x^2\,T' - x^2\,T''\right) = \frac{1}{2}(-\,\phi\phi'' + x\,\phi'\phi''); \tag{2.148}$$

and by multiplying this by $\mathrm{d}x$ and integrating between 0 and ∞, we find the expression for the kinetic energy:

$$T = \int_0^\infty 4\pi\,(T' + T'')\,\mathrm{d}x = -\frac{1}{2}\,\phi_0' - \frac{1}{4}\int_0^\infty \phi'^2\,\mathrm{d}x. \tag{2.149}$$

From Eqs. (2.143) and (2.149) we derive:

$$\epsilon = T + U = \frac{1}{2}\,\phi_0' + \frac{1}{4}\int_0^\infty \phi'^2\,\mathrm{d}x, \tag{2.150}$$

$$\frac{T}{U} = -\frac{1}{2}. \tag{2.151}$$

On comparing Eq. (2.142) with (2.150), we get[20]

$$\int_0^\infty \phi'^2\,\mathrm{d}x = -\frac{2}{7}\,\phi_0'. \tag{2.152}$$

We can then cast Eqs. (2.143), (2.144), and (2.149) in the very simple form

$$U = \frac{6}{7}\,\phi_0', \tag{2.153}$$

$$U_1 = \frac{5}{7}\,\phi_0', \tag{2.154}$$

$$T = -\frac{3}{7}\,\phi_0', \tag{2.155}$$

$$\frac{T}{U} = -\frac{1}{2}, \tag{2.156}$$

$$\frac{T}{U_1} = -\frac{3}{5}. \tag{2.157}$$

The sum of the energies of every electron, which we shall indicate by $\epsilon_1 = T + U_1$, is 2/3 of the total energy of the atom:

$$\epsilon = \frac{3}{7}\,\phi_0', \quad \epsilon = \frac{2}{7}\,\phi_0'; \tag{2.158}$$

[20]@ The original manuscript contains, at this point, the following paragraph: "I have not been able to prove directly this relation by following a mathematical procedure. However, I have directly verified that it is correct to within 1%. The following formulae are then exact if (2.152) is exact; otherwise they are only approximate relations." This paragraph has been subsequently canceled, while the note "proven" appears.

and, on going back to more conventional measure units, adopting the Rydberg (2.18×10^{-11} erg) as the measure unity for energy:

$$\epsilon = \frac{48\,\phi_0'}{7\,(6\pi)^{2/3}}\,Z^{7/3} \simeq -1.53\,Z^{7/3}, \tag{2.159}$$

$$\epsilon_1 = \frac{32\,\phi_0'}{7\,(6\pi)^{2/3}}\,Z^{7/3} \simeq -1.02\,Z^{7/3}. \tag{2.160}$$

On comparing the last expression with experiments, we note that the results obtained are slightly higher than the experimental values. This is due to the fact that the above arguments predict an infinite charge density in the vicinity of the nucleus, whereas for finite Z we should have a finite density. For not too heavy elements, for which experimental data are available, the error we make corresponds to the fundamental state energy for one of the innermost electrons. On the other hand, the relative error is much smaller for the heaviest elements, also because of the relativistic correction which affects the system in the opposite direction.

10. STATISTICAL BEHAVIOR OF THE FUNDAMENTAL TERMS IN NEUTRAL ATOMS

Having chosen, as usual, the unit length to be μ and the unit charge to be that of the nucleus, the number of electrons at a distance x to $x+\mathrm{d}x$ from the nucleus is

$$Z\,x\,\phi''\mathrm{d}x, \tag{2.161}$$

while the potential energy of any one of these electrons is

$$U = -\frac{1}{Z}\,\frac{\phi}{x}. \tag{2.162}$$

Out of the electrons (2.161), only

$$Z\,x\,\phi''\,k^{3/2}\,\mathrm{d}x \tag{2.163}$$

have a kinetic energy $T < -kU$ $(k < 1)$. It thus follows that there are

$$n = \int_0^\infty Z\,x\,\phi''\left(1 - \frac{x}{\phi}\,y\right)^{3/2}\mathrm{d}x, \quad y = \frac{T}{Z}, \tag{2.164}$$

electrons having an energy smaller than T; and, if we set $\alpha = n/Z$,

$$\begin{aligned} \alpha &= \int_0^\infty x\,\phi'' \left(1 - \frac{x}{\phi}\,y\right)^{3/2} \mathrm{d}x \\ &= \int_0^\infty \sqrt{x}\,(\phi - x\,y)^{3/2}\,\mathrm{d}x = F^{-1}(y), \qquad (2.165) \\ y &= F(\alpha). \qquad (2.166) \end{aligned}$$

Since $T = Zy$ clearly corresponds to the $(Z\alpha)$-th electron, the general expression for the term corresponding to the nth electron can be derived, having ordered the electrons by descending order, as follows:

$$T = Z\,F(\alpha), \quad \text{with} \quad \alpha = n/Z; \qquad (2.167)$$

and, using ordinary measure units, i.e., expressing the terms in Rydberg units, we find

$$T = \frac{16}{(6\pi)^{2/3}}\,Z^{4/3}\,F(\alpha) = 2.2590\,Z^{4/3}\,F(\alpha). \qquad (2.168)$$

11. NUMBERS TO THE FIFTH POWER[21]

x	x^5
3.1	286.29
3.2	335.54
3.3	391.35
3.4	454.35
3.5	525.22
3.6	604.66
3.7	693.44
3.8	792.35
3.9	902.24
4.0	1024.

x	x^5
4.1	1158.56
4.2	1306.91
4.3	1470.08
4.4	1649.16
4.5	1845.28
4.6	2059.63
4.7	2293.45
4.8	2548.04
4.9	2824.75
5.0	3125.

x	x^5
5.1	3450.25
5.2	3802.04
5.3	4181.95
5.4	4591.65
5.5	5032.84
5.6	5507.32
5.7	6016.92
5.8	6563.58
5.9	7149.24
6.0	7776.

[21]@ In the original manuscript, the fifth powers of numbers with odd second digit were missing, as well as those of the numbers from 8.5 to 10.0.

x	x^5
6.1	8445.96
6.2	9161.33
6.3	9924.36
6.4	10737.42
6.5	11602.91
6.6	12523.33
6.7	13501.25
6.8	14539.34
6.9	15640.31
7.0	16807.

x	x^5
7.1	18042.29
7.2	19349.18
7.3	20730.72
7.4	22190.07
7.5	23730.47
7.6	25355.25
7.7	27067.84
7.8	28871.74
7.9	30770.56
8.0	32768.

x	x^5
8.1	34867.84
8.2	37073.98
8.3	39390.41
8.4	41821.19
8.5	44370.53
8.6	47042.70
8.7	49842.09
8.8	52773.19
8.9	55840.59
9.0	59049.

x	x^5
9.1	62403.21
9.2	65908.15
9.3	69568.84
9.4	73390.40
9.5	77378.09
9.6	81537.27
9.7	85873.40
9.8	90392.08
9.9	95099.00
10.	100000

12. DIATOMIC MOLECULE WITH IDENTICAL NUCLEI

Let xy be a meridian cross section of the molecule, x its axis, and y the equator line. Also, let V_1 and V_2 be the potentials generated by each of the two atoms. Considering the effective potential in the form

$$V = V_1 + V_2 - \alpha \frac{2V_1V_2}{V_1+V_2}, \tag{2.169}$$

α must obey the differential equation

$$\nabla^2 V = V^{3/2}, \tag{2.170}$$

where we have neglected the proportionality constant by a suitable choice of measure units.

If we suppose, by approximation, that the value of α depends only on the distance from the center of the molecule, and constraining Eq. (2.170) to hold in the equatorial plane, we find the differential equation

$$V^{3/2} = \nabla^2 V, \tag{2.171}$$

where

$$V = (2 - \alpha)\, V_1, \tag{2.172}$$

$$\begin{aligned}\nabla^2 V &= 2\,\nabla^2 V_1 - \left(\nabla^2 \frac{2V_1V_2}{V_1+V_2}\right)\alpha \\ &\quad -2\left(\frac{V_1}{y} + \frac{\partial V_1}{\partial y}\right)\frac{\mathrm{d}\alpha}{\mathrm{d}y} - V_1\,\frac{\mathrm{d}^2\alpha}{\mathrm{d}y^2}.\end{aligned} \tag{2.173}$$

The constants are determined by requiring $\alpha(0)$ to be finite and the limit of α for $y = \infty$ to be equal to 1.

The hypothesis that α only depends on the distance from the center of the molecule is, however, too far from reality.

13. NUMBERS TO THE SIXTH POWER[22]

x	x^6
1.1	1.8
1.2	3.
1.3	4.8
1.4	7.5
1.5	11.4
1.6	16.8
1.7	24.1
1.8	34.
1.9	47.
2.	64.

x	x^6
2.1	85.8
2.2	113.4
2.3	148.
2.4	191.1
2.5	244.1
2.6	308.9
2.7	387.4
2.8	481.9
2.9	594.8
3.	729.

x	x^6
3.1	887.5
3.2	1073.7
3.3	1291.5
3.4	1544.8
3.5	1838.3
3.6	2176.8
3.7	2565.7
3.8	3010.9
3.9	3518.7
4.	4096.

x	x^6
4.1	4750.1
4.2	5489.
4.3	6321.4
4.4	7256.3
4.5	8303.8
4.6	9474.3
4.7	10779.2
4.8	12230.6
4.9	13841.3
5.	15625.

x	x^6
5.1	17596.3
5.2	19770.6
5.3	22164.4
5.4	24794.9
5.5	27680.6
5.6	30841.
5.7	34296.4
5.8	38068.7
5.9	42180.5
6.	46656.

x	x^6
6.1	51520.4
6.2	56800.2
6.3	62523.5
6.4	68719.5
6.5	75418.9
6.6	82654.
6.7	90458.4
6.8	98867.5
6.9	107918.2
7.	117649.

[22]@ In the original manuscript, the sixth powers of numbers with odd second digit were missing, as well as those for the numbers from 1.1 to 2.9 and from 8.5 to 10.0.

x	x^6
7.1	128100.3
7.2	139314.1
7.3	151334.2
7.4	164206.5
7.5	177978.5
7.6	192699.9
7.7	208422.4
7.8	225199.6
7.9	243087.5
8.	262144.

x	x^6
8.1	282429.5
8.2	304006.7
8.3	326940.4
8.4	351298.
8.5	377149.5
8.6	404567.2
8.7	433626.2
8.8	464404.1
8.9	496981.3
9.	531441.

x	x^6
9.1	567869.3
9.2	606355.
9.3	646990.2
9.4	689869.8
9.5	735091.9
9.6	782757.8
9.7	832972.
9.8	885842.4
9.9	941480.1
10.	1000000

14. NUMBERS TO THE SEVENTH POWER[23]

x	x^7
1.1	1.9
1.2	3.6
1.3	6.3
1.4	10.5
1.5	17.1
1.6	26.8
1.7	41.
1.8	61.2
1.9	89.4
2.	128.

x	x^7
2.1	180.1
2.2	249.4
2.3	340.5
2.4	458.6
2.5	610.4
2.6	803.2
2.7	1046.
2.8	1349.3
2.9	1725.
3.	2187.

x	x^7
3.1	2751.3
3.2	3436.
3.3	4261.8
3.4	5252.3
3.5	6433.9
3.6	7836.4
3.7	9493.2
3.8	11441.6
3.9	13723.1
4.	16384.

x	x^7
4.1	19475.4
4.2	23053.9
4.3	27181.9
4.4	31927.8
4.5	37366.9
4.6	43581.8
4.7	50662.3
4.8	58706.8
4.9	67822.3
5.	78125.

x	x^7
5.1	89741.1
5.2	102807.2
5.3	117471.1
5.4	133892.5
5.5	152243.5
5.6	172709.5
5.7	195489.7
5.8	220798.4
5.9	248865.1
6.	279936.

x	x^7
6.1	314274.3
6.2	352161.5
6.3	393898.1
6.4	439804.7
6.5	490222.8
6.6	545516.1
6.7	606071.2
6.8	672298.9
6.9	744635.3
7.	823543.

[23]@ In the original manuscript, the seventh powers of numbers with odd second digit were missing, as well as those for the numbers from 1.1 to 2.9 and from 8.5 to 10.0.

x	x^7
7.1	909512.
7.2	1003061.3
7.3	1104739.9
7.4	1215128.
7.5	1334838.9
7.6	1464519.5
7.7	1604852.3
7.8	1756556.9
7.9	1920390.9
8.	2097152.

x	x^7
8.1	2287679.2
8.2	2492854.7
8.3	2713605.1
8.4	2950903.5
8.5	3205770.9
8.6	3479278.2
8.7	3772547.9
8.8	4086756.
8.9	4423133.5
9.	4782969.

x	x^7
9.1	5167610.2
9.2	5578466.
9.3	6017008.7
9.4	6484775.9
9.5	6983373.
9.6	7514474.8
9.7	8079828.4
9.8	8681255.3
9.9	9320653.5
10.	10000000

15. SECOND APPROXIMATION FOR THE POTENTIAL INSIDE THE ATOM

From the statistical relation between the effective potential and the density,[24]

$$\rho = K\,(V - C)^{3/2}, \tag{2.174}$$

combined with the Poisson equation for the local potential:

$$\nabla^2 V_0 = -4\pi\,\rho, \tag{2.175}$$

and with the relation (only approximately verified)

$$\nabla^2 V = \frac{Z-n-1}{Z-n}\,\nabla^2 V_0 \tag{2.176}$$

for the atom with atomic number Z that has been ionized n times, we deduce

$$\nabla^2 V = -4\pi\,\rho\,\frac{Z-n-1}{Z-n}. \tag{2.177}$$

The potential inside the ion is

$$V = \frac{Ze}{r}\,\phi\left(\frac{r}{\mu}\right) + C, \tag{2.178}$$

where

$$\mu = 0.47\,Z^{-1/3}\left(\frac{Z-n}{Z-n-1}\right)^{2/3}\ \text{Å}, \tag{2.179}$$

[24]@ Here C is an integration constant; see below.

$$\begin{aligned} \phi'' &= \frac{\phi^{3/2}}{\sqrt{x}}, \quad \phi(0) = 1, && (2.180)\\ -x_0\,\phi'(x_0) &= \frac{n+1}{2}, \quad \phi(x_0) = 0, && (2.181)\\ C &= \frac{(n+1)e}{\mu x_0}. && (2.182) \end{aligned}$$

16. ATOMIC POLARIZABILITY

The potential inside an atom satisfies, to first and second order (as shown in the previous section), an equation of the kind

$$\nabla^2 V = K\,(V - C)^{3/2}. \tag{2.183}$$

Let us now consider the atom in a weak field E. Because of the mutual dependence between the variations of the atomic quantities and the applied field,[25] if the latter is weak, we deduce

$$\begin{aligned} \delta V &= -f(r)\,E\,r\,\cos(\boldsymbol{r}\cdot\mathbf{E}), && (2.184)\\ \delta C &= 0. && (2.185) \end{aligned}$$

Let us suppose that the field $-\mathbf{E}$ lies along the x axis. We then have

$$\begin{aligned} V_1 &= V + E\,x\,f(r), && (2.186)\\ \nabla^2 V_1 &= \nabla^2 V + E\left(x f''(r) + 3\,\frac{x}{r}\,f'(r)\right); && (2.187) \end{aligned}$$

$$\begin{aligned} (V_1 - C) &= (V - C) + E\,x\,f(r), && (2.188)\\ (V_1 - C)^{3/2} &= (V - C)^{3/2} + \frac{3}{2}\,(V - C)^{1/2}\,E\,x\,f(r) + \ldots; && (2.189) \end{aligned}$$

$$\begin{aligned} f''(r) + 3\,\frac{1}{r}\,f'(r) &= \frac{3}{2}\,K\,(V - C)^{1/2}\,f(r), && (2.190)\\ r^{3/2}\,f''(r) + 3\,r^{1/2}\,f'(r) &= \frac{3}{2}\,K\,(V - C)^{1/2}\,r^{3/2}\,f(r); && (2.191) \end{aligned}$$

[25]@ The original manuscript is corrupted, and our interpretation is only plausible.

and, having set

$$y = r^{3/2} f(r), \quad f(r) = \frac{y}{r^{3/2}}, \tag{2.192}$$

$$y'' = \frac{3}{2}\left(K\sqrt{V-C} + \frac{1}{2r^2}\right) y, \tag{2.193}$$

Eq. (2.186) becomes

$$V_1 = V + \frac{x}{r^{3/2}}\, y\, E. \tag{2.194}$$

The condition that $f(0)$ be finite allows us to obtain f or y up to a constant factor. This may then be determined by the requirement that the average value of $-\partial V/\partial x$ on the surface of the ion be equal to $-E$, that is, to the external field. This requirement reads

$$f(r_0) + \frac{1}{3}\, r_0\, f'(r_0) = 1. \tag{2.195}$$

The electric moment of the ion is then

$$M = E\, r_0^3\, (1 - f(r_0)). \tag{2.196}$$

17. FOURIER EXPANSIONS AND INTEGRALS

(1) For $x > 0$, we have

$$\begin{aligned} e^{-kx} &= \int_0^\infty \frac{4k}{k^2 + 4\pi\nu^2} \cos(2\pi\nu x)\, d\nu \\ &= \int_0^\infty \frac{8\pi\nu}{k^2 + 4\pi\nu^2} \sin(2\pi\nu x)\, d\nu \\ &= \int_{-\infty}^\infty \frac{2k}{k^2 + 4\pi\nu^2}\, e^{2\pi\nu i x}\, d\nu \\ &= \int_{-\infty}^\infty \frac{1}{2i} \frac{8\pi\nu}{k^2 + 4\pi\nu^2}\, e^{2\pi\nu i x}\, d\nu; \end{aligned}$$

for $x < 0$, on the other hand, the four integrals yield respectively the values e^{+kx}, $-e^{+kx}$, e^{+kx}, $-e^{+kx}$. For $x > -\alpha$, we have

$$\begin{aligned} e^{-kx} &= e^{k\alpha} \int_{-\infty}^\infty \frac{2k}{k^2 + 4\pi\nu^2}\, e^{2\pi\nu i(x+\alpha)}\, d\nu \\ &= e^{k\alpha} \int_{-\infty}^\infty \frac{1}{2i} \frac{8\pi\nu}{k^2 + 4\pi\nu^2}\, e^{2\pi\nu i(x+\alpha)}\, d\nu, \quad \text{etc.}; \end{aligned}$$

and, by letting $\alpha \to \infty$, the discontinuity at the point $x = -\alpha$ is shifted further and further to the left.

(2) We have:

$$\begin{aligned} e^{-kx^2} &= 2\sqrt{\frac{\pi}{k}} \int_0^\infty e^{-\pi^2\nu^2/k} \cos(2\pi\nu x)\, d\nu, \\ e^{-x^2} &= 2\sqrt{\pi} \int_0^\infty e^{-\pi^2\nu^2} \cos(2\pi\nu x)\, d\nu \\ &= \frac{1}{\sqrt{\pi}} \int_0^\infty e^{-w^2/4} \cos(wx)\, dw. \end{aligned}$$

(3) We have:

$$\frac{1}{\pi}\int_{-\infty}^{\infty} \frac{\sin[2\pi(\nu-\nu_0)a]}{\nu-\nu_0}\, e^{2\pi\nu ix}\, d\nu = \begin{cases} e^{2\pi\nu_0 ix}, & x^2 < a^2, \\ 0, & x^2 > a^2. \end{cases}$$

$$\frac{1}{\pi}\int_{-\infty}^{\infty} \left(\frac{\sin[2\pi(\nu-\nu_0)a]}{\nu-\nu_0} + \frac{\sin[2\pi(\nu+\nu_0)a]}{\nu+\nu_0}\right) \cos(2\pi\nu x)\, d\nu$$
$$= \begin{cases} \cos(2\pi\nu_0 x), & x^2 < a^2, \\ 0, & x^2 > a^2. \end{cases}$$

$$\frac{1}{\pi}\int_{-\infty}^{\infty} \left(\frac{\sin[2\pi(\nu-\nu_0)a]}{\nu-\nu_0} - \frac{\sin[2\pi(\nu+\nu_0)a]}{\nu+\nu_0}\right) \sin(2\pi\nu x)\, d\nu$$
$$= \begin{cases} \sin(2\pi\nu_0 x), & x^2 < a^2, \\ 0, & x^2 > a^2. \end{cases}$$

If $a = k/2\nu_0$, with integer k, the integrals become, respectively,

$$\begin{aligned} &(-1)^k \frac{1}{\pi}\int_{-\infty}^{\infty} \frac{\sin(k\pi\nu/\nu_0)}{\nu-\nu_0}\, e^{2\pi\nu ix}\, d\nu, \\ &(-1)^k \frac{1}{\pi}\int_{-\infty}^{\infty} \frac{2\nu_0 \sin(k\pi\nu/\nu_0)}{\nu^2-\nu_0^2} \cos(2\pi\nu x)\, d\nu, \\ &(-1)^k \frac{1}{\pi}\int_{-\infty}^{\infty} \frac{2\nu_0 \sin(k\pi\nu/\nu_0)}{\nu^2-\nu_0^2} \sin(2\pi\nu x)\, d\nu. \end{aligned}$$

18. BLACKBODY

Let E be the energy emitted per cm^2 and per unit time, while E_ν or E_λ denotes the same energy per unit frequency or wavelength. We have[26]

$$E(T) = \int_0^\infty E_\nu(\nu, T)\,\mathrm{d}\nu = \int_0^\infty E_\lambda(\lambda, T)\,\mathrm{d}\lambda, \qquad (2.197)$$

$$E_\nu = \frac{2\pi h\nu^3}{c^2}\,\frac{1}{\mathrm{e}^{h\nu/kT}-1}, \qquad (2.198)$$

$$E_\lambda = \frac{2\pi c^2 h}{\lambda^5}\,\frac{1}{\mathrm{e}^{hc/\lambda kT}-1}, \qquad (2.199)$$

where

$$E_\lambda = \frac{c}{\lambda^2}\,E_\nu, \qquad (2.200)$$

$$E_\nu = \frac{c}{\nu^2}\,E_\lambda. \qquad (2.201)$$

Then[27] (see Eq. (1.371)):

$$\begin{aligned} E(T) &= \int_0^\infty \frac{2\pi h\nu^3}{c^2}\,\frac{\mathrm{d}\nu}{\mathrm{e}^{h\nu/kT}-1} \\ &= \frac{2\pi k^4T^4}{c^2h^3}\int_0^\infty \frac{1}{\mathrm{e}^{h\nu/kT}-1}\left(\frac{h\nu}{kT}\right)^3 \mathrm{d}\left(\frac{h\nu}{kT}\right) \\ &= \frac{2\pi k^4T^4}{c^2h^3}\,\frac{\pi^4}{15} = \frac{2}{15}\,\frac{\pi^5}{c^2h^3}\,k^4T^4 \\ &\simeq 5.67\times 10^{-5}\,T^4\,\frac{\mathrm{erg}}{\mathrm{cm}^2\mathrm{s}} = 5.67\times 10^{-12}\,T^4\,\frac{\mathrm{W}}{\mathrm{cm}^2}. \end{aligned} \qquad (2.202)$$

The energy per unit volume is

$$E' = \frac{4}{c}\,E = \frac{8}{15}\,\frac{\pi^5}{c^3h^3}\,k^4T^4. \qquad (2.203)$$

If there is thermal equilibrium with the environment, the radiation pressure is[28]

$$p = \frac{1}{3}\,E' = \frac{4}{3}\,\frac{E}{c} = \frac{8}{45}\,\frac{\pi^5}{c^3h^3}\,k^4T^4 \simeq 2.52\times 10^{-15}\,T^4\,\frac{\mathrm{erg}}{\mathrm{cm}^3}. \qquad (2.204)$$

[26]@ In this section we use the Planck constant h, instead of the reduced constant $\hbar$, as in the original manuscript.

[27]@ The numerical value reported in the original manuscript was slightly different (5.55 instead of 5.67).

[28]@ The numerical value reported in the original manuscript is slightly different: 2.47 instead of 2.52.

If there are other blackbodies at zero temperature in the vicinity of our blackbody or if, more generally, our blackbody is in a space free of other radiation, we have to divide the above expression by 2.

19. RADIATION THEORY (PART 1)

The radiation within a region of volume Ω bounded by reflecting walls can be decomposed according to its characteristic frequencies. Since the radiation of a given frequency can be decomposed into two linearly polarized components, the number of such frequencies in the interval $(\nu,\ \nu + \mathrm{d}\nu)$ is

$$\mathrm{d}\mathcal{N} \;=\; \Omega\,\frac{8\pi\nu^2}{c^3}\,\mathrm{d}\nu. \tag{2.205}$$

This means that the density of waves in the volume–frequency space is $2/c^3$. Assuming that a stationary wave represents a stationary state of a light quantum of energy E with $E = h\nu$, we have[29]

$$\mathrm{d}\mathcal{N} \;=\; \Omega\,\frac{8\pi E^2}{c^3 h^3}\,\mathrm{d}E. \tag{2.206}$$

On the other hand, if $\alpha_1, \alpha_2, \alpha_3$ are the direction cosines of the trajectory of the quantum, we also have

$$p_x = \frac{E}{c}\cos\alpha_1, \quad p_y = \frac{E}{c}\cos\alpha_2, \quad p_z = \frac{E}{c}\cos\alpha_3, \tag{2.207}$$

and thus

$$\mathrm{d}\mathcal{N} \;=\; \frac{8\pi\Omega}{h3}(p_x^2 + p_y^2 + p_z^2)\,\mathrm{d}\sqrt{p_x^2 + p_y^2 + p_z^2}, \tag{2.208}$$

that is, the density of the stationary states in the phase space of a gas of light quanta is $2/h^3$, exactly as for an electron gas. The analogy cannot be carried further, since the former obeys Einstein's statistics, while the latter obeys Fermi's statistics. Let

$$\mathbf{C} \;=\; C_0 \sin(2\pi\nu t - \alpha)\,\mathbf{A} \tag{2.209}$$

be the vector potential corresponding to a given frequency, quantity $\mathbf{A}$ being a unit vector, C_0 a function of the position, and α a constant. We set

$$\mathbf{C} \;=\; u\,\mathbf{A}, \quad \text{with} \quad u \;=\; C_0 \sin(2\pi\nu t - \alpha)\,, \tag{2.210}$$

[29]@ In this section we use the Planck constant h instead of the reduced one $\hbar$, as in the original manuscript.

and denote by $\overline{u}$ and $\overline{C_0}$ the mean quadratic values of these quantities in the volume Ω:

$$\overline{u} = \overline{C}_0 \sin(2\pi\nu t - \alpha). \tag{2.211}$$

The total energy of the electric field at time t is

$$W_e = \frac{\Omega}{8\pi} \frac{4\pi^2\nu^2}{c^2} \overline{C}_0^2 \cos^2(2\pi\nu t - \alpha) = \frac{\Omega}{8\pi} \frac{\overline{u}'^2}{c^2}, \tag{2.212}$$

while the energy of the magnetic field is

$$W_m = \frac{\Omega}{8\pi} \frac{4\pi^2\nu^2}{c^2} \overline{C}_0^2 \sin^2(2\pi\nu t - \alpha) = = \frac{\Omega}{8\pi} \frac{4\pi^2\nu^2}{c^2} \overline{u}^2. \tag{2.213}$$

The total energy then becomes

$$W = \frac{\Omega}{8\pi c^2} \left(4\pi^2\nu^2 \overline{u}^2 + \overline{u}'^2\right). \tag{2.214}$$

Let us set

$$q = \overline{u} \sqrt{\frac{\Omega}{4\pi c^2}} = \frac{\overline{u}}{2c} \sqrt{\frac{\Omega}{\pi}}, \tag{2.215}$$

so that we infer

$$W = \pi^2\nu^2 q^2 + \frac{1}{2} \dot{q}^2; \tag{2.216}$$

and, by putting

$$p = \frac{\partial W}{\partial q} = \dot{q}, \tag{2.217}$$

we find

$$W = \frac{1}{2} \left(p^2 + 4\pi^2\nu^2 q^2\right). \tag{2.218}$$

This expression can be considered as the Hamiltonian of the system. On setting

$$H_s = \frac{1}{2} \left(p_s^2 + 4\pi^2\nu^2 q_s^2\right),$$

where the index $s = 1, 2, 3, \ldots$ labels all the possible stationary waves, and denoting by $H_0 = W_0$ the Hamiltonian of an atom inside the region Ω, the overall Hamiltonian —when mutual interaction is neglected— becomes

$$H = \sum_{s=0}^{\infty} H_s = W. \tag{2.219}$$

Let us now consider also the interaction and set $H_0' = H_0 +$ interaction. In first approximation, and when only one electron interacts with the radiation, quantity H_0' can be deduced from the relativistic Hamiltonian for the electron[30] (see Sec. 2.6):

$$W_0 = -e\,\phi + \frac{1}{2m}\,p_i^2 + \frac{e}{mc}\,p_i\,C_i = H_0 + \frac{e}{mc}\,p_i\,C_i. \qquad (2.220)$$

It does not make any difference to use $p_i\,C_i$ or $C_i\,p_i$ because $p_i\,C_i - C_i\,p_i = (h/2\pi i)\boldsymbol{\nabla}\cdot\mathbf{C} = 0$, as the potential ϕ inside the atom is constant and thus

$$\boldsymbol{\nabla}\cdot\mathbf{C} = -\frac{1}{c}\frac{\partial\phi}{\partial t} = 0.$$

The total Hamiltonian, including the interaction, then becomes

$$W = \sum_{s=0}^{\infty} H_s + \sum_{i=1}^{\infty} \frac{e}{mc}\,p_i\,C_i. \qquad (2.221)$$

Let us suppose that for $t = 0$ the region Ω is free of radiation. Then, classically, the electron will execute a damped motion. In a first approximation we can assume that such a motion is periodic; formally, this can be achieved by introducing, in the Hamiltonian, small terms depending only on time and on the p and q coordinates of the electron. Let us decompose its motion in harmonics and consider one of them, directed along the x axis with frequency ν_0. In the expansion of p_x, the term

$$p_{0x} = p_0 \sin(2\pi\,\nu_0\,t + \beta) \qquad (2.222)$$

will appear. Leaving out the other harmonics and focusing on the electromagnetic oscillator labeled by s, the Hamiltonian may be written as

$$\begin{aligned} W = {} & \frac{1}{2}\left(p_s^2 + 4\pi^2\nu_s^2\,q_s^2\right) + \frac{e}{mc}\,C_x^s\,p_{0x} \\ & + \text{terms independent of } q_s \text{ and } p_s. \end{aligned} \qquad (2.223)$$

C_x^s is the component of the vector potential along x and is proportional to q_s at a certain point. Let us set

$$C_s^x = b_s^x\,q_s. \qquad (2.224)$$

In general, b_s depends on the position. Assuming that the oscillations of the electron have a small amplitude with respect to the wavelength of the emitted waves, we may suppose that b_s is constant and equal to

[30]@ Obviously, the author has here assumed $H_0' = W_0$

its value at the center of the atom. Its mean squared value, obtained by averaging over many neighboring frequencies, is[31]

$$\overline{b_s^{x2}} = \frac{1}{3}\frac{\overline{u_s^2}}{q_s^2} = \frac{4}{3}\frac{\pi c^2}{\Omega}. \tag{2.225}$$

On substituting Eqs. (2.224) and (2.222) into Eq. (2.223), we obtain

$$\begin{aligned} W &= \frac{1}{2}\left(p_s^2 + 4\pi^2\nu_s^2\, q_s^2\right) + \frac{e}{mc}\, b_x^s\, p_0 \sin\left(2\pi\nu_0 t + \beta\right) q_s \\ &+ \text{terms independent of } q_s \text{ and } p_s. \end{aligned} \tag{2.226}$$

We then deduce

$$\dot{q}_s = p_s, \tag{2.227}$$

$$\dot{p}_s = -4\pi^2\,\nu_s^2\, q_s - \frac{e}{mc}\, b_x^s\, p_0 \sin\left(2\pi\nu_0 t + \beta\right), \tag{2.228}$$

$$\ddot{q}_s + 4\pi^2\,\nu_s^2\, q_s = -\frac{e}{mc}\, b_x^s\, p_0 \sin\left(2\pi\nu_0 t + \beta\right), \tag{2.229}$$

the last of which has the general integral

$$\begin{aligned} q_s &= A_s \sin 2\pi\nu_s t + B_s \cos 2\pi\nu_s t \\ &- \frac{e}{mc}\frac{b_x^s\, p_0 \sin\left(2\pi\nu_0 t + \beta\right)}{4\pi^2(\nu_s^2 - \nu_0^2)}. \end{aligned} \tag{2.230}$$

Let us suppose for simplicity that $\beta = 0$ and impose the restriction that at $t = 0$ there is no radiation:

$$q_s(0) = \dot{q}_s(0) = 0. \tag{2.231}$$

This is equivalent to choosing the origin of time at the instant $-\beta/2\pi\nu_0$ and assuming that the region Ω is empty at that time. On setting $t_1 = t + (\beta/2\pi\nu_0)$ and rewriting t in place of t_1, Eq. (2.230) becomes

$$\begin{aligned} q_s &= A_s \sin 2\pi\nu_s t + B_s \cos 2\pi\nu_s t \\ &- \frac{e}{mc}\frac{b_x^s\, p_0 \sin 2\pi\nu_0 t}{4\pi^2(\nu_s^2 - \nu_0^2)}, \end{aligned} \tag{2.232}$$

and the constraints (2.231) are to be satisfied with respect to the new independent variable. We thus find

$$B_s = 0, \quad A_s = \frac{\nu_0}{\nu_s}\frac{e}{mc}\frac{b_x^s\, p_0}{4\pi^2(\nu_s^2 - \nu_0^2)}, \tag{2.233}$$

[31]@ The following formula is derived from averaging the square of Eq. (2.224) and using Eqs. (2.211) and (2.215). Note that in the original manuscript the exponent 2 of $\overline{u_s}$ was missing.

$$
\begin{aligned}
q_s &= \frac{e}{mc}\,\frac{b_x^s\, p_0\, \sin 2\pi\nu_0 t}{4\pi^2(\nu_s^2-\nu_0^2)}\left(\frac{\nu_0}{\nu_s}\,\sin 2\pi\nu_s t \;-\; \sin 2\pi\nu_0 t\right) \\
&= \frac{e}{mc}\,\frac{b_x^s\, p_0\, \sin 2\pi\nu_0 t}{4\pi^2(\nu_s^2-\nu_0^2)}\left(\sin 2\pi\nu_s t \;-\; \sin 2\pi\nu_0 t\right. \\
&\qquad \left. -\,\frac{\nu_s-\nu_0}{\nu_s}\,\sin 2\pi\nu_s t\right) \\
&= \frac{e}{mc}\,\frac{b_x^s\, p_0\, \sin 2\pi\nu_0 t}{4\pi^2(\nu_s^2-\nu_0^2)}\left(2\,\cos 2\pi\,\frac{\nu_s+\nu_0}{2} t\,\sin 2\pi\,\frac{\nu_s-\nu_0}{2} t\right. \\
&\qquad \left. -\,\frac{\nu_s-\nu_0}{\nu_s}\,\sin 2\pi\nu_s t\right).
\end{aligned}
\tag{2.234}
$$

If W_s indicates the energy accumulated at time t in the sth oscillator, then the total energy will be

$$
\sum_s W_s \;=\; \int_{\nu_s=0}^{\nu_s=M} W_s\,\mathrm{d}\mathcal{N} \;=\; \frac{8\pi}{c^3}\int_0^M W_s\, n_s^2\,\Omega\,\mathrm{d}\nu_s. \tag{2.235}
$$

The integral must be evaluated over a large but finite range of frequencies. This is due to the fact that the previous equations were derived under the assumption that the excited wavelength were much larger than the amplitude of the oscillations of the electron. Thus the frequencies could not be arbitrarily large. However, $\sum_s W_s$ tends to grow linearly with time and can thus exceed such a limit; on the other hand, each W_s has a maximum (as one can see from Eq. (2.234)), but this does not contradict what was said above since, if we replace W_s with its maximum values, the integral in Eq. (2.235) diverges. However, by removing from the integration domain a small region containing ν_0, the integral becomes convergent. This means that, after a long enough time, almost all the emitted radiation will be contained in an arbitrarily small region around ν_0. The frequencies ν_s of interest are thus very close to ν_0, so that in Eq. (2.234) we can replace $(\nu_s+\nu_0)/2$ with ν_0.[32] Hence:

$$
\begin{aligned}
q_s &= \frac{e}{mc}\,\frac{b_s^2 p_0 t}{8\pi^2\nu_0(\nu_s-\nu_0)} \\
&\quad \times \left[2\,\sin\pi(\nu_s-\nu_0)t\,\cos 2\pi\nu_0 t \;-\; \frac{\nu_s-\nu_0}{\nu_0}\,\sin 2\pi\nu_0 t\right].
\end{aligned}
\tag{2.237}
$$

[32]@ The original manuscript then continued as follows: "We can also replace $\sin 2\pi(\nu_s-\nu_0)/2$ with $2\pi(\nu_s-\nu_0)/2$. Hence

$$
q_s = \frac{e}{mc}\,\frac{b_s p_0}{4\pi^2(\nu_s^2-\nu_0^2)}\left(2\pi(\nu_s-\nu_0)t\,\cos 2\pi\nu_0 t \;-\; \frac{\nu_s-\nu_0}{\nu_0}\,\sin 2\pi\nu_0\right). \tag{2.236}
$$

[It continues in the next page.]

As we shall see, for large t the frequency range in which the radiation has a relevant effect is of the order $1/t$. Thus the first term in brackets is, in general, of order unity, whereas the second one is arbitrarily small as t tends to infinity. In this limit, it therefore follows that

$$q_s = \frac{e}{mc}\frac{b_s^x p_0}{4\pi^2\nu_0}\frac{\sin\pi(\nu_s-\nu_0)t}{\nu_s-\nu_0}\cos 2\pi\nu_0 t, \tag{2.238}$$

$$p_s = -2\pi\nu_0\frac{e}{mc}\frac{b_s^x p_0}{4\pi^2\nu_0}\frac{\sin\pi(\nu_s-\nu_0)t}{\nu_s-\nu_0}\cos 2\pi\nu_0 t, \tag{2.239}$$

$$\begin{aligned} W_s &= \frac{1}{2}\left(p_s^2 + 4\pi^2\nu_s^2 q_s^2\right) \\ &= 2\pi^2\nu_0^2\frac{e^2}{m^2c^2}\frac{b_s^{x2}p_0^2}{16\pi^4\nu_0^2}\frac{\sin^2\pi(\nu_s-\nu_0)t}{(\nu_s-\nu_0)^2}, \end{aligned} \tag{2.240}$$

$$\begin{aligned} \sum W_s &= \int 2\pi^2\nu_0^2\frac{e^2}{m^2c^2}\frac{b_s^{x2}p_0^2}{16\pi^4\nu_0^2}\frac{\sin^2\pi(\nu_s-\nu_0)t}{(\nu_s-\nu_0)^2}\frac{8\pi\nu_0^2}{c^3}\,\Omega\,\mathrm{d}\nu_s \\ &= \frac{\Omega}{\pi c^3}\frac{e^2\nu_0^2}{m^2c^2}p_0^2\,\overline{b_s^{x2}}\int\frac{\sin^2\pi(\nu_s-\nu_0)t}{(\nu_s-\nu_0)^2}\,\mathrm{d}\nu_s \\ &= \frac{\Omega}{\pi c^3}\frac{e^2\nu_0^2}{m^2c^2}p_0^2\,\frac{4}{3}\frac{\pi c^2}{\Omega}\,\pi^2 t \\ &= \frac{4}{3}\frac{e^2\nu_0^2}{m^2c^3}\pi^2 p_0^2\, t. \end{aligned} \tag{2.241}$$

For the electron motion we have

$$p_x = \dot{x}\, m, \tag{2.246}$$

---

[32]We can assume that the quantity $\nu_0 t$ is large, that is, we can consider a large enough time with respect to the oscillation period. In this case, the second term in the brackets is negligible and we obtain

$$q_s = \frac{e}{mc}\frac{b_s p_0 t}{2\pi(\nu_s-\nu_0)}\cos 2\pi\nu_0 t, \tag{2.242}$$

$$p_s = \frac{e}{mc}\frac{b_s p_0 t}{2\pi(\nu_s-\nu_0)}\left(-2\pi\nu_0\sin 2\pi\nu_0 t + \frac{\cos 2\pi\nu_0 t}{t}\right). \tag{2.243}$$

Neglecting the last term in the expression for p_s for large t,

$$p_s = -2\pi\nu_0\frac{e}{mc}\frac{b_s p_0 t}{2\pi(\nu_s-\nu_0)}\sin 2\pi\nu_0 t; \tag{2.244}$$

we thus obtain

$$W_s = \frac{1}{2}\left(p_s^2 + 4\pi^2\nu_s^2 q_s^2\right) = 2\pi^2\nu_s^2\frac{e^2}{m^2c^2}\frac{b_s^2 p_0^2 t^2}{4\pi^2(\nu_s^2-\nu_0^2)}. \tag{2.245}$$

However, the previous expressions do not hold when the quantity $(\nu_s-\nu_0)t$ is large, since we have replaced $\sin\pi(\nu_s-\nu_0)t$ with $\pi(\nu_s-\nu_0)t$."

However, this part was crossed out by the author.

$$\overline{p_x^2} = \frac{m^2}{4\pi^2\nu_0^2}\,\overline{\ddot{x}^2}, \tag{2.247}$$

$$p_0^2 = 2\,\overline{p_x^2} = \frac{m^2}{2\pi^2\nu_0^2}\,\overline{\ddot{x}^2}, \tag{2.248}$$

so that

$$\sum W_s = \frac{2}{3}\frac{e^2\overline{\ddot{x}^2}}{c^3}\,t, \tag{2.249}$$

while the energy radiated per unit time is given by

$$E = \sum \dot{W}_s = \frac{2}{3}\frac{e^2\overline{\ddot{x}^2}}{c^3}, \tag{2.250}$$

in agreement with Balmer's formula.

20. MOMENT OF INERTIA OF THE EARTH

Let m be the mass of the Earth (using measure units such that the Newton gravitational constant is equal to 1), $\mathcal{I}_p$ the polar moment of inertia, and $\mathcal{I}_e$ the equatorial moment of inertia. The potential of the gravitational force in an external point at a distance R from the center O of the Earth, and such that the vector $\boldsymbol{R}$ makes an angle θ with the equator, is (see Sec. 1.7):

$$V = \frac{m}{R} + \frac{1}{R^3}\left(\mathcal{I}_0 - \frac{3}{2}\mathcal{I}_\theta\right), \tag{2.251}$$

where $\mathcal{I}_0$ is the central moment of inertia and $\mathcal{I}_\theta$ the moment of inertia with respect to an axis forming an angle θ with the equator. Since

$$\mathcal{I}_0 = \mathcal{I}_e + \frac{1}{2}\mathcal{I}_p = \frac{3}{2}\mathcal{I}_e + \frac{1}{2}\left(\mathcal{I}_p - \mathcal{I}_e\right), \tag{2.252}$$

$$\mathcal{I}_\theta = \mathcal{I}_e\cos^2\theta + \mathcal{I}_p\sin^2\theta = \mathcal{I}_e + \left(\mathcal{I}_p - \mathcal{I}_e\right)\sin^2\theta, \tag{2.253}$$

it follows that

$$V = \frac{m}{R} + \frac{1}{R^3}\left(\mathcal{I}_p - \mathcal{I}_e\right)\left(\frac{1}{2} - \frac{3}{2}\sin^2\theta\right). \tag{2.254}$$

To evaluate $\mathcal{I}_p$ and $\mathcal{I}_e$, we can use the fact that the potential on the Earth surface takes the same value at the pole and at the equator, once the centrifugal force is taken into account. Denoting by r_e and r_p the equatorial and polar radii, respectively, to first order the potentials at the equator and at the pole read

$$V_e = \frac{m}{r_e} + \frac{1}{2}\frac{\mathcal{I}_p - \mathcal{I}_e}{r^3} + \frac{\beta}{2}\frac{m}{r}, \tag{2.255}$$

$$V_p = \frac{m}{r_p} - \frac{\mathcal{I}_p - \mathcal{I}_e}{r^3}, \tag{2.256}$$

where quantity r is the mean radius of the Earth, which replaces r_e, r_p and similar quantities in the correction terms to first order. Using (again in first approximation)

$$\frac{1}{r_p} - \frac{1}{r_e} = \frac{s}{r}, \tag{2.257}$$

where s is the flattening factor of the Earth, we get

$$\frac{m}{r}\left(s - \frac{\beta}{2}\right) = \frac{3}{2}\frac{\mathcal{I}_p - \mathcal{I}_e}{r^3}, \tag{2.258}$$

$$\mathcal{I}_p - \mathcal{I}_e = \frac{2}{3}\left(s - \frac{\beta}{2}\right) mr^2, \tag{2.259}$$

or, setting $s = 1/297$ and $\beta = 1/289$,

$$\mathcal{I}_p - \mathcal{I}_e = \frac{1}{916} mr^2. \tag{2.260}$$

Substituting this results in Eq. (2.254), we obtain

$$V = \frac{m}{R} + \frac{1}{R^3}\left[\frac{1}{916} mr^2 \left(\frac{1}{2} - \frac{3}{2}\sin^2\theta\right)\right]. \tag{2.261}$$

The potential acting on a celestial body of mass M therefore is

$$MV = \frac{Mm}{R} + \frac{M}{R^3}\left[\frac{1}{916} mr^2 \left(\frac{1}{2} - \frac{3}{2}\sin^2\theta\right)\right]. \tag{2.262}$$

Thus, a component of the force along the radius vector is present; its magnitude is

$$F = -\frac{3}{916}\frac{Mmr^2}{R^4}\sin\theta\cos\theta, \tag{2.263}$$

and on Earth the following torque acts:

$$C = \frac{3}{916}\frac{Mmr^2}{R^3}\sin\theta\cos\theta. \tag{2.264}$$

This torque would move the terrestrial axis towards the celestial meridian of the perturbing star. So, if the latter were the Sun, at the solstices the axis of the Earth would be drawn towards the pole of the ecliptic. At other times of the year, the meridian in which the torque lies would be at some angle with respect to the meridian normal to the ecliptic. Denoting by ϵ the latter angle, and with α and β the inclination of the terrestrial axis and the angle that the Sun determines on the ecliptic after the spring equinox, respectively, we have

$$\epsilon = 90 + \phi, \tag{2.265}$$

where ϕ is the longitude measured, as is customary, from the meridian normal to that containing the pole of the ecliptic (that is. from the meridian where the Sun is at the equinox), and

$$\tan\phi = \tan\beta\cos\alpha. \tag{2.266}$$

If we describe the Earth's motion as that of a gyroscope, its axis will move all the time perpendicularly to the meridian containing the star, with the angular velocity:

$$\eta = \frac{C}{\mathcal{I}_p\omega}, \tag{2.267}$$

ω being the angular speed of the Earth. The components normal to the meridian containing the pole of the ecliptic are

$$\eta_1 = \frac{C}{\mathcal{I}_p\omega}\cos\epsilon = -\frac{C}{\mathcal{I}_p\omega}\sin\phi = \frac{C}{\mathcal{I}_p\omega}\frac{\tan\beta\cos\alpha}{\sqrt{1+\tan^2\beta\cos^2\alpha}}, \tag{2.268}$$

$$\eta_2 = \frac{C}{\mathcal{I}_p\omega}\sin\epsilon = \frac{C}{\mathcal{I}_p\omega}\cos\phi = \frac{C}{\mathcal{I}_p\omega}\frac{1}{\sqrt{1+\tan^2\beta\cos^2\alpha}}. \tag{2.269}$$

On replacing C with Eq. (2.264) and recalling that $\sin\theta = \sin\alpha\sin\beta$, we find

$$\eta_1 = \frac{3}{916}\frac{Mmr^2}{R^3\mathcal{I}_p\omega}\sin\alpha\cos\alpha\frac{\tan\beta\sin\beta\sqrt{1-\sin^2\alpha\sin^2\beta}}{\sqrt{1+\tan^2\beta\cos^2\alpha}}, \tag{2.270}$$

$$\eta_2 = \frac{3}{916}\frac{Mmr^2}{R^3\mathcal{I}_p\omega}\sin\alpha\frac{\sin\beta\sqrt{1-\sin^2\alpha\sin^2\beta}}{\sqrt{1+\tan^2\beta\cos^2\alpha}}. \tag{2.271}$$

Neglecting the eccentricity of the orbit, the average value of η_2 is zero since on changing β into $-\beta$, quantity η_2 changes sign.

If α is very small, the foregoing formulae become

$$\phi = \beta, \tag{2.272}$$

$$\theta = \alpha\sin\beta, \tag{2.273}$$

$$\eta = \frac{3}{916} \frac{Mmr^2}{R^3 \mathcal{I}_p \omega} \alpha \sin \beta, \tag{2.274}$$

$$\eta_1 = \frac{3}{916} \frac{Mmr^2}{R^3 \mathcal{I}_p \omega} \alpha \sin^2 \beta, \tag{2.275}$$

$$\eta_2 = \frac{3}{916} \frac{Mmr^2}{R^3 \mathcal{I}_p \omega} \alpha \sin \beta \cos \beta. \tag{2.276}$$

Assuming a circular orbit, the average values of η_1 and η_2 are

$$\overline{\eta}_1 = \frac{1}{2} \frac{3}{916} \frac{Mmr^2}{R^3 \mathcal{I}_p \omega} \alpha, \tag{2.277}$$

$$\overline{\eta}_2 = 0. \tag{2.278}$$

The axis of the Earth rotates around the axis of the ecliptic with an angular velocity $n = \overline{\eta}_1 / \sin \alpha$ which, for small α, is

$$n = \frac{\overline{\eta}_1}{\alpha} = \frac{1}{2} \frac{3}{916} \frac{Mmr^2}{R^3 \mathcal{I}_p \omega} = \frac{3}{2} \frac{M}{R^3 \omega} \frac{\mathcal{I}_p - \mathcal{I}_e}{\mathcal{I}_p}. \tag{2.279}$$

After having added the effect of the Moon and having neglected the nutation, we get

$$n = \frac{3}{2} \frac{\mathcal{I}_p - \mathcal{I}_e}{\mathcal{I}_p} \left(\frac{M}{R^3 \omega} + \frac{M'}{R'^3 \omega} \right) \tag{2.280}$$

and, from this,

$$\mathcal{I}_p = \frac{3}{2} (\mathcal{I}_p - \mathcal{I}_e) \left(\frac{M}{R^3} + \frac{M'}{R'^3} \right) \frac{1}{n} \frac{1}{\omega}. \tag{2.281}$$

If time is measured in years/2π, we have

$$\frac{M}{R^3} = 1, \quad \frac{M'}{R'^3} = 2.25, \quad \frac{1}{n} = 25800, \quad \omega = 366. \tag{2.282}$$

It follows that

$$\mathcal{I}_p \simeq 344 \, (\mathcal{I}_p - \mathcal{I}_e) \tag{2.283}$$

and, since $\mathcal{I}_p - \mathcal{I}_e = mr^2/916$,

$$\mathcal{I}_p = \frac{344}{916} m r^2 = 0.375 \, m r^2, \tag{2.284}$$

which is too high a value, since $\mathcal{I}_p/(\mathcal{I}_p - \mathcal{I}_e) = 305$.

21. RADIATION THEORY (PART 2)

Let us consider once more the Hamiltonian

$$H_0 + \sum_{s=1}^{\infty} \left(\frac{1}{2} p_s^2 + 2\pi^2 \nu_s^2 q_s^2 \right) + \sum_{s=1}^{\infty} \frac{e}{mc} \boldsymbol{p}_s \cdot \boldsymbol{b}_s q_s, \tag{2.285}$$

$\boldsymbol{b}_s$ being a vector[33] that depends on position and such that its average value is

$$\overline{|\boldsymbol{b}_s|^2} = \frac{4\pi c^2}{\Omega}. \tag{2.286}$$

Let ψ_n be the eigenfunction of the nth stationary state of the unperturbed atom and $\psi_s^{r_s}$ the eigenfunction of the rth state of the sth unperturbed oscillator. Neglecting any interaction, the eigenfunction of the whole system will be

$$\psi = \sum_{n, r_1, r_2, \dots} a_{n, r_1, r_2, r_3, r_4, \dots r_i \dots} \psi_n \psi_1^{r_1} \psi_2^{r_2} \dots \psi_r^{r_i} \dots, \tag{2.287}$$

where the quantities a are constant. However, due to interactions, these a will depend on time, in accordance with the following differential equations

$$\frac{\hbar}{i} \dot{a}_{n, r_1, r_2, \dots} = \sum a_{n', r_1', r_2', \dots} A_{n, r_1, r_2, \dots, n', r_1', r_2', \dots}, \tag{2.288}$$

A denoting the interaction matrix. It is immediately clear that the only terms that can be different from zero are those corresponding to an atom changing its state and to the quantum number of the oscillators changing by one unit. For $r_s' = r_s \pm 1$ we have

$$\begin{aligned} & A_{n, r_1, r_2, \dots r_s \dots, n', r_1, r_2, \dots r_s' \dots} \\ = \; & \frac{e}{c} 2\pi i \left(\nu_n - \nu_{n'} \right) \left(b_s^x \eta_{nn'}^x + b_s^y \eta_{nn'}^y + b_s^z \eta_{nn'}^z \right) \\ & \times \sqrt{\frac{\hbar (r_s + 1/2 \pm 1/2)}{4\pi \nu_s}} \exp \{ 2\pi i (\nu_n - \nu_{n'} \pm \nu_s) t \}, \end{aligned} \tag{2.289}$$

wherein η_x, η_y, η_z are the polarization matrices along the x, y, z axis of the unperturbed atom, respectively, and ν_n are the levels of the atom, that is, $\nu_n = E_n / h$. We have assumed $\boldsymbol{b}_s$ to be constant.

---

[33]@ Note that here the author considered a sort of generalization of what had been done in Sec. 2.19: The vector potential is written as $\mathbf{C} = \boldsymbol{b} q$, treating q as a scalar quantity (in a certain sense).

Let us suppose that the atom is initially in the state n, while the oscillators are in the ground state. It will then suffice to take all the quantities a to be zero, with the exception of

$$a_{n,0,0,0,\dots} = 1. \tag{2.290}$$

After a quite short time we shall have

$$\begin{aligned} \dot{a}_{n',0,\dots,0,1,0,\dots} &= \frac{i}{\hbar} A_{n',0,\dots,0,1,0,\dots,n,0,\dots,0,\dots} \\ &= -\frac{e}{c}\frac{2\pi}{\hbar}\nu_{nn'}\,\boldsymbol{b}_s\cdot\boldsymbol{\eta}_{nn'}\sqrt{\frac{\hbar}{4\pi\nu_s}}\exp\{2\pi i(\nu_{nn'}-\nu_s)t\}, \end{aligned} \tag{2.291}$$

setting $\nu_{nn'} = \nu_{n'} - \nu_n$. Thus

$$a_{n',0,\dots,0,1,0,\dots} = \frac{e}{c}\frac{i}{\hbar}\nu_{nn'}\,\boldsymbol{b}_s\cdot\boldsymbol{\eta}_{nn'}\sqrt{\frac{\hbar}{4\pi\nu_s}}\,\frac{e^{2\pi i(\nu_{nn'}-\nu_s)t}-1}{\nu_{nn'}-\nu_s}, \tag{2.292}$$

so that

$$\begin{aligned} |a_{n',0,\dots,0,1,0,\dots}|^2 &= \frac{e^2}{c^2}\frac{1}{\hbar^2}\nu_{nn'}^2\,|\boldsymbol{b}_s\cdot\boldsymbol{\eta}_{nn'}|^2\sqrt{\frac{\hbar}{4\pi\nu_s}}\,\frac{4\sin^2\pi(\nu_{nn'}-\nu_s)t}{(\nu_{nn'}-\nu_s)^2} \\ &= \frac{e^2}{c^2}\frac{\nu_{nn'}^2}{\pi\hbar\nu_s}\,|\boldsymbol{b}_s\cdot\boldsymbol{\eta}_{nn'}|^2\,\frac{\sin^2\pi(\nu_{nn'}-\nu_s)t}{(\nu_{nn'}-\nu_s)^2}. \end{aligned} \tag{2.293}$$

Since the average value of $|\boldsymbol{b}_s\cdot\boldsymbol{\eta}_{nn'}|^2$ is

$$\overline{|\boldsymbol{b}_s\cdot\boldsymbol{\eta}_{nn'}|^2} = \frac{4}{3}\frac{\pi c^2}{\Omega}|\eta|^2 \tag{2.294}$$

and the value of ν_s is close to $\nu_{nn'}$, the probability to find the atom in the state n' is

$$\begin{aligned} P &= \frac{4}{3}\frac{e^2}{c^2}\frac{\pi c^2}{\Omega}|\eta|^2\,\nu_{nn'}\frac{1}{\pi\hbar}\int\frac{8\pi\nu_{nn'}^2\Omega}{c^3}\,\frac{\sin^2\pi(\nu_{nn'}-\nu_s)t}{(\nu_{nn'}-\nu_s)^2}\,\mathrm{d}\nu_s \\ &= \frac{64}{3}\frac{2\pi^5}{\hbar}\frac{e^2|\eta|^2\nu_{nn'}^3}{c^3}\,t, \end{aligned} \tag{2.295}$$

while the depopulation rate due to the $n \to n'$ transition is[34]

$$\frac{\mathrm{d}P}{\mathrm{d}t} = \frac{64\pi^4 e^2\nu_{nn'}^3|\eta|^2}{3hc^3} = \frac{16\pi^4 e^2\nu_{nn'}^4|2\eta|^2}{3c^3}\,\frac{1}{h\nu_{nn'}}. \tag{2.296}$$

[34]@ In the last formula we have restored the Planck constant h, as in the original manuscript.

22. ABOUT MATRICES

A physical quantity A can be represented by a linear operator that transforms vectors into vectors, in a space with infinite dimensions. Let us fix an arbitrary reference frame and denote by $\psi_1, \psi_2, \ldots, \psi_n, \ldots$ the unit vectors along the various axes. They may also be complex. In this case the orthogonality relations will be

$$\psi_i \cdot \psi_k^* = \delta_{ik}. \tag{2.297}$$

A matrix A_{rs} can be associated with the operator A, but this matrix will depend on the choice of the reference frame. Its elements[35] are defined by the relation[36]

$$A \psi_s = A_{rs} \psi_r. \tag{2.300}$$

Let us introduce a new set of reference axes and let $\chi_1, \chi_2, \ldots, \chi_n, \ldots$ be the unit vectors along these new axes. For real axes, the operator S transforming the vectors ψ into the vectors χ reduces to a rotation. Its matrix is defined by the relation

$$\chi_k = S_{ik} \psi_i, \tag{2.301}$$

and, due to Eq. (2.297), which also holds for χ,

$$S_{ik} S_{il}^* = \delta_{kl}. \tag{2.302}$$

If S^{-1} is the inverse operator of S, we have

$$\psi_j = S_{kj}^{-1} \chi_k \tag{2.303}$$

and, substituting in Eq. (2.301),

$$\chi_k = S_{ik} S_{li}^{-1} \chi_l. \tag{2.304}$$

Thus

$$S_{ik} S_{li}^{-1} = \delta_{kl}, \tag{2.305}$$

[35]@ Notice that the author denoted by the same symbol the operator and its representative matrix. However, any confusion is avoided by noting that a matrix has always two subscripts labelling explicitly the row and the column.

[36]The multiplication rule can be deduced as follows

$$A B \psi_s = A B_{rs} \psi_r = A_{tr} B_{rs} \psi_r, \tag{2.298}$$

that is,

$$(A B)_{ts} = A_{tr} B_{rs}. \tag{2.299}$$

and this is satisfied if

$$S^{-1}_{rs} = S^*_{sr}. \tag{2.306}$$

In fact, in this case,

$$S_{ik}\,S^{-1}_{li} = S_{ik}\,S^*_{il}. \tag{2.307}$$

Equation (2.305) can be immediately derived from the relation

$$S^{-1}\,S = 1, \tag{2.308}$$

which can be written in the form

$$S^{-1}_{li}\,S_{ik} = S_{ik}\,S^{-1}_{li} = \delta_{kl}. \tag{2.309}$$

Then, from Eq. (2.308), it follows that

$$S_{ki}\,S^{-1}_{il} = S_{ki}\,S^*_{li} = \delta_{kl}. \tag{2.310}$$

This relation is similar to Eq. (2.302), but it refers to the rows rather than to the columns.

Let us go back to Eq. (2.300) and replace the vectors $\boldsymbol{\psi}$ with their expressions from Eq. (2.303):

$$A\,S^{-1}_{rs}\,\boldsymbol{\chi}_r = A_{rs}\,S^{-1}_{ir}\,\boldsymbol{\chi}_i; \tag{2.311}$$

on setting[37]

$$A\,\boldsymbol{\chi}_s = A'_{rs}\,\boldsymbol{\chi}_r, \tag{2.312}$$

$$A'_{ir}\,S^{-1}_{rs}\,\boldsymbol{\chi}_i = A_{rs}\,S^{-1}_{ir}\,\boldsymbol{\chi}_i, \tag{2.313}$$

they become

$$A'_{ir}\,S^{-1}_{rs} = S^{-1}_{ir}\,A_{rs}, \tag{2.314}$$

$$A'_{ir}\,S^{-1}_{rs}\,S_{sj} = S^{-1}_{ir}\,A_{rs}\,S_{sj}, \tag{2.315}$$

$$A'_{ij} = S^{-1}_{ir}\,A_{rs}\,S_{sj}. \tag{2.316}$$

In the same way, on substituting Eq. (2.301) into Eq. (2.312), we find

$$A\,S_{rs}\,\boldsymbol{\psi}_r = A'_{rs}\,S_{ir}\,\boldsymbol{\psi}_i, \tag{2.317}$$

$$A_{ir}\,S_{rs}\,\boldsymbol{\psi}_i = A'_{rs}\,S_{ir}\,\boldsymbol{\psi}_i, \tag{2.318}$$

$$A_{ir}\,S_{rs} = S_{ir}\,A'_{rs}, \tag{2.319}$$

$$A_{ij} = S_{ir}\,A'_{rs}\,S_{sj}. \tag{2.320}$$

---

[37]@ Note that Eq. (2.312) *defines* the matrix (A'_{rs}) representing the operator A in the basis $\{\boldsymbol{\chi}\}$.

These formulae are analogous to the ones written above and can be directly derived from them. For example, from Eq. (2.316) it follows that

$$S_{ai}\, A'_{ij}\, S^{-1}_{jb} \;=\; S_{ai}\, S^{-1}_{ir}\, A_{rs}\, S_{sj}\, S^{-1}_{jb}\,, \tag{2.321}$$

$$A_{rs} \;=\; S_{ai}\, A'_{ij}\, S^{-1}_{jb}\,, \tag{2.322}$$

which is identical to Eq. (2.320).

Denoting by $[A]$ and $[A']$ the matrices corresponding to the operator A in the two reference frames, and by $[S]$ and $[S^{-1}]$ the matrices having elements S_{rs} and S^{-1}_{rs}, we then have

$$[A]\,[S] \;=\; [S]\,[A'], \tag{2.323}$$

$$[A'] \;=\; [S^{-1}]\,[A]\,[S]. \tag{2.324}$$

In all the previous derivations we have assumed S to be the operator that transforms the vectors $\boldsymbol{\psi}$ into the vectors $\boldsymbol{\chi}$, which means that we can write

$$\boldsymbol{\chi}_i \;=\; S\,\boldsymbol{\psi}_i. \tag{2.325}$$

This requires the vectors $\boldsymbol{\chi}$ and $\boldsymbol{\psi}$ to be numbered by the same set of indices; on the other hand, such a condition is not actually needed. We can therefore associate the matrix $[S]$ not really with an operator, but simply with a function S_{rs} of two variables, namely, of the indices r and s of vectors $\boldsymbol{\psi}_r$ and $\boldsymbol{\chi}_s$: a function satisfying the relation

$$\boldsymbol{\chi}_s \;=\; S_{rs}\,\boldsymbol{\psi}_r. \tag{2.326}$$

23. RADIATION THEORY (PART 3)

Let us consider again that the atom is initially in the nth state and that the oscillators are at rest. If an inner state n' exists, to first order we can neglect any other stationary state of the atom. While the volume in which the atom is enclosed tends to infinity, the probability of exciting only one frequency ν_s tends to zero. This means that we can treat nearly all the oscillators as if they were at rest for the duration of the emission.[38] From Eq. (2.289), we have

$$\dot{a}_{n',0,\ldots 0,1,0\ldots} \;=\; -\,\frac{e}{c}\,\frac{2\pi}{\hbar}\,\nu_{nn'}\,\boldsymbol{b}_s\!\cdot\!\boldsymbol{\eta}_{n'n}$$

[38] More precisely, we exclude the quantum states corresponding to two or more excited oscillators.

$$\times \sqrt{\frac{\hbar}{4\pi\nu_s}}\, \mathrm{e}^{2\pi i(\nu_{nn'}-\nu_s)t}\, a_{n,0,\ldots 0,0,0\ldots}, \tag{2.327}$$

$$\dot{a}_{n,0,\ldots 0,0,0\ldots} = \sum_s \frac{e}{c}\frac{2\pi}{\hbar}\,\nu_{nn'}\,\boldsymbol{b}_s\cdot\boldsymbol{\eta}_{nn'} \times \sqrt{\frac{\hbar}{4\pi\nu_s}}\, \mathrm{e}^{-2\pi i(\nu_{nn'}-\nu_s)t}\, a_{n',0,\ldots 0,1,0\ldots}. \tag{2.328}$$

We can suppose that $\boldsymbol{\eta}_{nn'}$ is real and thus $\boldsymbol{\eta}_{nn'} = \boldsymbol{\eta}_{n'n}$. We shall now try to satisfy these equations by setting $a_{n,0,\ldots 0,0,0\ldots} = \exp\{-\gamma t/2\}$. We then get

$$\dot{a}_{n',0,\ldots 0,1,0\ldots} = -\frac{e}{c}\frac{2\pi}{\hbar}\,\nu_{nn'}\,\boldsymbol{b}_s\cdot\boldsymbol{\eta}_{nn'} \times \sqrt{\frac{\hbar}{4\pi\nu_s}}\, \mathrm{e}^{2\pi i(\nu_{nn'}-\nu_s)t}\, \mathrm{e}^{-\gamma t/2} \tag{2.329}$$

and thus

$$a_{n',0,\ldots 0,1,0\ldots} = -\frac{e}{c}\frac{2\pi}{\hbar}\,\nu_{nn'}\,\boldsymbol{b}_s\cdot\boldsymbol{\eta}_{nn'} \times \sqrt{\frac{\hbar}{4\pi\nu_s}}\, \frac{\mathrm{e}^{2\pi i(\nu_{nn'}-\nu_s-\gamma/2)t} - 1}{2\pi i(\nu_{nn'}-\nu_s)-\gamma/2}, \tag{2.330}$$

$$\dot{a}_{n,0,\ldots 0,0,0\ldots} = -\sum_s \frac{e^2}{c^2}\frac{4\pi^2}{\hbar^2}\frac{\hbar}{4\pi\nu_s}\,\nu_{nn'}^2\,|\boldsymbol{b}_s\cdot\boldsymbol{\eta}_{nn'}|^2 \times \frac{\mathrm{e}^{-\gamma t/2} - \mathrm{e}^{2\pi i(\nu_{nn'}-\nu_s)t}}{2\pi i(\nu_{nn'}-\nu_s)-\gamma/2}\,\mathrm{e}^{-\gamma t/2}. \tag{2.331}$$

Assuming, as usual, that ν_s is very close to $\nu_{nn'}$ and that Eq. (2.294) holds, and on transforming the sum into an integral, we find

$$\begin{aligned}\dot{a}_{n,0,\ldots 0,0,0\ldots} &= -\frac{e^2}{c^2}\frac{4\pi^2}{\hbar^2}\frac{\hbar}{4\pi\nu_s}\,\nu_{nn'}^2\,\frac{4}{3}\frac{\pi c^2}{\Omega}\,|\eta_{nn'}|^2 \\ &\times \frac{8\pi\nu_{nn'}^2}{c^3}\,\Omega\,\mathrm{e}^{-\gamma t/2}\int \frac{\mathrm{e}^{-\gamma t/2} - \mathrm{e}^{2\pi i(\nu_{nn'}-\nu_s)t}}{2\pi i(\nu_{nn'}-\nu_s)-\gamma/2}\,\mathrm{d}\nu_s \\ &= -\frac{32\pi^3 e^2\nu_{nn'}^3|\eta_{nn'}|^2}{3\hbar c^3}\,\mathrm{e}^{-\gamma t/2}\int \frac{\mathrm{e}^{-\gamma t/2} - \mathrm{e}^{2\pi i(\nu_{nn'}-\nu_s)t}}{2\pi i(\nu_{nn'}-\nu_s)-\gamma/2}\,\mathrm{d}\nu_s.\end{aligned} \tag{2.332}$$

Then, since

$$\dot{a}_{n,0,\ldots 0,0,0\ldots} = -\frac{\gamma}{2}\mathrm{e}^{-\gamma t/2},$$

we deduce that

$$\frac{\gamma}{2} = \frac{32\pi^3 e^2 \nu_{nn'}^3 |\eta_{nn'}|^2}{3\hbar c^3} \int \frac{\mathrm{e}^{-\gamma t/2} - \mathrm{e}^{2\pi i(\nu_{nn'} - \nu_s)t}}{2\pi i(\nu_{nn'} - \nu_s) - \gamma/2} \, \mathrm{d}\nu_s. \tag{2.333}$$

It can be proven that the integral on the r.h.s. equals 1/2, and thus[39]

$$\gamma = \frac{32\pi^3 e^2 \nu_{nn'}^3 |\eta_{nn'}|^2}{3\hbar c^3}. \tag{2.334}$$

24. PERTURBED KEPLERIAN MOTION IN A PLANE

Let us consider a point-particle of unitary mass attracted by a force M/r^2 acting towards a fixed center O. The equation of the trajectory is

$$r = \frac{k}{1 + e\cos(\theta - \alpha)}, \tag{2.335}$$

where k, e and α are constants. Indeed, putting

$$\begin{aligned} k &= \frac{r^2 V_t^2}{M} \qquad e = \sqrt{\frac{(k-r)^2 V_t^2 + k^2 V_r^2}{kM}}, \\ \alpha &= \theta - \arctan \frac{k V_r}{(k-r) V_t} \\ &= \theta - \arcsin \frac{k}{re} \frac{V_r}{V_t} = \theta - \arccos \frac{k-r}{re}, \end{aligned} \tag{2.336}$$

with V_r and V_t denoting the radial and the transverse velocities, Eq. (2.335) is identically satisfied when Eqs. (2.337) are substituted in it. Moreover, the quantities k, e, α given in Eq. (2.336) are constant. In-

[39]@ In the original manuscript, the beginning of an attempt to prove this result is reported (the complex exponential is expanded into trigonometric functions). Here we only quote the following words: "The imaginary part of the integral is undetermined, but we are only interested in the real part of γ, since only this quantity enters in the expression of $|a_n|^2$, which is what has physical meaning."

deed,

$$
\begin{aligned}
\dot{k} &= \frac{\mathrm{d}}{\mathrm{d}t}\frac{r^4\dot{\theta}^2}{M} = \frac{2r^3\dot{\theta}}{M}\left(2\dot{r}\dot{\theta} + r\ddot{\theta}\right) = \frac{2r^3\dot{\theta}}{M}\,a_t = 0, \\
\dot{e} &= \frac{1}{e}\left[-k\,\frac{\dot{r}}{r^2}\left(\frac{k}{r} - 1\right) + \frac{k}{M}\,\dot{r}\,\ddot{r}\right] \\
&= \frac{k\dot{r}}{eM}\left(\ddot{r} - r\,\dot{\theta}^2 + \frac{M}{r^2}\right) = \frac{k\dot{r}}{eM}\left(a_r + \frac{M}{r^2}\right) = 0, \\
\dot{\alpha} &= \dot{\theta} - \frac{r\,e\,V_t}{k\dot{r}}\,\frac{k\dot{r}}{er^2} = \dot{\theta} - \frac{V_t}{r} = 0.
\end{aligned}
\tag{2.337}
$$

An expression for the semi-major axis can be deduced:

$$
\begin{aligned}
a &= \frac{k}{1-e^2} = \frac{k^2M}{kM - (k-r)^2V_t^2 - k^2V_r^2} \\
&= \frac{k^2M}{kM - k^2V^2 + 2krV_t^2 - kM} = \frac{M}{2M/r - V^2} \\
&= \frac{r}{2 - rV^2/M} = \frac{r}{2 - V^2/V_0^2} = r\,\frac{V_0^2}{2V_0^2 - V^2},
\end{aligned}
\tag{2.338}
$$

in which $V = \sqrt{V_r^2 + V_t^2}$ is the total speed and $V_0 = \sqrt{M/r}$ the speed corresponding to circular motion. The semi-minor axis will be

$$
\begin{aligned}
b &= \frac{k}{\sqrt{1-e^2}} = \sqrt{k\,a} = \frac{r^2V_t}{\sqrt{M}\,\sqrt{2 - rV^2/M}} \\
&= \frac{rV_t}{\sqrt{2M/r - V^2}} = r\,\frac{V_t}{\sqrt{2V_0^2 - V^2}}.
\end{aligned}
\tag{2.339}
$$

The radius k normal to the major axis can also be written as

$$
k = \frac{r^2V_t^2}{M} = r\,\frac{V_t^2}{V_0^2}. \tag{2.340}
$$

The distance of the moving point-particle from the second focus is

$$
r' = 2a - r = \frac{r^2V^2/M}{2 - rV^2/M} = r\,\frac{V^2}{2V_0^2 - V^2}, \tag{2.341}
$$

and the period of revolution is

$$
T = \frac{2\pi ab}{rV_t} = \frac{2\pi M}{\left(2M/r - V^2\right)^{3/2}} = \frac{2\pi}{\sqrt{M}}\,a^{3/2}. \tag{2.342}
$$

Let us now assume that an arbitrary field is superimposed on the Newtonian field, and let us denote by χ_r and χ_t the radial and transverse components of this additional force. Equation (2.335) will still hold, but now k, e, α are no longer constant. They are varying functions that depend on r, θ, V_r and V_t, and are defined by Eqs. (2.336). We'll obviously have

$$\begin{aligned}
\dot{k} &= \frac{\partial k}{\partial V_r}\,\chi_\mathrm{r} + \frac{\partial k}{\partial V_t}\,\chi_\mathrm{t} = 2\,\frac{r^2 V_t}{M}\,\chi_\mathrm{t} \;=\; 2k\,\frac{\chi_\mathrm{t}}{V_t}, && (2.343)\\
\dot{e} &= \frac{\partial e}{\partial V_r}\,\chi_\mathrm{r} + \frac{\partial e}{\partial V_t}\,\chi_\mathrm{t} \\
&= \left(2\,\frac{k-r}{eM}\,V_t^2 + \frac{k}{eM}\,V_r^2\right)\frac{\chi_\mathrm{t}}{V_t} + \frac{k}{eM}\,V_r\chi_\mathrm{r}, && (2.344)\\
\dot{\alpha} &= \frac{k+r}{e^2 M}\,V_r\,\chi_\mathrm{t} - \frac{k-r}{e^2 M}\,V_t\,\chi_\mathrm{r}, && (2.345)\\
\dot{a} &= \frac{2}{M}\,a^2\,(V_r\,\chi_\mathrm{r} + V_t\,\chi_\mathrm{t})\,. && (2.346)
\end{aligned}$$

Then, let us assume we know χ_r and χ_t as functions of r, θ and t. From Eq. (2.335), they can be expressed as functions of k, e, α, θ, t, as well as V_r and V_t:

$$\begin{aligned}
V_t &= \frac{\sqrt{kM}}{r} = V_t(k, e, \alpha, \theta), && (2.347)\\
V_r &= V_t\,\frac{re}{M}\,\sin(\theta-\alpha) \;=\; V_r(k, e, \alpha, \theta). && (2.348)
\end{aligned}$$

On substituting in the three independent equations (2.343),(2.344),(2.345) [Eq. (2.346) actually can be derived from the previous ones], we find

$$\begin{aligned}
\dot{k} &= \dot{k}(k, e, \alpha, \theta, t), && (2.349)\\
\dot{e} &= \dot{e}(k, e, \alpha, \theta, t), && (2.350)\\
\dot{\alpha} &= \dot{\alpha}(k, e, \alpha, \theta, t). && (2.351)
\end{aligned}$$

Thus another equation is needed in order to determine the motion. It comes from the first of Eqs. (2.336):

$$\dot{\theta} = \frac{\sqrt{kM}}{r^2} = \dot{\theta}(k, e, \alpha, \theta). \qquad (2.352)$$

From the initial values $k_0, e_0, \alpha_0, \theta_0$ of the four variables at time t_0, the values of k, e, α and θ at any instant can be evaluated by using Eqs. (2.349)-(2.352). If the perturbation is small, the problem is solved by successive approximations. Denoting by θ' the value of θ at any instant of time in the case of no perturbation, as derived from Kepler's equation, in the zeroth-order approximation we'll have

$$\theta = \theta', \quad k = k_0, \quad e = e_0, \quad \alpha = \alpha_0. \tag{2.353}$$

In first approximation, we instead have $k = k_1$, $e = e_1$, $\alpha = \alpha_1$, $\theta = \theta_1$, with

$$\begin{aligned} k_1 &= k_0 + \int_{t_0}^{\infty} \dot{k}(k_0, e_0, \alpha_0, \theta', t)\, \mathrm{d}t, \\ e_1 &= e_0 + \int_{t_0}^{\infty} \dot{e}(k_0, e_0, \alpha_0, \theta', t)\, \mathrm{d}t, \\ \alpha_1 &= \alpha_0 + \int_{t_0}^{\infty} \dot{\alpha}(k_0, e_0, \alpha_0, \theta', t)\, \mathrm{d}t. \end{aligned} \tag{2.354}$$

It is not possible to write a similar expression for θ, since in the exact expression

$$\theta = \theta_0 + \int_{t_0}^{\infty} \dot{\theta}(k, e, \alpha, \theta, t)\, \mathrm{d}t \tag{2.355}$$

the two terms on the r.h.s. are of the same order of magnitude, so that, if we set an approximate value for θ on the r.h.s., we wouldn't get a more approximate value on the l.h.s. However, we can try to transform Eq. (2.352). To this end, let us note that the form of the function $\dot{\theta}(k, e, \alpha, \theta)$ does not depend on the perturbing forces, and it thus is the same if no perturbation is present. We shall then have

$$\dot{\theta}' = \dot{\theta}(k_0, e_0, \alpha_0, \theta'); \tag{2.356}$$

and, setting

$$\theta = \theta' + \gamma, \tag{2.357}$$

we shall get

$$\begin{aligned} \dot{\gamma} &= \dot{\theta}(k, e, \alpha, \theta) - \dot{\theta}(k_0, e_0, \alpha_0, \theta') \\ &= \dot{\theta}(k, e, \alpha, \theta) - \dot{\theta}' = \dot{\gamma}(k, e, \alpha, \theta, t). \end{aligned} \tag{2.358}$$

If instead of Eq. (2.352) we wanted to use the equation

$$\dot{\theta} = \dot{\theta}' + \dot{\gamma}(k, e, \alpha, \theta, t), \tag{2.359}$$

this would not have been adequate for our calculation by successive approximations, since, by setting an approximate value of θ in $\dot{\gamma}(k, e, \alpha, \theta, t)$, we do not get an approximate value for $\dot{\gamma}$ (because $\dot{\gamma}$ is not zero when perturbative forces are absent, unless we use for θ its exact value θ').

In order to evaluate θ_1 it is necessary to use the expression[40]

$$t = t_0 + \int_{\theta_0}^{2\pi} \frac{\mathrm{d}\theta_1}{\dot{\theta}(k_1', e_1', \alpha_1', \theta_1)}, \tag{2.360}$$

in which we will have set

$$k_1' = k_1[\overline{\theta}'(\theta_a)], \quad e_1' = e_1[\overline{\theta}'(\theta_a)], \quad \alpha_1' = \alpha_1[\overline{\theta}'(\theta_a)], \tag{2.361}$$

with $\theta' = \theta'(t)$, $t = \overline{\theta}'(\theta')$, $k_1 = k_1(r)$, and so on. In general, for the expressions approximated to order n ($n > 1$), the following formulae hold:

$$\begin{aligned} k_n &= k_0 + \int_{t_0}^{\infty} \dot{k}\,(k_{n-1}, e_{n-1}, \alpha_{n-1}, \theta_{n-1}, t)\; \mathrm{d}t, \\ e_n &= e_0 + \int_{t_0}^{\infty} \dot{e}\,(k_{n-1}, e_{n-1}, \alpha_{n-1}, \theta_{n-1}, t)\; \mathrm{d}t, \\ \alpha_n &= \alpha_0 + \int_{t_0}^{\infty} \dot{\alpha}\,(k_{n-1}, e_{n-1}, \alpha_{n-1}, \theta_{n-1}, t)\; \mathrm{d}t, \\ t &= t_0 + \int_{\theta_0}^{2\pi} \frac{\mathrm{d}\theta_n}{\dot{\theta}(k_n', e_n', \alpha_n', \theta_n)}, \end{aligned} \tag{2.362}$$

where

$$k_n' = k_n\left(\overline{\theta}_{n-1}(\theta_n)\right), \quad e_n' = e_n\left(\overline{\theta}_{n-1}(\theta_n)\right), \quad \alpha_n' = \alpha_n\left(\overline{\theta}_{n-1}(\theta_n)\right);$$

$$k_n = k_n(t), \quad e_n = e_n(t), \quad \alpha_n = \alpha_n(t).$$

The last of Eqs. (2.362) is justified by the fact that, knowing k, e, α as functions of t to order n (that is to say up to infinitesimals of order n, when the perturbing forces tend to zero) and knowing t as a function of θ to order $n-1$, it is possible to derive k, e and α as functions of θ to order n, since $dk/\mathrm{d}t, de/\mathrm{d}t, \mathrm{d}\alpha/\mathrm{d}t$ are themselves infinitesimals of the first order.

Let us now suppose that the perturbing forces are constant in time or, more precisely, that they can be considered as such on time scales longer than the revolution period. Let us also assume that they are small enough so that k, e and α change little during a period. We shall denote by $\underline{\dot{k}}, \underline{\dot{e}}$ and $\underline{\dot{\alpha}}$ the secular variations of these quantities, i.e., the average

[40]@ In the original manuscript, the upper limit of the integral is not explicitly given.

values of the quantities $\dot{k}, \dot{e}, \dot{\alpha}$ over the entire period. Clearly, we shall have

$$\underline{\dot{k}} = \underline{\dot{k}}(k, e, \alpha, t), \quad \underline{\dot{e}} = \underline{\dot{e}}(k, e, \alpha, t), \quad \underline{\dot{\alpha}} = \underline{\dot{\alpha}}(k, e, \alpha, t). \tag{2.363}$$

The explicit form of Eqs. (2.363) depends on the form of the functions

$$\chi_{\rm r} = \chi_{\rm r}(r, \theta, t), \quad \chi_{\rm t} = \chi_{\rm t}(r, \theta, t), \tag{2.364}$$

and the time dependence obtains only if $\chi_{\rm r}$ and $\chi_{\rm t}$ depend on time, with the constraint (as assumed) that the time variation is small.

Let us now examine the specific case in which $\chi_{\rm t} = 0$; $\chi_{\rm r} = \epsilon\, r^n$. From Eqs. (2.343), (2.344), (2.345), we deduce

$$\begin{aligned} \dot{k} &= 0, \quad \dot{e} = \frac{k}{eM} V_r\, \epsilon\, r^n, \\ \dot{\alpha} &= \frac{r-k}{e^2 M} V_t\, \epsilon\, r^n = \frac{1}{e^2}\sqrt{\frac{k}{M}}\left(r^n - k\, r^{n-1}\right)\epsilon, \end{aligned} \tag{2.365}$$

and thus

$$\underline{\dot{k}} = 0, \quad \underline{\dot{e}} = 0, \quad \underline{\dot{\alpha}} = \frac{1}{e^2}\sqrt{\frac{k}{M}}\left(\overline{r^n} - k\,\overline{r^{n-1}}\right)\epsilon, \tag{2.366}$$

with

$$\overline{r^n} = \frac{(1-e^2)^{3/2}}{2\pi} k^n \int_0^{2\pi} \frac{\mathrm{d}\theta}{(1+e\cos\theta)^{n+r}}. \tag{2.367}$$

It follows that

$$\begin{aligned} \overline{r^{-1}} &= \left(1 - e^2\right) k^{-1}, \\ \overline{r^{-2}} &= \left(1 - e^2\right)^{3/2} k^{-2}, \\ \overline{r^{-3}} &= \left(1 - e^2\right)^{3/2} k^{-3}, \\ \overline{r^{-4}} &= \left(1 - e^2\right)^{3/2} \left(1 + \frac{1}{2} e^2\right) k^{-4}, \\ \overline{r^{-5}} &= \left(1 - e^2\right)^{3/2} \left(1 + \frac{3}{2} e^2\right) k^{-5}, \\ \overline{r^{-6}} &= \left(1 - e^2\right)^{3/2} \left(1 + 3\, e^2 + \frac{3}{8} e^4\right) k^{-6}, \end{aligned} \tag{2.368}$$

and[41]

$$
\begin{aligned}
\overline{r} &= \left(1 - e^2\right)^{-1} \left(1 + \frac{1}{2}\, e^2\right) k, \\
\overline{r^2} &= \left(1 - e^2\right)^{-2} \left(1 + \frac{3}{2}\, e^2\right) k^2, \\
\overline{r^3} &= \left(1 - e^2\right)^{-3} \left(1 + 3\, e^2 + \frac{3}{8}\, e^4\right) k^3, \\
\overline{r^4} &= \left(1 - e^2\right)^{-4} \left(1 + 5\, e^2 + \frac{15}{8}\, e^4\right) k^4, \\
\overline{r^5} &= \left(1 - e^2\right)^{-5} \left(1 + \frac{15}{2}\, e^2 + \frac{45}{8}\, e^4 + \frac{5}{16}\, e^6\right) k^5, \\
\overline{r^6} &= \left(1 - e^2\right)^{-6} \left(1 + \frac{21}{2}\, e^2 + \frac{105}{8}\, e^4 + \frac{35}{16}\, e^6\right) k^6, \\
\overline{r^7} &= \left(1 - e^2\right)^{-7} \left(1 + 14\, e^2 + \frac{105}{4} e^4 \right. \\
&\qquad \left. + \frac{35}{4}\, e^6 + \frac{35}{128}\, e^8\right) k^7, \\
\overline{r^8} &= \left(1 - e^2\right)^{-8} \left(1 + 18\, e^2 + \frac{189}{4}\, e^4 \right. \\
&\qquad \left. + \frac{105}{4}\, e^6 + \frac{315}{128}\, e^8\right) k^8.
\end{aligned}
\tag{2.369}
$$

[41]@ In the original manuscript, the explicit expressions for $\overline{r}, \overline{r^2}, \ldots, \overline{r^8}$ are missing.

We then infer that

$$n = 0, \qquad \underline{\dot{\alpha}} = \sqrt{\frac{k}{M}}\,\epsilon,$$

$$n = -1, \qquad \underline{\dot{\alpha}} = \sqrt{\frac{k}{M}}\left(1 - e^2\right)\frac{1-\sqrt{1-e^2}}{e^2}\,\frac{\epsilon}{k},$$

$$n = -2, \qquad \underline{\dot{\alpha}} = 0,$$

$$n = -3, \qquad \underline{\dot{\alpha}} = -\sqrt{\frac{k}{M}}\left(1 - e^2\right)^{3/2}\frac{1}{2}\,\frac{\epsilon}{k^3},$$

$$n = -4, \qquad \underline{\dot{\alpha}} = -\sqrt{\frac{k}{M}}\left(1 - e^2\right)^{3/2}\frac{\epsilon}{k^4},$$

$$n = -5, \qquad \underline{\dot{\alpha}} = -\sqrt{\frac{k}{M}}\left(1 - e^2\right)^{3/2}\left(\frac{3}{2} + \frac{3}{8}e^2\right)\frac{\epsilon}{k^5}.$$

25. RADIATION THEORY (PART 4)

Let us consider two atomic quantum states labeled by indices 1 and 2 and let ν be the transition frequency. Let A_{21} be the probability that an atom in the state 2 will spontaneously make a transition to state 1 in unit time, $B_{21}U$ the probability for the same transition due to radiation of frequency ν, quantity U being the radiation energy per unit frequency and unit volume. Let also $B_{12}U$ denote the probability of the inverse transition, and N_1, N_2 the number of atoms in states 1 and 2, respectively. At the equilibrium we shall have

$$\frac{N_1}{N_2} = \frac{A_{21} + B_{21}U}{B_{12}U}. \qquad (2.370)$$

If the background temperature is T and Boltzmann's law is assumed, we find[42]

$$\frac{N_2}{N_1} = \mathrm{e}^{-h\nu/kT}, \qquad (2.371)$$

[42]@ In this section we use Planck's constant h as in the original manuscript.

$$U = \frac{8\pi}{c^3}\,\frac{\nu^3 h}{\mathrm{e}^{h\nu/kT}-1}, \tag{2.372}$$

from which

$$\begin{aligned} & B_{12}\,\frac{8\pi}{c^3}\,\frac{\nu^3 h}{\mathrm{e}^{h\nu/kT}-1} \\ & = A_{21}\,\mathrm{e}^{-h\nu/kT} + B_{21}\,\frac{8\pi}{c^3}\,\frac{\nu^3 h}{\mathrm{e}^{h\nu/kT}-1}\,\mathrm{e}^{-h\nu/kT}, \end{aligned} \tag{2.373}$$

which is always satisfied only if

$$B_{12} = B_{21}, \tag{2.374}$$

$$A_{21} = \frac{8\pi}{c^3}\,\nu^3\,h\,B_{12}. \tag{2.375}$$

Let us now try to obtain these results using the radiation theory developed above. Let

$$\psi_0 = \sum_{n,r_1,\dots,r_s,\dots} \psi_{n,r_1,\dots,r_s,\dots}\, a_{n,r_1,\dots,r_s,\dots} \tag{2.376}$$

be the eigenfunction at an arbitrary time. If we neglect all the quantum states other than 1 and 2, we shall get

$$\begin{aligned} \dot{a}_{1,\dots,n_s+1,\dots} = & -\sum \frac{e}{c}\,\frac{4\pi^2}{h}\,\nu\,\boldsymbol{b}_s\cdot\boldsymbol{\eta}_{12}\,\sqrt{\frac{h(n_s+1)}{8\pi^2\nu_s}} \\ & \times \exp\left\{2\pi i(\nu-\nu_s)t\right\}\, a_{2,\dots,n_s-1,\dots}, \end{aligned} \tag{2.377}$$

because of the fact that we can assume that in the transition $2 \to 1$ the emitted energy has a characteristic frequency close to ν. In the same way we can write

$$\begin{aligned} \dot{a}_{2,\dots,n_s-1,\dots} = & \frac{e}{c}\,\frac{4\pi^2}{h}\,\nu\,\boldsymbol{b}_s\cdot\boldsymbol{\eta}_{12}\,\sqrt{\frac{hn_s}{8\pi^2\nu}} \\ & \times \exp\left\{2\pi i(\nu_s-\nu)t\right\}\, a_{1,\dots,n_s,\dots}. \end{aligned} \tag{2.378}$$

Since

$$N_1 = \sum |a_{1,\dots}|^2, \tag{2.379}$$

$$N_2 = \sum |a_{2,\dots}|^2, \tag{2.380}$$

it follows that

$$\dot{N}_1 = \sum \left(a_{1,\dots}\,\dot{a}^*_{1,\dots} + \dot{a}_{1,\dots}\,a^*_{1,\dots}\right). \tag{2.381}$$

In order to simplify our calculations, let us suppose that all the a_1s are initially zero. The previous equation would seem to yield $\dot{N}_i = 0$, but this is wrong, since it results from an incorrect use of a limit procedure with an infinite number of frequencies. The calculation must be performed in the same manner as in Sec. 2.21. The only difference with respect to what was done there is that now the argument of the square root is $n_s + 1$ instead of 1. Since in the final result there appears the square of such square root, all we have to do is multiplying the result by the average value of $n_s + 1$. Denoting by n the average value of n_s, we find

$$\dot{N}_1 = N_2 \frac{64\pi^4\nu^3e^2|\eta_{12}|^2}{3hc^3} (n + 1) . \tag{2.382}$$

In the same way, on assuming all the atoms to be initially in state 1, we find the same formula, apart from changing N_1 into N_2, and vice versa, and having n instead of $n + 1$, due to the fact that in Eq. (2.378) we have n_s and not $n_s + 1$:

$$\dot{N}_2 = N_1 \frac{64\pi^4\nu^3e^2|\eta_{12}|^2}{3hc^3} n. \tag{2.383}$$

From the foregoing, we can now derive Einstein's A and B coefficients:

$$A_{21} = \frac{64\pi^4\nu^3e^2|\eta_{12}|^2}{3hc^3}, \tag{2.384}$$

$$B_{21} = B_{12} = \frac{nA_{21}}{U} = \frac{A_{21}}{(8\pi/c^3)\nu^3h}, \tag{2.385}$$

which agree with Eqs. (2.374) and (2.375).

26. DEFINITE INTEGRALS[43]

(13) We have

$$\int_0^1 \left(1 - x^2\right)^n = \frac{1}{2n+1} \frac{2^{2n}n!^2}{(2n)!}. \tag{2.386}$$

For large n, the l.h.s. is approximated by

$$\int_0^\infty \mathrm{e}^{-nx^2}\,\mathrm{d}x = \frac{1}{2}\sqrt{\frac{\pi}{n}},$$

[43]See Sec. 1.37.

from which, for large n, we find

$$\frac{n!^2 2^{2n}}{(2n)!} = \sqrt{\pi n} + \ldots, \tag{2.387}$$

as it can immediately be seen from Stirling's formula (see Sec. 1.27).

(14) We have

$$\int_{-\infty}^{\infty} \frac{\mathrm{e}^{ix}}{a+ix}\,\mathrm{d}x = \begin{cases} 2\pi\,\mathrm{e}^{-a}, & a > 0, \\ 0, & a < 0. \end{cases} \tag{2.388}$$

(15) We have

$$\int_{-\infty}^{\infty} \frac{\cos x}{a^2+x^2}\,\mathrm{d}x = \frac{\pi}{a}\,\mathrm{e}^{-a}. \tag{2.389}$$

(16) We have

$$\int_{-\infty}^{\infty} \frac{x\sin x}{a^2+x^2}\,\mathrm{d}x = \pi\,\mathrm{e}^{-a}. \tag{2.390}$$

(17) We have

$$\int_{-\infty}^{\infty} \frac{\mathrm{e}^{iax}}{1+k^2x^2}\,\mathrm{d}x = \frac{\pi}{k}\,\mathrm{e}^{-a/k}, \quad k > 0, \quad a > 0. \tag{2.391}$$

(18) We have [44]

$$\int_{-\infty}^{\infty} \frac{x\mathrm{e}^{iax}}{1+k^2x^2}\,\mathrm{d}x = \frac{i\pi}{k^2}\,\mathrm{e}^{-a/k}, \quad \frac{a}{k} > 0. \tag{2.392}$$

(19) We have

$$\int_{-\infty}^{\infty} \mathrm{e}^{ix^2}\,\mathrm{d}x = \frac{1+i}{\sqrt{2}}\sqrt{\pi}. \tag{2.393}$$

(14bis) We have

$$\int_{-\infty}^{\infty} \frac{\mathrm{e}^{ikx}}{1+ix}\,\mathrm{d}x = \begin{cases} 2\pi\,\mathrm{e}^{-k}, & k > 0, \\ 0, & k < 0. \end{cases} \tag{2.394}$$

[44]@ More precisely, this result holds for $a > 0$ (keeping $a/k > 0$), while for $a < 0$ we simply get the opposite.

(20) On setting $\mathrm{d}q_1 = \mathrm{d}x_1\mathrm{d}y_1\mathrm{d}z_1$, and $r_1 = \sqrt{x_1^2 + y_1^2 + z_1^2}$, one gets:[45]

$$\int \mathrm{e}^{-ar_1}\,\mathrm{d}q_1 = \frac{8\pi}{a^3}, \qquad a > 0, \tag{2.395}$$

$$\int \frac{1}{r_1}\,\mathrm{e}^{-ar_1}\,\mathrm{d}q_1 = \frac{8\pi}{a^3}\,\frac{a}{2}, \qquad a > 0. \tag{2.396}$$

(21) On setting

$$\begin{aligned}
\mathrm{d}\tau &= \mathrm{d}x_1\mathrm{d}y_1\mathrm{d}z_1\mathrm{d}x_2\mathrm{d}y_2\mathrm{d}z_2, \\
r_{12} &= \sqrt{(x_1 - x_2)^2 + (y_1 - y_2)^2 + (z_1 - z_2)^2}, \\
r_1 &= \sqrt{x_1^2 + y_1^2 + z_1^2}, \quad r_2 = \sqrt{x_2^2 + y_2^2 + z_2^2}, \quad a, b > 0,
\end{aligned}$$

one gets:

$$\int \mathrm{e}^{-ar_1}\,\mathrm{e}^{-br_2}\,\mathrm{d}\tau = \frac{64\pi^2}{a^3b^3}, \tag{2.397}$$

$$\int \frac{1}{r_1}\,\mathrm{e}^{-ar_1}\,\mathrm{e}^{-br_2}\,\mathrm{d}\tau = \frac{64\pi^2}{a^3b^3}\,\frac{a}{2}, \tag{2.398}$$

$$\int \frac{1}{r_{12}}\,\mathrm{e}^{-ar_1}\,\mathrm{e}^{-br_2}\,\mathrm{d}\tau = \frac{64\pi^2}{a^3b^3}\,\frac{a^2 + 3ab + b^2}{2(a+b)^3}\,ab. \tag{2.399}$$

27. SERIES EXPANSIONS[46]

(1) Let us consider the following function of x:

$$y = \sum_{n=0}^{\infty} \frac{f(n)x^n}{n!}\,(-1)^n. \tag{2.400}$$

Under certain constraints, we have

$$\lim_{x\to\infty} \frac{x^r y}{\mathrm{e}^x} = 0 \tag{2.401}$$

for any value of r. If $f(n)$ =constant, then Eq. (2.401) is surely satisfied. If $f(n) = n$, then $y = -x\mathrm{e}^{-x}$ and Eq. (2.401) is again satisfied. In the same way we can prove that it is satisfied when

[45]@ The following integrals are evaluated over the entire real axis for each variable.

[46]See Secs. 1.22 and 3.1.

$f(n) = n(n-1)$ or $f(n) = n(n-1)(n-2)$, and so on. Moreover, the same holds for $f(n) = 1/(n+1)$ or $1/(n+1)(n+2)$ or $1/(n+1)(n+2)(n+3)$, and so on. It follows that Eq. (2.401) is satisfied when $f(n)$ is any rational function of n or, more generally, when $f(n)$ *can be expanded in decreasing powers of* n*, the first one being an arbitrary power* n^k *(with integer* k*).*
Equation (2.401) is also satisfied if $f(n)$ can be expanded in decreasing powers of n, each power step being unity, for example, and the first power n^c being rational or irrational. In this case, in fact, the function $f_1(n)$ in the series

$$y + y' = \sum \frac{f_1(n)x^n}{n!}$$

can be expanded starting from n^{c-1}. Thus, for a given arbitrary value of r, Eq. (2.401) will be satisfied when y is replaced by

$$y + k\,y' + \frac{k(k-1)}{2}\,y'' + \ldots + y^{(k)}, \tag{2.402}$$

quantity k depending on r. Now, if we set

$$\lim_{x\to\infty} \frac{x^r(z+z')}{\mathrm{e}^x} = 0, \tag{2.403}$$

which is equivalent to

$$z + z' = \alpha\,x^{-r}\,\mathrm{e}^x, \tag{2.404}$$

with an infinitesimal α, then it follows that

$$z = \mathrm{e}^{-x}\int \alpha\,x^{-r}\,\mathrm{e}^{2x}\,\mathrm{d}x = \mathrm{e}^{-x}\,\beta\,x^{-r}\,\mathrm{e}^{2x} = \beta\,x^{-r}\,\mathrm{e}^x, \tag{2.405}$$

where β is another infinitesimal. By repeating this procedure k times, we find that Eq. (2.401) is also satisfied by y.
Equation (2.401) also holds if $f(n)$ is the product of $\log n$ and an algebraic function of n, and this can be proven as we did above. More in general, replacing $f(n)$ by

$$\begin{aligned} &f(n) - k\,f(n-1) + \frac{k(k-1)}{2}\,f(n-2) \\ &- \frac{k(k-1)(k-2)}{6}\,f(n-3) + \ldots \pm f(n-k), \end{aligned} \tag{2.406}$$

it is possible, by appropriately choosing k, to make y infinitesimal of an arbitrarily large order (for large n), and Eq. (2.401) gets satisfied.

(2) In first approximation, for large n and small ϵ/n, we have

$$\binom{n}{n/2+\epsilon} = \frac{n!}{(n/2+\epsilon)!\,(n/2-\epsilon)!} = 2^n\sqrt{\frac{2}{\pi n}}\,\mathrm{e}^{-2\epsilon^2/n}. \tag{2.407}$$

(3) For large x, the series expansion of θ (which is, however, always divergent) reads

$$\begin{aligned}\theta(x) &= \frac{2}{\sqrt{\pi}}\int_0^x \mathrm{e}^{-x^2}\,\mathrm{d}x \\ &= 1-\frac{1}{\sqrt{\pi}}\mathrm{e}^{-x^2}\left(\frac{1}{x}-\frac{1}{2x^3}+\frac{3}{4x^5}-\frac{15}{8x^7}+\frac{105}{16x^9}-\dots\right).\end{aligned}$$

Although this is a divergent expansion, it nevertheless can be used, since it supplies values that approximate the true ones by excess or by defect, alternately.

28. RADIATION THEORY (PART 5): FREE ELECTRON SCATTERING

We have considered the stationary waves that may be present in a volume Ω without any assumption about their form or about the shape of the volume. For simplicity, here we assume that the potential $\mathbf{C}_s$ relative to the radiation of frequency ν_s is of the following form, which is compatible with Eq. (2.215):

$$\mathbf{C}_s = \sqrt{4\pi c^2/\Omega}\,q_s\,\mathrm{e}^{2\pi i(\gamma_s' x+\gamma_s'' y+\gamma_s''' z)}\,\mathbf{A}_s, \tag{2.408}$$

where $\mathbf{A}_s$ is a unit vector normal to the propagation direction. The frequency is given by

$$\nu_s = c\sqrt{\gamma_s'^2+\gamma_s''^2+\gamma_s'''^2}, \tag{2.409}$$

and the number of oscillators, relative to the wave numbers, in the intervals $\gamma_s' - \gamma_s' + \mathrm{d}\gamma_s'$, $\gamma_s'' - \gamma_s'' + \mathrm{d}\gamma_s''$, and $\gamma_s''' - \gamma_s''' + \mathrm{d}\gamma_s'''$, will be

$$\mathrm{d}N = 2\Omega\,\mathrm{d}\gamma_s'\,\mathrm{d}\gamma_s''\,\mathrm{d}\gamma_s'''. \tag{2.410}$$

We shall assume the following eigenfunctions for the free electron:

$$U_n = \frac{1}{\sqrt{\Omega}}\exp\left\{2\pi i(\delta_n' x+\delta_n'' y+\delta_n''' z)\right\}, \tag{2.411}$$

and there will be

$$dn = \Omega \, d\delta'_n \, d\delta''_n \, d\delta'''_n \tag{2.412}$$

of them. The electron momenta corresponding to Eq. (2.411) will be

$$p^n_x = -h\,\delta'_n, \quad p^n_y = -h\delta''_n, \quad p^n_z = -h\delta'''_n. \tag{2.413}$$

In the same way, the momenta of the light quanta are, from Eq. (2.408),

$$p^s_x = -h\,\gamma'_s, \quad p^s_y = -h\gamma''_s, \quad p^s_z = -h\gamma'''_s. \tag{2.414}$$

The first- and second-order interaction terms in the total Hamiltonian (see Sec. 2.6) are

$$\begin{aligned} &\sum_s \frac{e}{mc} \boldsymbol{p} \cdot \mathbf{A}_s \sqrt{\frac{4\pi c^2}{\Omega}}\, q_s \, e^{2\pi i(\gamma'_s x + \gamma''_s y + \gamma'''_s z)} \\ &+ \frac{e^2}{2mc^2} \frac{4\pi c^2}{\Omega} \sum_{r,s} q_r\, q_s\, \mathbf{A}_r \cdot \mathbf{A}_s \, e^{2\pi i[(\gamma'_r + \gamma'_s)x + (\gamma''_r + \gamma''_s)y + (\gamma'''_r + \gamma'''_s)z]}. \end{aligned} \tag{2.415}$$

Since only the second-order perturbation terms will enter in the perturbation matrix, the first-order terms are either small or changing too fast. By neglecting the first-order terms, the only non-zero elements in the matrix will be the ones corresponding to an arbitrary exchange of electrons and to a one-level transition (in either directions) of two and only two oscillators. Since we are only interested in large and slowly varying terms, let us suppose that one of these oscillators, say the rth, undergoes the transition from the quantum number k_r to $k_r + 1$, while the other, say the sth, undergoes the transition k_s to $k_s - 1$. The matrix element corresponding to such a transition is

$$\begin{aligned} B_{n,k_r,k_s;n',k_r+1,k_s-1} &= \frac{4\pi e^2}{2m\Omega^2} \mathbf{A}_r \cdot \mathbf{A}_s \sqrt{\frac{\hbar(r+1)}{4\pi\nu_r}} \sqrt{\frac{\hbar s}{4\pi\nu_s}} \\ &\times \int e^{2\pi i[(\gamma'_r - \gamma'_s - \delta'_n + \delta'_{n'})x + \ldots]} \, d\tau \, e^{2\pi i(\nu_{n'} - \nu_n + \nu_r - \nu_s)t}. \end{aligned} \tag{2.416}$$

Let us suppose the volume Ω is a cube of side length a. Then the absolute value of the integral becomes

$$\begin{aligned} &\frac{\sin\pi(\gamma'_r - \gamma'_s - \delta'_n + \delta'_{n'})a}{\pi(\gamma'_r - \gamma'_s - \delta'_n + \delta'_{n'})} \frac{\sin\pi(\gamma''_r - \gamma''_s - \delta''_n + \delta''_{n'})a}{\pi(\gamma''_r - \gamma''_s - \delta''_n + \delta''_{n'})} \\ &\times \frac{\sin\pi(\gamma'''_r - \gamma'''_s - \delta'''_n + \delta'''_{n'})a}{\pi(\gamma'''_r - \gamma'''_s - \delta'''_n + \delta'''_{n'})}. \end{aligned} \tag{2.417}$$

Furthermore let us assume that at time $t = 0$ all the atoms are in the state n and the oscillators in the state 0, apart from the oscillator s

which is in the state k_s. We shall associate the coefficient 1 with the eigenfunction corresponding to this state and the coefficient 0 to all the others. For a small period of time we have the following: If $a_{n',1,k_s-1}$ is the coefficient of the eigenfunction corresponding to the atom in the state n', the rth oscillator in state 1, and the oscillator s in the state $k_s - 1$, then

$$\dot{a}_{n',1,k_s-1} \;=\; \frac{i}{\hbar}\, B_{n',1,k_s-1;n,0,k_s}\,, \tag{2.418}$$

that is, neglecting a phase factor,

$$\begin{aligned}\dot{a}_{n',1,k_s-1} \;&=\; \mathbf{A}_r\cdot\mathbf{A}_s\,\frac{e^2}{2m\Omega^2}\sqrt{\frac{k_s}{\nu_r\nu_s}}\;\mathrm{e}^{2\pi i(\nu_n-\nu_{n'}+\nu_s-\nu_r)t}\\ &\times\frac{\sin\pi(\gamma_r'-\gamma_s'-\delta_n'+\delta_{n'}')a}{\pi(\gamma_r'-\gamma_s'-\delta_n'+\delta_{n'}')}\,\frac{\sin\pi(\gamma_r''-\gamma_s''-\delta_n''+\delta_{n'}'')a}{\pi(\gamma_r''-\gamma_s''-\delta_n''+\delta_{n'}'')}\\ &\times\frac{\sin\pi(\gamma_r'''-\gamma_s'''-\delta_n'''+\delta_{n'}''')a}{\pi(\gamma_r'''-\gamma_s'''-\delta_n'''+\delta_{n'}''')}\,,\end{aligned} \tag{2.419}$$

so that

$$\begin{aligned}\left|a_{n',1,k_s-1}\right|^2 \;&=\; \left|\mathbf{A}_r\cdot\mathbf{A}_s\right|^2\,\frac{e^4}{4m^2\Omega^4}\,\frac{k_s}{\nu_r\nu_s}\\ &\times\frac{\sin\pi(\gamma_r'-\gamma_s'-\delta_n'+\delta_{n'}')a}{\pi(\gamma_r'-\gamma_s'-\delta_n'+\delta_{n'}')}\,\frac{\sin\pi(\gamma_r''-\gamma_s''-\delta_n''+\delta_{n'}'')a}{\pi(\gamma_r''-\gamma_s''-\delta_n''+\delta_{n'}'')}\\ &\times\frac{\sin\pi(\gamma_r'''-\gamma_s'''-\delta_n'''+\delta_{n'}''')a}{\pi(\gamma_r'''-\gamma_s'''-\delta_n'''+\delta_{n'}''')}\,\frac{\sin\pi(\nu_n-\nu_{n'}+\nu_s-\nu_r)t}{\pi^2(\nu_n-\nu_{n'}+\nu_s-\nu_r)^2}\,.\end{aligned} \tag{2.420}$$

Summing over all the values of r and n' and transforming the sum into an integral, we get

$$\begin{aligned}\sum_{n',r}\left|a_{n',1,k_s-1}\right|^2 \;&=\; \frac{e^4}{\Omega^2 m^2}\,\frac{k_s}{\nu_s}\int \mathrm{d}\gamma_s'\mathrm{d}\gamma_s''\mathrm{d}\gamma_s'''\,\mathrm{d}\delta_n'\mathrm{d}\delta_n''\,\mathrm{d}\delta_n'''\,\frac{\left|\mathbf{A}_r\cdot\mathbf{A}_s\right|^2}{\nu_s}\\ &\times\frac{\sin\pi(\gamma_r'-\gamma_s'-\delta_n'+\delta_{n'}')a}{\pi(\gamma_r'-\gamma_s'-\delta_n'+\delta_{n'}')}\,\frac{\sin\pi(\gamma_r''-\gamma_s''-\delta_n''+\delta_{n'}'')a}{\pi(\gamma_r''-\gamma_s''-\delta_n''+\delta_{n'}'')}\\ &\times\frac{\sin\pi(\gamma_r'''-\gamma_s'''-\delta_n'''+\delta_{n'}''')a}{\pi(\gamma_r'''-\gamma_s'''-\delta_n'''+\delta_{n'}''')}\,\frac{\sin\pi(\nu_n-\nu_{n'}+\nu_s-\nu_r)t}{\pi^2(\nu_n-\nu_{n'}+\nu_s-\nu_r)^2}\,.\end{aligned} \tag{2.421}$$

Let us suppose that a is very large. The integrand will be significantly different from zero only for those transitions that satisfy the momentum conservation, that is,

$$\gamma_r' - \gamma_s' - \delta_n' + \delta_{n'}' \;=\; 0 \tag{2.422}$$

(and similarly for the other components). On integrating over $\mathrm{d}\gamma_r'$, $\mathrm{d}\gamma_r''$ and $\mathrm{d}\gamma_r'''$, we find

$$\begin{aligned}\sum_{n',r} \left|a_{n',1,k_s-1}\right|^2 &= \frac{e^4}{\Omega^2 m^2} \frac{k_s}{\nu_s} \int \mathrm{d}\delta_n' \mathrm{d}\delta_n'' \, \mathrm{d}\delta_n''' \\ &\times \frac{\overline{|\mathbf{A}_r \cdot \mathbf{A}_s|^2}}{\nu_s} \, \Omega \, \frac{\sin \pi(\nu_n - \nu_{n'} + \nu_s - \nu_r)t}{\pi^2(\nu_n - \nu_{n'} + \nu_s - \nu_r)^2}, \qquad (2.423)\end{aligned}$$

where

$$\nu_r = c\sqrt{\gamma_r'^2 + \gamma_r''^2 + \gamma_r'''^2}, \qquad (2.424)$$

the components γ_r being given by Eq. (2.422). If t is large enough, the following must hold:

$$\nu_n - \nu_{n'} + \nu_s - \nu_r \simeq 0. \qquad (2.425)$$

We can then restrict the integration to those values of $\delta_{n'}$ that, through Eqs. (2.422) and (2.424), satisfy Eq. (2.425). To calculate the intensity, let us consider low-energy light quanta and slow electrons. Then we have

$$\nu_r \simeq \nu_s; \qquad (2.426)$$

and, denoting by θ the angle between the incident and the scattered quantum, we moreover have

$$\begin{aligned}\overline{|\mathbf{A}_r \cdot \mathbf{A}_s|^2} &= \frac{1}{2} - \frac{1}{4}\sin^2\frac{\theta}{2}, \qquad (2.427)\\ \frac{1}{\hbar}\left|p_{n'} - p_n\right| &= -4\pi\frac{\nu_s}{c}\sin\frac{\theta}{2}, \qquad (2.428)\end{aligned}$$

so that the integral in Eq. (2.423) becomes

$$\begin{aligned}\sum_{n',r} \left|a_{n',1,k_s-1}\right|^2 &= \frac{4e^4 k_s}{m^2 \Omega c^2} \int \pi \cos\frac{\theta}{2} \mathrm{d}\theta \sin^2\frac{\theta}{2} \\ &\times \left(\frac{1}{2} - \frac{1}{4}\sin^2\frac{\theta}{2}\right) \frac{\sin^2 \pi c r t \sin\theta/2}{\pi^2 c^2 r^2 \sin^2\theta/2} \, \mathrm{d}r \\ &= \pi t \frac{4e^4 k_s}{m^2 \Omega c^3} \int_0^\pi \cos\frac{\theta}{2} \sin\frac{\theta}{2} \left(\frac{1}{2} - \frac{1}{4}\sin^2\frac{\theta}{2}\right) \mathrm{d}\theta \\ &= \pi t \frac{4e^4 K_S}{3m^2 c^3 \Omega} = \frac{4}{3}\frac{\pi e^4 t}{m^2 c^3}\frac{u}{h\nu_s}, \qquad (2.429)\end{aligned}$$

u being the energy per unit volume.

29. DE BROGLIE WAVES

The expression

$$\psi = \int_{-\infty}^{\infty} \mathrm{e}^{-i2\pi\gamma x}\, \mathrm{e}^{i2\pi\nu t}\, \frac{\mathrm{d}\gamma}{\alpha + i\,2\pi(\gamma - \gamma_0)} \tag{2.430}$$

represents a wavepacket with phase velocity

$$v_{\mathrm{ph}} = \nu/\gamma.$$

If α tends to zero, Eq. (2.430) reduces to

$$\begin{aligned} \psi &= \mathrm{e}^{-i2\pi\gamma_0 x}\, \mathrm{e}^{i2\pi\nu_0 t} \int_{-\infty}^{\infty} \exp\{i2\pi(\gamma-\gamma_0)\,(t\,\mathrm{d}\nu_0/\mathrm{d}\gamma_0 - x)\}\, \frac{\mathrm{d}(\gamma-\gamma_0)}{\alpha + i\,2\pi(\gamma - } \\ &= \mathrm{e}^{-i2\pi\gamma_0 x}\, \mathrm{e}^{i2\pi\nu_0 t} \int_{-\infty}^{\infty} \frac{\mathrm{e}^{iy}\,\mathrm{d}y}{2\pi\,[(t\,\mathrm{d}\nu_0/\mathrm{d}\gamma_0 - x)\,\alpha + iy]}. \qquad (2.4 \end{aligned}$$

If $\alpha \to 0$, we get the following (see Eq. (2.388)). For $\alpha > 0$:

$$\psi = \begin{cases} \mathrm{e}^{-i2\pi\gamma_0 x}\, \mathrm{e}^{i2\pi\nu_0 t}, & \text{for } t\,\mathrm{d}\nu_0/\mathrm{d}\gamma_0 - x > 0, \\ 0, & \text{for } t\,\mathrm{d}\nu_0/\mathrm{d}\gamma_0 - x < 0. \end{cases} \tag{2.432}$$

This represents a plane wave extending from $x = -\infty$ to $x = (\mathrm{d}\nu_0/\mathrm{d}\gamma_0)t$ whose (forward) wavefront moves with the group velocity

$$v_{\mathrm{gr}} = \frac{\mathrm{d}\nu_0}{\mathrm{d}\gamma_0}. \tag{2.433}$$

For $\alpha < 0$:

$$\psi = \begin{cases} 0, & \text{for } t\,\mathrm{d}\nu_0/\mathrm{d}\gamma_0 - x > 0, \\ \mathrm{e}^{-i2\pi\gamma_0 x}\, \mathrm{e}^{i2\pi\nu_0 t}, & \text{for } t\,\mathrm{d}\nu_0\mathrm{d}\gamma_0 - x < 0. \end{cases} \tag{2.434}$$

This represents a plane wave extending from $x = (\mathrm{d}\nu_0/\mathrm{d}\gamma_0)t$ to $x = +\infty$ whose (backward) wavefront moves with the group velocity (2.433).

30. $e^2 \simeq hc$?

Let us consider two electrons A and B placed at a distance ℓ from each other. In some sense, the surrounding region[47] will be quantized. As a first approximation we can describe the situation in terms of a pointlike mass moving with a group velocity equal to the light speed c. Let us also make the arbitrary assumption that such a point-particle moves periodically between A and B and back. Let us further suppose that it is free of interactions while travelling between A and B, whereas in A and in B it inverts its velocity due to the collision with the electrons sitting there. If its motion is quantized, we have

$$|p| = nh/2\ell \tag{2.435}$$

and, assuming $n = 1$,

$$|p| = h/2\ell. \tag{2.436}$$

At every collision the electron receives a "kick" equal to

$$2\,|p| = h/\ell, \tag{2.437}$$

and the number of collisions per unit time is

$$\frac{1}{T} = \frac{c}{2\ell}, \tag{2.438}$$

so that a continuous force will be acting on each electron, whose magnitude is

$$F = \frac{2p}{T} = \frac{hc}{2\ell^2}. \tag{2.439}$$

If we identify Eq. (2.439) with Coulomb's law

$$F = \frac{e^2}{\ell^2}, \tag{2.440}$$

we find

$$e = \sqrt{\frac{hc}{2}}. \tag{2.441}$$

Such a value is, however, 21 times greater than the real one.

[47]@ The word used in the original manuscript literally means "aether."

31. THE EQUATION $y'' + Py = 0$

Let us consider the equation

$$y'' + Py = 0. \tag{2.442}$$

On setting

$$y = u \exp\left\{i \int (k/u^2)\,\mathrm{d}x\right\}, \tag{2.443}$$

we obtain

$$y' = \left(u' + i\frac{k}{u}\right) \exp\left\{i \int (k/u^2)\,\mathrm{d}x\right\}, \tag{2.444}$$

$$y'' = \left(u'' - \frac{k^2}{u^3}\right) \exp\left\{i \int (k/u^2)\,\mathrm{d}x\right\}, \tag{2.445}$$

$$u'' - \frac{k^2}{u^3} + uP = 0. \tag{2.446}$$

Given the initial conditions y_0 and y_0' at $x = x_0$, we set

$$u_0 = |y_0|, \tag{2.447}$$

so that the arbitrary (real) additive constant of the integral in Eq. (2.443) is determined modulo 2π. For $y_0 \neq 0$, according to Eq. (2.444), we then put

$$y_0' = \frac{y_0}{|y_0|}\left(u_0' + i\frac{k}{|y_0|}\right). \tag{2.448}$$

We can suppose that both u_0' and k are real. Then, if P is also real, the integration of Eq. (2.442) with a complex variable is equivalent to the integration of Eq. (2.446) with a real variable.

Note that, if y_0'/y_0 is real, then $k = 0$ and Eq. (2.446) reduces to (2.442).

Given an arbitrary solution of Eq. (2.446) with an arbitrary value of k, not only the function y in Eq. (2.443) but also its complex conjugate

$$y = u \exp\left\{-i \int (k/u^2)\,\mathrm{d}x\right\}, \tag{2.449}$$

satisfy Eq. (2.442), so that the general solution to Eq. (2.442) is

$$y = u\left[A \exp\left\{i \int (k/u^2)\,\mathrm{d}x\right\} + B \exp\left\{-i \int (k/u^2)\,\mathrm{d}x\right\}\right]. \tag{2.450}$$

On setting

$$u_1 = u/\sqrt{k}, \tag{2.451}$$

we obtain

$$u_1'' - \frac{1}{u_1^3} + u_1 P = 0, \tag{2.452}$$

and the general solution can still be written in the form

$$y = u_1 \left[A \exp\left\{ i \int \mathrm{d}x/u_1^2 \right\} + B \exp\left\{ -i \int \mathrm{d}x/u_1^2 \right\} \right]. \tag{2.453}$$

From this it follows that we can always reduce the problem to the $k = 1$ case. When the initial values for y and y' are given, it is possible to proceed as described above, since Eq. (2.443) now becomes

$$y = \sqrt{k}\, u_1 \left[i \int \mathrm{d}x/u_1^2 \right], \tag{2.454}$$

and the constant $\sqrt{k}$, the integration constants, and the initial values $u_{1\,0}$ and $u'_{1\,0}$ can be determined from Eqs. (2.447), (2.448), and (2.451). Alternatively, we can find an arbitrary solution of Eq. (2.452), fix arbitrarily the integration constants, and then the coefficients A and B can be determined in such a way that the initial conditions on y_0 and y_0' are met.

Assuming P to be a slowly varying function, in first approximation the function

$$u = P^{-1/4} \tag{2.455}$$

is a solution of Eq. (2.452). The general solution, to first order, then is

$$\begin{cases} y = \dfrac{1}{\sqrt[4]{P}} \left(A \cos \int \sqrt{P} \mathrm{d}x + B \sin \int \sqrt{P} \mathrm{d}x \right), & P > 0, \\ y = \dfrac{1}{\sqrt[4]{-P}} \left[A \exp\left\{ \int \sqrt{-P} \mathrm{d}x \right\} \right. \\ \qquad \left. + B \exp\left\{ - \int \sqrt{-P} \mathrm{d}x \right\} \right], & P < 0. \end{cases} \tag{2.456}$$

The condition that P be a slowly varying function is expressed as

$$\left| \frac{P'}{P} \right| \ll 1.$$

In order to derive the second-order approximation, we can replace u'' in Eq. (2.452) with the value obtained from Eq. (2.455). Thus, in second approximation, we get

$$-\frac{1}{4} P'' P^{-5/4} + \frac{5}{16} P'^2 P^{-9/4} - \frac{1}{u^3} + u P = 0. \tag{2.457}$$

On replacing u with $P^{-1/4} + \Delta u$ and setting

$$\frac{1}{u^3} \simeq P^{3/4} - 3\,P\,\Delta u, \tag{2.458}$$

Eq. (2.457) becomes

$$-\frac{1}{4}\,P''\,P^{-5/4} + \frac{5}{16}\,P'^2\,P^{-9/4} + 4\,P\,\Delta u = 0, \tag{2.459}$$

from which it follows that

$$\Delta u = \frac{1}{16}\,P''\,P^{-9/4} - \frac{5}{64}\,P'^2\,P^{-13/4} \tag{2.460}$$

and

$$u = P^{-1/4}\left(1 + \frac{P\,P'' - (5/4)\,P'^2}{16\,P^3}\right). \tag{2.461}$$

In this order of approximation, the following will be valid:

$$\frac{1}{u^2} = P^{1/2}\left(1 - \frac{P\,P'' - (5/4)\,P'^2}{8\,P^3}\right), \tag{2.462}$$

$$\int \frac{\mathrm{d}x}{u^2} = -\frac{P'}{8P^{3/2}} + \int \sqrt{P}\left(1 - \frac{P'^2}{32P^3}\right)\mathrm{d}x, \tag{2.463}$$

and the solutions for y will be of the kind

$$y = \frac{1}{\sqrt[4]{P}}\left(1 + \frac{P\,P'' - (5/4)\,P'^2}{16\,P^3}\right) \times \left\{ \begin{array}{c} \sin \\ \cos \end{array} \left[-\frac{P'}{8P^{3/2}} + \int \sqrt{P}\left(1 - \frac{P'^2}{32P^3}\right)\mathrm{d}x\right]\right., \tag{2.464}$$

for $P > 0$. Similar solutions hold for $P < 0$:

$$y = \frac{1}{\sqrt[4]{-P}}\left(1 + \frac{P\,P'' - (5/4)\,P'^2}{16\,P^3}\right) \times \exp\left\{\pm\left[-\frac{P'}{8(-P)^{3/2}} + \int \sqrt{-P}\left(1 - \frac{P'^2}{32P^3}\right)\mathrm{d}x\right]\right\}, \tag{2.465}$$

or, by setting $P_1 = -P$,

$$y = \frac{1}{\sqrt[4]{P_1}}\left(1 - \frac{P_1\,P_1'' - (5/4)\,P_1'^2}{16\,P_1^3}\right) \times \exp\left\{\pm\left[\frac{P_1'}{8P_1^{3/2}} + \int \sqrt{P_1}\left(1 + \frac{P_1'^2}{32P_1^3}\right)\mathrm{d}x\right]\right\}. \tag{2.466}$$

Let us consider, in general, the case for which $P < 0$, so that $P_1 > 0$. Equation (2.442) can then be written as

$$y'' - P y = 0. \tag{2.467}$$

Let us set

$$y = z \exp\left\{\int \sqrt{P_1} \mathrm{d}x\right\}. \tag{2.468}$$

We then have

$$y' = \left(z' + z\sqrt{P_1}\right) \exp\left\{\int \sqrt{P_1}\mathrm{d}x\right\}, \tag{2.469}$$

$$y'' = \left[z'' + 2z'\sqrt{P_1} + z\left(P_1 + \frac{P_1'}{2\sqrt{P_1}}\right)\right] \exp\left\{\int \sqrt{P_1}\mathrm{d}x\right\}, \tag{2.470}$$

$$z'' + 2z'\sqrt{P_1} + z\frac{P_1'}{2\sqrt{P_1}} = 0, \tag{2.471}$$

$$\frac{z''}{2\sqrt{P_1}\,z} + \frac{z'}{z} + \frac{1}{4}\frac{P_1'}{P_1} = 0. \tag{2.472}$$

If P is slowly varying, we can in first approximation set

$$z = P_1^{-1/4}; \tag{2.473}$$

and, by considering both signs of $\sqrt{P_1}$, we find again Eq. (2.456).

If y_1 is a solution of Eq. (2.443), the general solution reads

$$y = A\,y_1 + B\,y_1 \int \frac{\mathrm{d}x}{y_1^2}. \tag{2.474}$$

Indeed, on setting

$$y_2 = y_1 \int \frac{\mathrm{d}x}{y_1^2},$$

we get

$$y_2' = y_1' \int \frac{\mathrm{d}x}{y_1^2} + \frac{1}{y_1},$$

$$y_2'' = y_1'' \int \frac{\mathrm{d}x}{y_1^2},$$

and thus

$$0 = y_2'' - \frac{y_1''}{y_1}\,y_2 = y_2'' + P\,y_2. \tag{2.475}$$

32. INDETERMINACY OF VECTOR AND SCALAR POTENTIALS

Let us consider a magnetic and an electric field in a given spacetime region. The potentials ϕ and $\mathbf{C}$ are rather undetermined since we can set

$$\mathbf{H} = \nabla \times \mathbf{C} = \nabla \times \mathbf{C}_1, \tag{2.476}$$

$$\mathbf{E} = -\nabla\phi - \frac{1}{c}\frac{\partial \mathbf{C}}{\partial t} = \nabla\phi_1 - \frac{1}{c}\frac{\partial \mathbf{C}_1}{\partial t}, \tag{2.477}$$

with $\phi_1 \neq \phi$ and $\mathbf{C}_1 \neq \mathbf{C}$. Correspondingly, we could write two wave equations for an electron:

$$\left[-\left(\frac{W}{c} + \frac{e}{c}\,\phi\right)^2 + \sum_i \left(p_i + \frac{e}{c}\,C_i\right)^2 + m^2c^2\right]^2 \psi = 0, \tag{2.478}$$

$$\left[-\left(\frac{W}{c} + \frac{e}{c}\,\phi_1\right)^2 + \sum_i \left(p_i + \frac{e}{c}\,C_{1\,i}\right)^2 + m^2c^2\right]^2 \psi_1 = 0. \tag{2.479}$$

We can always put

$$\mathbf{C}_1 - \mathbf{C} = \nabla A, \tag{2.480}$$

$$\phi_1 - \phi = -\frac{1}{c}\frac{\partial A}{\partial t}, \tag{2.481}$$

with A an arbitrary function of space and time. This function is no longer arbitrary if we impose the so-called continuity constraint[48]

$$\nabla \cdot \mathbf{C} + \frac{1}{c}\frac{\partial \phi}{\partial t} = \nabla \cdot \mathbf{C}_1 + \frac{1}{c}\frac{\partial \phi_1}{\partial t} = 0, \tag{2.482}$$

since, in this case,

$$\nabla^2 A - \frac{1}{c^2}\frac{\partial^2 A}{\partial t^2}. \tag{2.483}$$

From

$$W \exp\left\{\frac{i}{\hbar}\frac{e}{c}A\right\} = \exp\left\{\frac{i}{\hbar}\frac{e}{c}A\right\}\,\left[W + e\,(\phi_1 - \phi)\right], \tag{2.484}$$

$$p_i \exp\left\{\frac{i}{\hbar}\frac{e}{c}A\right\} = \exp\left\{\frac{i}{\hbar}\frac{e}{c}A\right\}\,\left[p_i + \frac{e}{c}\,(C_{1\,i} - C_i)\right], \tag{2.485}$$

[48]@ This is more widely known as the Lorenz gauge condition.

we deduce that

$$\left(\frac{W}{c} + \frac{e}{c}\phi_1\right) \exp\left\{\frac{i}{\hbar}\frac{e}{c}A\right\} = \exp\left\{\frac{i}{\hbar}\frac{e}{c}A\right\} \left(\frac{W}{c} + \frac{e}{c}\phi\right), \tag{2.486}$$

$$\left(\frac{W}{c} + \frac{e}{c}\phi_1\right)^2 \exp\left\{\frac{i}{\hbar}\frac{e}{c}A\right\} = \exp\left\{\frac{i}{\hbar}\frac{e}{c}A\right\} \left(\frac{W}{c} + \frac{e}{c}\phi\right)^2, \tag{2.487}$$

and that

$$\left(p_i + \frac{e}{c}C_{1\,i}\right) \exp\left\{\frac{i}{\hbar}\frac{e}{c}A\right\} = \exp\left\{\frac{i}{\hbar}\frac{e}{c}A\right\} \left(p_i + \frac{e}{c}C_i\right), \tag{2.488}$$

$$\left(p_i + \frac{e}{c}C_{1\,i}\right)^2 \exp\left\{\frac{i}{\hbar}\frac{e}{c}A\right\} = \exp\left\{\frac{i}{\hbar}\frac{e}{c}A\right\} \left(p_i + \frac{e}{c}C_i\right)^2. \tag{2.489}$$

It follows that, if ψ is a solution of Eq. (2.478), then the quantity

$$\psi_1 = \psi \exp\left\{\frac{i}{\hbar}\frac{e}{c}A\right\} \tag{2.490}$$

will be a solution of Eq. (2.479). Since the phase shift of ψ given by Eq. (2.490) is physically irrelevant, due to the fact that it is identical for all the eigenfunctions evaluated at the same point and at the same time, we have then proven that the two Hamiltonians considered by us are equivalent.

33. ON THE SPONTANEOUS IONIZATION OF A HYDROGEN ATOM PLACED IN A HIGH POTENTIAL REGION

Let us consider a hydrogen atom placed at the common center of two spheres of radii R and $R + \mathrm{d}R$, respectively. On the first of these spheres there is a charge $-Q'/\mathrm{d}R$ and on the second a charge $(Q'/\mathrm{d}R) - e$ (we set $Q' = QR$); then we take $\mathrm{d}R$ to be infinitesimal. The atomic electron will experience a potential:

$$\begin{cases} V = e/x - A, & x < R, \\ V = 0, & x > R, \end{cases} \tag{2.491}$$

x being the distance from the center and A a constant.[49] For the sake of simplicity, we adopt the radius of the first Bohr orbit as the unit length,

[49]@ In the original manuscript, the value $A = Q^2/R^2$ is given. But, for dimensional reasons, this is wrong. The constant A can be fixed by requiring the continuity of the potential; in this case we have $A = e/R$.

e as the unit charge and $\hbar$ as the unit action. Our unit for energy will be $me^4/\hbar^2 = 4\pi R_y \hbar$, and then $1/(4\pi R_y)$ will be the unit time, where R_y is Rydberg's frequency. Furthermore, we choose the electron mass as the unit mass. The Schrödinger equation corresponding to zero azimuthal quantum number, when and setting $\chi = \psi/x$, will be

$$\begin{cases} \chi'' + 2\,(E - A + 1/x)\,\chi = 0, & x < R, \\ \chi'' + 2E\,\chi = 0, & x > R. \end{cases} \tag{2.492}$$

Let us set $E - A = E_1$. If the atom is in its ground state, then E_1 is approximately equal to $-1/2$.[50] We then set

$$-E_1 = \frac{1}{2} + \frac{1}{2}\,\alpha, \tag{2.493}$$

so that Eqs. (2.492) become

$$\begin{cases} \chi'' + (1 - \alpha + 2/x)\,\chi = 0, & x < R, \\ \chi'' + (2A - 1 - \alpha)\,\chi = 0, & x > R. \end{cases} \tag{2.494}$$

A solution to the first of these equations for $\alpha = 0$ is

$$\chi = x\,\mathrm{e}^{-x}. \tag{2.495}$$

Let us cast the solution for $\alpha \neq 0$ in the form

$$\chi = x\,\mathrm{e}^{-x} + \alpha\,y, \tag{2.496}$$

with the constraints $y(0) = 0$, $y'(0) = 0$. On substituting into Eq. (2.494), we find

$$y'' = x\,\mathrm{e}^{-x} + (1 + \alpha - 2/x)\,y, \tag{2.497}$$

showing that y depends on α. Since the initial conditions on y have been fixed, y is completely determined. For large values of x, this takes the asymptotic form

$$y = k_\alpha\,\mathrm{e}^{x\sqrt{1+\alpha}}/x^{1/\sqrt{1+\alpha}}. \tag{2.498}$$

Since we have assumed that α is small, as an approximation we could set $k_\alpha = k_0$, and k_0 will be evaluated from the asymptotic form for y

[50]@ The ground state energy of a hydrogen atom is $-e^2/2a_B$, where a_B is the Bohr radius. In the adopted units, this energy equals $-1/2$.

with the constraints $y(0) = 0,\ y'(0) = 0$. The differential equation for such y is

$$y'' = x\,\mathrm{e}^{-x} + (1 - 2/x)\,y, \tag{2.499}$$

and the asymptotic expression for the solution will be of the form

$$y = k_0\,\mathrm{e}^{x}/x. \tag{2.500}$$

Let us then evaluate k_0. The function y can be expanded in increasing powers of x:

$$y = \frac{1}{6}x^3 - \frac{1}{9}x^4 + \ldots + a_n x^n + \ldots, \tag{2.501}$$

where the coefficients a_r can be obtained from the recursive relation

$$a_n = -(-1)^n\frac{n-2}{n!} + \frac{a_{n-2} - 2a_{n-1}}{(n-1)n}. \tag{2.502}$$

Starting from a_2, these can be written in the form

$$a_{2n+1} = \frac{1}{(2n)!}\left[n\frac{2n+2}{2n+1} - \left(1 + \frac{1}{3} + \frac{1}{5} + \ldots + \frac{1}{2n-1}\right)\right], \tag{2.503}$$

$$a_{2n} = -\frac{1}{(2n-1)!}\left[n - \left(1 + \frac{1}{3} + \frac{1}{5} + \ldots + \frac{1}{2n-1}\right)\right]. \tag{2.504}$$

Indeed, if Eqs. (2.503) and (2.504) hold for some value of n (and it can be directly verified that they hold for $n = 1$), then Eq. (2.504) will still be valid if we replace n with $n+1$ since, from Eq. (2.502), we derive

$$\begin{aligned}
-(2n+1)!\,a_{2n+2} &= \frac{n}{n+1} - \frac{n}{n+1}\left(1 + \frac{1}{3} + \frac{1}{5} + \ldots + \frac{1}{2n-1}\right) \\
&- \frac{1}{n+1}\left(1 + \frac{1}{3} + \frac{1}{5} + \ldots + \frac{1}{2n-1}\right) + \frac{n^2}{n+1} + \frac{2n}{2n+1} \\
&= n + 1 - \left(1 + \frac{1}{3} + \frac{1}{5} + \ldots + \frac{1}{2n+1}\right),
\end{aligned}$$

and thus Eq. (2.504) holds for a_{2n+2} too. In the same way, by using again Eq. (2.502), we find

$$\begin{aligned}
(2n+2)!\,a_{2n+3} &= \frac{2n+1}{2n+3} - \frac{2n+1}{2n+3}\left(1 + \frac{1}{3} + \frac{1}{5} + \ldots + \frac{1}{2n-1}\right) \\
&- \frac{2}{2n+3}\left(1 + \frac{1}{3} + \frac{1}{5} + \ldots + \frac{1}{2n-1}\right) + n\frac{2n+2}{2n+3} + \frac{2n+2}{2n+3} \\
&= (n+1)\frac{2n+2}{2n+3} - \left(1 + \frac{1}{3} + \frac{1}{5} + \ldots + \frac{1}{2n+1}\right),
\end{aligned}$$

and thus Eq. (2.503) holds for a_{2n+3} as well. Equations (2.503) and (2.504) then always apply.

The expansion (2.501) can be indefinitely derived term by term. We can thus set

$$y''' + 3y'' + 3y' + y = \sum_0^\infty b_r\, x^r; \tag{2.505}$$

and, in general, we have

$$\begin{aligned} b_r &= a_r + 2(r+1)a_{r+1} + 3(r+1)(r+2)a_{r+2} \\ &\quad + (r+1)(r+2)(r+3)a_{r+3}, \end{aligned} \tag{2.506}$$

or, due to Eqs. (2.503) and (2.504):

$$\begin{aligned} &(2n+1)!\, b_{2n} = -2n^2(2n+1) + 3n(2n+1)(2n+2) \\ &- 3(n+1)(2n+1)(2n+2) + (n+1)(2n+1)(2n+4) \\ &+ 2n(2n+1)\left(1 + \frac{1}{3} + \ldots + \frac{1}{2n-1}\right) \\ &- 3(2n+1)^2\left(1 + \frac{1}{3} + \ldots + \frac{1}{2n-1}\right) \\ &+ 3(2n+1)(2n+2)\left(1 + \frac{1}{3} + \ldots + \frac{1}{2n+1}\right) \\ &- (2n+1)(2n+3)\left(1 + \frac{1}{3} + \ldots + \frac{1}{2n+1}\right) = 1, \end{aligned} \tag{2.507}$$

$$\begin{aligned} &(2n+2)!\, b_{2n+1} = n(2n+2)^2 - 3(n+1)(2n+2)^2 \\ &+ 3(n+1)(2n+2)(2n+4) - (n+2)(2n+2)(2n+4) \\ &- (2n+1)(2n+2)\left(1 + \frac{1}{3} + \ldots + \frac{1}{2n-1}\right) \\ &+ 3(2n+1)^2\left(1 + \frac{1}{3} + \ldots + \frac{1}{2n+1}\right) \\ &- 3(2n+2)(2n+3)\left(1 + \frac{1}{3} + \ldots + \frac{1}{2n+1}\right) \\ &+(2n+2)(2n+4)\left(1 + \frac{1}{3} + \ldots + \frac{1}{2n+3}\right) = 1 - \frac{1}{2n+3}. \end{aligned} \tag{2.508}$$

It follows that

$$\begin{aligned} \sum_0^\infty b_r\, x^r &= \sum_1^\infty \frac{x^{s-1}}{s!} - \sum_0^\infty \frac{x^{2s+1}}{(2s+3)!} \\ &= \frac{1}{x}\sum_1^\infty \frac{x^s}{s!} - \frac{1}{x^2}\sum_1^\infty \frac{x^{2s+1}}{(2s+1)!} \end{aligned}$$

$$
\begin{aligned}
&= \frac{\mathrm{e}^x - 1}{x} - \frac{\mathrm{e}^x - \mathrm{e}^{-x} - 2x}{2x^2} \\
&= \frac{\mathrm{e}^x}{x} \qquad \frac{\mathrm{e}^x - \mathrm{e}^{-x}}{2x^2} \\
&= \mathrm{e}^x \left(\frac{1}{x} - \frac{1}{2x^2} \right) + \mathrm{e}^{-\frac{1}{2x^2}} .
\end{aligned} \tag{2.509}
$$

By substituting into Eq. (2.505) and rewriting the l.h.s., we find

$$
\begin{aligned}
&(y'' + 2y' + y) + \frac{\mathrm{d}}{\mathrm{d}x} (y'' + 2y' + y) \\
&= \mathrm{e}^x \left(\frac{1}{x} - \frac{1}{2x^2} \right) + \mathrm{e}^{-x} \frac{1}{2x^2},
\end{aligned} \tag{2.510}
$$

and thus

$$
y'' + 2y' + y = \mathrm{e}^{-x} \left\{ \int \left[\mathrm{e}^{2x} \left(\frac{1}{x} - \frac{1}{2x^2} \right) + \frac{1}{2x^2} \right] \mathrm{d}x + C \right\}. \tag{2.511}
$$

Since, for $x = 0$, we have $y = 0$, $y' = 0$, $y'' = 0$, this becomes

$$
y'' + 2y' + y = \mathrm{e}^x \frac{1}{2x} - \mathrm{e}^{-x} \left(\frac{1}{2x} + 1 \right), \tag{2.512}
$$

that is,

$$
(y' + y) + \frac{\mathrm{d}}{\mathrm{d}x} (y' + y) = \mathrm{e}^x \frac{1}{2x} - \mathrm{e}^{-x} \left(\frac{1}{2x} + 1 \right), \tag{2.513}
$$

so that

$$
y' + y = \mathrm{e}^{-x} \left\{ \int \left[\frac{\mathrm{e}^{2x}}{2x} - \left(\frac{1}{2x} + 1 \right) \right] \mathrm{d}x + C \right\}. \tag{2.514}
$$

Taking into account the initial conditions, we get

$$
y' + y = -x\,\mathrm{e}^{-x} + \mathrm{e}^{-x} \int_0^x \frac{\mathrm{e}^{2x} - 1}{2x} \mathrm{d}x \tag{2.515}
$$

and

$$
y = \mathrm{e}^{-x} \left[\int \left(-x + \int_0^x \frac{\mathrm{e}^{2x} - 1}{2x} \mathrm{d}x \right) \mathrm{d}x + C \right]; \tag{2.516}
$$

and finally, using again the initial conditions, we find

$$
y = -\frac{1}{2} x^2 \mathrm{e}^{-x} + \mathrm{e}^{-x} \int_0^x \mathrm{d}x_1 \int_0^{x_1} \frac{\mathrm{e}^{2x_2} - 1}{2x_2} \mathrm{d}x_2. \tag{2.517}
$$

As a further check, let us evaluate y' and y'':

$$\begin{aligned} y' &= -x\,\mathrm{e}^{-x} + \frac{1}{2}x^2\,\mathrm{e}^{-x} - \mathrm{e}^{-x}\int_0^x \mathrm{d}x_1 \int_0^{x_1} \frac{\mathrm{e}^{2x_2}-1}{2x_2}\,\mathrm{d}x_2 \\ &\quad + \mathrm{e}^{-x}\int_0^x \frac{\mathrm{e}^{2x}-1}{2x}\,\mathrm{d}x, \qquad (2.518) \\ y'' &= -\mathrm{e}^{-x} + 2x\,\mathrm{e}^{-x} - \frac{1}{2}x^2\,\mathrm{e}^{-x} + \mathrm{e}^{-x}\int_0^x \mathrm{d}x_1 \int_0^{x_1} \frac{\mathrm{e}^{2x_2}-1}{2x_2}\,\mathrm{d}x_2 \\ &\quad - 2\,\mathrm{e}^{-x}\int_0^x \frac{\mathrm{e}^{2x}-1}{2x}\,\mathrm{d}x + \frac{\mathrm{e}^{x}-\mathrm{e}^{-x}}{2x}. \qquad (2.519) \end{aligned}$$

Since

$$\int_0^x \mathrm{d}x_1 \int_0^{x_1} \frac{\mathrm{e}^{2x_2}-1}{2x_2}\,\mathrm{d}x_2 = x\int_0^x \frac{\mathrm{e}^{2x}-1}{2x}\,\mathrm{d}x - \frac{1}{4}\mathrm{e}^{2x} + \frac{1}{2}x + \frac{1}{4},$$

the previous formulae become

$$\begin{aligned} y &= x\,\mathrm{e}^{-x}\int_0^x \frac{\mathrm{e}^{2x}-1}{2x}\,\mathrm{d}x - \frac{1}{4}\mathrm{e}^{x} + \left(\frac{1}{4} + \frac{1}{2}x - \frac{1}{2}x^2\right)\mathrm{e}^{-x}, \qquad (2.520) \\ y' &= (1-x)\,\mathrm{e}^{-x}\int_0^x \frac{\mathrm{e}^{2x}-1}{2x}\,\mathrm{d}x + \frac{1}{4}\mathrm{e}^{x} \\ &\quad + \left(-\frac{1}{4} - \frac{3}{2}x + \frac{1}{2}x^2\right)\mathrm{e}^{-x}, \qquad (2.521) \\ y'' &= (x-2)\,\mathrm{e}^{-x}\int_0^x \frac{\mathrm{e}^{2x}-1}{2x}\,\mathrm{d}x + \mathrm{e}^{x}\left(\frac{1}{2x} - \frac{1}{4}\right) \\ &\quad + \mathrm{e}^{-x}\left(-\frac{1}{2x} - \frac{3}{4} + \frac{5}{2}x - \frac{1}{2}x^2\right). \qquad (2.522) \end{aligned}$$

From these it follows that

$$y'' = \left(1 - \frac{2}{x}\right)y + x\,\mathrm{e}^{-x}, \qquad (2.523)$$

that is, the differential equation (2.499) is satisfied. Moreover, clearly

$$y(0) = y'(0) = 0, \qquad (2.524)$$

as we required. For $x \to \infty$ we get

$$\int_0^x \frac{\mathrm{e}^{2x}-1}{2x}\,\mathrm{d}x = \mathrm{e}^{2x}\left(\frac{1}{4x} + \frac{1}{\gamma x^2} + \text{higher order terms}\right), \qquad (2.525)$$

and the asymptotic expression for y is

$$y = \frac{1}{8}\frac{\mathrm{e}^{x}}{x}, \qquad (2.526)$$

from which we can obtain the constant k_0 in Eq. (2.500):

$$k_0 = \frac{1}{8}. \tag{2.527}$$

For large values of x, the solution of Eqs. (2.491) and (2.492) then is approximately

$$\chi = x\,\mathrm{e}^{-x} + \frac{\alpha}{8}\,\mathrm{e}^{x\sqrt{1+\alpha}}/x^{1/\sqrt{1+\alpha}}. \tag{2.528}$$

We now suppose that R is large in our units; this means that R has to be large with respect to atomic dimensions. We then have

$$\chi(R) = R\,\mathrm{e}^{-R} + \frac{\alpha}{8}\,\frac{\mathrm{e}^{R\sqrt{1+\alpha}}}{R^{1/\sqrt{1+\alpha}}}, \tag{2.529}$$

$$\chi''(R) = (1-R)\,\mathrm{e}^{-R} + \frac{\alpha}{8}\,\frac{\mathrm{e}^{R\sqrt{1+\alpha}}}{R^{1/\sqrt{1+\alpha}}}\left(\sqrt{1+\alpha} - \frac{1}{R\sqrt{1+\alpha}}\right). \tag{2.530}$$

For reasons that will be clear later on, we are only interested in very small values of α, so that the second term in the expression of $\chi(R)$ is of the same order as the first one. This means that α has to be of the order $R^2\mathrm{e}^{-2R}$. It is then possible to replace $\sqrt{1+\alpha}$ with 1 everywhere. Neglecting also 1 relative to R, the equations above become

$$\begin{aligned} \chi(R) &= R\,\mathrm{e}^{-R} + \frac{\alpha}{8}\,\frac{\mathrm{e}^{R}}{R}, \\ \chi''(R) &= -R\,\mathrm{e}^{-R} + \frac{\alpha}{8}\,\frac{\mathrm{e}^{R}}{R}. \end{aligned} \tag{2.531}$$

Equation (2.528) takes a simple form for large values of x smaller than R:

$$\chi = x\,\mathrm{e}^{-x} + \frac{\alpha}{8}\,\frac{\mathrm{e}^{x}}{x}. \tag{2.532}$$

For $x > R$ the second equation in (2.492) must be satisfied. Let us suppose that $E < 0$, because otherwise no spontaneous ionization would take place. Since

$$E = A - \frac{1}{2} - \frac{1}{2}\alpha, \tag{2.533}$$

A has to be much larger than $\frac{1}{2}$. The second equation in (2.492) thus has sinusoidal solutions. For x greater than R we then have

$$\begin{aligned} \chi = {} & \left(R\,\mathrm{e}^{-R} + \frac{\alpha}{8}\,\frac{\mathrm{e}^{R}}{R}\right)\cos\sqrt{2E}(x-R) \\ & + \frac{1}{\sqrt{2E}}\left(-R\,\mathrm{e}^{-R} + \frac{\alpha}{8}\,\frac{\mathrm{e}^{R}}{R}\right)\sin\sqrt{2E}(x-R). \end{aligned} \tag{2.534}$$

Let us set

$$B = A - \frac{1}{2} + 4R^2\,e^{-2R}\,\frac{A-1}{A}. \tag{2.535}$$

The quantities E, B and $A - 1/2$ have all values that are very close one another so that, when they appear as factors in some expression, it is definitely possible to substitute one for the other in order to simplify the formulae. In expressions that are arguments of sines and cosines, another approximation is required. Since

$$E = B - \frac{4(A-1)}{A}\,R^2\,e^{-2R} - \frac{1}{2}\,\alpha, \tag{2.536}$$

we'll set

$$\sqrt{2E} = \sqrt{2B} - \frac{1}{\sqrt{2E}}\left(\frac{4(A-1)}{A}\,R^2\,e^{-2R} + \frac{1}{2}\,\alpha\right). \tag{2.537}$$

We denote by γ the second term above divided by 2π:[51]

$$\gamma = -\frac{1}{2\pi}\,\frac{1}{\sqrt{2E}}\left(\frac{4(A-1)}{A}\,R^2\,e^{-2R} + \frac{1}{2}\,\alpha\right), \tag{2.538}$$

so that

$$\alpha = -\,4\pi\,\sqrt{2E}\,\gamma - \frac{8(A-1)}{A}\,R^2\,e^{-2R}. \tag{2.539}$$

If this is substituted into Eq. (2.534) and the approximations described above are used, we find

$$\begin{aligned} \chi = & \left(\frac{1}{A}\,R\,e^{-R} - \frac{\sqrt{2B}}{4}\,\frac{e^R}{R}\,2\pi\gamma\right)\,\cos\left(\sqrt{2B} + 2\pi\gamma\right)(x-R) \\ & + \left(-\,\frac{\sqrt{2B}}{A}\,R\,e^{-R} - \frac{1}{4}\,\frac{e^R}{R}\,2\pi\gamma\right)\,\sin\left(\sqrt{2B} + 2\pi\gamma\right)(x-R), \end{aligned} \tag{2.540}$$

or, again as an approximation,

$$\begin{aligned} \chi = & \sqrt{\frac{2}{A}\,R^2\,e^{-2R} + \frac{A}{8}\,R^{-2}\,e^{2R}\,4\pi^2\gamma^2} \\ & \times\,\cos\left[\left(\sqrt{2B} + 2\pi\gamma\right)(x-R) + z\right], \end{aligned} \tag{2.541}$$

where z is an angle that depends on γ. If we choose χ to be normalized with respect to dx, then it must be multiplied by a factor N:

$$u = N\,\chi \tag{2.542}$$

[51]@ The author considers this γ as the (correction to) the momentum of the system under consideration (in the adopted units).

such that

$$N\sqrt{\frac{2}{A}\,R^2\,\mathrm{e}^{-2R} \;+\; \frac{A}{8}\,R^{-2}\,\mathrm{e}^{2R}\,4\pi^2\gamma^2} \;=\; 2. \tag{2.543}$$

Indeed,

$$\lim_{\epsilon\to 0}\int_0^\infty \chi(\gamma)\,\mathrm{d}x\int_{-\epsilon}^{+\epsilon}\chi(\gamma+\epsilon)\,\mathrm{d}\epsilon \;=\; 1/N^2. \tag{2.544}$$

This yields the following normalized eigenfunctions for $x < R$ and $x > R$, respectively:

$$\begin{aligned} u &= \frac{\sqrt{\dfrac{A}{2}\,\dfrac{\mathrm{e}^R}{R}}}{\sqrt{1+\dfrac{\pi^2A^2}{4}\dfrac{\mathrm{e}^{4R}}{R^4}\gamma^2}}\left(2x\,\mathrm{e}^{-x} \;+\; \frac{\alpha}{4}\,\frac{\mathrm{e}^x}{x}\right)\,\mathrm{e}^{iBt}\,\mathrm{e}^{2\pi i\sqrt{2B}\gamma t}, \\ u &= 2\,\cos\left[\left(\sqrt{2B}+2\pi\gamma\right)(x-R)+z\right]\,\mathrm{e}^{iBt}\,\mathrm{e}^{2\pi i\sqrt{2B}\gamma t}. \end{aligned} \tag{2.545}$$

Here we have taken into account the time dependence and the fact that $E = B + 2\pi\sqrt{2B}\gamma$. Note that in Eq. (2.545) the term $2x\mathrm{e}^{-x} + (\alpha/4)(\mathrm{e}^x/x)$ has been factored out since

$$\int_0^R\left(2x\,\mathrm{e}^{-x} \;+\; \frac{\alpha}{4}\,\frac{\mathrm{e}^x}{x}\right)^2\,\mathrm{d}x \;\simeq\; \int_0^R\left(2x\,\mathrm{e}^{-x}\right)^2\,\mathrm{d}x \;\simeq\; 1, \tag{2.546}$$

so that, for small values of x, it represents the eigenfunction of the quasi-stationary state $1s$, normalized in the usual way.

Let us now suppose that initially the electron is in the ground state. Its eigenfunction is approximately spherically symmetric, so that we can write

$$\psi \;=\; U(x)/x, \tag{2.547}$$

where at time 0 we have

$$U_0 \;\simeq\; 2x\,\mathrm{e}^{-x}. \tag{2.548}$$

Denoting by u_0 the functions u defined above at time $t = 0$, let us expand U_0 in series of u_0 :

$$U_0 \;=\; \int_{-\infty}^{\infty} c\,u_0\,\mathrm{d}\gamma. \tag{2.549}$$

We shall have

$$c \;=\; \int_0^\infty U_0\,u_0\,\mathrm{d}x \;\simeq\; \sqrt{\frac{A}{2}\,\frac{\mathrm{e}^R}{R}}\bigg/\sqrt{1+\frac{\pi^2A^2}{4}\,\frac{\mathrm{e}^{4R}}{R^4}\gamma^2}\;; \tag{2.550}$$

and, since, for an arbitrary t,

$$U = \int_{-\infty}^{\infty} c\, u\, \mathrm{d}\gamma, \tag{2.551}$$

on substituting Eqs. (2.547) and (2.550), we find for x less than R:

$$U = \mathrm{e}^{iBt} A \frac{\mathrm{e}^{2R}}{R^2} \left(x\,\mathrm{e}^{-x} + \frac{\overline{\alpha}}{8} \frac{\mathrm{e}^{x}}{x} \right) \int_{-\infty}^{\infty} \frac{\mathrm{e}^{2\pi i\sqrt{2B}\gamma t}\, \mathrm{d}\gamma}{1 + (\pi^2 A^2 \mathrm{e}^{4R}/4R^4)\gamma^2}, \tag{2.552}$$

where it is simple to show (from Eq. (2.539)) that

$$\overline{\alpha} = -\frac{A-1}{A} 8R^2\,\mathrm{e}^{-2R} - \frac{\sqrt{2B}}{A} 8R^2\,\mathrm{e}^{-2R} \tag{2.553}$$

coincides with the α relative to the stationary ground state considered here; the demonstration is similar to the one exposed in what follows, which will lead to Eq. (2.560)).[52] Now

$$\begin{aligned}
&\int_{-\infty}^{\infty} \frac{\mathrm{e}^{2\pi i\sqrt{2B}\gamma t}\, \mathrm{d}\gamma}{1 + (\pi^2 A^2 \mathrm{e}^{4R}/4R^4)\gamma^2} \\
&= \frac{4}{R} R^2\,\mathrm{e}^{-2R} \int_{-\infty}^{\infty} \frac{\mathrm{e}^{2\pi i\sqrt{2B}\gamma t}\, \mathrm{d}\left(A\mathrm{e}^{2R}\gamma/4R^2\right)}{1 + 4\pi^2 \left(A\mathrm{e}^{2R}\gamma/4R^2\right)^2} \\
&= \begin{cases} \dfrac{2}{A} R^2\,\mathrm{e}^{-2R} \exp\left\{ \dfrac{4R^2\sqrt{2B}}{A\mathrm{e}^{2R}} t \right\}, & \text{for } t < 0, \\[2ex] \dfrac{2}{A} R^2 \exp\left\{ -\dfrac{4R^2\sqrt{2B}}{A\mathrm{e}^{2R}} t \right\}, & \text{for } t > 0. \end{cases}
\end{aligned} \tag{2.554}$$

We are only interested in the solution for $t > 0$, since we shall set the initial conditions regardless of the way the system reached such initial state. Thus, for $t > 0$ and $x < R$, we'll have

$$U = \left(x\,\mathrm{e}^{-x} + \frac{\overline{\alpha}}{8} \frac{\mathrm{e}^{x}}{x} \right) \mathrm{e}^{iBt} \exp\left\{ -4R^2\sqrt{2B}\, t/A\mathrm{e}^{2R} \right\}, \tag{2.555}$$

while, for $x > R$,

$$U = \mathrm{e}^{iBt}\, 2\sqrt{\frac{A}{2} \frac{\mathrm{e}^{R}}{R}} \int_{-\infty}^{\infty} \frac{\cos\left[\left(\sqrt{2B} + 2\pi\gamma\right)(x-R) + z \right] \mathrm{e}^{2\pi i\sqrt{2B}\gamma t}}{\sqrt{1 + (\pi^2 A^2 \mathrm{e}^{4R}/4R^4)\gamma^2}}\, \mathrm{d}\gamma, \tag{2.556}$$

[52]@ This paragraph is added as a postponed footnote in the original manuscript.

$$
\begin{aligned}
U &= \mathrm{e}^{iBt}\,\frac{\mathrm{e}^R}{R}\left[\int_{-\infty}^{\infty}\frac{1-(\pi A\sqrt{2B}\mathrm{e}^{2R}/2R^2)\gamma}{1+(\pi^2A^2\mathrm{e}^{4R}/4R^4)\gamma^2}\right.\\
&\times\cos\left(\sqrt{2B}+2\pi\gamma\right)(x-R)\ \mathrm{e}^{2\pi i\sqrt{2B}\gamma t}\,\mathrm{d}\gamma\\
&\left.-\int_{-\infty}^{\infty}\frac{\sqrt{2B}+(\pi A\mathrm{e}^{2R}/2R^2)\gamma}{1+(\pi^2A^2\mathrm{e}^{4R}/4R^4)\gamma^2}\sin\left(\sqrt{2B}+2\pi\gamma\right)(x-R)\ \mathrm{e}^{2\pi i\sqrt{2B}\gamma t}\,\mathrm{d}\gamma\right]\\
&= \mathrm{e}^{iBt}\,\frac{\mathrm{e}^R}{R}\left[\mathrm{e}^{i\sqrt{2B}(x-R)}\int_{-\infty}^{\infty}\frac{M+Ni}{2}\,\mathrm{e}^{2\pi i[\sqrt{2B}t+(x-R)]\gamma}\,\mathrm{d}\gamma\right.\\
&\left.+\ \mathrm{e}^{-i\sqrt{2B}(x-R)}\int_{-\infty}^{\infty}\frac{M-Ni}{2}\,\mathrm{e}^{2\pi i[\sqrt{2B}t-(x-R)]\gamma}\,\mathrm{d}\gamma\right],
\end{aligned}
\tag{2.557}
$$

with

$$
M = \frac{1-(\pi A\sqrt{2B}\mathrm{e}^{2R}/2R^2)\gamma}{1+(\pi^2A^2\mathrm{e}^{4R}/4R^4)\gamma^2}, \tag{2.558}
$$

$$
N = \frac{\sqrt{2B}+(\pi A\mathrm{e}^{2R}/2R^2)\gamma}{1+(\pi^2A^2\mathrm{e}^{4R}/4R^4)\gamma^2}. \tag{2.559}
$$

We have

$$
\int_{-\infty}^{\infty}\mathrm{e}^{2\pi i[\sqrt{2B}t\pm(x-R)]\gamma}\Bigg/\left(1+\frac{\pi^2A^2}{4}\frac{\mathrm{e}^{4R}}{R^4}\gamma^2\right)\,\mathrm{d}\gamma
$$

$$
= \begin{cases}
\dfrac{2}{A}\dfrac{R^2}{\mathrm{e}^{2R}}\exp\left\{+\dfrac{4}{A}\dfrac{R^2}{\mathrm{e}^{2R}}[\sqrt{2B}t\pm(x-R)]\right\}, \\
\text{for } \sqrt{2B}\,t\pm(x-R)<0, \\[2ex]
\dfrac{2}{A}\dfrac{R^2}{\mathrm{e}^{2R}}\exp\left\{-\dfrac{4}{A}\dfrac{R^2}{\mathrm{e}^{2R}}[\sqrt{2B}t\pm(x-R)]\right\}, \\
\text{for } \sqrt{2B}\,t\pm(x-R)>0.
\end{cases} \tag{2.560}
$$

Moreover:[53]

$$
\int_{-\infty}^{\infty}\frac{\gamma\,\mathrm{e}^{2\pi i[\sqrt{2B}t\pm(x-R)]\gamma}\,\mathrm{d}\gamma}{1+(\pi^2A^2\mathrm{e}^{4R}/4R^4)\gamma^2}
$$

---

[53] @ In the original manuscript the signs of the results of the integral were reversed, but here the correct expressions are used.

$$
= \begin{cases} -i \, \dfrac{4}{\pi A^2} \, \dfrac{R^4}{\mathrm{e}^{4R}} \, \exp\left\{ + \dfrac{4}{A} \, \dfrac{R^2}{\mathrm{e}^{2R}} [\sqrt{2B}t \pm (x-R)] \right\}, \\ \text{for } \sqrt{2B}\, t \pm (x-R) < 0, \\ \\ i \, \dfrac{4}{\pi A^2} \, \dfrac{R^4}{\mathrm{e}^{4R}} \, \exp\left\{ - \dfrac{4}{A} \, \dfrac{R^2}{\mathrm{e}^{2R}} [\sqrt{2B}t \pm (x-R)] \right\}, \\ \text{for } \sqrt{2B}\, t \pm (x-R) > 0. \end{cases} \tag{2.561}
$$

Thus, although we will be concerned only with the solutions for $x > R$ and $t > 0$, we will have to consider two separate cases, depending on whether $\sqrt{2B}t - (x - R)$ is positive or negative, while $\sqrt{2B}t + (x - R)$ is always positive. Since the quantity

$$
\int_{-\infty}^{\infty} (1/2)(M + Ni) \, \mathrm{e}^{2\pi i [\sqrt{2B}t + (x-R)]\gamma} \, \mathrm{d}\gamma \tag{2.562}
$$

is identically zero when $\sqrt{2B} + x - R > 0$, due to Eqs. (2.560) and (2.561), we have, respectively[54]

$$
U = \begin{cases} \sqrt{\dfrac{8}{A}} \, \dfrac{R}{\mathrm{e}^{R}} \, \mathrm{e}^{iBt} \, \exp\left\{ -i \arcsin \sqrt{(2A-1)/2A} - i\sqrt{2B}(x-R) \right\} \\ \times \exp\left\{ 4R^2(x-R)/(A\mathrm{e}^{2R}) - 4R^2\sqrt{2B}\, t/(A\mathrm{e}^{2R}) \right\}, \\ \text{for } \sqrt{2B}t - (x-R) > 0, \\ \\ 0, \quad \text{for } \sqrt{2B}t - (x-R) < 0, \end{cases} \tag{2.563}
$$

having again made use of the approximation $2B = 2A - 1$ where possible. For $\sqrt{2B}t - (x - R) > 0$, independently of the small time-dependent damping factor and of the space-dependent growth factor, Eq. (2.563) represents a progressive plane wave moving towards high values of x. For sufficiently small values of t and $x - R$, the electron flux per unit time is

$$
F = \frac{8R^2\sqrt{2B}}{A\mathrm{e}^{2R}}. \tag{2.564}
$$

On the other hand, the damping factor can be written as

$$
\mathrm{e}^{-t/2T}, \tag{2.565}
$$

where T is the time-constant.[55] It follows, as it is natural, that

$$
F = \frac{1}{T}, \quad T = \frac{A\mathrm{e}^{2R}}{8R^2\sqrt{2B}}. \tag{2.566}
$$

[54]@ Note that the author missed a factor 2 in front of the following expression.

[55]@ Below, in this section, we shall call T the "mean-life", following the author's terminology.

In his notes on the half-life of α particles in radioactive nuclei, Gamov[56] assumed an exponential time dependence and postulated that, at large distances from the nucleus, the eigenfunction of the α particle is a spherical progressive wave, thus determining T by using Eq. (2.566). The previous discussion shows how well grounded his arguments were and also that Gudar's objections on alleged inconsistencies arising from the space dependent growth factor in U from Eq. (2.563) were not true. Indeed, the first equation in (2.563) holds only up to a distance

$$x - R = \sqrt{2B}\, t, \tag{2.567}$$

while beyond it we have $U = 0$. This means that, for times close to $t = 0$, Eq. (2.563) is verified only in a region close to the nucleus, whereas, with the passing of time (taking into account the approximations used), it is valid within a radius $\sqrt{2B}t = vt$, where v is precisely the velocity of the emitted particles. Notice that, even if the finite life of the quasi-stationary state induces a small uncertainty in the emission velocity, the wavefront appears sharp due to the approximations we have made in the computation. We shall shortly show how, by reducing the approximation further, it is possible to highlight such an uncertainty of v, and to determine the velocity curve independently of the general statistical principles of quantum mechanics.

The formulae that we have just derived suggest some interesting observations:

I. Having verified that the first of Eqs. (2.563) holds at short distances almost since the beginning, we can try to derive directly a solution of the required form, without worrying about what happens at larger distances. This is Gamov's method. In other words, let us assume that the time dependence is, at any distance, given by

$$\mathrm{e}^{2\pi i \nu t}\, \mathrm{e}^{-t/2T} = \mathrm{e}^{2\pi i t(\nu - 1/4\pi i T)}, \tag{2.568}$$

so that ψ represents formally a stationary state with a complex eigenvalue. Now, from Eqs. (2.532) and (2.534) and taking account of the time dependence, with a suitable approximate normalization, the general

[56]@ The author refers here to G.Gamov, *Z. Phys.* **41** (1928) 204. He had already worked on Gamov's theory also in connection with his Thesis work [E.Majorana, *The Quantum Theory of Radioactive Nuclei* (in Italian), E. Fermi supervisor, unpublished].

solution for the stationary states becomes to be

$$U = \begin{cases} e^{iEt}\, 2\left(x\,e^{-x} + \dfrac{\alpha}{8}\dfrac{e^{x}}{x}\right), \quad \text{for } x < R, \\[2ex] e^{iEt}\, 2\left[\left(\dfrac{R}{e^R} + \dfrac{\alpha}{8}\dfrac{e^R}{R}\right)\cos\sqrt{2E}(x-R) \right. \\ \qquad \left. + \dfrac{1}{\sqrt{2E}}\left(-\dfrac{R}{e^R} + \dfrac{\alpha}{8}\dfrac{e^R}{R}\right)\sin\sqrt{2E}(x-R)\right], \\ \text{for } x > R. \end{cases} \tag{2.569}$$

For $x > R$, we can also write

$$U = e^{iEt}\left\{\left[\frac{R}{e^R} + \frac{\alpha}{8}\frac{e^R}{R} - \frac{i}{\sqrt{2E}}\left(-\frac{R}{e^R} + \frac{\alpha}{8}\frac{e^R}{R}\right)\right] e^{i\sqrt{2E}(x-R)}\right.$$
$$\left. + \left[\frac{R}{e^R} + \frac{\alpha}{8}\frac{e^R}{R} + \frac{i}{\sqrt{2E}}\left(-\frac{R}{e^R} + \frac{\alpha}{8}\frac{e^R}{R}\right)\right] e^{-i\sqrt{2E}(x-R)}\right\}. \tag{2.570}$$

The requirement for having no ingoing wave is

$$\frac{R}{e^R} + \frac{\alpha}{8}\frac{e^R}{R} - \frac{i}{\sqrt{2E}}\left(-\frac{R}{e^R} + \frac{\alpha}{8}\frac{e^R}{R}\right) = 0, \tag{2.571}$$

whence

$$\alpha = -\frac{\sqrt{2E}+i}{\sqrt{2E}-i}\, 8R^2\, e^{-2R}, \tag{2.572}$$

and, setting in first approximation $\sqrt{2E} = \sqrt{2A-1}$,

$$\alpha = -\frac{A-1}{A}\, 8R^2\, e^{-2R} - i\,\frac{\sqrt{2A-1}}{A}\, 8R^2\, e^{-2R}, \tag{2.573}$$

so that

$$E = A - \frac{1}{2} - \frac{1}{2}\alpha = B + i\,\frac{\sqrt{2A-1}}{A}\, 8R^2\, e^{-2R}, \tag{2.574}$$

or, to the same order of approximation,

$$E = B + i\frac{\sqrt{2B}}{A}\, 4R^2\, e^{-2R}. \tag{2.575}$$

It follows that, for $x < R$,

$$U = e^{iBt}\, 2\left[x\,e^{-x} - \left(\frac{A-1}{A}\, R^2\, e^{-2R} + i\frac{\sqrt{2B}}{A}\, R^2\, e^{-2R}\right)\frac{e^x}{x}\right]$$
$$\times\, e^{-\frac{\sqrt{2B}}{A} 4R^2 e^{-2R}}, \tag{2.576}$$

as already found. Then, by this method one can also determine the "mean-life" T:

$$T = \frac{A}{\sqrt{2B}} \frac{e^{2R}}{8R^2}. \tag{2.577}$$

II. The "mean-life" T is proportional to $A/\sqrt{2B}$, where B is the mean energy of the electron or, which is the same, the mean kinetic energy of the electron when it crosses the spherical surface of radius R. Since, using a very rough approximation, $B \simeq A - 1/2$, the mean-life is proportional to $(B+1/2)/\sqrt{2B}$. If we take $A = 1/2$, that is to say that A is equal exactly to the ionization potential, then $B = 0$, and the mean-life naturally becomes infinite. What may be surprising is that the ionization probability per unit time increases with increasing B until it reaches a maximum value, and then starts to decrease and eventually fall off to zero as $B \to \infty$. The maximum is achieved when $B = 1/2$ and thus $A = 1$, that is to say, at twice the value of the ionization potential. The minimum mean-life is then

$$T = \frac{e^{2R}}{8R^2}. \tag{2.578}$$

The explanation of this paradox is the following. Whenever there is a surface that sharply separates two regions with different potentials, it behaves as a reflecting surface not only for the particles coming from the region with lower potential energy, but also for the ones coming from the opposite side, provided that the absolute value of the kinetic energy (positive or negative) is small with respect to the abrupt potential energy jump.

III. We have seen that the energy of the electron has been determined with inaccurately. We can talk in terms of probability that it lies between E and $E + \mathrm{d}E$ or, analogously, of probability that the speed of the emitted electron falls between v and $v + \mathrm{d}v$. From Eq. (2.537), we have

$$v = \sqrt{2E} \simeq \sqrt{2B} + 2\pi\,\gamma, \tag{2.579}$$

$$\mathrm{d}v \simeq 2\pi\,\mathrm{d}\gamma. \tag{2.580}$$

The probability that γ has a value between γ and $\gamma + \mathrm{d}\gamma$ is $c^2\mathrm{d}\gamma$; from Eq. (2.550), the probability of v lying in the interval $v, v + \mathrm{d}v$ is

$$\frac{(Ae^{2R}/4\pi R^2)\mathrm{d}v}{1 + (\pi^2 A^2 e^{4R}/4R^4)\gamma^2} \simeq \frac{(Ae^{2R}/4\pi R^2)\mathrm{d}v}{1 + (A^2 e^{4R}/16R^4)\,(v - \sqrt{2B})^2}. \tag{2.581}$$

In first approximation, the same holds for the energy. The probability for unit energy is

$$\frac{(Ae^{2R}/4\pi\sqrt{2B}R^2}{1+(A^2e^{4R}/32BR^4)\,(E-B)^2} = \frac{K_1/\pi}{1+K_2^2\,(E-B)^2} = \frac{1/\pi K}{1+(E-B)^2/K^2}. \tag{2.582}$$

As we shall show later, when considering radioactive phenomena, if we deal with quasi-stationary states, we always find the same probability, independently of the form of the potential (provided it has spherical symmetry). The parameter K which defines the probability amplitude is related to the mean-life T by the relation

$$K = 1/2T = \tau, \tag{2.583}$$

or, going back to the usual units,

$$K = \hbar/2T. \tag{2.584}$$

Note that such expressions agree with the general uncertainty relations.

IV. Let us push further the approximation in the case $x > R$, while keeping the definition (2.538) for γ. We now have

$$E = B + 2\pi\sqrt{2B}\,\gamma, \tag{2.585}$$

and, instead of Eq. (2.537), in second approximation we get

$$\sqrt{2E} = \sqrt{2B} + 2\pi\gamma - \frac{2\pi^2}{\sqrt{2B}}\gamma^2. \tag{2.586}$$

Equation (2.557) now becomes

$$\begin{aligned} U = {}& e^{iEt}\,\frac{e^R}{R}\left[e^{i\sqrt{2B}(x-R)}\int_{-\infty}^{\infty}\frac{M+Ni}{2}\right. \\ &\times\exp\left\{2\pi i[\sqrt{2B}t+(x-R)]\gamma-\frac{2\pi^2 i\gamma^2}{\sqrt{2B}}(x-R)\right\}\,d\gamma \\ &+e^{-i\sqrt{2B}(x-R)}\int_{-\infty}^{\infty}\frac{M-Ni}{2} \\ &\left.\times\exp\left\{2\pi i[\sqrt{2B}t-(x-R)]\gamma+\frac{2\pi^2 i\gamma^2}{\sqrt{2B}}(x-R)\right\}\,d\gamma\right]. \end{aligned} \tag{2.587}$$

34. SCATTERING OF AN α PARTICLE BY A RADIOACTIVE NUCLEUS

Let us consider the emission of an α particle by a radioactive nucleus and assume that such a particle is described by a quasi-stationary wave. As Gamov has shown, after some time this wave scatters at infinity. In other words, the particle spends some time near the nucleus but eventually ends up far from it. We now begin to study the features of such a quasi-stationary wave, and then address the *inverse* of the problem studied by Gamov.[57] Namely, we want to determine the probability that an α particle, colliding with a nucleus that has just undergone an α radioactive transmutation, will be captured by the nucleus so as to reconstruct a nucleus of the element preceding the original one in the radioactive genealogy. This issue has somewhat been addressed by Gudar, although not deeply enough. It is directly related to our hypothesis according to which, under conditions rather different from the ones we are usually concerned with, a process can take place that reconstitutes the radioactive element.

Following Gamov, let us suppose that spherical symmetry is realized, so that the azimuthal quantum of the particle near the nucleus is zero. For simplicity, we neglect for the moment the overall motion of the other nuclear components. The exact formulae will have to take account of that motion, and thus the formulae that we shall now derive will have to be modified; but this does not involve any major difficulty. For the spherically symmetric stationary states, setting, as usual, $\psi = \chi/x$, we shall have

$$\frac{\mathrm{d}^2\chi}{\mathrm{d}x^2} + \frac{2m}{\hbar^2}\,(E - U)\,\chi = 0. \tag{2.588}$$

Beyond a given distance R, which we can assume to be of the order of the atomic dimensions, the potential U practically vanishes. The functions χ will then be symmetric for $E > 0$. For definiteness, we require U to be exactly zero for $x > R$, but it will be clear that no substantial error is really introduced in this way in our calculations. For the time being, let us consider the functions χ to depend only on position, and —as it is allowed— to be real. Furthermore, we use the normalization condition

$$\int_0^R \chi^2\,\mathrm{d}x = 1. \tag{2.589}$$

[57] @ Once more, the author is referring to G.Gamov, *Z. Phys.* **41** (1928) 204: See the previous footnote.

Let us now imagine that it exists a quasi-stationary state such that it is possible to construct a function u_0 which vanishes for $x > R$, satisfies the constraint

$$\int_0^R |u_0|^2 \, \mathrm{d}x \;=\; 1, \tag{2.590}$$

and approximately obeys[58] the differential equation (2.588) at the points where its value is large. This function u_0 will be suited to represent the α particle at the initial time. It is possible to expand it in terms of the functions χ that are obtained by varying E within a limited range. Let us then set

$$E \;=\; E_0 + W. \tag{2.591}$$

The existence of such a quasi-stationary state is revealed by the fact that for $x < R$ the functions χ, normalized according to Eq. (2.589), and their derivatives are small for small W.

In first approximation, we can set, for $x < R$,

$$\begin{aligned} \chi_W \;&=\; \chi_0 + W\, y(x), \\ \chi_W' \;&=\; \chi_0' + W\, y'(x), \end{aligned} \tag{2.592}$$

and these are valid (as long as U has a reasonable behavior) with great accuracy and for all values of W in the range of interest. In particular, for $x = R$:

$$\begin{aligned} \chi_W(R) \;&=\; \chi_0(R) + W\, y(R), \\ \chi_W'(R) \;&=\; \chi_0'(R) + W\, y'(R). \end{aligned} \tag{2.593}$$

Bearing in mind that Eq. (2.588) simply reduces for $x > R$ to

$$\frac{\mathrm{d}^2\chi_W}{\mathrm{d}x^2} + \frac{2m}{\hbar^2}\,(E_0 + W)\,\chi_W \;=\; 0, \tag{2.594}$$

for $x > R$ we get

$$\begin{aligned} \chi_W \;=\;& (a + bW)\,\cos\frac{1}{\hbar}\sqrt{2m(E_0+W)}(x-R) \\ & + (a_1 + b_1 W)\,\sin\frac{1}{\hbar}\sqrt{2m(E_0+W)}(x-R), \end{aligned} \tag{2.595}$$

[58] For an approximately determined value of t, while being almost real.

having set

$$\begin{aligned} a &= \chi_0(R), \quad b = y(R), \\ a_1 &= \frac{\hbar\,\chi_0'(R)}{\sqrt{2m(E_0+W)}}, \quad b_1 = \frac{\hbar\,y'(R)}{\sqrt{2m(E_0+W)}}. \end{aligned} \tag{2.596}$$

Note that a_1 and b_1 are not strictly constant but, to the order of approximation for which our problem is determined, we can consider them as constant and replace them with

$$a_1 = \frac{\hbar\,\chi_0'(R)}{\sqrt{2mE_0}}, \quad b_1 = \frac{\hbar\,y'(R)}{\sqrt{2mE_0}}. \tag{2.597}$$

Moreover, since E_0 is not completely determined, we shall fix it in order to simplify Eq. (2.595); with this aim, we can shift R by a fraction of wavelength $h/\sqrt{2mE_0}$. It will then be found that Eq. (2.595) can always be replaced with the simpler one

$$\begin{aligned} \chi_W &= \alpha\,\cos\sqrt{2m(E_0+W)}\,(x-R)/\,\hbar \\ &\quad + \beta W\,\sin\sqrt{2m(E_0+W)}\,(x-R)/\,\hbar. \end{aligned} \tag{2.598}$$

We set

$$\sqrt{2m(E_0+W)}\,/\,\hbar = \sqrt{2mE_0}\,/\,\hbar + 2\pi\,\gamma = C + 2\pi\,\gamma, \tag{2.599}$$

and, in first approximation, the following will hold:

$$2\pi\,\gamma \simeq \frac{W}{\hbar\sqrt{2E_0/m}} = \frac{W}{\hbar v}, \tag{2.600}$$

v being the (average) speed of the emitted α particles. On substituting into Eq. (2.598), we approximately find

$$\begin{aligned} \chi_W &= \alpha\,\cos(C+2\pi\gamma)(x-R) \\ &\quad + \beta'\,\gamma\,\sin(C+2\pi\gamma)(x-R), \end{aligned} \tag{2.601}$$

with

$$\beta' = \beta\,2\pi\hbar\,\sqrt{2E_0/m}. \tag{2.602}$$

For the moment, the χ_W functions are normalized as follows:

$$\int_0^R \chi_W^2\,\mathrm{d}x = 1.$$

We denote by η_W the same eigenfunctions normalized with respect to $\mathrm{d}\gamma$. For $x > R$, we then get

$$\begin{aligned}\eta_W &= \frac{2}{\sqrt{\alpha^2+\beta'^2\gamma^2}}\,[\alpha\cos(C+2\pi\gamma)(x-R) \\ &+ \beta'\gamma\sin(C+2\pi\gamma)(x-R)] = \frac{2}{\sqrt{\alpha^2+\beta'^2\gamma^2}}\,\chi_W.\end{aligned} \tag{2.603}$$

We expand u_0, which represents the α particle at the initial time, as a series in η_W, and get

$$u_0 = \int_{-\infty}^{\infty} K_\gamma\,\eta_W\,\mathrm{d}\gamma. \tag{2.604}$$

Now, since $u_0 = \chi_W$ for $x \leq R$ and therefore

$$K_\gamma = \int_0^\infty \eta_W\,u_0\,\mathrm{d}x = \frac{2}{\sqrt{\alpha^2+\beta'^2\gamma^2}}\int_0^R \chi_W^2\,\mathrm{d}x = \frac{2}{\sqrt{\alpha^2+\beta'^2\gamma^2}}, \tag{2.605}$$

on substituting into Eq. (2.604), we obtain

$$u_0 = \int_{-\infty}^{\infty} \frac{4\,\chi_W}{\alpha^2+\beta'^2\gamma^2}\,\mathrm{d}\gamma. \tag{2.606}$$

For small values of x, the different functions χ_W actually coincide and are also equal to u_0; it must then be true that

$$1 = \int_{-\infty}^{\infty} \frac{4}{\alpha^2+\beta'^2\gamma^2}\,\mathrm{d}\gamma = -\frac{4\pi}{\alpha\beta'}, \tag{2.607}$$

and, consequently,

$$\beta' = -\frac{4\pi}{\alpha} \tag{2.608}$$

must necessarily hold. Because of Eq. (2.600), if we introduce the time dependence, we approximately get

$$u = \mathrm{e}^{iEt/\hbar}\int_{-\infty}^{\infty} \frac{4\chi_W\,\exp\left\{2\pi i\sqrt{2E_0/m}\;\gamma t\right\}}{\alpha^2+16\pi^2\gamma^2/\alpha^2}\,\mathrm{d}\gamma. \tag{2.609}$$

For small values of x the χ_W's can be replaced with u_0, and we have

$$u = u_0\,\mathrm{e}^{iE_0t/\hbar}\exp\left\{-\alpha^2\sqrt{2E_0/m}\,t/2\right\}. \tag{2.610}$$

This can be written as

$$u = u_0\,\mathrm{e}^{iE_0t/\hbar}\,\mathrm{e}^{-t/2T}, \tag{2.611}$$

quantity T denoting the time-constant ("mean-life")

$$T = \frac{1}{\alpha^2\sqrt{2E_0/m}} = \frac{1}{\alpha^2 v}. \qquad (2.612)$$

In this way, and using also Eq. (2.608), both α and β' can be expressed in terms of T:

$$\alpha = \frac{\pm 1}{\sqrt{vT}} = \frac{\pm 1}{\sqrt[4]{2(E/m)T^2}}, \qquad (2.613)$$

$$\beta' = \mp 4\pi\sqrt{vT} = \mp 4\pi\sqrt[4]{2(E/m)T^2}. \qquad (2.614)$$

It will be clear that only one stationary state corresponds to a hyperbolic-like orbit in the classical theory. The revolution period or, more precisely, the time interval between two intersections of the orbit with the spherical surface of radius r, is given by

$$P_W = \frac{4}{(\alpha^2 + \beta'^2\gamma^2)v}, \qquad (2.615)$$

and the maximum value is reached for $W = 0$:

$$P_W = \frac{4}{\alpha^2 v} = 4T. \qquad (2.616)$$

As a purely classical picture suggests, the probabilities for the realization of single stationary states are proportional to the revolution periods (see Eq. (2.605)), and T itself can be derived from classical arguments. Indeed, if a particle is on an orbit W and inside the sphere of radius R, on average it will stay in this orbit for a time $T_W = (1/2)P_W = (2/v)/(\alpha^2 + \beta'^2\gamma^2)$, and the mean value of T_W will be

$$\overline{T_W} = \int_0^\infty T_W^2 \,\mathrm{d}\gamma \Big/ \int_0^\infty T_W \,\mathrm{d}\gamma = \frac{1}{\alpha^2 v} = T. \qquad (2.617)$$

However, we must caution that, by pushing the analogy even further to determine the expression for the survival probability, we would eventually get a wrong result.

The eigenfunction u takes the form in Eq. (2.610) only for small values of x. Neglecting what happens for values of x that are not too small, but still lower than R, and considering, moreover, even the case $x > R$, from Eqs. (2.602) and (2.606) we have

$$\begin{aligned} u = \mathrm{e}^{iE_0t/\hbar} \Bigg[& \int_0^\infty \frac{4\alpha\cos(C + 2\pi\gamma)(x - R)}{\alpha^2 + \beta'^2\gamma^2}\,\mathrm{e}^{2\pi i v\gamma t}\,\mathrm{d}\gamma \\ & - \int_0^\infty \frac{4\beta'\gamma\sin(C + 2\pi\gamma)(x - R)}{\alpha^2 + \beta'^2\gamma^2}\,\mathrm{e}^{2\pi i v\gamma t}\,\mathrm{d}\gamma \Bigg], \end{aligned} \qquad (2.618)$$

where α and β' depend on T according to Eqs. (2.613), (2.614). Equation (2.618) can be written as

$$\begin{aligned} u &= \mathrm{e}^{iE_0t/\hbar} \left[\mathrm{e}^{iC(x-R)} \int_0^\infty \frac{(2\alpha - 2i\beta'\gamma)}{\alpha^2 + \beta'^2\gamma^2} \, \mathrm{e}^{2\pi i(vt+x-R)\gamma} \, \mathrm{d}\gamma \right. \\ & \left. + \, \mathrm{e}^{-iC(x-R)} \int_0^\infty \frac{(2\alpha + 2i\beta'\gamma)}{\alpha^2 + \beta'^2\gamma^2} \, \mathrm{e}^{2\pi i[vt-(x-R)]\gamma} \, \mathrm{d}\gamma \right]. \end{aligned} \tag{2.619}$$

Since α and β' have opposite signs and, for $t > 0$ and $x > R$, one has $vt + x - R > 0$, the first integral is zero, while the second one equals

$$\int_0^\infty \frac{(2\alpha + 2i\beta'\gamma)}{\alpha^2 + \beta'^2\gamma^2} \, \mathrm{e}^{2\pi i[vt-(x-R)]\gamma} \, \mathrm{d}\gamma = 2 \int_0^\infty \frac{\mathrm{e}^{2\pi i[vt-(x-R)]\gamma}}{\alpha - i\beta'\gamma} \, \mathrm{d}\gamma$$

$$= \left\{ \begin{array}{ll} -\dfrac{4\pi}{\beta'} \, \mathrm{e}^{2\pi(\alpha/\beta')[vt-(x-R)]} = -\dfrac{4\pi}{\beta'} \, \mathrm{e}^{-(\alpha^2/2)[vt-(x-R)]}, \\ \\ 0, \end{array} \right. \tag{2.620}$$

for $vt - (x - R) > 0$ and $vt - (x - R) < 0$, respectively. On substituting into Eq. (2.621) and recalling that, from Eq. (2.599), $C = mv/\hbar$, we finally find

$$u = \left\{ \begin{array}{l} \alpha \, \mathrm{e}^{iE_0t/\hbar} \, \mathrm{e}^{-imv(x-R)/\hbar} \, \mathrm{e}^{-t/2T} \, \mathrm{e}^{(x-R)/(2vT)}, \\ \\ 0, \end{array} \right. \tag{2.621}$$

for $vt - (x - R) > 0$ and $vt - (x - R) < 0$, respectively. Let us now assume that the nucleus has lost the α particle; this means that, initially, it is $u_0 = 0$ near the nucleus. We now evaluate the probability that such a nucleus will re-absorb an α particle when bombarded with a parallel beam of particles. To characterize the beam we'll have to give the intensity per unit area, the energy per particle, and the duration of the bombardment. The only particles with a high absorption probability are those having energy close to E_0, with an uncertainty of the order h/T. On the other hand, in order to make clear the interpretation of the results, the duration τ of the bombardment must be small compared to T. Then it follows that, from the uncertainty relations, the energy of the incident particles will be determined with an error greater than h/T. Thus, instead of fixing the intensity per unit area, it is more appropriate to give the intensity per unit area and unit energy close to E_0; so, let N be the total number of particles incident on the nucleus during the entire duration of the bombardment, per unit area and unit energy.

Suppose that, initially, the incident plane wave is confined between two parallel planes at distance d_1 and $d_2 = d_1 + \ell$ from the nucleus, respectively. Since we have assumed that the initial wave is a plane wave, it will be

$$u_0 = u_0(\xi), \tag{2.622}$$

ξ being the abscissa (distance from the nucleus) of a generic plane that is parallel to the other two. Then, for $\xi < d_1$ or $\xi > d_2$, it is $u_0 = 0$. Furthermore, we'll suppose $d_1 > R$ and, without introducing any further constraint,

$$\ell = \frac{h\rho}{m\sqrt{2E_0/m}} = \frac{h\rho}{mv} = \rho\,\lambda, \tag{2.623}$$

with ρ an integer number and λ the wavelength of the emitted α particle. We can now expand ψ_0 between d_1 and d_2 in a Fourier series and thus as a sum of terms of the kind

$$k_\sigma\,\mathrm{e}^{\sigma 2\pi i(\xi - d_1)/\ell}, \tag{2.624}$$

with integer σ. The terms with negative σ roughly represent outgoing particles, and thus we can assume them to be zero. Let us concentrate on the term

$$k_\rho\,\mathrm{e}^{\rho 2\pi i(\xi - d_1)/\ell} = k_\rho\,\mathrm{e}^{imv(\xi - d_1)/\hbar} \tag{2.625}$$

and let us set[59]

$$u_0 = \psi_0 + k_\rho\,\mathrm{e}^{imv(\xi - d_1)/\hbar}. \tag{2.626}$$

The eigenfunctions of a free particle moving perpendicularly to the incoming wave, normalized with respect to $\mathrm{d}E$, are[60]

$$\psi_\sigma = \frac{1}{\sqrt{2hE/m}}\,\mathrm{e}^{i\sqrt{2mE}(\xi - d_1)/\hbar}. \tag{2.627}$$

Note that the square root at the exponent must be considered once with the positive sign and once with the negative sign, and E runs twice between 0 and ∞. However, only the eigenfunctions with the positive square root sign are of interest to us, since they represent the particles moving in the direction of decreasing ξ. We can set

$$\psi_0 = \int_0^\infty c_E\,\psi_\rho\,\mathrm{d}E, \tag{2.628}$$

[59]@ Note that the author split the wavefunction of the incident particles into a term related to the principal energy E_0 (the second term in Eq. (2.626)) plus another term which will be expanded according to Eq. (2.628).

[60]@ In the original manuscript, these eigenfunctions are denoted by ψ_ρ, but here, for clarity, they will be denoted by ψ_σ

wherein

$$c_E = \int_{d_1}^{d_2} \psi_0 \, \psi_\rho^* \, \mathrm{d}\xi. \tag{2.629}$$

In particular, we put

$$c_{E_0} = \int_{d_1}^{d_2} \psi_0 \, \frac{1}{\sqrt{hv}} \, \mathrm{e}^{-imv(\xi-d_1)/\hbar} \, \mathrm{d}\xi = \frac{k_\rho \ell}{\sqrt{hv}}. \tag{2.630}$$

Since, evidently,

$$N = c_{E_0}^2, \tag{2.631}$$

one finds

$$N = \frac{k_\rho^2 \ell^2}{hv}. \tag{2.632}$$

Let us now expand u_0 in terms of the eigenfunctions associated with the central field produced by the remaining nuclear constituents. Since only the spherically symmetric eigenfunctions having eigenvalues very close to E_0 are significantly different from zero near the nucleus, we shall concentrate only on these. For $x > R$, the expression of these eigenfunctions is given in Eqs. (2.603), (2.613), (2.614). Actually, the η_W given by Eq. (2.603) are the eigenfunctions relative to the problem reduced to one dimension. In order to have the spatial eigenfunctions, normalized with respect to γ, we must consider

$$g_W = \frac{\eta_W}{\sqrt{4\pi} x}. \tag{2.633}$$

In this way we will set

$$\psi_0 = \int_0^\infty p_\gamma \, g_W \, \mathrm{d}\gamma + \ldots, \tag{2.634}$$

wherein

$$p_\gamma = \iiint \mathrm{d}S \, g_W \, \psi_0 = \int_{d_1}^{d_2} 2\pi \, x \, g_W \, \mathrm{d}x \int_{d_1}^{x} \psi_0 \, \mathrm{d}\xi. \tag{2.635}$$

We can set

$$g_W = \frac{1}{\sqrt{4\pi} x} \left[A_\gamma \, \mathrm{e}^{i(C+2\pi\gamma)(x-d_1)} + B_\gamma \, \mathrm{e}^{-i(C+2\pi\gamma)(x-d_1)} \right], \tag{2.636}$$

and, from Eq. (2.603),

$$\begin{aligned} A_\gamma &= \frac{\alpha - i\beta'\gamma}{\sqrt{\alpha^2 + \beta'^2\gamma^2}} \, \mathrm{e}^{i(C+2\pi\gamma)(d_1-R)}, \\ B_\gamma &= \frac{\alpha + i\beta'\gamma}{\sqrt{\alpha^2 + \beta'^2\gamma^2}} \, \mathrm{e}^{-i(C+2\pi\gamma)(d_1-R)}. \end{aligned} \tag{2.637}$$

We can now assume that d_1, and thus d_2, is arbitrarily large; but $\ell = d_2 - d_1$ has to be small because the duration of the bombardment, which is of the order ℓ/v, must be negligible with respect to T. Since $2\pi\gamma$ is of the same order as α^2, that is to say, of the same order as $1/vT$ (see Eq. (2.612)), $2\pi\gamma\ell$ is absolutely negligible. For $d_1 < x < d_2$ it is then possible to rewrite Eq. (2.636) as

$$g^W = \frac{1}{\sqrt{4\pi}x}\left[A_\gamma\,\mathrm{e}^{imv(x-d_1)/\hbar} + B_\gamma\,\mathrm{e}^{-imv(x-d_1)/\hbar}\right], \tag{2.638}$$

given Eqs. (2.637).

Let us now substitute this into Eq. (2.635), taking into account Eqs. (2.626) and (2.632). We'll simply have

$$\begin{aligned} p_\gamma &= \frac{2\pi B_\gamma}{\sqrt{4\pi}}\int_{d_1}^{d_2}\mathrm{e}^{-imv(x-d_1)/\hbar}\,\mathrm{d}x\int_{d_1}^{x}\mathrm{e}^{imv(\xi-d_1)/\hbar}\,\mathrm{d}\xi \\ &= \frac{hB_\gamma k_\rho\ell}{i\sqrt{4\pi}\,m\,v} = \frac{B_\gamma h^{3/2}\sqrt{N}}{i\sqrt{4\pi}\,m\,\sqrt{v}} = q\,B_\gamma, \end{aligned} \tag{2.639}$$

with

$$q = \frac{h^{3/2}N^{1/2}}{i\,m\,v^{1/2}\sqrt{4\pi}}. \tag{2.640}$$

Substituting into Eq. (2.634), one gets

$$\psi_0 = q\int_0^\infty B_\gamma g_W\,\mathrm{d}\gamma + \ldots \tag{2.641}$$

and, at an arbitrary time,

$$\psi = \mathrm{e}^{iE_0t/\hbar}\,q\int_0^\infty B_\gamma\,g_W\,\mathrm{e}^{2\pi iv\gamma t}\,\mathrm{d}\gamma + \ldots, \tag{2.642}$$

or, taking into account Eqs. (2.633) and (2.603),

$$\psi = \mathrm{e}^{iE_0t/\hbar}\,\frac{q}{\sqrt{4\pi}x}\int_0^\infty \frac{2B_\gamma}{\sqrt{\alpha^2+\beta'^2\gamma^2}}\,\chi_W\,\mathrm{e}^{2\pi iv\gamma t}\,\mathrm{d}\gamma + \ldots. \tag{2.643}$$

We now want to investigate the behavior of ψ near the nucleus. There, assuming that other quasi-stationary state different from the one we are considering do not exist, the terms we have not written down in the expansion of ψ can contribute significantly only during a short time interval after the scattering of the wave. If this is the case, ψ will have spherical symmetry near the nucleus. We set

$$\psi = \frac{u}{\sqrt{4\pi}x}, \tag{2.644}$$

so that the number of particles that will eventually be captured is

$$\int |u^2| \, \mathrm{d}x \tag{2.645}$$

(the integration range should extend up to a reasonable distance, for example up to R). Substituting into Eq. (2.643), and noting that for small values of x we approximately have $\chi_W = \chi_0$, one obtains

$$u = q \, \chi_0 \, \mathrm{e}^{iE_0 t/\hbar} \int_0^\infty \frac{2}{\alpha - i\beta'\gamma} \, \mathrm{e}^{2\pi i[vt-(d_1-R)]\gamma} \, \mathrm{d}\gamma. \tag{2.646}$$

Since, as we already noted, $\alpha\beta' < 0$, and setting $d = d_1 - R$, from Eqs. (2.613), we find

$$u = \begin{cases} q \, \alpha \, \chi_0 \, \mathrm{e}^{iE_0 t/\hbar} \, \mathrm{e}^{-\dfrac{t - d/v}{2T}} = q \, \alpha \, \mathrm{e}^{-iCd} \, \mathrm{e}^{-\dfrac{t - d/v}{2T}}, & \text{for } t > \dfrac{d}{v}, \\ 0, \quad \text{for } t < \dfrac{d}{v}. \end{cases} \tag{2.647}$$

The meaning of these formulae is very clear: The α-particle beam, which by assumption does not last for a long time, reaches the nucleus at the time $t = d/v$, and there is a probability $|q\alpha|^2$ that a particle is captured (obviously, $q^2\alpha^2 \ll 1$). The effect of the beam then ceases and, if a particle has been absorbed, it is re-emitted on the time scale predicted by the laws of radioactive phenomena. If we set $n = |q\alpha|^2$, then from Eqs. (2.612) and (2.640) we get

$$n = \frac{2\pi^2\hbar^3}{m^2 v^2 T} \, N, \tag{2.648}$$

which tells us that the absorption probabilities are completely independent of any hypothesis on the form of the potential near the nucleus, and that they only depend on the time-constant T.[61]

Equation (2.648) has been derived using only mechanical arguments

---

[61]@ The original manuscript then continues with two large paragraphs which have however been crossed out by the author. The first one reads as follows:
"Since only the particles with energy near E_0 are absorbed, we can think, with some imagination, that every energy level $E_0 + W$ is associated with a different absorption coefficient ℓ_W, and that such ℓ_W is proportional to the probability that a particle in the quasi-stationary state has energy $E_0 + W$. From (2.600), (2.608), (2.612), and (2.605), we then have

$$\ell_W = \frac{D}{1 + 4T^2 W^2/\hbar^2}. \tag{2.649}$$

but, as a matter of fact, we can get the same result using thermodynamics. Let us consider one of our radioactive nuclei in a bath of α particles in thermal motion. To the degree of approximation we have treated the problem so far, we can consider the nucleus to be at rest. Due to the assumed spherical symmetry of the system, a particle in contact with the nucleus is in a quantum state with a simple statistical weight. Such a state, of energy E_0, is not strictly stationary, but has a finite half-life; this should be considered, as in all similar cases, as a second-order effect. Assuming that the density and the temperature of the gas of α particles is such that there exist D particles per unit volume and unit energy near E_0, then, in an energy interval $\mathrm{d}E$, we will find

$$D\,\mathrm{d}E \tag{2.654}$$

particles per unit volume. Let us denote by p the momentum of the particles, so that we have

$$p = \sqrt{2mE_0}, \tag{2.655}$$

$$\mathrm{d}p = \sqrt{m/2E_0}\,\mathrm{d}E. \tag{2.656}$$

E_0 appears instead of E in the previous equations because we are considering particles with energy close to E_0. The $D\mathrm{d}E$ particles fill a unitary volume in ordinary space, and in momentum space they fill the volume between two spheres of radii p and $p + \mathrm{d}p$, respectively. Thus, in phase space they occupy a volume

$$4\pi\, p^2\, \mathrm{d}p = 4\pi\, m^2 \sqrt{2E_0/m}\,\mathrm{d}E = 4\pi\, m^2\, v\, \mathrm{d}E. \tag{2.657}$$

Since the number of incident particles per unit area and unit energy with energy between $(E_0 + W)$ and $(E_0 + W) + \mathrm{d}W$ is $N\mathrm{d}W$, we must have

$$n = N \int_{-\infty}^{\infty} \ell_W\, \mathrm{d}W = N\, D\, \frac{\pi\hbar}{2T}, \tag{2.650}$$

from which, comparing with (2.648),

$$D = \frac{1}{\pi}\, \frac{h^2}{m^2 v^2} = \frac{\lambda^2}{\pi}. \tag{2.651}$$

This is a very simple expression for the absorption cross section of particles with energy E_0, i.e., the particles with the greatest absorption coefficient. If we set

$$N' = N\, \frac{\pi\hbar}{2T}, \tag{2.652}$$

then Eq. (2.648) becomes

$$n = \frac{\lambda^2}{\pi}\, N', \tag{2.653}$$

which means that the absorption of N' particles of energy E_0 is equivalent to the absorption of N particles per unit energy." The second paragraph is not reproduced here since it appears to be incomplete.

This volume contains

$$\frac{m^2 v}{2\pi^2 \hbar^3}\, dE \tag{2.658}$$

quantum states. Therefore, on average, we have

$$D\, \frac{2\pi^2 \hbar^3}{m^2 v} \tag{2.659}$$

particles in every quantum state with energy close to E_0. This is also the mean number of particles inside the nucleus, provided that the expression (2.659) is much smaller than 1, so that we can neglect the interactions between the particles. Since the time-constant ("mean-life") of the particles in the nucleus is T, then

$$n \;=\; \frac{2\pi^2 \hbar^3 D}{m^2 v T} \tag{2.660}$$

particles will be emitted per unit time and, in order to maintain the equilibrium, the same number of particles will be absorbed. Concerning the collision probability with a nucleus, and then the absorption probability, D particles per unit volume and energy are equivalent to a parallel beam of $N = Dv$ particles per unit area, unit energy and unit time. On substituting, we then find

$$n \;=\; \frac{2\pi^2 \hbar^3}{m^2 v^2 T}\, N, \tag{2.661}$$

which coincides with Eq. (2.648).

35. RETARDED POTENTIAL[62]

Let us consider a periodic solution of Eq. (1.21) and let

$$H \;=\; u \sin \omega t, \tag{2.662}$$

with u a time-independent function. The equation

$$\nabla^2 u + \frac{\omega^2}{c^2}\, u \;=\; 0 \tag{2.663}$$

will hold; and, setting $k^2 = \omega^2/c^2$, we find

$$\nabla^2 u + k^2 u \;=\; 0. \tag{2.664}$$

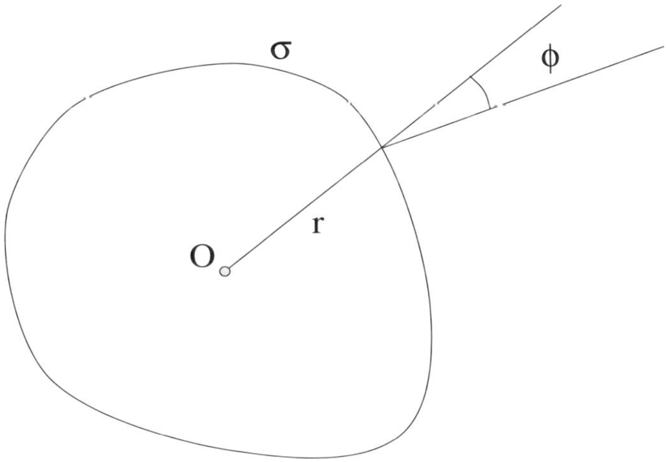

Fig. 2.1. Definition of some quantities used in the text.

Equation (1.33) then becomes

$$u \sin \omega t = \frac{1}{4\pi} \int \left[\sin \omega(t - r/c) \left(u \cos \phi + r \frac{\partial u}{\partial n} \right) + \frac{\omega r}{c} u \cos \phi \cos \omega(t - r/c) \right] \frac{d\sigma}{r^2}, \qquad (2.665)$$

and thus

$$u = \frac{1}{4\pi} \int \left[\cos \frac{\omega r}{c} \left(u \cos \phi + r \frac{\partial u}{\partial n} \right) + \frac{\omega r}{c} u \cos \phi \sin \frac{\omega r}{c} \right] \frac{d\sigma}{r^2}. \qquad (2.666)$$

If the distances r are large with respect to the wavelength, we will simply have

$$u = \frac{1}{4\pi} \int \frac{1}{r} \left(\frac{\partial u}{\partial n} \cos \frac{\omega r}{c} + u \frac{\omega}{c} \cos \phi \sin \frac{\omega r}{c} \right) d\sigma, \qquad (2.667)$$

or, in terms of the wavelength,

$$u = \frac{1}{2\lambda} \int \frac{1}{r} \left(\frac{\lambda}{2\pi} \frac{\partial u}{\partial n} \cos \frac{2\pi r}{\lambda} + u \cos \phi \sin \frac{2\pi r}{\lambda} \right) d\sigma; \qquad (2.668)$$

note that we are dealing with stationary waves.

[62] See Sec. 1.2.

Table 2.3. Matching values for the solutions of the equation $y'' = xy$ (see the text).

x	M	M'	N	N'
-4	0.2199	-1.2082	0.5732	1.3972
0	1	0	0	1
4	68.1777	131.6581	93.5172	180.6092

36. THE EQUATION $y'' = xy$

It is easy to find approximate solutions to this equation with Wentzel's method (see Sec. 2.31 and also 2.5). However, these do not hold anymore when x approaches zero. Therefore the problem that arises is how to connect the asymptotic expressions for $x > 0$ (by a few units, at least) with those for $x < 0$. Since the equation is homogenous, we need only to know how to match two particular solutions, to be able to perform the matching for *any* solution. Let us consider the following two particular solutions:

$$\begin{aligned} M &= 1 + \frac{x^3}{2{\cdot}3} + \frac{x^6}{2{\cdot}3{\cdot}5{\cdot}6} + \frac{x^9}{2{\cdot}3{\cdot}5{\cdot}6{\cdot}8{\cdot}9} + \ldots, \\ N &= x + \frac{x^4}{3{\cdot}4} + \frac{x^7}{3{\cdot}4{\cdot}6{\cdot}7} + \frac{x^{10}}{3{\cdot}4{\cdot}6{\cdot}7{\cdot}9{\cdot}10} + \ldots. \end{aligned} \tag{2.669}$$

For $|x| > 4$ the first and, even better, second-approximation asymptotic expressions are practically exact. It is then enough to compute, from Eq. (2.669), the values of M, N, M', N' for $x = \pm 4$. These can be found in Table 2.3.[63] In Fig.2.2 we report the functions M and N in the interval $-4 < x < 0$.

37. RESONANCE DEGENERACY FOR MANY-ELECTRON ATOMS

Let us consider n electrons $q_1, q_2, \ldots, q_n$ in n orbits described by the eigenfunctions $\psi_1, \psi_2, \ldots, \psi_n$ with different eigenvalues. If we neglect the

[63]@ Note that the numerical values reported in Table 2.3, as written in the original manuscript, were obtained from Eqs. (2.669) by taking the expansions up to the non-vanishing *tenth* term (and the same is true for the derivatives), which means up to the x^{27} and x^{28} power terms for M and N, respectively (and x^{29}, x^{30} for M' and N', respectively).

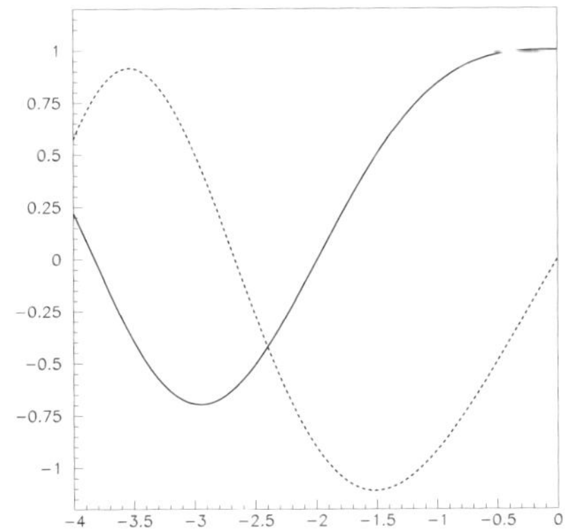

Fig. 2.2. The functions M (solid line) and N (dashed line) in the interval $-4 < x < 0$.

interaction, in the zeroth-order approximation, we can assume that the eigenfunction of the system is the product of eigenfunctions of the single electrons. Since the different electron eigenfunctions can be ordered in $n!$ different ways, we shall have $n!$ independent eigenfunctions of the kind

$$\Psi_r = \psi_1(q_{r_1})\,\psi_2(q_{r_2})\cdots\psi_n(q_{r_n}), \tag{2.670}$$

$r_1, r_2, \ldots, r_n$ being an arbitrary permutation of the first n numbers. Let us denote by P_r the substitution

$$\left(\begin{array}{ccccc} 1 & 2 & 3 & \ldots & n \\ a_1 & a_2 & a_3 & \ldots & a_n \end{array}\right), \tag{2.671}$$

and define also P_r as the operators acting on functions of n variables or groups of variables, which we briefly write as q:

$$P_r\, f(q) = f(P_r\, q). \tag{2.672}$$

In the previous equation, P_r on the l.h.s. must be considered as an operator, and on the r.h.s. as a substitution that alters the order of the independent variables. Clearly, this double meaning cannot lead to misunderstandings. It is also understood that P_1 is the identity permutation. From Eq. (2.670) it follows that

$$\Psi_1 = \psi_1(q_1)\,\psi_2(q_2)\cdots\psi_n(q_n), \tag{2.673}$$

and, from Eqs. (2.670), (2.672), and (2.673),

$$\Psi_r = P_r\,\Psi_1. \tag{2.674}$$

As a perturbation term, let us introduce in the Hamiltonian the interaction H, which we'll have to consider symmetric with respect to q, so that

$$P_r\, H(q) \;=\; H(q), \quad r = 1, 2, \ldots, n!. \tag{2.675}$$

The H_{rs} term of the perturbation matrix will be

$$H_{rs} \;=\; \int \Psi_r^*\, H\, \Psi_s\, \mathrm{d}q \;=\; \int P_r\, \psi_1^*\; H\; P_s\, \psi_1\, \mathrm{d}q, \tag{2.676}$$

where $\mathrm{d}q$ obviously denotes the volume element in the space of the q variables. Note that the last integral extends from $-\infty$ to ∞ for all the variables, and thus it does not depend on q. Consequently the operator P_r will reduce to unity when it is applied to it.[64]

38. VARIOUS FORMULAE

38.1 Schwarz Formula

The Schwarz formula is

$$\left|\sum_{i=1}^{n} a_i\, b_i\right|^2 \;\leq\; \sum_{i=1}^{n} a_i^2 \cdot \sum_{i=1}^{n} b_i^2. \tag{2.677}$$

Indeed,

$$\sum_{i=1}^{n} a_i^2 \cdot \sum_{i=1}^{n} b_i^2 \,-\, \left|\sum_{i=1}^{n} a_i\, b_i\right|^2 \;=\; \frac{1}{2} \sum_{i,j=1}^{n} \left(a_i\, b_j \;-\; a_j\, b_i\right)^2. \tag{2.678}$$

If it is understood that every couple i, j must be taken only once, and, since the terms $i = j$ are zero, by introducing for example the condition $i < j$, the term on the l.h.s. can be rewritten as

$$\sum_{i<j} \left(a_i\, b_j \;-\; a_j\, b_i\right)^2.$$

There is also another Schwarz formula:

$$\left|\int_a^b y\, z\, \mathrm{d}x\right|^2 \;\leq\; \int_a^b y^2\, \mathrm{d}x \int_a^b z^2\, \mathrm{d}x \tag{2.679}$$

[64]@ This section was evidently left incomplete by the author.

(with $b > a$). Indeed,

$$\int_a^b y^2\,\mathrm{d}x \int_a^b z^2\,\mathrm{d}x - \left|\int_a^b y\,z\,\mathrm{d}x\right|^2$$
$$= \frac{1}{2}\int_{x=a}^{x=b}\int_{\xi=a}^{\xi=b} [y(x)\,z(\xi) - y(\xi)\,z(x)]^2\,\mathrm{d}x\,\mathrm{d}\xi. \quad (2.680)$$

38.2 Maximum Value of Random Variables

Let $x_1, x_2, \ldots, x_n$ be n random independent variables, following the same normal distribution law

$$P_x = \frac{1}{\sqrt{\pi}}\,\mathrm{e}^{-x^2}, \quad (2.681)$$

which can also be viewed as n independent realizations of the same random variable x. Let y be the largest (in the algebraic sense) of these. Its distribution is clearly

$$P_y = \frac{\mathrm{d}}{\mathrm{d}y}\left(\frac{1-\theta(y)}{2}\right)^n, \quad (2.682)$$

with

$$\theta(y) = \frac{2}{\sqrt{\pi}}\int_0^y \mathrm{e}^{-y^2}\,\mathrm{d}y. \quad (2.683)$$

For large n, the values of y for which P_y is significantly different from zero are large themselves. Limiting our analysis to such a part of the curve representing P_y, we can then derive its asymptotic behavior for large n by the Equation in item 3) of Sec. 2.27. As a first approximation, we have

$$P_y = \frac{\mathrm{d}}{\mathrm{d}y}\left(1 - \frac{1}{2\sqrt{\pi}}\frac{\mathrm{e}^{-y^2}}{y}\right)^n, \quad (2.684)$$

which approximately is

$$P_y = \frac{\mathrm{d}}{\mathrm{d}y}\exp\left\{-\frac{n\mathrm{e}^{-y^2}}{2\sqrt{2}y}\right\}; \quad (2.685)$$

and, with a further approximation, becomes

$$P_y = \frac{n}{\sqrt{\pi}}\exp\left\{-\left(\frac{n\mathrm{e}^{-y^2}}{2\sqrt{2}y} + y^2\right)\right\}. \quad (2.686)$$

Let y_0 be the value for which P_y has a maximum. Since

$$\frac{\mathrm{d}}{\mathrm{d}y}\left(\frac{n}{2\sqrt{\pi}}\frac{\mathrm{e}^{-y^2}}{y} + y^2\right) = -\frac{n}{\sqrt{\pi}}\mathrm{e}^{-y^2} - \frac{n}{2\sqrt{\pi}}\frac{\mathrm{e}^{-y^2}}{y^2} + 2y, \quad (2.687)$$

in first approximation one gets

$$\begin{aligned} y_0 &= \sqrt{\log n}, \quad \text{absolute error} \to 0, \\ \frac{n}{2\sqrt{\pi}}\,\mathrm{e}^{-y_0^2} &= \sqrt{\log n}, \quad \text{absolute error} \to 0, \\ \frac{n}{2\sqrt{\pi}}\,\frac{\mathrm{e}^{-y_0^2}}{y_0} &= 1, \\ \mathrm{e}^{-y_0^2} &= \frac{2\sqrt{\pi \log n}}{n}, \quad \text{relative error} \to 0. \end{aligned}$$

It follows that

$$P_{y_0} = \frac{n}{\sqrt{\pi}}\,\mathrm{e}^{-y_0^2} \exp\left\{-\frac{n}{2\sqrt{\pi}}\,\frac{\mathrm{e}^{-y_0^2}}{y}\right\} = \frac{2\sqrt{\log n}}{e}. \tag{2.688}$$

Thus we have obtained both the maximum value of P_y and the corresponding value of y:

$$y_0 = \sqrt{\log n}, \tag{2.689}$$

$$P_{y_0} = \frac{2\sqrt{\log n}}{e} = \frac{2y_0}{e}. \tag{2.690}$$

Moreover, the width of P_y (the interval in which P_y is large enough) is of the order $1/y_0$. We haven't yet succeeded in establishing whether the value of y_0 is given by $\sqrt{\log n}$ with a precision greater than $1/y_0$, as would be desirable. It is then convenient to follow another procedure. Since

$$P_y = \frac{\mathrm{d}}{\mathrm{d}y}\left(\frac{1}{\sqrt{\pi}}\int_{-\infty}^{y} \mathrm{e}^{-y^2}\,\mathrm{d}y\right)^n, \tag{2.691}$$

if we require $P'_{y_0} = 0$, we have

$$(n-1)\,\mathrm{e}^{-y_0^2} = 2y_0 \int_{-\infty}^{y} \mathrm{e}^{-y^2}\,\mathrm{d}y, \tag{2.692}$$

i.e., with a relative error tending to 0,

$$\frac{n}{2\sqrt{\pi}}\,\mathrm{e}^{-y_0^2} = y_0. \tag{2.693}$$

Taking the logarithm of the previous expression, up to infinitesimals, we get

$$\log n - \log 2\sqrt{\pi} - y^2 = \log y. \tag{2.694}$$

On setting $y_0 = \sqrt{\log n} + \epsilon$, we obtain, in first approximation,

$$- \log 2\sqrt{\pi} \; - \; 2\epsilon \sqrt{\log n} \; = \; \log \sqrt{\log n}, \tag{2.695}$$

which yields

$$\epsilon \; = \; - \frac{\log 2\sqrt{\pi \log n}}{2\sqrt{\log n}}. \tag{2.696}$$

Thus, the second-order approximation for y_0 becomes

$$y_0 \; = \; \log n \; - \; \frac{\log 2\sqrt{\pi \log n}}{2\sqrt{\log n}}. \tag{2.697}$$

It follows that the correction term goes to zero less rapidly than the amplitude of the P_y curve, which behaves as $1/\sqrt{\log n}$. This is something we have to bear in mind.

Pushing further the approximation would not yield corrections comparable with $1/\sqrt{\log n}$. Then, we use the following as a first-order approximation values for y_0 and P_{y_0}:

$$y_0 \; = \; \sqrt{\log n} - \frac{\log 2\sqrt{\pi \log n}}{2\sqrt{\log n}}, \tag{2.698}$$

$$P_{y_0} \; = \; (2/e) \; \sqrt{\log n}, \tag{2.699}$$

or

$$P_{y_0} \; = \; 2y_0/e. \tag{2.700}$$

38.3 Binomial Coefficients

n											
0	1										
1	1	1									
2	1	2	1								
3	1	3	3	1							
4	1	4	6	4	1						
5	1	5	10	10	5	1					
6	1	6	15	20	15	6	1				
7	1	7	21	35	35	21	7	1			
8	1	8	28	56	70	56	28	8	1		
9	1	9	36	84	126	126	84	36	9	1	
10	1	10	45	120	210	252	210	120	45	10	1

n						
11	1	11	55	165	330	462
	462	330	165	55	11	1
12	1	12	66	220	495	792
	924	792	495	220	66	12
	1					
13	1	13	78	286	715	1287
	1716	1716	1287	715	286	78
	13	1				
14	1	14	91	364	1001	2002
	3003	3432	3003	2002	1001	364
	91	14	1			
15	1	15	105	455	1365	3003
	5005	6435	6435	5005	3003	1365
	455	105	15	1		

n						
16	1	16	120	560	1820	4368
	8008	11440	12870	11440	8008	4368
	1820	560	120	16	1	
17	1	17	136	680	2380	6188
	12376	19448	24319	24310	19448	12376
	6188	2380	680	136	17	1
18	1	18	153	816	3060	8568
	18564	31824	43758	48620	43758	31824
	18564	8568	3060	816	153	18
	1					
19	1	19	171	969	3876	11628
	27132	50388	75582	92378	92378	75582
	50388	27132	11628	3876	969	171
	19	1				
20	1	20	190	1140	4845	15504
	38760	77520	125970	167960	184756	167960
	125970	77520	38760	15504	4845	1140
	190	20	1			

38.4 Expansion of $1/(1-x)^n$

We have

$$\frac{1}{(1-x)^n} = \sum_{r=0}^{\infty} \binom{n+r-1}{r} x^r = \sum_{r=0}^{\infty} \binom{n+r-1}{n-1} x^r. \qquad (2.701)$$

It follows that

$$\binom{n+r-1}{r} = \sum_{r=0}^{r} \binom{n+r-2}{r}, \qquad (2.702)$$

Table 2.4. Coefficients of the expansion of the function $1/(1-x)^n$.

	$r=0$	1	2	3	4	5	6	7	8	9
$n=0$	1	0	0	0	0	0	0	0	0	0
1	1	1	1	1	1	1	1	1	1	1
2	1	2	3	4	5	6	7	8	9	10
3	1	3	6	10	15	21	28	36	45	55
4	1	4	10	20	35	56	84	120	165	220
5	1	5	15	35	70	126	210	330	495	715
6	1	6	21	56	126	252	462	792	1287	2002
7	1	7	28	84	210	462	924	1716	3003	5005
8	1	8	36	120	330	792	1716	3432	6435	11440
9	1	9	45	165	495	1287	3003	6435	12870	24310
10	1	10	55	220	715	2002	5005	11440	24310	48620

or

$$\sum_{r=0}^{r} \binom{k-1+r}{r} = \binom{k+r}{r}. \tag{2.703}$$

In Table 2.4 we report some coefficients of the expansion of $1/(1-x)^n$.

38.5 Relations between the Binomial Coefficients

$$\binom{n-1}{r-1} + \binom{n-1}{r} = \binom{n}{r}, \tag{2.704}$$

$$\sum_{r=0}^{n} \frac{1}{2^{n+r}} \binom{n+r}{n} = 1 \tag{2.705}$$

(see Sec. 1.32) ;

$$\sum_{r=0}^{\infty} \frac{1}{2^{n+r}} \binom{n+r}{n} = 2 \tag{2.706}$$

(see Sec. 1.32). It follows that

$$\sum_{r=1}^{\infty} \frac{1}{2^{2n+r}} \binom{2n+r}{n} = 1, \tag{2.707}$$

$$\sum_{r=0}^{l} \binom{n+r}{r} = \binom{n+l+1}{l} \tag{2.708}$$

(see previous subsection); or

$$\sum_{r=0}^{l} \binom{n+r}{n} = \binom{n+l+1}{n+1}, \tag{2.709}$$

$$\sum_{r=0}^{2r>n} \frac{1}{2r+1} \binom{n}{2r} = \frac{2^n}{n+1}. \qquad (2.710)$$

38.6 Mean Values of r^n between Concentric Spherical Surfaces[65]

Let P be a point with coordinates $\alpha = 0$, $\beta = 0$, $\gamma = 1$, and P_1 a point on the spherical surface whose equation is

$$\alpha^2 + \beta^2 + \gamma^2 = x^2 < 1. \qquad (2.711)$$

If r is the distance between P and P_1, we'll denote by S_n the mean value[66] of r^n:

$$S_n = \frac{1}{4\pi x^2} \int_0^{4\pi x^2} r^n \, d\sigma = \frac{1}{4\pi} \int_0^{4\pi} r^n \, d\omega; \qquad (2.712)$$

it follows that

$$\frac{dS_n}{dx} = \frac{1}{4\pi} \int_0^{4\pi} \nabla r^n \cdot \boldsymbol{u} \, d\omega, \qquad (2.713)$$

$$x^2 \frac{dS_n}{dx} = \frac{1}{4\pi} \int_0^{4\pi x^2} \nabla r^n \cdot \boldsymbol{u} \, d\sigma, \qquad (2.714)$$

$\boldsymbol{u}$ being a unitary vector normal to the sphere. From Eq. (2.714), we have

$$\begin{aligned} x^2 \frac{dS_n}{dx} &= \frac{1}{4\pi} \int_0^{4\pi x^3} \nabla^2 r^n \, dS \\ &= \frac{1}{4\pi} \int_0^{4\pi x^3} n(n+1)\, r^{n-2} \, dS, \qquad (2.715) \\ \frac{d}{dx}\left(x^2 \frac{dS_n}{dx}\right) &= \frac{n(n+1)}{4\pi} \int_0^{4\pi x^2} r^{n-2} \, d\sigma \\ &= n(n+1)\, x^2 \, S_{n-2}. \qquad (2.716) \end{aligned}$$

We thus get

$$\frac{d^2 S_n}{dx^2} + \frac{2}{x} \frac{dS_n}{dx} = n(n+1)\, S_{n-2}, \qquad (2.717)$$

[65] See Sec. 1.21.

[66] @ In what follows, the author denoted by $d\sigma$, $d\omega$, and dS the surface element, the solid angle element, and the volume element, respectively.

which can also be written in the form

$$\frac{1}{x}\frac{\mathrm{d}^2(xS_n)}{\mathrm{d}x^2} = n(n+1)\,S_{n-2}. \tag{2.718}$$

On the other hand, Eq. (2.713) can be rewritten as

$$\frac{\mathrm{d}S_n}{\mathrm{d}x} = \frac{1}{4\pi}\int_0^{4\pi} n\,r^{n-1}\,\frac{r^2+x^2-1}{2xr}\,\mathrm{d}\omega, \tag{2.719}$$

that is,

$$\frac{\mathrm{d}S_n}{\mathrm{d}x} = \frac{n}{2x}\,S_n - n\,\frac{1-x^2}{2x}\,S_{n-2}. \tag{2.720}$$

Taking the derivative of the last expression and substituting it into Eq. (2.717), we finally get

$$(n+2)\,S_n - 2n\,(1+x^2)\,S_{n-2} + (n-2)\,(1-x^2)^2\,S_{n-4} = 0. \tag{2.721}$$

Equations (2.718) and (2.721), with the obvious relations

$$S_0 = 1, \quad S_{-1} = 1, \quad S_n(0) = 1, \tag{2.722}$$

make it possible to evaluate all the S_n.

Let us evaluate S_1; from Eqs. (2.718) and (2.722), we have

$$\begin{aligned}
\frac{\mathrm{d}^2(xS_1)}{\mathrm{d}x^2} &= 2x, \\
\frac{\mathrm{d}(xS_1)}{\mathrm{d}x} &= 1 + x^2, \\
x\,S_1 &= x + \frac{1}{3}\,x^3, \\
S_1 &= 1 + \frac{1}{3}\,x^2.
\end{aligned} \tag{2.723}$$

Substituting $n=0$ into Eq. (2.721), one gets

$$2 - 2(1-x^2)^2\,S_{-4} = 0, \tag{2.724}$$

from which

$$S_{-4} = \frac{1}{(1-x^2)^2}. \tag{2.725}$$

From Eq. (2.718) one obtains

$$\frac{\mathrm{d}^2(xS_{-2})}{\mathrm{d}x^2} = \frac{2x}{(1-x^2)^2},$$

$$
\begin{aligned}
\frac{\mathrm{d}(x S_{-2})}{\mathrm{d}x} &= \frac{1}{1-x^2}, \\
x\, S_{-2} &= \frac{1}{2} \log \frac{1+x}{1-x}, \\
S_{-2} &= \frac{1}{2x} \log \frac{1+x}{1-x}.
\end{aligned} \tag{2.726}
$$

Since we now know the values of S_1, S_{-4}, and S_{-2} from Eqs. (2.723), (2.725), and (2.726), all the remaining S_n can be evaluated using only Eq. (2.721). For example, setting $n = 2$, one gets

$$
4 S_2 \; - \; 4(1+x^2) \; = \; 0, \tag{2.727}
$$

from which

$$
S_2 \; = \; 1 \, + \, x^2, \tag{2.728}
$$

as can be directly checked. Here we report the values of the first S_n with positive n:

$$
\begin{aligned}
S_0 &= 1, & S_0(1) &= 1, \\
S_1 &= 1 + \frac{1}{3} x^2 = \frac{(1+x)^3 - (1-x)^3}{6x}, & S_1(1) &= \frac{4}{3}, \\
S_2 &= 1 + x^2 = \frac{(1+x)^4 - (1-x)^4}{8x}, & S_2(1) &= 2, \\
S_3 &= 1 + 2x^2 + \frac{1}{5} x^4 = \ldots, & S_3(1) &= \frac{16}{5}, \\
S_4 &= 1 + \frac{10}{3} x^2 + x^4 = \ldots, & S_4(1) &= \frac{16}{3}.
\end{aligned}
$$

In general, for $n > -2$, we have

$$
S_n(1) \; = \; \frac{2^{n+1}}{n+2}. \tag{2.729}
$$

Clearly, in this formula $S_n(1)$ is the mean value of the nth powers of the distances between two surface elements on a sphere of unit radius (see Sec. 1.21 for the analogous formulae corresponding to surface elements on a circle).

For negative n, we instead have

$$
\begin{aligned}
S_0 &= 1, & S_0(1) &= 1, \\
S_{-1} &= 1, & S_{-1}(1) &= 1, \\
S_{-2} &= \frac{1}{2x}\log\frac{1+x}{1-x}, & S_{-2}(1) &= \infty, \\
S_{-3} &= \frac{1}{1-x^2} = \frac{1}{2x}\left(\frac{1}{1-x} - \frac{1}{1+x}\right), \\
S_{-4} &= \frac{1}{(1-x^2)^2} = \frac{1}{4x}\left(\frac{1}{(1-x)^2} - \frac{1}{(1+x)^2}\right), \\
S_{-5} &= \frac{1+\frac{1}{3}x^2}{(1-x^2)^3} = \frac{1}{6x}\left(\frac{1}{(1-x)^3} - \frac{1}{(1+x)^3}\right).
\end{aligned}
$$

Notice that, with the exception of S_{-2}, the quantities S_n (with integer n) are rational functions. Let us set

$$S_n = \sum_{r=0}^{\infty} a_n^r\, x^{2r}, \tag{2.730}$$

with (see Eq. (2.722))

$$a_n^0 = 1. \tag{2.731}$$

Equation (2.718) can be written, more in general, as

$$\frac{1}{x}\frac{\mathrm{d}^{2k}(xS_n)}{\mathrm{d}x^{2k}} = (n+1)n(n-1)\cdots(n-2k+2)\,S_{n-2k}. \tag{2.732}$$

Thus, from Eqs. (2.722), it follows that

$$a_n^r\,(2r+1)! = (n+1)n(n-1)\cdots(n-2r+2), \tag{2.733}$$

so that

$$a_n^r = \frac{(n+1)n(n-1)\cdot(n-2r+2)}{(2r+1)!}, \tag{2.734}$$

$$S_n = \sum_{r=0}^{\infty} \frac{(n+1)n(n-1)\cdots(n-2r+2)}{(2r+1)!}\, x^{2r}. \tag{2.735}$$

The last equation can also be written in the form

$$S_n = \sum_{r=0}^{\infty} \binom{n+1}{2r} \frac{x^{2r}}{2r+1}. \tag{2.736}$$

For integer $n > -2$, the sum reduces to a finite polynomial. In particular, we recover Eq. (2.729) (cf. Eq. (2.710)):

$$S_n(1) = \sum_{r=0}^{2r \geq n+1} \binom{n+1}{2r} \frac{1}{2r+1} = \frac{2^{n+1}}{n+2}. \tag{2.737}$$

Then we get

$$\begin{aligned}
S_6 &= 1 + 7x^2 + 7x^4 + x^6, & S_6(1) &= 16, \\
S_5 &= 1 + 5x^2 + 3x^4 + \frac{1}{7}x^6, & S_5(1) &= \frac{64}{7}, \\
S_4 &= 1 + \frac{10}{3}x^2 + x^4 = \ldots, & S_4(1) &= \frac{16}{3}, \\
S_3 &= 1 + 2x^2 + \frac{1}{5}x^4 = \ldots, & S_3(1) &= \frac{16}{5}, \\
S_2 &= 1 + x^2 = \frac{(1+x)^4 - (1-x)^4}{8x}, & S_2(1) &= 2, \\
S_1 &= 1 + \frac{1}{3}x^2 = \frac{(1+x)^3 - (1-x)^3}{6x}, & S_1(1) &= \frac{4}{3}, \\
S_0 &= 1, & S_0(1) &= 1, \\
S_{-1} &= 1, & S_{-1}(1) &= 1, \\
S_{-2} &= 1 + \frac{1}{3}x^2 + \frac{1}{5}x^4 + \frac{1}{7}x^6 + \ldots, \\
S_{-3} &= 1 + x^2 + x^4 + x^6 + \ldots, \\
S_{-4} &= 1 + 2x^2 + 3x^4 + 4x^6 + \ldots, \\
S_{-5} &= 1 + \frac{2\cdot5}{3}x^2 + \frac{3\cdot7}{3}x^4 + \frac{4\cdot9}{3}x^6 + \ldots, \\
S_{-6} &= 1 + \frac{4\cdot5\cdot6}{4!}x^2 + \frac{6\cdot7\cdot8}{4!}x^4 + \frac{8\cdot9\cdot10}{4!}x^6 + \ldots,
\end{aligned}$$

and so on. Equation (2.735) can be rewritten for the two cases $n > -2$ and $n < -2$, respectively, as follows:[67]

$$S_n = \sum_{r=0}^{2r=n+1/2\pm1/2} \binom{n+1}{2r} \frac{x^{2r}}{2r+1}, \qquad n > -2, \tag{2.738}$$

$$S_n = \sum_{r=0}^{\infty} \frac{1}{-n-2} \binom{-n-2+2r}{-n-3} x^{2r}, \qquad n < -2. \tag{2.739}$$

Now, let $y\,\mathrm{d}r$ be the probability that r lies between r and $r + \mathrm{d}r$. We have

$$y = 0, \quad \text{for } |r-1| > x. \tag{2.740}$$

Otherwise, let us consider the point with coordinates $\alpha = 0$, $\beta = 0$, $\gamma = x$ on the internal sphere. Take the spherical surface with center at this point and with radius r,

$$\alpha^2 + \beta^2 + (\gamma - x)^2 = r^2, \tag{2.741}$$

and let it intersect the external spherical surface

$$\alpha^2 + \beta^2 + \gamma^2 = 1. \tag{2.742}$$

For the circumference common to the two spherical surfaces,

$$\begin{aligned} 2\gamma x - x^2 &= 1 - x^2, \\ \gamma &= \frac{1+x^2}{2x} - \frac{r^2}{2x}, \end{aligned} \tag{2.743}$$

we'll have

$$y = \frac{1}{2}\left|\frac{\mathrm{d}\gamma}{\mathrm{d}r}\right| = \frac{r}{2x}. \tag{2.744}$$

In conclusion,

$$y = \begin{cases} 0, & \text{for } r < 1-x, \\ \dfrac{r}{2x}, & \text{for } 1-x < r < 1+x, \\ 0, & \text{for } 1+x < r; \end{cases} \tag{2.745}$$

and, in particular,

$$\begin{aligned} y(1-x) &= \frac{1-x}{2x}, \\ y(1+x) &= \frac{1+x}{2x}. \end{aligned} \tag{2.746}$$

[67]@ The + sign in the upper limit for the sum refers to odd n, the − sign to even n.

We then infer that

$$
\begin{aligned}
S_n &= \int_{-\infty}^{\infty} r^n \, y \, \mathrm{d}r \;=\; \int_{1-x}^{1+x} \frac{r^{n+1}}{2x} \, \mathrm{d}r \\
&= \frac{(1+x)^{n+2} - (1-x)^{n-2}}{2(n+2)x},
\end{aligned}
\tag{2.747}
$$

which contains Eqs. (2.729), (2.735), (2.738), and (2.739). Equation (2.747) does not hold for $n = -2$, in which case one has

$$
S_{-2} = \int_{1-x}^{1+x} \frac{1}{2rx} \, \mathrm{d}r = \frac{1}{2x} \log \frac{1+x}{1-x}, \tag{2.748}
$$

as already obtained above.

3

VOLUMETTO III: 28 JUNE 1929

1. EVALUATION OF SOME SERIES

(17)
$$\sum_{r=1}^{\infty} \frac{1}{r} e^{-ry} \sin rx = \arctan \frac{\sin x}{e^y - \cos x}$$

$$= \arctan \frac{\tan x/2}{\tanh y/2} - \frac{x}{2}, \qquad (3.1)$$

or, setting $K = e^{-y}$,

$$\sum_{r=1}^{\infty} \frac{K^r}{r} \sin rx = \arctan \frac{K \sin x}{1 - K \cos x}$$

$$= \arctan \left(\frac{1+K}{1-K} \tan \frac{x}{2} \right) - \frac{x}{2}. \qquad (3.2)$$

Special cases:

(a) K=1:

$$\sum_{r=1}^{\infty} \frac{\sin rx}{r} = \frac{\pi}{2} - \frac{x}{2}; \qquad (3.3)$$

see (12).

(b) $x = \pi/2$:

$$K - \frac{1}{3}K^3 + \frac{1}{5}K^5 + \ldots = \arctan K. \qquad (3.4)$$

(c) Setting $x = \pi/4$ in (a), after some simple algebra we obtain

$$\frac{2}{3{\cdot}5} - \frac{2}{7{\cdot}9} + \frac{2}{11{\cdot}13} - \frac{2}{15{\cdot}17} + \ldots = \frac{\pi}{2\sqrt{2}} - 1. \qquad (3.5)$$

(18)
$$\frac{2}{1{\cdot}3}+\frac{2}{3{\cdot}5}+\frac{2}{5{\cdot}7}+\frac{2}{7{\cdot}9}+\ldots=1,$$
$$\frac{2}{1{\cdot}3}+\frac{2}{5{\cdot}7}+\frac{2}{9{\cdot}11}+\ldots=\frac{\pi}{4},$$
$$\frac{2}{3{\cdot}5}+\frac{2}{7{\cdot}9}+\frac{2}{11{\cdot}13}+\frac{2}{15{\cdot}17}+\ldots=1-\frac{\pi}{4},$$
$$\frac{2}{3{\cdot}5}+\frac{2}{11{\cdot}13}+\frac{2}{19{\cdot}21}+\frac{2}{27{\cdot}29}+\ldots=\pi\,\frac{\sqrt{2}-1}{8},$$
$$\frac{2}{7{\cdot}9}+\frac{2}{15{\cdot}17}+\frac{2}{23{\cdot}25}+\frac{2}{31{\cdot}33}+\ldots=1-\pi\,\frac{\sqrt{2}+1}{8}.$$

(19)
$$\frac{2}{8^2(8^2-1)}+\frac{2}{16^2(16^2-1)}+\frac{2}{24^2(24^2-1)}+\ldots$$
$$=1-\pi\,\frac{\sqrt{2}+1}{8}-\frac{\pi^2}{192}.$$

(20)
$$\frac{2}{8^4(8^4-1)}+\frac{2}{16^4(16^4-1)}+\frac{2}{24^4(24^4-1)}+\ldots$$
$$=1-\pi\,\frac{\sqrt{2}+1}{8}-\frac{\pi^2}{192}-\frac{\pi^4}{90{\cdot}2048}.$$

(21)
$$\sum_{r=0}^{\infty}\binom{n+1}{2r}\frac{x^{2r}}{2r+1}=\frac{(1+x)^{n+2}-(1-x)^{n+2}}{2(n+2)x}$$

for $x\leq 1$; see Sec. 2.38.6. If n is a positive integer, the series reduces to a finite sum up to $2r=n+1/2\pm 1/2$.

Special cases:

(a) $x=1$:
$$\sum_{r=0}^{\infty}\binom{n+1}{2r}\frac{1}{2r+1}=\frac{2^{2n+1}}{n+2}. \tag{3.6}$$

(b) The above formula fails for $n=-2$; in this limit:
$$1+\frac{1}{3}x^2+\frac{1}{5}x^4+\frac{1}{7}x^6+\ldots=\frac{1}{2x}\log\frac{1+x}{1-x}. \tag{3.7}$$

(c) For other interesting expressions with integer n, see Sec. 2.38.6.

(22) $$\sum_{r=1}^{\infty} \frac{\cos rx}{r} = -\log 2 - \log \sin \frac{x}{2}$$

for $0 < x < 2\pi$.

(23) Changing k (a non-odd integer) in $-k$ in Eq. (3.274) and summing the two expressions, from $y(k) + y(-k) = 0$, we obtain

$$\frac{1}{1-k^2} - \frac{3}{9-k^2} + \frac{5}{25-k^2} - \frac{7}{49-k^2} + \ldots \pm \frac{2n+1}{(2n+1)^2-k^2} + \ldots = \frac{\pi}{4\cos k\pi/2} \tag{3.8}$$

2. THE EQUATION $\Box H = r$

We first give a formula related to the simpler equation:

$$\nabla^2 V = p. \tag{3.9}$$

Since $1/r$ is a harmonic function, we have

$$\frac{1}{r}\nabla^2 V = \frac{1}{r}\nabla^2 V - V\nabla^2\frac{1}{r} = \nabla\cdot\left(\frac{1}{r}\nabla V - V\nabla\frac{1}{r}\right); \tag{3.10}$$

and, from (3.9),

$$\nabla\cdot\left(\frac{1}{r}\nabla V - V\nabla\frac{1}{r}\right) = \frac{p}{r}. \tag{3.11}$$

If r is the distance between P_0 and an arbitrary point P, on integrating over a region S' between a closed surface σ around P_0 and a sphere of radius ϵ centered in P_0, we get

$$\int_{S'} \frac{p}{r}\,\mathrm{d}S = \int_\sigma \left(V\cos\alpha + r\frac{\partial V}{\partial n}\right)\frac{\mathrm{d}\sigma}{r^2} - \int_0^{4\pi\epsilon^2}\left(V + \epsilon\frac{\partial V}{\partial n}\right)\frac{\mathrm{d}\sigma}{\epsilon^2}, \tag{3.12}$$

n being the outward normal and α the angle between this normal and the position vector. For $\epsilon \to 0$, S' tends to the whole region S enclosed by σ and Eq. (3.12) becomes

$$V(P_0) = -\frac{1}{4\pi}\int_S \frac{p}{r}\,\mathrm{d}S + \frac{1}{4\pi}\int_\sigma\left(V\cos\alpha + r\frac{\partial V}{\partial n}\right)\frac{\mathrm{d}\sigma}{r^2}. \tag{3.13}$$

Let us now consider the differential equation

$$\nabla^2 H - \frac{1}{c^2}\frac{\partial^2 H}{\partial t^2} = r, \tag{3.14}$$

r being a known function of space and time. If r is again the distance from a reference point P_0, and we define the function H_1 to be

$$H_1(P,t) = H(P, t - \frac{r}{c}), \tag{3.15}$$

then it follows:

$$\begin{aligned}
H(P,t) &= H_1(P, t+r/c), \\
H'_x(P,t) &= H'_{1\,x}(P, t+r/c) + \frac{x}{rc} H'_{1\,t}(P, t+r/c), \\
H''_{xx}(P,t) &= H''_{1\,xx}(P, t+r/c) \frac{2x}{rc} H''_{1\,xt}(P, t+r/c) \\
&\quad + \frac{x^2}{r^2c^2} H''_{1\,tt}(P, t+r/c) + \frac{r^2 - x^2}{r^3 c} H'_{1\,t}(P, t+r/c), \\
\nabla^2 H(P,t) &= \nabla^2 H_1(P, t+r/c) + \frac{1}{c^2} H''_{1\,tt}(P, t+r/c) \\
&\quad + \frac{2}{c} H''_{1\,tr}(P, t+r/c) + \frac{2}{rc} H'_{1\,t}(P, t+r/c), \\
\frac{1}{c^2} H''_{tt}(P,t) &= \frac{1}{c^2} H''_{tt}(P, t+r/c).
\end{aligned}$$

From Eq. (3.14), it follows:

$$r(P,t) = \nabla^2 H_1(P, t+r/c) + \frac{2}{c} H''_{1\,tr}(P, t+r/c) + \frac{2}{rc} H'_{1\,t}(P, t+r/c), \tag{3.16}$$

or, changing $t + r/c$ into t:

$$\nabla^2 H_1(P, t) + \frac{2}{c} H''_{1\,tr}(P, t) + \frac{2}{rc} H'_{1\,t}(P, t) = r(P, t - r/c). \tag{3.17}$$

If A is an arbitrary function of space and time, we write

$$\overline{A}(P,t) = A(P, t - r/c), \tag{3.18}$$

and Eq. (3.17) becomes

$$\nabla^2 H_1 + \frac{2}{c}\frac{\partial^2 H_1}{\partial t \partial r} + \frac{2}{rc}\frac{\partial H_1}{\partial t} = \overline{r}. \tag{3.19}$$

Let us set

$$p = \overline{r} - \frac{2}{c}\frac{\partial^2 H_1}{\partial t \partial r} - \frac{2}{rc}\frac{\partial H_1}{\partial t}; \tag{3.20}$$

then Eq. (3.19) becomes

$$\nabla^2 H_1 = p. \tag{3.21}$$

For a given value of t, H_1 and p are space-dependent functions, and we can use Eq. (3.13). We then get

$$\begin{aligned} H_1(P_0,t) &= -\frac{1}{4\pi}\int_S \frac{p}{r}\,\mathrm{d}S + \frac{1}{4\pi}\int_\sigma \left(H_1 \cos\alpha + r\frac{\partial H_1}{\partial n}\right)\frac{\mathrm{d}\sigma}{r^2} \\ &= -\frac{1}{4\pi}\int_S \frac{\overline{r}}{r}\,\mathrm{d}S + \frac{1}{4\pi}\frac{2}{c}\int_S \left(\frac{\partial^2 H_1}{\partial t\partial r} + \frac{1}{r}\frac{\partial H_1}{\partial t}\right)\frac{\mathrm{d}S}{r} \\ &\quad + \frac{1}{4\pi}\int_\sigma \left(H_1 \cos\alpha + r\frac{\partial H_1}{\partial n}\right)\frac{\mathrm{d}\sigma}{r^2}. \end{aligned} \tag{3.22}$$

Furthermore,

$$\begin{aligned} \int_S \left(\frac{\partial^2 H_1}{\partial t\partial r} + \frac{1}{r}\frac{\partial H_1}{\partial t}\right)\frac{\mathrm{d}S}{r} &= \int \mathrm{d}\omega \int \left(r\frac{\partial^2 H_1}{\partial t\partial r} + \frac{\partial H_1}{\partial t}\right)\mathrm{d}r \\ &= \int \mathrm{d}\omega \int \frac{\partial}{\partial r}\left(r\frac{\partial H_1}{\partial t}\right)\mathrm{d}r \\ &= \int_\sigma r\frac{\partial H_1}{\partial t}\,\mathrm{d}\omega \\ &= \int_\sigma r\frac{\partial H_1}{\partial t}\cos\alpha\,\frac{\mathrm{d}\sigma}{r^2}. \end{aligned} \tag{3.23}$$

Substituting this expression in Eq. (3.22), we find

$$\begin{aligned} H_1(P_0,t) &= -\frac{1}{4\pi}\int_S \frac{\overline{r}}{r}\,\mathrm{d}S \\ &\quad + \frac{1}{4\pi}\int_\sigma \left(H_1\cos\alpha + r\frac{\partial H_1}{\partial n} + \frac{2r}{c}\frac{\partial H_1}{\partial t}\cos\alpha\right)\frac{\mathrm{d}\sigma}{r^2}. \end{aligned} \tag{3.24}$$

However,

$$\begin{aligned} H_1(P_0,t) &= H(P_0,t), \\ H_1(P,t) &= H(P,t-r/c) = \overline{H}(P,t), \\ \frac{\partial H_1(P,t)}{\partial n} &= \frac{\partial H(P,t-r/c)}{\partial n} = \overline{\frac{\partial H(P,t)}{\partial n}} - \frac{\cos\alpha}{c}\overline{\frac{\partial H(P,t)}{\partial t}}; \end{aligned}$$

and, on substitution into Eq. (3.24):

$$\begin{aligned} H(P_0,t) &= -\frac{1}{4\pi}\int_S \frac{\overline{r}}{r}\,\mathrm{d}S \\ &\quad + \frac{1}{4\pi}\int_\sigma \left(\overline{H}\cos\alpha + r\overline{\frac{\partial H}{\partial n}} + \frac{r}{c}\overline{\frac{\partial H}{\partial t}}\cos\alpha\right)\frac{\mathrm{d}\sigma}{r^2}, \end{aligned} \tag{3.25}$$

which clearly expresses, setting $r = 0$, a more general principle than that of Huygens.

Let us consider a periodic solution of Eq. (3.14):

$$H = u\,\mathrm{e}^{i\sigma t}. \tag{3.26}$$

On setting

$$k = \frac{\sigma}{c}, \tag{3.27}$$

Eq. (3.14) becomes

$$\nabla^2 u + k^2 u = r\,\mathrm{e}^{-i\sigma t}. \tag{3.28}$$

Let us consider

$$r = y\,\mathrm{e}^{i\sigma t}, \tag{3.29}$$

with y a function depending only on space variables; it follows that

$$\nabla^2 u + k^3 u = y. \tag{3.30}$$

If Eqs. (3.26) and (3.29) are satisfied, then, to every function that is solution of Eq. (3.14), there corresponds a function that is solution of Eq. (3.30); the same holds if we change i in $-i$ in Eqs. (3.26) and (3.29). If u satisfies Eq. (3.30), then, from Eqs. (3.25), (3.26), and (3.29), we obtain

$$\begin{aligned} u(P_0) &= -\frac{1}{4\pi}\int_S \frac{\mathrm{e}^{-ikr}}{r}\,y\,\mathrm{d}S \\ &\quad + \frac{1}{4\pi}\int_\sigma \left(u\,(1+ikr)\cos\alpha + r\,\frac{\partial u}{\partial n}\right)\mathrm{e}^{-ikr}\,\frac{\mathrm{d}\sigma}{r^2}. \end{aligned} \tag{3.31}$$

Changing i into $-i$, we obtain a second expression for u:

$$\begin{aligned} u(P_0) &= -\frac{1}{4\pi}\int_S \frac{\mathrm{e}^{ikr}}{r}\,y\,\mathrm{d}S \\ &\quad + \frac{1}{4\pi}\int_\sigma \left(u\,(1-ikr)\cos\alpha + r\,\frac{\partial u}{\partial n}\right)\mathrm{e}^{ikr}\,\frac{\mathrm{d}\sigma}{r^2}. \end{aligned} \tag{3.32}$$

Taking the sum of these two expressions and dividing the result by 2, we get a third expression for u, which is explicitly real:

$$\begin{aligned} u(P_0) = &-\frac{1}{4\pi}\int_S \frac{\cos kr}{r}\,y\,\mathrm{d}S + \frac{1}{4\pi}\int_\sigma \Big(u\cos kr\cos\alpha \\ &+ u\,kr\sin kr\cos\alpha + r\,\frac{\partial u}{\partial n}\cos kr\Big)\,\frac{\mathrm{d}\sigma}{r^2}. \end{aligned} \tag{3.33}$$

If instead we take the difference of the two expressions and divide by $2i$, we obtain the remarkable identity

$$\begin{aligned} 0 \;=\; & -\frac{1}{4\pi}\int_S \frac{\sin kr}{r}\, y\,\mathrm{d}S \\ & +\frac{1}{4\pi}\int_\sigma \left(u \sin kr \cos\alpha - u\,kr \cos kr \cos\alpha + r\,\frac{\partial u}{\partial n}\sin kr\right)\frac{\mathrm{d}\sigma}{r^2}, \end{aligned}$$

that is,

$$\begin{aligned} & \int_S \frac{\sin kr}{r}\, y\,\mathrm{d}S \\ & = \int_\sigma \left(u \sin kr \cos\alpha - u\,kr \cos kr \cos\alpha + r\,\frac{\partial u}{\partial n}\sin kr\right)\frac{\mathrm{d}\sigma}{r^2}. \end{aligned} \tag{3.34}$$

On taking the limit $k \to 0$, Eq. (3.30) reduces to Eq. (3.9) and Eq. (3.33) to Eq. (3.13). Substituting Eq. (3.30) into (3.34), we have:

$$\begin{aligned} & \int_S \frac{\sin kr}{r}\left(\boldsymbol{\nabla}^2 u + k^2 u\right)\mathrm{d}S \\ & = \int_\sigma \left(u \sin kr \cos\alpha - u\,kr \cos kr \cos\alpha + r\,\frac{\partial u}{\partial n}\sin kr\right)\frac{\mathrm{d}\sigma}{r^2}, \end{aligned} \tag{3.35}$$

which is a true identity holding for an arbitrary function u. In particular in Eq. (3.35) we can take k arbitrarily small and expand each term as a power series of k. On equating the first-order terms on the two sides we find

$$\int_S \boldsymbol{\nabla}^2 u\,\mathrm{d}S = \int_\sigma \frac{\partial u}{\partial n}\,\mathrm{d}\sigma, \tag{3.36}$$

which is the well-known divergence theorem. Other identities can be obtained by equating higher order terms; for example, from third-order terms:

$$\int_S \left(u - \frac{1}{6}r^2\,\boldsymbol{\nabla}^2 u\right)\mathrm{d}S = \int_\sigma \left(\frac{1}{3}u\,r\cos\alpha - \frac{1}{6}r^2\,\frac{\partial u}{\partial n}\right)\mathrm{d}\sigma, \tag{3.37}$$

which can be directly proven by observing that

$$u - \frac{1}{6}r^2\,\boldsymbol{\nabla}^2 u = \frac{1}{6}\left(u\,\boldsymbol{\nabla}^2 r^2 - r^2\,\boldsymbol{\nabla}^2 u\right). \tag{3.38}$$

Let us now consider again Eq. (3.31) within some approximations. First, let us suppose that r is large with respect to the wavelength, so that we can neglect 1 compared to ikr; furthermore, let σ be the surface of a progressive wave that has a curvature radius smaller than its wavelength. At small distances such a wave can be treated as a plane wave, and approximately we shall have

$$\frac{\partial u}{\partial n} = \pm i\, k\, u, \tag{3.39}$$

where the signs $\pm$ correspond to a wave approaching or leaving P_0. Within these approximations, Eq. (3.31) reduces to

$$u(P_0) = \frac{k\, i}{4\pi} \int_\sigma u\, (\cos\, \alpha \pm 1)\, \frac{\mathrm{e}^{-i\, kr}}{r}\, \mathrm{d}\sigma \tag{3.40}$$

or, introducing the wavelength from the relation

$$k = \frac{2\pi}{\lambda}, \tag{3.41}$$

$$u(P_0) = \frac{i}{\lambda} \int_\sigma \frac{\cos\, \alpha \pm 1}{2}\, \frac{u\, \mathrm{e}^{-\frac{2\pi i}{\lambda}\, r}}{r}\, \mathrm{d}\sigma. \tag{3.42}$$

If α is small and the wave is approaching P_0 :

$$u(P_0) = \frac{i}{\lambda} \int_\sigma \frac{\mathrm{e}^{-\frac{2\pi i}{\lambda}\, r}}{r}\, u\, \mathrm{d}\sigma. \tag{3.43}$$

3. EQUILIBRIUM OF A ROTATING HETEROGENEOUS LIQUID BODY (CLAIRAUT PROBLEM)

Suppose that a rotating body is a superposition of incompressible liquid layers each of a given density. Assuming a small angular velocity ω, the deformations of the body are of order ω^2, and we'll take ω^2 as the reference small quantity.

The liquid elements attract one another according to Newton's law, in which we keep, with some convenient choice of the unit system, the attraction coefficient to be unity. When the body is at rest, the density is a never-increasing function of the distance from the center:

$$\rho = \rho(r), \quad \rho' \leq 0. \tag{3.44}$$

In the same way, Newton's potential (generated by the forces) depends on r:

$$V_0 = V_0(r). \tag{3.45}$$

We denote by D the average density of the part of the body at a distance smaller than r:

$$D = \frac{\int_0^r \rho r^2 \,\mathrm{d}r}{r^3/3}. \tag{3.46}$$

It follows:

$$r^3 D = 3 \int_0^r \rho r^2 \,\mathrm{d}r, \tag{3.47}$$

$$3r^2 D + r^3 D' = 3 \rho r^2, \tag{3.48}$$

that is,

$$3\rho = 3D + r D'; \tag{3.49}$$

and, taking the derivative,

$$3\rho' = 4D' + r D'', \tag{3.50}$$

which will be useful later on.

The force acting on a unit mass at a distance r will be

$$\left(\frac{1}{r^2}\right) \int_0^r 4\pi r^2 \rho \,\mathrm{d}r = \frac{4}{3}\pi r D, \tag{3.51}$$

so that

$$V_0' = -\frac{4}{3}\pi r D. \tag{3.52}$$

Now let us set the body in rotation; in the new equilibrium configuration, an element in P will have moved to P' . Let us put

$$\eta = \overline{PP'} \cos\left(r, \overline{PP'}\right). \tag{3.53}$$

The normal shift η can be expanded in spherical functions Y:

$$\eta = \sum H Y, \tag{3.54}$$

with H being functions that depend on the radius. If the rotation takes place about the z axis, only the spherical functions that are symmetric with respect to the z axis will appear in the expansion (3.54), and these can be expressed in terms of the Legendre polynomials

$$P_n(\cos\theta). \tag{3.55}$$

Furthermore, η should not be sensitive to changes of z in $-z$. Thus we can restrict our calculations to the spherical functions of even order. Moreover, on the surface of the sphere of radius r, we have

$$\int \eta \, \mathrm{d}\sigma = 0. \tag{3.56}$$

Hence the zeroth-order spherical function should not be considered. The first contribution comes from the second-order function, which we take in the form

$$Y = (x^2 + y^2 - 2z^2)/r^2. \tag{3.57}$$

Here we suppose that for all the other functions we have $H = 0$. This corresponds to assuming that, in a first approximation, the equal-density surfaces are ellipsoids. This hypothesis is clearly satisfied by the free surface. Equation (3.54) then reduces to

$$\eta = H Y, \tag{3.58}$$

with Y given by Eq. (3.57).
The flattening of the equal-density surfaces with average radius equal to r clearly is

$$s = 3H/r. \tag{3.59}$$

In the same way we suppose that the Newtonian potential is, in first approximation,

$$V = V_0 + L Y. \tag{3.60}$$

Adding the contribution from the centrifugal force, we obtain the total potential to be considered for the local equilibrium:

$$\begin{aligned} U &= V + \frac{1}{2}\omega^2 \left(x^2 + y^2\right) \\ &= V_0 + \frac{1}{3}\omega^2 r^2 + \left(L + \frac{1}{6}\omega^2 r^2\right) Y. \end{aligned} \tag{3.61}$$

In first approximation, from Eqs. (3.53) and (3.58), the density ρ of the rotating fluid is

$$\rho_1 = \rho - \eta \rho' = \rho - H \rho' Y. \tag{3.62}$$

To determine H and L, which are unknowns in the present problem, we have to use the Poisson equation and impose the condition that the equal-density surfaces coincide with the equipotential surfaces. The Poisson equation gives

$$\nabla^2 V = -4\pi \rho_1, \tag{3.63}$$

or, using

$$\begin{aligned} \nabla^2 V_0 &= 4\pi\,\rho, && (3.64)\\ V - V_0 &= L\,Y, && (3.65)\\ \rho_1 - \rho &= -H\,\rho'\,Y, && (3.66) \end{aligned}$$

simply

$$\nabla^2 L\,Y = 4\pi\,H\,\rho'\,Y. \tag{3.67}$$

Dividing by Y, we get

$$4\pi\,H\,\rho' = L'' + \frac{2}{r}\,L' - \frac{6}{r^2}\,L. \tag{3.68}$$

In a first approximation the equipotential surfaces (U = const.) are ellipsoids of revolution about z. The flattening of the meridian section will, in first approximation, be

$$s_U = -3\,\frac{L + (1/6)\,\omega^2\,r^2}{r\,V_0'} = +3\,\frac{L + (1/6)\,\omega^2\,r^2}{(4/3)\pi\,r^2\,D}. \tag{3.69}$$

If, as we have seen, the equal-density surfaces are ellipsoids of revolution as well, with flattening given by Eq. (3.59), in order for the two families of surfaces to coincide, we should have

$$s = s_U, \tag{3.70}$$

that is

$$H = \frac{L + (1/6)\,\omega^2\,r^2}{(4/3)\pi\,r\,D}. \tag{3.71}$$

Extracting L from Eq. (3.71), we get

$$\begin{aligned} L &= \frac{4}{3}\pi\,r\,D\,H - \frac{1}{6}\,\omega^2\,r^2, && (3.72)\\ L' &= \frac{4}{3}\pi\,D\,H + \frac{4}{3}\pi\,r\,D'\,H + \frac{4}{3}\pi\,r\,D\,H' - \frac{1}{3}\,\omega^2\,r, && (3.73)\\ L'' &= \frac{8}{3}\pi\,D'\,H + \frac{8}{3}\pi\,D\,H' + \frac{8}{3}\pi\,r\,D'\,H' \\ &\quad + \frac{4}{3}\pi\,r\,D''\,H + \frac{4}{3}\pi\,r\,D\,H'' - \frac{1}{3}\,\omega^2. && (3.74) \end{aligned}$$

Substituting in Eq. (3.68), we eliminate L:

$$\begin{aligned} 3\,H\,\rho' &= -\frac{4\,D\,H}{r} + 4\,D'\,H + 4\,D\,H' + 2r\,D'\,H' \\ &\quad + r\,D''\,H + r\,D\,H''; && (3.75) \end{aligned}$$

and, using Eq. (3.50),

$$3\,H\,\rho' \;=\; 4\,D'\,H \,+\, r\,D''\,H, \tag{3.76}$$

we finally obtain

$$0 \;=\; -\,\frac{4\,D\,H}{r} \,+\, 4\,D\,H' \,+\, 2r\,D' \,+\, r\,D\,H''H', \tag{3.77}$$

or

$$D\left(-4 \,+\, 4r\,\frac{H'}{H} \,+\, r^2\,\frac{H''}{H}\right) \,+\, 2r\,D'\,\frac{r\,H'}{H} \;=\; 0. \tag{3.78}$$

Let us set

$$q \;=\; r\,s'/s; \tag{3.79}$$

remembering that $s = 3H/r$, we get

$$\frac{s'}{s} \;=\; \frac{H'}{H} \,-\, \frac{1}{r}, \tag{3.80}$$

$$q \;=\; r\,\frac{H'}{H} \,-\, 1, \tag{3.81}$$

from which we deduce:

$$r\,\frac{H'}{H} \;=\; 1 \,+\, q, \tag{3.82}$$

$$\frac{H'}{H} \,+\, r\,\frac{H''}{H} \,-\, r\left(\frac{H'}{H}\right)^2 \;=\; q' \tag{3.83}$$

$$r\,\frac{H'}{H} \,+\, r^2\,\frac{H''}{H} \,-\, r^2\left(\frac{H'}{H}\right)^2 \;=\; r\,q' \tag{3.84}$$

$$1 \,+\, q \,+\, r^2\,\frac{H''}{H} \,-\, (1 \,+\, q)^2 \;=\; r\,q' \tag{3.85}$$

$$r^2\,\frac{H''}{H} \;=\; r\,q' \,+\, q \,+\, q^2. \tag{3.86}$$

Substituting Eqs. (3.82) and (3.86) into Eq. (3.78), we find

$$D\left(r\,q' \,+\, 5\,q \,+\, q^2\right) \,+\, 2r\,D'\,(1 \,+\, q) \;=\; 0, \tag{3.87}$$

which is the Clairaut equation.

If rD'/D tends to 0 as r tends to 0, for $r = 0$ we should have

$$5\,q \,+\, q^2 \;=\; 0, \tag{3.88}$$

that is,

$$q \;=\; 0,\; -5. \tag{3.89}$$

Now, expanding V around the center of the rotating body, we have

$$V = V(0) + A\left(x^2 + y^2\right) + B\,z^2 + \ldots. \tag{3.90}$$

Assuming $\rho'(0)$ to be finite (and, in particular, zero), V can be expanded in series of x, y, z and, for symmetry reasons, the odd power terms vanish. Denoting by ϵ a small function of 4th order in r, we get

$$U = V(0) + A\left(x^2 + y^2\right) + \frac{1}{2}\omega^2\left(x^2 + y^2\right) + B\,z^2 + \epsilon, \tag{3.91}$$

with, obviously,

$$4A + 2B = -4\pi\,\rho(0). \tag{3.92}$$

Let us consider

$$U = \text{const.}; \tag{3.93}$$

we find ($A_1 = -A$, $B_1 = -B$)

$$\left(A_1 - \frac{1}{2}\omega^2\right)\left(x^2 + y^2\right) + B_1\,z^2 + \epsilon = \text{const.}, \tag{3.94}$$

and, neglecting second-order terms:

$$s = \frac{1/\sqrt{A_1 - \omega^2/2} - 1/\sqrt{B_1}}{1/\sqrt{A_1 - \omega^2/2}} = 1 - \sqrt{\frac{A_1 - \omega^2/2}{B_1}}. \tag{3.95}$$

We then have

$$s'(0) = 0 \tag{3.96}$$

and, *a fortiori*,

$$q(0) = \frac{r\,s'(0)}{s(0)} = 0. \tag{3.97}$$

The point

$$(r, q) = (0, 0). \tag{3.98}$$

belongs to the integral curve defined by Eq. (3.87).

Let us assume that D can be expanded in an even power series of r:

$$D = D(0) + ar^2 + br^4 + cr^6 + \ldots \tag{3.99}$$

and, in the same way,

$$q = q_0 + \alpha r^2 + \beta r^4 + \gamma r^6 + \ldots, \tag{3.100}$$

or, by setting

$$a_0 = D_0, \tag{3.101}$$
$$a_2 = a, \tag{3.102}$$
$$a_4 = b, \tag{3.103}$$
$$a_6 = c, \tag{3.104}$$
$$\dots,$$
$$\alpha_0 = q_0 = 0, \tag{3.105}$$
$$\alpha_2 = \alpha, \tag{3.106}$$
$$\alpha_4 = \beta, \tag{3.107}$$
$$\alpha_6 = \gamma, \tag{3.108}$$
$$\dots,$$

$$D = \sum a_{2n}\, r^{2n}, \tag{3.109}$$
$$q = \sum \alpha_{2n}\, r^{2n}; \tag{3.110}$$

and, substituting in Eq. (3.87),

$$\begin{aligned}&\left(\sum a_{2n}\, r^{2n}\right)\left[\sum\left(2n\,\alpha_{2n}\, r^{2n} + 5\alpha_{2n}\, r^{2n}\right) + \left(\sum \alpha_{2n}\, r^{2n}\right)^2\right] \\ &+ 2\sum 2n\,\alpha_{2n}\, r^{2n}\left(1 + \sum \alpha_{2n}\, r^{2n}\right) = 0.\end{aligned} \tag{3.111}$$

Let us evaluate the initial coefficients of the expansions of q. We have:

$$D = D_0 + ar^2 + br^4 + cr^6 + \dots, \tag{3.112}$$
$$r\,D' = 2ar^2 + 4br^4 + 6cr^6 + \dots, \tag{3.113}$$
$$q = \alpha r^2 + \beta r^4 + \gamma r^6 + \dots, \tag{3.114}$$
$$r\,q' = 2\alpha r^2 + 4\beta r^4 + 6\gamma r^6 + \dots, \tag{3.115}$$
$$q^2 = \alpha^2 r^4 + 2\alpha\beta r^6 + \dots, \tag{3.116}$$
$$\begin{aligned} r\,q' + 5\,q + q^2 &= 7\alpha r^2 + (9\beta + \alpha^2) r^4 + \dots \\ &= (11\gamma + 2\alpha\beta) r^6 + \dots,\end{aligned} \tag{3.117}$$
$$1 + q = 1 + \alpha r^2 + \beta r^4 + \gamma r^6 + \dots, \tag{3.118}$$
$$\begin{aligned} D\left(r\,q' + 5\,q + q^2\right) &= 7\alpha\, D_0\, r^2 + \Big((9\beta + \alpha^2) D_0 + 7\alpha a\Big)\, r^4 \\ &+ \Big((11\gamma + 2\alpha\beta) D_0 + (9\beta + \alpha^2) a + \\ &+ 7\alpha b\Big)\, r^6 + \dots,\end{aligned} \tag{3.119}$$
$$\begin{aligned} 2\,r\,D'\,(1 + q) &= 4ar^2 + (4\alpha a + 8b) r^4 \\ &+ (4\beta a + 8\alpha b + 12c) r^6 + \dots.\end{aligned} \tag{3.120}$$

It follows that

$$\begin{aligned} 7\alpha D_0 + 4a &= 0, && (3.121)\\ 9\beta + \alpha^2)D_0 + 7\alpha a + 4\alpha a + 8b &= 0, && (3.122)\\ (11\gamma + 2\alpha\beta)D_0 + (9\beta + \alpha^2)a + 7\alpha b & \\ + 4\beta a + 8\alpha b + 12c &= 0. && (3.123) \end{aligned}$$

The force of attraction on the surface of a planet with mass M is given by a potential that in first approximation reads

$$V = \frac{M}{r} + \frac{\mathcal{I}_0 - 3\mathcal{I}/2}{r^3}, \tag{3.124}$$

where $\mathcal{I}$ is the moment of inertia with respect to the line OP [1] and $\mathcal{I}_0$ is the (polar, not axial) moment of inertia with respect to the center of mass.

On the surface we have

$$U = \frac{M}{r} + \frac{\mathcal{I}_0 - 3\mathcal{I}/2}{r^3} + \frac{1}{2}\omega^2\left(x^2 + y^2\right) = \text{const.} \tag{3.125}$$

by denoting with R_p the polar radius and with R_e the equatorial radius, we get

$$\frac{M}{R_p} - \frac{C - A}{R_p^3} = \frac{M}{R_e} + \frac{1}{2}\frac{C - A}{R_e^3} + \frac{1}{2}\omega^2 R_e^2, \tag{3.126}$$

where C is the moment of inertia with respect to the polar axis [2] and A the moment of inertia with respect to any equatorial axis, so that

$$\mathcal{I}_0 = A + \frac{1}{2}C. \tag{3.127}$$

Let us denote by f the ratio between the centrifugal force and the gravitational force at the equator and with r_1 the average radius of the planet; in first approximation we then have

$$f = \frac{\omega^2 r_1^3}{M}. \tag{3.128}$$

From Eq. (3.126), it follows, in first approximation, that

$$M\left(\frac{1}{R_p} - \frac{1}{R_e}\right) = \frac{3}{2}\frac{C - A}{r_1^3} + \frac{1}{2}\frac{f}{r_1}M; \tag{3.129}$$

[1]@ Joining the center of the planet with the considered external point.

[2]@ Above, this quantity was denoted by I.

and, setting as usual

$$s_1 = \frac{R_e - R_p}{R_e}, \tag{3.130}$$

we have, again in first approximation,

$$s_1 = \frac{3}{2} \frac{C - A}{M r_1^2} + \frac{1}{2} f, \tag{3.131}$$

or, denoting by D_1 the average density inside the planet

$$s_1 - \frac{1}{2} f = \frac{9 (C - A)}{8\pi \, r_1^5 \, D_1}. \tag{3.132}$$

The average moment of inertia of the Earth becomes

$$\mathcal{I} = \frac{8\pi}{3} \int_0^{r_1} \rho \, r^4 \, \mathrm{d}r. \tag{3.133}$$

Now, we have

$$\begin{aligned} \int_0^{r_1} \rho \, r^4 \, \mathrm{d}r &= \int_0^{r_1} r^2 \rho \, r^2 \, \mathrm{d}r = \int_0^{r_1} r^2 \, \mathrm{d} \left(\frac{1}{3} r^3 D \right) \\ &= \frac{1}{3} r_1^5 \, D_1 - \frac{2}{3} \int_0^{r_1} r^4 \, D \, \mathrm{d}r, \end{aligned} \tag{3.134}$$

from which it follows that

$$\mathcal{I} = \frac{8\pi}{9} \left(r_1^5 \, D_1 - 2 \int_0^{r_1} r^4 \, D \, \mathrm{d}r \right). \tag{3.135}$$

In first approximation, we get

$$C \simeq \mathcal{I}; \tag{3.136}$$

and, substituting in Eq. (3.132), we find

$$s_1 - \frac{1}{2} f = \frac{C - A}{C} \left(1 - \frac{2}{r_1^5 \, D_1} \int r^4 \, D \, \mathrm{d}r \right). \tag{3.137}$$

Let us consider again the Clairaut equation (3.87) and evaluate the expression

$$\begin{aligned} \frac{\mathrm{d}}{\mathrm{d}r} \left(r^5 \, D \sqrt{1+q} \right) &= 5 \, r^4 \, D \sqrt{1+q} + r^5 \, D' \sqrt{1+q} \\ &\quad + r^5 \, D \frac{q'}{2 \sqrt{1+q}} \\ &= \frac{5 \, r^4 \, D}{\sqrt{1+q}} \left(1 + q + \frac{r D'}{5D} (1 + q) + \frac{r q'}{10} \right) \end{aligned} \tag{3.138}$$

From Eq. (3.87), we find

$$\frac{r D'}{5D}(1+q) = -\frac{r q'}{10} - \frac{q}{2} - \frac{q^2}{10}, \tag{3.139}$$

so that, substituting in Eq. (3.138), we get

$$\frac{\mathrm{d}}{\mathrm{d}r}\left(r^5 D \sqrt{1+q}\right) = \frac{5 r^4 D}{\sqrt{1+q}}\left(1 + \frac{1}{2} q - \frac{1}{10} q^2\right), \tag{3.140}$$

from which it follows that [3]

$$r_1^5 D_1 \sqrt{1+q_1} = \int_0^{r_1} \frac{5 r^4 D}{\sqrt{1+q}}\left(1 + \frac{1}{2} q - \frac{1}{10} q^2\right) \mathrm{d}r. \tag{3.141}$$

Let us set

$$K = \frac{1 + q/2 - q^2/10}{\sqrt{1+q}} \tag{3.142}$$

so that Eq. (3.141) becomes [4]

$$r_1^5 D_1 \sqrt{1+q_1} = \int_0^{r_1} 5 r^4 D K \, \mathrm{d}r. \tag{3.143}$$

If q is sufficiently small, K is approximately unity: [5]

q	k
0	1
0.1	1.00018
0.2	1.00051
0.3	1.00072
0.4	1.00066
0.5	1.00021
0.6	0.99928
0.7	0.99782
0.8	0.99580
0.9	0.99317
1	0.98995
2	0.92376
3	0.8

[3]@ In the original manuscript, the upper limit of the integral is r; however, it is evident that the appropriate limit is r_1.

[4]@ See the previous footnote.

[5]@ In the table below, the author reported only the values for $K = 1, 1.00074, 1.00021, 0.98995$ corresponding to $q = 0, 0.3, 0.5, 0.9$, respectively. Particular emphasis is given to the value for $q = 0.3$.

The maximum of q occurs on the surface $q = q_1$, and its minimum at the center $(r\,,\,q) = (0\,,\,0)$.

Let us evaluate q_1; as we have seen in Eq. (3.131), on an equipotential surface outside the planet,

$$s = \frac{1}{2}f + \frac{3}{2}\frac{C - A}{Mr^2}. \tag{3.144}$$

In first approximation, the first term on the l.h.s. increases as r^3. Then, deriving the above equation, we get

$$r\,s' = \frac{3}{2}f - 3\,\frac{C - A}{M\,r^2}. \tag{3.145}$$

Comparison with Eq. (3.144) gives

$$\begin{aligned} r\,s' + 2s &= \frac{5}{2}f, \\ r\,s' &= \frac{5}{2}f - 2s, \\ q = \frac{r\,s'}{s} &= \frac{5}{2}\frac{f}{s} - 2. \end{aligned} \tag{3.146}$$

In particular, if f is evaluated with respect to the surface of the planet, one gets

$$q_1 = \frac{5}{2}\frac{f}{s_1} - 2. \tag{3.147}$$

For the Earth $q_1 = 0.57$. Thus K takes a value very close to 1. Assuming $K = 1$ (this hypothesis can always be made when the planet's density is not exceedingly inhomogeneous), Eq. (3.143) becomes

$$r_1^5\,D_1\sqrt{\frac{5}{2}\frac{f}{s_1} - 1} \simeq \int 5\,r^4\,D\,\mathrm{d}r, \tag{3.148}$$

or, using Eq. (3.137),

$$\sqrt{\frac{5}{2}\frac{f}{s_1} - 1} \simeq \frac{5}{2} - \frac{5}{2}\frac{C}{C - A}\left(s_1 - \frac{1}{2}f\right). \tag{3.149}$$

For the Earth

$$f = 1/288, \tag{3.150}$$

$$\frac{C}{C - A} = 305, \tag{3.151}$$

and thus

$$s_1 = 1/297, \tag{3.152}$$

which is in excellent agreement with experiment. Substituting these values in Eq. (3.145) we find

$$\frac{1}{297} = \frac{1}{2}\frac{1}{288} + \frac{3}{2}\frac{C - A}{Mr_1^2}, \tag{3.153}$$

from which

$$C - A = \frac{1}{920} M\, r_1^2, \tag{3.154}$$

and, using Eq. (3.151),

$$C = 0.332\, M\, r_1^2 \tag{3.155}$$

while for a constant density we would have $I = 0.4Mr_1^2$.

Values reported at the Madrid Conference [6] are

$$\begin{aligned} R_e &\simeq 6378, && (3.156)\\ R_p &= 6357, && (3.157)\\ s &= 1/297, && (3.158)\\ D_1 &= 5.515. && (3.159) \end{aligned}$$

Let us suppose that the density of the Earth can be expressed in the form:

$$\rho = a + b\,r^2 + c\,r^4. \tag{3.160}$$

We want to find the unknown coefficients using the conditions

$$\begin{aligned} D_1 &= 5.515,\\ \rho_1 &= 2.5,\\ \mathcal{I} &= 0.332\, M\, r_1^2. \end{aligned} \tag{3.161}$$

We have

$$\rho_1 = a + b\,r_1^2 + c\,r_1^4, \tag{3.162}$$

$$\begin{aligned} \frac{1}{3} r_1^3 D_1 &= \int_0^{r_1} \left(a\,r^2 + b\,r^4 + c\,r^6\right) \mathrm{d}r\\ &= \frac{1}{3} a\,r_1^3 + \frac{1}{5} b\,r_1^5 + \frac{1}{7} c\,r_1^7, \end{aligned} \tag{3.163}$$

that is,

$$D_1 = a + \frac{3}{5} b\,r_1^2 + \frac{3}{7} c\,r_1^4. \tag{3.164}$$

[6] @ The author does not provide details about this reference.

Moreover,

$$\begin{aligned} \mathcal{I} &= \frac{8\pi}{3} \int_0^{r_1} \left(a\,r^4 + b\,r^6 + c\,r^8 \right) \mathrm{d}r \\ &= \frac{8\pi}{3} \left(\frac{1}{5}\,a\,r_1^5 + \frac{1}{7}\,b\,r_1^7 + \frac{1}{9}\,c\,r_1^9 \right), \end{aligned} \tag{3.165}$$

that is,

$$\frac{\mathcal{I}}{r_1^5} = \frac{8\pi}{3} \left(\frac{1}{5}\,a + \frac{1}{7}\,b\,r_1^2 + \frac{1}{9}\,c\,r_1^4 \right). \tag{3.166}$$

On the other hand,

$$\frac{M}{r_1^3} = \frac{4}{3}\pi\, D_1 + \frac{4}{3}\,M \left(a + \frac{3}{5}\,b\,r_1^2 + \frac{3}{7}\,c\,r_1^4 \right), \tag{3.167}$$

from which it follows that

$$\frac{\mathcal{I}}{M r_1^2} = \frac{2}{5}\, \frac{a + \frac{5}{7}\,b\,r_1^2 + \frac{5}{9}\,c\,r_1^4}{a + \frac{3}{5}\,b\,r_1^2 + \frac{3}{7}\,c\,r_1^4}. \tag{3.168}$$

The l.h.s. of Eqs. (3.162), (3.164), (3.168) involve known quantities, so that the set of linear equations in the unknowns a, br_1^2, cr_1^4 is

$$\begin{gathered} a + b\,r_1^2 + c\,r_1^4 = \rho_1, \\ a\,(1 - \delta) + br_1^2 \left(\frac{5}{7} - \delta\,\frac{3}{5} \right) + cr_1^4 \left(\frac{5}{9} - \delta\,\frac{3}{7} \right) = 0, \\ a + \frac{3}{5}\,b\,r_1^2 + \frac{3}{7}\,c\,r_1^4 = D_1, \end{gathered} \tag{3.169}$$

where we have set

$$\delta = \frac{5}{2}\,\frac{\mathcal{I}}{M r_1^2}. \tag{3.170}$$

Moreover let us put

$$\epsilon = \rho_1 / D_1. \tag{3.171}$$

From Eqs. (3.169) it then follows that

$$\begin{aligned} a\,(1 - \delta) + br_1^2 \left(\frac{5}{7} - \frac{3}{5}\,\delta \right) + cr_1^4 \left(\frac{5}{9} - \frac{3}{7}\,\delta \right) &= 0, \\ a\,(1 - \epsilon) + br_1^2 \left(1 - \frac{3}{5}\,\epsilon \right) + cr_1^4 \left(1 - \frac{3}{7}\,\delta \right) &= 0, \end{aligned} \tag{3.172}$$

from which

$$br_1^2 = -\frac{\dfrac{4}{9} - \dfrac{4}{7}\delta + \dfrac{8}{63}\epsilon}{\dfrac{10}{63} - \dfrac{6}{35}\delta + \dfrac{4}{147}\epsilon}\, a, \tag{3.173}$$

$$cr_1^4 = \frac{\dfrac{2}{7} - \dfrac{2}{5}\delta + \dfrac{4}{35}\epsilon}{\dfrac{10}{63} - \dfrac{6}{35}\delta + \dfrac{4}{147}\epsilon}\, a. \tag{3.174}$$

It follows that

$$\begin{aligned} a &= \ell\,(175 - 189\,\delta + 30\,\epsilon)\,, \\ br_1^2 &= -\ell\,(490 - 630\,\delta + 140\,\epsilon)\,, \\ cr_1^4 &= \ell\,(315 - 441\,\delta + 126\,\epsilon)\,. \end{aligned} \tag{3.175}$$

Substituting in Eq. (3.169), we get

$$\begin{aligned} 16\,\epsilon\,\ell &= \rho_1, \\ 0 &= 0, \\ 16\,\ell &= D_1. \end{aligned} \tag{3.176}$$

From which, invoking Eq. (3.171)

$$l = D_1/16. \tag{3.177}$$

Finally, we get

$$\begin{aligned} \rho = \; & \frac{(175 - 189\,\delta + 30\,\epsilon)\;D_1}{16} \\ & - \frac{(490 - 630\,\delta + 140\,\epsilon)\;D_1}{16}\,\frac{r^2}{r_1^2} \\ & + \frac{(315 - 441\,\delta + 126\,\epsilon)\;D_1}{16}\,\frac{r^4}{r_1^4}, \end{aligned} \tag{3.178}$$

with δ and ϵ defined in Eqs. (3.170) and (3.171). For the Earth, from Eqs. (3.175), it follows that

$$\delta = 0.83, \quad \epsilon = 0.45. \tag{3.179}$$

Substituting these values into Eq. (3.178), we find

$$\rho_1 = D_1 \left(1.977 - 1.881\,\frac{r^2}{r_1^2} + 0.354\,\frac{r^4}{r_1^4} \right). \tag{3.180}$$

The maximum value for the density (at the center of the Earth) then would be

$$\rho_0 = 1.977\, D_1 = 1.977 \cdot 5.515 = 10.90. \qquad (3.181)$$

From

$$1.977 - 1.881 + 0.354 = 0.45, \qquad (3.182)$$

$$1.977 - 1.881 \cdot \frac{3}{5} + 0.354 \cdot \frac{3}{7} = 1.000, \qquad (3.183)$$

$$1.977 \cdot \frac{2}{5} - 1.881 \cdot \frac{2}{7} + 0.354 \cdot \frac{2}{9} = 0.332, \qquad (3.184)$$

setting

$$\rho = D_1 \left(\alpha + \beta \frac{r^2}{r_1^2} + \gamma \frac{r^4}{r_1^4} \right), \qquad (3.185)$$

the coefficients α, β, γ are seen to satisfy the equations

$$\begin{aligned} \alpha + \beta + \gamma &= \frac{\rho_1}{D_1}, \\ \alpha + \frac{3}{5}\beta + \frac{3}{7}\gamma &= 1, \\ \frac{2}{5}\alpha + \frac{2}{7}\beta + \frac{2}{9}\gamma &= \frac{\mathcal{I}}{M r_1^2}, \end{aligned} \qquad (3.186)$$

which are simpler than Eqs. (3.169).

4. DETERMINATION OF A FUNCTION FROM ITS MOMENTS

Let y be a function of x:

$$y = y(x), \qquad (3.187)$$

and suppose that, for $x^2 > a^2$, we have $y = 0$ and an integral

$$\int_{-\infty}^{\infty} |y| \, \mathrm{d}x \qquad (3.188)$$

that is finite. Let us define the moments μ_0, μ_1,..., μ_n of order $0, 1, 2, ..., n$ as

$$\mu_0 = \int y \, \mathrm{d}x,$$

$$\mu_1 = \int x\, y \, \mathrm{d}x,$$

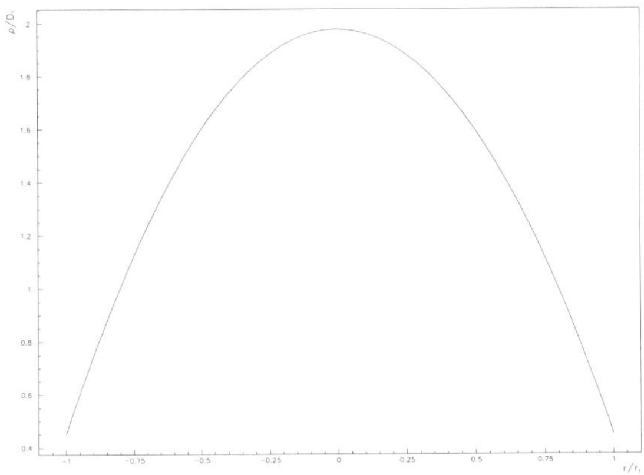

Fig. 3.1. The density of the Earth ρ as a function of the distance from its center (see the text for notation).

$$\begin{aligned} &\ldots, \\ \mu_n &= \int x^n\, y\, \mathrm{d}x, \end{aligned} \tag{3.189}$$

and set

$$z(t) = \int y\, \mathrm{e}^{ixt}\, \mathrm{d}x, \tag{3.190}$$

so that

$$y = \frac{1}{2\pi} \int_{-\infty}^{\infty} \mathrm{e}^{-ixt}\, z\, \mathrm{d}t. \tag{3.191}$$

From Eq. (3.190), it then follows that

$$\begin{aligned} \frac{\mathrm{d}z}{\mathrm{d}t} &= i \int x\, y\, \mathrm{e}^{ixt}\, \mathrm{d}x, \\ &\ldots, \\ \frac{\mathrm{d}^n z}{\mathrm{d}t^n} &= i^n \int x^n\, y\, \mathrm{e}^{ixt}\, \mathrm{d}x. \end{aligned} \tag{3.192}$$

For $t = 0$, we have $z(0) = \mu_0$ and, in general,

$$\left(\frac{\mathrm{d}^n z}{\mathrm{d}t^n}\right)_0 = i^n\, \mu_n. \tag{3.193}$$

From the assumptions mentioned above, z can be expanded in a absolutely convergent Mac-Laurin series:

$$z = \sum_0^\infty \mu_n \frac{(it)^n}{n!}. \tag{3.194}$$

Substituting in Eq. (3.191), we then have

$$y = \frac{1}{2\pi} \int_{-\infty}^{\infty} \mathrm{e}^{-ixt} \sum_0^\infty \mu_n \frac{(it)^n}{n!} \, \mathrm{d}t, \tag{3.195}$$

where, obviously, the integral and the series cannot be inverted. We can also write

$$\begin{aligned} y &= \frac{1}{\pi} \int_0^\infty \cos xt \sum_0^\infty (-1)^r \mu_{2r} \frac{t^{2r}}{(2r)!} \, \mathrm{d}t \\ &+ \frac{1}{\pi} \int_0^\infty \sin xt \sum_0^\infty (-1)^r \mu_{2r+1} \frac{t^{2r+1}}{(2r+1)!} \, \mathrm{d}t. \end{aligned} \tag{3.196}$$

Example 1. Let $y = 1$ for $0 < x < 1$ and $y = 0$ for $x < 0$ and $x > 1$. The moments of this function will be

$$\mu_0 = 1, \quad \mu_1 = \frac{1}{2}, \quad \ldots \quad \mu_n = \frac{1}{n+1}.$$

Let us substitute them in Eq. (3.196) and note that, in the present case,

$$\begin{aligned} \sum_0^\infty (-1)^r \mu_{2r} \frac{t^{2r}}{(2r)!} &= \sum_0^\infty (-1)^r \frac{t^{2r}}{(2r+1)!} \\ = \frac{1}{t} \sum_0^\infty (-1)^r \frac{t^{2r+1}}{(2r+1)!} &= \frac{\sin t}{t}, \end{aligned} \tag{3.197}$$

$$\begin{aligned} \sum_0^\infty (-1)^r \mu_{2r+1} \frac{t^{2r+1}}{(2r+1)!} &= \frac{1}{t} \sum_0^\infty (-1)^r \frac{t^{2r+2}}{(2r+2)!} \\ = \frac{1}{t} \left(1 - \sum_0^\infty (-1)^r \frac{t^{2r}}{(2r)!} \right) &= \frac{1 - \cos t}{t}. \end{aligned} \tag{3.198}$$

We have

$$\begin{aligned} y &= \frac{1}{\pi} \int_0^\infty \frac{\cos xt \sin t + \sin xt \, (1 - \cos t)}{t} \, \mathrm{d}t \\ &= \frac{1}{\pi} \int_0^\infty \frac{\sin (1-x)t}{t} \, \mathrm{d}t + \frac{1}{\pi} \int_0^\infty \frac{\sin xt}{t} \, \mathrm{d}t. \end{aligned} \tag{3.199}$$

The first integral takes the values $\pi/2$ for $x < 1$ and $-\pi/2$ for $x > 1$. The second integral takes the values $-\pi/2$ for $x < 0$ and $\pi/2$ for $x > 0$. We then have:

$$\begin{aligned} &\text{for } x < 0, && y = \frac{1}{2} - \frac{1}{2} = 0; \\ &\text{for } 0 < x < 1, && y = \frac{1}{2} + \frac{1}{2} = 1; \\ &\text{for } x > 1, && y = -\frac{1}{2} + \frac{1}{2} = 0; \end{aligned} \tag{3.200}$$

as we supposed.

Example 2. Let $y = 0$ for $x < 0$ and $y = \mathrm{e}^{-x}$ for $x > 0$. Now the conditions above are not satisfied, and we have to abandon mathematical rigor somewhat. We have

$$\mu_n = n!. \tag{3.201}$$

Substituting in Eq. (3.195), we get

$$\sum_0^\infty \mu_n \frac{(it)^n}{n!} = \sum_0^\infty (it)^n = \frac{1}{1 - it}, \tag{3.202}$$

which is valid only for $t^2 < 1$, since otherwise the expansion does not converge. However, let us suppose that we can always write

$$\sum_0^\infty (it)^n = \frac{1}{1 - it}, \tag{3.203}$$

so that Eq. (3.195) becomes

$$y = \frac{1}{2\pi} \int_{-\infty}^{\infty} \frac{\mathrm{e}^{-ixt}}{1 - it}\, \mathrm{d}t. \tag{3.204}$$

If in the formula (14bis) in Sec. 2.26),

$$\int_{-\infty}^{\infty} \frac{\mathrm{e}^{ix}\,\mathrm{d}x}{a + ix} = \begin{cases} 2\pi\,\mathrm{e}^{-a}, & a > 0, \\ 0, & a < 0, \end{cases} \tag{3.205}$$

we replace a with x and x with $-tx$, we get

$$\begin{aligned} \int_{-\infty}^{\infty} \frac{\mathrm{e}^{-ixt}\, x\,\mathrm{d}t}{x - itx} &= \int_{-\infty}^{\infty} \frac{\mathrm{e}^{-ixt}\,\mathrm{d}t}{1 - it} \\ &= \begin{cases} 2\pi\,\mathrm{e}^{-x}, & \textit{for } x > 0, \\ 0, & \textit{for } x < 0, \end{cases} . \end{aligned} \tag{3.206}$$

resulting in

$$y = \begin{cases} e^{-x}, & \textit{for } x > 0, \\ 0, & \textit{for } x < 0, \end{cases} \tag{3.207}$$

as we supposed.

Example 3. Let $y = e^{-x^2}$. We then have

$$\mu_{2r+1} = 0, \tag{3.208}$$

$$\begin{aligned} \mu_{2r} &= \left(\frac{2r-1}{2}\right)! \\ &= \sqrt{\pi}\cdot\frac{1}{2}\cdot\frac{3}{2}\cdot\ldots\cdot\frac{2r-1}{2}, \\ &= \sqrt{\pi}\,\frac{(2r)!}{r!\cdot 2^{2r}}. \end{aligned} \tag{3.209}$$

It follows that

$$\sum_0^\infty (-1)^r\,\mu_{2r}\,\frac{t^{2r}}{(2r)!} = \sqrt{\pi}\sum_0^\infty (-1)^r\,\frac{t^{2r}}{2^{2r}\,r!} = \sqrt{\pi}e^{-\frac{t^2}{4}}. \tag{3.210}$$

Substituting in Eq. (3.195),

$$\begin{aligned} y &= \frac{1}{2\sqrt{\pi}}\int_{-\infty}^{\infty} e^{-ixt}\,e^{-\frac{t^2}{4}}\,dt \\ &= \frac{1}{2\sqrt{\pi}}e^{-x^2}\int_{-\infty}^{\infty} e^{-\left(\frac{t}{2}+ix\right)^2}\,dt \\ &= \frac{1}{\sqrt{\pi}}e^{-x^2}\int_{-\infty}^{\infty} e^{-\left(\frac{t}{2}+ix\right)^2}\,d\left(\frac{t}{2}+ix\right) \\ &= e^{-x^2}. \end{aligned} \tag{3.211}$$

as we supposed.

Example 4. Let us consider the problem of finding the function whose moments are

$$\mu_0 = 1, \tag{3.212}$$

$$\mu_1 = \frac{1}{4}, \tag{3.213}$$

$$\mu_2 = \frac{1}{9}, \tag{3.214}$$

$$\ldots,$$

$$\mu_n = \frac{1}{(n+1)^2}. \tag{3.215}$$

We have

$$\begin{aligned}
z &= \sum_0^\infty \mu_n \frac{(it)^n}{n!} = \sum_0^\infty \frac{(it)^n}{(n+1)!\,(n+1)}, && (3.216)\\
z\,it &= \sum_1^\infty \frac{(it)^q}{q!\,q}, && (3.217)\\
i\,(z't + z) &= i \sum_1^\infty \frac{(it)^{q-1}}{q!}, && (3.218)\\
it\,(z't + z) &= \sum_1^\infty \frac{(it)^q}{q!} = \mathrm{e}^{it} - 1, && (3.219)\\
z' &= -\frac{z}{t} + \frac{\mathrm{e}^{it} - 1}{it^2}. && (3.220)
\end{aligned}$$

Then, noting that $z = 1$ for $t = 0$, we get

$$z = \frac{1}{t} \int_0^t \frac{\mathrm{e}^{it} - 1}{it}\,\mathrm{d}t. \qquad (3.221)$$

Substitution into Eq. (3.195) or (3.191) gives

$$\begin{aligned}
y &= \frac{1}{2\pi} \int_{-\infty}^{\infty} \mathrm{e}^{-ixt}\,\mathrm{d}t \cdot \frac{1}{t} \int_0^t \frac{\mathrm{e}^{it_1} - 1}{it_1}\,\mathrm{d}t_1\\
&= \frac{1}{\pi} \left(\int_0^\infty \frac{\cos xt}{t}\,\mathrm{d}t \int_0^t \frac{\sin t_1}{t_1}\,\mathrm{d}t_1 \right.\\
&\quad \left. + \int_0^\infty \frac{\sin xt}{t}\,\mathrm{d}t \int_0^t \frac{1 - \cos t_1}{t_1}\,\mathrm{d}t_1 \right)\\
&= \frac{1}{\pi} \int_0^{\frac{\pi}{4}} \frac{\mathrm{d}\theta}{\sin\theta\,\cos\theta} \int_0^\infty \left(\frac{1}{r}\right) [\cos(xr\cos\theta)\,\sin(r\sin\theta)\\
&\quad + \sin(xr\cos\theta)\,(1 - \cos(r\sin\theta))]\,\mathrm{d}r\\
&= \frac{1}{\pi} \int_0^{\frac{\pi}{4}} \frac{\mathrm{d}\theta}{\sin\theta\,\cos\theta} \cdot\\
&\quad \cdot \int_0^\infty \frac{1}{r}\,\{\sin[r(\sin\theta - x\cos\theta)] + \sin(rx\cos\theta)\}\,\mathrm{d}r. \qquad (3.222)
\end{aligned}$$

For $0 < \theta < \pi/4$, the second integral takes the value

$$\begin{array}{ll} 0, & \text{if } \; x < 0,\\ \pi, & \text{if } \; 0 < x < \tan\theta. \end{array} \qquad (3.223)$$

For $0 < x < 1$, we then have

$$y = \int_{\arctan x}^{\frac{\pi}{4}} \frac{\mathrm{d}\theta}{\sin\theta\,\cos\theta} = [\log\tan\theta]_{\frac{\pi}{4}}^{\arctan x} = -\log x. \qquad (3.224)$$

The function is thus determined for all values of x:

$$\begin{aligned} &\text{for } x < 0, \quad y = 0, \\ &\text{for } 0 < x < 1, \quad y = -\log x, \\ &\text{for } x > 1, \quad y = 0. \end{aligned} \tag{3.225}$$

It is easy to check that the function y defined above satisfies the required conditions.

Example 5. Let $y\,dx$ be the probability that the distance between two points (i.e., surface elements) belonging to a circle of unitary radius lies between x and $x + \mathrm{d}x$. From (7) in Sec. 1.21, the moments of y are

$$\mu_n = \frac{4}{n+4} \frac{(n+1)!}{(1+n/2)!\,(1+n/2)!}. \tag{3.226}$$

In particular,

$$\mu_0 = 1, \tag{3.227}$$

$$\mu_1 = \frac{128}{45\pi}, \tag{3.228}$$

$$\mu_2 = 1, \tag{3.229}$$

$$\ldots.$$

Substitution into (3.194) gives

$$\begin{aligned} z &= \sum_0^\infty \mu_n \frac{(it)^n}{n!} = \sum_0^\infty 4 \frac{n+1}{n+4} \frac{(it)^n}{(1+n/2)!\,(1+n/2)!} \\ &= \sum_0^\infty \frac{2\,(n+1)(it)^n}{(1+n/2)!\,(2+n/2)!}. \end{aligned} \tag{3.230}$$

Let us set $z = z_1 + iz_2$. We get

$$z_1 = \sum_0^\infty (-1)^r\, 2\,(2r+1) \frac{t^{2r}}{(r+1)!\,(r+2)!}, \tag{3.231}$$

$$\int_0^t z_1\,\mathrm{d}t = \sum_0^\infty (-1)^r\, 2 \frac{t^{2r+1}}{(r+1)!\,(r+2)!}, \tag{3.232}$$

$$\begin{aligned} t^2 \int_0^t z_1\,\mathrm{d}t &= \sum_0^\infty (-1)^r\, 2 \frac{t^{2r+3}}{(r+1)!\,(r+2)!} \\ &= -\sum_1^\infty (-1)^s\, 2 \frac{t^{2s+1}}{s!\,(s+1)!} \\ &= -\sum_1^\infty (-1)^s\, 2 \frac{(2t/2)^{2s+1}}{s!\,(s+1)!} \\ &= 2t - 2I_1(2t), \end{aligned} \tag{3.233}$$

whence

$$\int_0^t z_1\,\mathrm{d}t = \frac{2}{t} - 2\,\frac{I_1(2t)}{t^2}, \tag{3.234}$$

$$z_1 = -\frac{2}{t^2} - 4\,\frac{I_1'(2t)}{t^2} + \frac{I_1(2t)}{t^3}; \tag{3.235}$$

and, since

$$I_1'(2t) = I_0(2t) - \frac{1}{2t}\,I_1(2t), \tag{3.236}$$

it follows that

$$z_1 = -\frac{2 + 4\,I_0(2t)}{t^2} + \frac{I_1(2t)}{t^3}. \tag{3.237}$$

For z_2 we have instead

$$z_2 = \sum_0^\infty (-1)^r\,4\,(r+1)\,\frac{t^{2r+1}}{(r+3/2)!\,(r+5/2)!}, \tag{3.238}$$

$$\int_0^t z_2\,\mathrm{d}t = \sum_0^\infty (-1)^r\,2\,\frac{t^{2r+2}}{(r+3/2)!\,(r+5/2)!}, \tag{3.239}$$

$$t^2 \int_0^t z_2\,\mathrm{d}t = \sum_0^\infty (-1)^r\,2\,\frac{t^{2r+4}}{(r+3/2)!\,(r+5/2)!}, \tag{3.240}$$

and so on.

Example 6. Let $y(r)\,dr$ be the probability that the distance between two points belonging to two concentric spherical surfaces, one with unit radius and the other with a radius $a < 1$, lies between r and $r + dr$. In this case the moments of $y(r)$ are

$$\mu_0 = 1, \tag{3.241}$$

$$\mu_1 = 1 + \frac{1}{3}\,a^2, \tag{3.242}$$

$$\ldots,$$

$$\mu_n = \frac{(1+a)^{n+2} - (1-a)^{n+2}}{2(n+2)a}, \tag{3.243}$$

$$\ldots;$$

see Sec. 2.38.6. Substituting in Eq. (3.194) we get

$$z = \frac{(1+a)^2}{2a} \sum_0^\infty \frac{(1+a)^n\,(it)^n}{n!\,(n+2)} - \frac{(1-a)^2}{2a} \sum_0^\infty \frac{(1-a)^n\,(it)^n}{n!\,(n+2)}, \tag{3.244}$$

$$\begin{aligned}
\int_0^t z\,\mathrm{d}t &= \frac{1+a}{2a\,i}\sum_0^\infty \frac{(1+a)^{n+1}\,(it)^{n+1}}{(n+2)!} \\
&\quad - \frac{1-a}{2a\,i}\sum_0^\infty \frac{(1-a)^{n+1}\,(it)^{n+1}}{(n+2)!},
\end{aligned} \tag{3.245}$$

$$\begin{aligned}
t\int_0^t z\,\mathrm{d}t &= -\frac{1}{2a}\sum_0^\infty \frac{(1+a)^{n+2}\,(it)^{n+2}}{(n+2)!} \\
&\quad + \frac{1}{2a}\sum_0^\infty \frac{(1-a)^{n+2}\,(it)^{n+2}}{(n+2)!}, \\
&= -\frac{1}{2a}\left(\mathrm{e}^{i\,(1+a)\,t} - 1 - i\,(1+a)\,t\right) \\
&\quad + \frac{1}{2a}\left(\mathrm{e}^{i\,(1-a)\,t} - 1 - i\,(1-a)\,t\right) \\
&= \frac{\mathrm{e}^{i\,(1-a)\,t} - \mathrm{e}^{i\,(1+a)\,t}}{2a} + i\,t,
\end{aligned} \tag{3.246}$$

$$\int_0^t z\,\mathrm{d}t = \frac{\mathrm{e}^{i\,(1-a)\,t} - \mathrm{e}^{i\,(1+a)\,t}}{2at} + i, \tag{3.247}$$

$$\begin{aligned}
z &= \frac{\mathrm{e}^{i\,(1+a)\,t} - \mathrm{e}^{i\,(1-a)\,t}}{2at^2} \\
&\quad - i\,\frac{(1+a)\,\mathrm{e}^{i\,(1+a)\,t} - (1-a)\,\mathrm{e}^{i\,(1-a)\,t}}{2at}.
\end{aligned} \tag{3.248}$$

However,

$$\begin{aligned}
\int_{-\infty}^{\infty} z\,\mathrm{e}^{-i\,r\,t}\,\mathrm{d}t &= \left[\mathrm{e}^{-i\,r\,t}\int_0^t z\,\mathrm{d}t\right]_{-\infty}^{\infty} \\
&\quad - \int_{-\infty}^{\infty} -i\,r\,\mathrm{e}^{-i\,r\,t}\,\mathrm{d}t \int_0^t z(t_1)\,\mathrm{d}t_1 \\
&= \left[\mathrm{e}^{-i\,r\,t}\left(\frac{\mathrm{e}^{i\,(1-a)\,t} - \mathrm{e}^{i\,(1+a)\,t}}{2at} + i\right)\right]_{-\infty}^{\infty} \\
&\quad + \frac{i\,r}{2a}\int_{-\infty}^{\infty} \frac{\mathrm{e}^{i\,(1-a-r)\,t} - 1}{t}\,\mathrm{d}t \\
&\quad - \frac{i\,r}{2a}\int_{-\infty}^{\infty} \frac{\mathrm{e}^{i\,(1+a-r)\,t} - 1}{t}\,\mathrm{d}t \\
&\quad - r\int_{-\infty}^{\infty} \mathrm{e}^{-i\,r\,t}\,\mathrm{d}t.
\end{aligned} \tag{3.249}$$

The first and fourth terms on the l.h.s. are indeterminate, and their mean values are zero. The second term takes the values

$$\begin{aligned} &\pi\,\frac{r}{2a}, \quad \text{for} \quad r < 1 - a, \\ &\pi\,\frac{r}{2a}, \quad \text{for} \quad r > 1 - a. \end{aligned} \tag{3.250}$$

The third term assumes the values

$$\begin{aligned} &\pi\,\frac{r}{2a}, \quad \text{for} \quad r < 1 + a, \\ &-\pi\,\frac{r}{2a}, \quad \text{for} \quad r > 1 + a. \end{aligned} \tag{3.251}$$

Then, from Eq. (3.191), changing x into r, it follows that

$$\begin{aligned} y &= 0, \quad \text{for} \quad r < 1 - a, \\ y &= \frac{r}{2a}, \quad \text{for} \quad 1 - a < r < 1 + a, \\ y &= 0, \quad \text{for} \quad r > 1 + a; \end{aligned} \tag{3.252}$$

see Sec. 2.38.6.

5. PROBABILITY CURVES

(1) Probability that two points belonging to two concentric spherical surfaces with radii a and $b < a$ are at a distance r from each other: The probability density $y = \mathrm{d}P/\mathrm{d}r$ is given by

$$\begin{aligned} y &= 0, \quad \text{for} \quad r < a - b, \\ y &= \frac{r}{2ab}, \quad \text{for} \quad a - b < r < a + b, \\ y &= 0, \quad \text{for} \quad r > a + b; \end{aligned} \tag{3.253}$$

see Secs. 2.39.6 and 3.4. Its moments then are

$$\mu_n = \int r^n\, y\, \mathrm{d}r = \frac{(a+b)^{n+2} - (a-b)^{n+2}}{2ab(n+2)}. \tag{3.254}$$

In particular,

$$\mu_{-1} = \frac{1}{a}, \tag{3.255}$$

$$\mu_0 = 1, \tag{3.256}$$

$$\mu_1 = a + \frac{1}{3}\frac{b^2}{a}, \tag{3.257}$$

$$\mu_2 = a^2 + b^2, \tag{3.258}$$

$$\ldots.$$

(2) Probability density that two points on a length ℓ segment are at a distance r from each other:

$$\begin{aligned} y(r) &= \frac{2(\ell - r)}{l^2}, \quad 0 < r < \ell, \\ y(r) &= 0, \quad \text{otherwise.} \end{aligned} \tag{3.259}$$

Its moments are

$$\mu_n = \frac{2\ell^n}{(n+1)\,(n+2)}. \tag{3.260}$$

In particular

$$\mu_0 = \ell, \tag{3.261}$$

$$\mu_1 = \frac{\ell}{3}, \tag{3.262}$$

$$\mu_2 = \frac{\ell^2}{6}, \tag{3.263}$$

$$\ldots.$$

(3) Probability density that two points on two co-planar and concentric circumferences with radii a and $b < a$ are at a distance r from each other:

$$y = \frac{2r}{\pi\sqrt{-(a^2 - b^2)^2 + 2(a^2 + b^2)r^2 - r^4}}. \tag{3.264}$$

For $b = a$, we have simply

$$y = \frac{2r}{\pi\sqrt{4a^2r^2 - r^4}}. \tag{3.265}$$

6. EVALUATION OF THE INTEGRAL $\int_0^{\pi/2} \frac{\sin kx}{\sin x}\,\mathrm{d}x$

From the relation

$$\sin\,(k+2)\,x \;-\; \sin\,kx \;=\; 2\,\cos\,(k+1)\,x\,\sin\,x, \tag{3.266}$$

we deduce

$$\frac{\sin\,(k+2)\,x}{\sin\,x} \;=\; \frac{\sin\,kx}{\sin\,x} + 2\,\cos\,(k+1)\,x. \tag{3.267}$$

Integrating between 0 and $\pi/2$ and setting

$$y(k) \;=\; \int_0^{\pi/2} \frac{\sin\,kx}{\sin\,x}\,\mathrm{d}x, \tag{3.268}$$

we find

$$y(k) \;-\; y(k+2) \;=\; -\frac{2}{k+1}\,\sin\,\frac{\sin\,(k+1)\,\pi}{2}. \tag{3.269}$$

Considering this relation with k replaced by $k+2,\ k+4,\ \ldots,\ k+2n$, respectively, and noting that

$$\begin{aligned}\lim_{k\to\infty}\int_0^{\pi/2}\frac{\sin\,kx}{\sin\,x}\,\mathrm{d}x \;&=\; \lim_{k\to\infty}\int_0^{\infty}\frac{\sin\,kx}{x}\,\mathrm{d}x\\ &=\; \int_0^{\infty}\frac{\sin\,x}{x}\,\mathrm{d}x \;=\; \frac{\pi}{2},\end{aligned} \tag{3.270}$$

that is,

$$y(\infty) \;=\; \frac{\pi}{2}, \tag{3.271}$$

we get

$$y(k) \;=\; \frac{\pi}{2} - \sum_0^{\infty} \frac{2}{k+1+2r}\,\sin\,\frac{(k+1+2r)\,\pi}{2}, \tag{3.272}$$

that is,

$$y(k) \;=\; \frac{\pi}{2} - \sin\,\frac{(k+1)\,\pi}{2}\sum_0^{\infty}\frac{2(-1)^r}{(k+1)+2r}. \tag{3.273}$$

In other words, we have

$$y(k) \;=\; \frac{\pi}{2} - 2\left(\frac{1}{k+1} - \frac{1}{k+3} + \frac{1}{k+5} - \ldots\right)\sin\,\frac{(k+1)\pi}{2}. \tag{3.274}$$

Let us consider the following function defined for positive α:

$$\phi(\alpha) = 1 - \frac{1}{1+\alpha} + \frac{1}{1+2\alpha} - \frac{1}{1+3\alpha} + \ldots, \tag{3.275}$$

while, for $\alpha = 0$, we define

$$\phi(0) = \lim_{\alpha \to 0} \phi(\alpha) = \frac{1}{2}. \tag{3.276}$$

This is always an increasing function of its argument taking values between 1/2 and 1.

Notice that the identity

$$\phi(\alpha) + \frac{1}{1+\alpha}\,\phi\left(\frac{\alpha}{1+\alpha}\right) = 1, \tag{3.277}$$

enables us to determine the series expansion of $\phi(\alpha)$ for $\alpha \to 0$. We can also give an integral formula for $\phi(\alpha)$:

$$\phi(\alpha) = \int_0^1 \frac{\mathrm{d}x}{1 + x^\alpha}, \tag{3.278}$$

from which we deduce the following particular values:

$$\phi(0) = \lim_{\alpha \to 0} \phi(\alpha) = \frac{1}{2}, \quad \phi(1) = \log 2, \quad \phi(2) = \frac{\pi}{4}. \tag{3.279}$$

For integer α, we have

$$\frac{1}{1 + x^\alpha} = \left(\frac{1}{1 - x\,\delta} + \frac{1}{1 - x\,\delta^3} + \ldots + \frac{1}{1 - x\,\delta^{2\alpha-1}}\right)\frac{1}{\alpha}, \tag{3.280}$$

where $\delta = \mathrm{e}^{i\pi/\alpha}$, that is the first αth root of -1. Equation (3.280) can be immediately verified by expanding the two sides of it in a power series of x or of $1/\alpha$, depending on whether α is less than or greater than 1, respectively.

Substituting Eq. (3.280) in Eq. (3.278), for integer α we have

$$\begin{aligned}\phi(\alpha) = {} & - \frac{1}{\alpha}\left[\frac{1}{\delta}\log(1 - \delta) + \frac{1}{\delta^3}\log(1 - \delta^3) + \ldots\right.\\ & \left. + \frac{1}{\delta^{2\alpha-1}}\log(1 - \delta^{2\alpha-1})\right];\end{aligned} \tag{3.281}$$

or, noting that $\delta^{2\alpha} = 1$,

$$\begin{aligned}\phi(\alpha) = {} & - \frac{1}{\alpha}\left[\delta^{2\alpha-1}\log(1 - \delta) + \delta^{2\alpha-3}\log(1 - \delta^3) + \ldots\right.\\ & \left. + \delta\log(1 - \delta^{2\alpha-1})\right].\end{aligned} \tag{3.282}$$

Note that the imaginary part of each logarithm on the r.h.s. of Eq. (3.282) is univocally determined when it belongs to the interval $(-i\pi/2, i\pi/2)$. In view of the fact that the terms on the r.h.s. of Eq. (3.282) are complex conjugate in pairs and that

$$\log\,(1-\delta^r) = \log\left(2\sin\frac{r\,\pi}{2}\right) + i\,\frac{r\,\pi}{2\alpha} - i\,\frac{\pi}{2}, \tag{3.283}$$

$$\delta^{2\alpha-r} = \cos\frac{r\,\pi}{\alpha} - i\,\sin\frac{r\,\pi}{\alpha}, \tag{3.284}$$

Eq. (3.282) becomes

$$\begin{aligned}\phi(\alpha) &= \cos\frac{\pi}{\alpha}\log\left(2\sin\frac{\pi}{2\alpha}\right)\\ &+ \cos\frac{3\pi}{\alpha}\log\left(2\sin\frac{3\pi}{2\alpha}\right) + \ldots\\ &+ \cos\frac{(2\alpha-1)\pi}{\alpha}\log\left(2\sin\frac{(2\alpha-1)\pi}{2\alpha}\right)\\ &+ \frac{\pi}{2\alpha}\sin\frac{\pi}{\alpha} + \frac{3\pi}{2\alpha}\sin\frac{3\pi}{\alpha} + \ldots\\ &+ \frac{(2\alpha-1)\pi}{\alpha}\sin\frac{(2\alpha-1)\pi}{\alpha}.\end{aligned} \tag{3.285}$$

We deduce the following particular cases:

$$\begin{aligned}\phi(0) &= \frac{1}{2},\quad \phi(1) = \log 2,\quad \phi(2) = \frac{\pi}{4},\\ \phi(3) &= \frac{1}{3}\log 2 + \frac{\sqrt{3}}{9}\pi,\quad \phi(4) = \ldots.\end{aligned} \tag{3.286}$$

We can then evaluate $\phi(\alpha)$ for integer α; however, the repeated use of Eq. (3.277) allows us to evaluate this function for arbitrary values of the independent variable in the form $\alpha/(1+n\alpha)$, with integer α and n. Excluding the trivial case with $\alpha = 0$, for each value of n and varying α between 1 and ∞, we have a discrete group of possible values of the independent variable for which the function can be evaluated; the lowest value is $1/(n+1)$, while the upper limit is $1/n$. The set of values for which the function can be evaluated is then made of a discrete set of points of the form $1/n$; hence it is not possible to evaluate the whole function based on the considerations above and continuity properties.

If we know the value of $\phi(\alpha)$ for an arbitrary value of α, using Eq. (3.277) we can always reduce the problem to the case with $\alpha < 1$, since $\alpha/(1+\alpha) < 1$. Using Eq. (3.277) twice, we can also restrict the problem

Table 3.1. Some values of the function $\phi(\alpha)$.

α	$\phi(\alpha)$	α	$\phi(\alpha)$
0	0.50000	0.26	0.56304
0.02	0.50500	0.28	0.56759
0.04	0.51000	0.30	0.57201
0.06	0.51498	0.32	0.57652
0.08	0.51994	0.34	0.58089
0.10	0.52488	0.36	0.58521
0.12	0.52979	0.38	0.58946
0.14	0.53467	0.40	0.59366
0.16	0.53951	0.42	0.59779
0.18	0.54431	0.44	0.60186
0.20	0.54907	0.46	0.60587
0.22	0.55378	0.48	0.60982
0.24	0.55843	0.50	0.61371

to the case with $\alpha < 1/2$. We then have the results listed in the Table 3.1. [7]

For small α, the following expansion is useful:

$$\phi(\alpha) = \frac{1}{2} + \frac{1}{4}\alpha - \frac{1}{8}\alpha^3 + \frac{1}{4}\alpha^5 - \frac{17}{16}\alpha^7 + \ldots. \tag{3.287}$$

Substituting Eq. (3.275) in Eq. (3.274), we get

$$y(k) = \frac{\pi}{2} - \phi\left(\frac{2}{k+1}\right)\left(\frac{2}{k+1}\right)\sin\frac{(k+1)\pi}{2}. \tag{3.288}$$

We infer

$$\begin{aligned}
y(0) &= 0, \\
y(2) &= 2, \\
y(4) &= 2\left(1 - \frac{1}{3}\right) = \frac{4}{3}, \\
y(6) &= 2\left(1 - \frac{1}{3} + \frac{1}{5}\right), \\
y(8) &= 2\left(1 - \frac{1}{3} + \frac{1}{5} - \frac{1}{7}\right), \\
y(10) &= 2\left(1 - \frac{1}{3} + \frac{1}{5} - \frac{1}{7} + \frac{1}{9}\right), \\
y(1) &= y(3) = y(5) = \ldots = y(2n+1) = \frac{\pi}{2}.
\end{aligned}$$

[7]@ In the original manuscript, only the values corresponding to $\alpha = 0$, $\alpha = 0.40$, and $\alpha = 0.50$ were reported in the Table.

This relation is immediately verified when $k = 1$, while it follows from Eq. (3.269) when k is an arbitrary odd integer. We deduce:

$$\int_0^\infty \frac{\sin x}{x}\,\mathrm{d}x = \lim_{k\to\infty}\int_0^{\pi/2}\frac{\sin kx}{\sin x}\,\mathrm{d}x = \frac{\pi}{2}. \tag{3.289}$$

7. INFINITE PRODUCTS

(1) We have

$$\frac{2}{1}\cdot\frac{2}{3}\cdot\frac{4}{3}\cdot\frac{4}{5}\cdot\frac{6}{5}\cdot\frac{6}{7}\cdot\ldots = \frac{\pi}{2}. \tag{3.290}$$

(2) For $k > 0$:

$$(1-k)\left(1-\frac{k}{4}\right)\left(1-\frac{k}{9}\right)\left(1-\frac{k}{16}\right)\ldots = \frac{\sin\pi\sqrt{k}}{\pi\sqrt{k}} \tag{3.291}$$

On setting $x = \pi\sqrt{k}$, $k = x^2/\pi^2$, we can write

$$\frac{\sin x}{x} = (1-k)\left(1-\frac{k}{4}\right)\left(1-\frac{k}{9}\right)\left(1-\frac{k}{16}\right)\ldots. \tag{3.292}$$

For $x = \pi/2$ $(k = 1/4)$ we recover the Wallis formula (1).

(3) We have

$$\frac{1}{2}\cdot\frac{4^2{\cdot}7}{5^3}\cdot\frac{7^2{\cdot}10}{8^3}\cdot\frac{10^2{\cdot}13}{11^3}\cdot\frac{13^2{\cdot}16}{14^3}\cdot\ldots = \left(\lim_{x\to\infty}\frac{P_1}{P_2}\right)^3 = \lambda^3 \tag{3.293}$$

(see Sec. 2.5), with

$$\begin{aligned} P_1''(x) &= x\,P_1(x), & P_1(0) &= 1, & P_1'(0) &= 0,\\ P_2''(x) &= x\,P_2(x), & P_2(0) &= 0, & P_2'(0) &= 1. \end{aligned}$$

From

$$P_2 = P_1\int_0^\infty \frac{\mathrm{d}x}{P_1^2}, \tag{3.294}$$

it follows that

$$\frac{1}{\lambda} = \int_0^\infty \frac{\mathrm{d}x}{P_1^2}. \tag{3.295}$$

8. BERNOULLI NUMBERS AND POLYNOMIALS

Bernoulli polynomials can be derived from the generating function

$$\psi(x,t) = \frac{t\,\mathrm{e}^{xt}}{\mathrm{e}^t - 1} = \sum_0^\infty \frac{B_n(x)}{n!}\,t^n. \tag{3.296}$$

Bernoulli numbers are the constant terms in the polynomials $B_n(x)$:

$$B_n = B_n(0). \tag{3.297}$$

On setting $x = 0$ in Eq. (3.296), we directly deduce the definition of Bernoulli numbers:

$$\frac{t}{\mathrm{e}^t - 1} = \sum_0^\infty \frac{B_n}{n!}\,t^n. \tag{3.298}$$

We list the first few Bernoulli numbers and polynomials:

$$B_0 = 1, \quad B_1 = -\frac{1}{2}, \quad B_2 = \frac{1}{6}, \quad B_3 = 0,$$

$$B_4 = -\frac{1}{30}, \quad B_5 = 0, \quad B_6 = \frac{1}{42}, \quad B_7 = 0,$$

$$B_8 = -\frac{1}{30}, \quad B_9 = 0, \quad B_{10} = \frac{5}{66}, \quad B_{11} = 0.$$

$$B_0(x) = 1,$$

$$B_1(x) = x - \frac{1}{2},$$

$$B_2(x) = x^2 - x + \frac{1}{6},$$

$$B_3(x) = x^3 - \frac{3}{2}x^2 + \frac{1}{2}x,$$

$$B_4(x) = x^4 - 2x^3 + x^2 - \frac{1}{30},$$

$$B_5(x) = x^5 - \frac{5}{2}x^4 + \frac{5}{3}x^3 - \frac{1}{6}x,$$

$$B_6(x) = x^6 - 3x^5 + \frac{5}{2}x^4 - \frac{1}{2}x^2 + \frac{1}{42},$$

$$B_7(x) = x^7 - \frac{7}{2}x^6 + \frac{7}{2}x^5 - \frac{7}{6}x^3 + \frac{1}{6}x,$$

$$B_8(x) = x^8 - 4x^7 + \frac{14}{3}x^6 - \frac{7}{3}x^4 + \frac{2}{3}x^2 - \frac{1}{30},$$

$$B_9(x) = x^9 - \frac{9}{2}x^8 + 6x^7 - \frac{21}{5}x^5 + 2x^3 - \frac{3}{10}x,$$

$$B_{10}(x) = x^{10} - 5x^9 + \frac{15}{2}x^8 - 7x^6 + 5x^4 - \frac{3}{2}x^2 + \frac{5}{66},$$

$$B_{11}(x) = x^{11} - \frac{11}{2}x^{10} + \frac{55}{6}x^9 - 11x^7 + 11x^5 - \frac{11}{2}x^3 + \frac{5}{6}x.$$

9. POISSON BRACKETS

In quantum mechanics, the Poisson bracket of two quantities a and b is defined as the following expression [8]:

$$[a\,,\,b] = \frac{i}{\hbar}(a\,b - b\,a) = -\,[b\,,\,a]\,. \tag{3.299}$$

Denoting by q and p the canonical variables and observing that $p = -(\hbar/i)\partial/\partial q$, we have

$$[q_i\,,\,p_i] = 1, \tag{3.300}$$

$$[a\,,\,b] = \sum_i \left(\frac{\partial a}{\partial q_i}\frac{\partial b}{\partial p_i} - \frac{\partial a}{\partial p_i}\frac{\partial b}{\partial q_i}\right), \tag{3.301}$$

$$\begin{aligned}[x\,,\,H] &= \sum_i \left(\frac{\partial x}{\partial q_i}\frac{\partial H}{\partial p_i} - \frac{\partial x}{\partial p_i}\frac{\partial H}{\partial q_i}\right) \\ &= \sum_i \left(\frac{\partial x}{\partial q_i}\dot{q}_i + \frac{\partial x}{\partial p_i}\dot{p}_i\right) = \dot{x}.\end{aligned} \tag{3.302}$$

Let us cite the expressions for the Poisson brackets of some quantities:

[8]@ In the original manuscript, the old notation $h/2\pi$ is used for the quantity we here denote by $\hbar$.

(1) Given

$$u_x = q_y\,p_z - q_z\,p_y, \tag{3.303}$$
$$u_y = q_z\,p_x - q_x\,p_z, \tag{3.304}$$
$$u_z = q_x\,p_y - q_y\,p_x, \tag{3.305}$$
$$\tag{3.306}$$

it follows that

$$[u_x\,,\,u_y] = -\,[u_y\,,\,u_x] = u_z, \tag{3.307}$$
$$[u_y\,,\,u_z] = -\,[u_z\,,\,u_y] = u_x, \tag{3.308}$$
$$[u_z\,,\,u_x] = -\,[u_x\,,\,u_z] = u_y, \tag{3.309}$$

$$\left[u_x^2\,,\,u_y\right] = u_x\,u_z + u_z\,u_x, \tag{3.310}$$
$$\left[u_x^2\,,\,u_z\right] = -\,u_x\,u_y - u_y\,u_x, \quad \text{etc.}, \tag{3.311}$$
$$\left[u_x\,,\,u_y^2\right] = u_y\,u_z + u_z\,u_y, \tag{3.312}$$
$$\left[u_x\,,\,u_z^2\right] = -\,u_x\,u_y - u_y\,u_x, \quad \text{etc.}, \tag{3.313}$$

$$\left[u_x^2 + u_y^2 + u_z^2\,,\,u_x\right] = 0, \quad \text{etc..} \tag{3.314}$$

(2) Given

$$q_x = r\,\sin\theta\,\cos\phi, \tag{3.315}$$
$$q_y = r\,\sin\theta\,\sin\phi, \tag{3.316}$$
$$q_z = r\,\cos\theta, \tag{3.317}$$

we have

$$[r\,,\,p_x] = \frac{q_x}{r} = \sin\theta\,\cos\phi, \tag{3.318}$$
$$[r\,,\,p_y] = \frac{q_y}{r} = \sin\theta\,\sin\phi, \tag{3.319}$$
$$[r\,,\,p_z] = \frac{q_z}{r} = \cos\theta, \tag{3.320}$$
$$[\cos\theta\,,\,p_x] = -\,\frac{q_x\,q_z}{r^3} = -\,\frac{\sin\theta\,\cos\theta\,\cos\phi}{r}, \tag{3.321}$$
$$[\cos\theta\,,\,p_y] = -\,\frac{q_y\,q_z}{r^3} = -\,\frac{\sin\theta\,\cos\theta\,\sin\phi}{r}, \tag{3.322}$$
$$[\cos\theta\,,\,p_z] = -\,\frac{r^2 - q_z^2}{r^3} = \frac{\sin^2\theta}{r}, \tag{3.323}$$

$$
\begin{aligned}
[\sin\theta\,,\,p_x] &= \frac{\cos^2\theta\,\cos\phi}{r}, && (3.324)\\
[\sin\theta\,,\,p_y] &= \frac{\cos^2\theta\,\sin\phi}{r}, && (3.325)\\
[\sin\theta\,,\,p_z] &= -\frac{\sin\theta\,\cos\theta}{r}, && (3.326)\\
[\theta\,,\,p_x] &= \frac{\cos\theta\,\cos\phi}{r}, && (3.327)\\
[\theta\,,\,p_y] &= \frac{\cos\theta\,\sin\phi}{r}, && (3.328)\\
[\theta\,,\,p_z] &= -\frac{\sin\theta}{r}, && (3.329)\\
[\cos\phi\,,\,p_x] &= \frac{\sin^2\phi}{r\,\sin\theta}, && (3.330)\\
[\cos\phi\,,\,p_y] &= -\frac{\cos\phi\,\sin\phi}{r\,\sin\theta}, && (3.331)\\
[\cos\phi\,,\,p_z] &= 0, && (3.332)\\
[\sin\phi\,,\,p_x] &= -\frac{\cos\phi\,\sin\phi}{r\,\sin\theta}, && (3.333)\\
[\sin\phi\,,\,p_y] &= \frac{\cos^2\phi}{r\,\sin\theta}, && (3.334)\\
[\sin\phi\,,\,p_z] &= 0, && (3.335)\\
[\phi\,,\,p_x] &= -\frac{\sin\phi}{r\,\sin\theta}, && (3.336)\\
[\phi\,,\,p_y] &= \frac{\cos\phi}{r\,\sin\theta}, && (3.337)\\
[\phi\,,\,p_z] &= 0. && (3.338)
\end{aligned}
$$

(3) Let us set, for simplicity:

$$
\begin{aligned}
X &= u_x, && (3.339)\\
Y &= u_y, && (3.340)\\
Z &= u_z. && (3.341)
\end{aligned}
$$

Denoting by k the azimuthal quantum number and m (or n) the equatorial quantum number, we have the following matrices of degeneration in units $\hbar$:

$$
\begin{aligned}
Z_{m,n} &= \delta_{m,n}\,m, && (3.342)\\
X_{m,n} &= \frac{1}{2}\,(\delta_{m+1,n} + \delta_{m,n+1})\,\sqrt{k(k+1) - mn}, && (3.343)
\end{aligned}
$$

$$Y_{m,n} = \frac{i}{2}\left(\delta_{m+1,n} - \delta_{m,n+1}\right)\sqrt{k(k+1)-mn}. \quad (3.344)$$

For example, for $k = 2$, we have

$$Z = \begin{pmatrix} 2 & 0 & 0 & 0 & 0 \\ 0 & 1 & 0 & 0 & 0 \\ 0 & 0 & 0 & 0 & 0 \\ 0 & 0 & 0 & -1 & 0 \\ 0 & 0 & 0 & 0 & -2 \end{pmatrix}, \quad (3.345)$$

$$X = \begin{pmatrix} 0 & 1 & 0 & 0 & 0 \\ 1 & 0 & \sqrt{\frac{3}{2}} & 0 & 0 \\ 0 & \sqrt{\frac{3}{2}} & 0 & \sqrt{\frac{3}{2}} & 0 \\ 0 & 0 & \sqrt{\frac{3}{2}} & 0 & 1 \\ 0 & 0 & 0 & 1 & 0 \end{pmatrix}, \quad (3.346)$$

$$Y = \begin{pmatrix} 0 & i & 0 & 0 & 0 \\ -i & 0 & i\sqrt{\frac{3}{2}} & 0 & 0 \\ 0 & -i\sqrt{\frac{3}{2}} & 0 & i\sqrt{\frac{3}{2}} & 0 \\ 0 & 0 & -i\sqrt{\frac{3}{2}} & 0 & i \\ 0 & 0 & 0 & -i & 0 \end{pmatrix}, \quad (3.347)$$

$$Z^2 = \begin{pmatrix} 2 & 0 & 0 & 0 & 0 \\ 0 & 1 & 0 & 0 & 0 \\ 0 & 0 & 0 & 0 & 0 \\ 0 & 0 & 0 & -1 & 0 \\ 0 & 0 & 0 & 0 & 4 \end{pmatrix}, \quad (3.348)$$

$$X^2 = \begin{pmatrix} 1 & 0 & \sqrt{\frac{3}{2}} & 0 & 0 \\ 0 & \frac{5}{2} & 0 & \frac{3}{2} & 0 \\ \sqrt{\frac{3}{2}} & 0 & 3 & 0 & \sqrt{\frac{3}{2}} \\ 0 & \frac{3}{2} & 0 & \frac{5}{2} & 0 \\ 0 & 0 & \sqrt{\frac{3}{2}} & 0 & 1 \end{pmatrix}, \quad (3.349)$$

$$Y^2 = \begin{pmatrix} 1 & 0 & -\sqrt{\frac{3}{2}} & 0 & 0 \\ 0 & \frac{5}{2} & 0 & -\frac{3}{2} & 0 \\ -\sqrt{\frac{3}{2}} & 0 & 3 & 0 & -\sqrt{\frac{3}{2}} \\ 0 & -\frac{3}{2} & 0 & \frac{5}{2} & 0 \\ 0 & 0 & -\sqrt{\frac{3}{2}} & 0 & 1 \end{pmatrix}, \quad (3.350)$$

$$X^2 + Y^2 + Z^2 = \begin{pmatrix} 6 & 0 & 0 & 0 & 0 \\ 0 & 6 & 0 & 0 & 0 \\ 0 & 0 & 6 & 0 & 0 \\ 0 & 0 & 0 & 6 & 0 \\ 0 & 0 & 0 & 0 & 6 \end{pmatrix}, \tag{3.351}$$

$$X\,Y = \begin{pmatrix} -i & 0 & 0i\sqrt{\frac{3}{2}} & 0 & 0 \\ 0 & -\frac{1}{2}i & 0 & \frac{3}{2}i & 0 \\ -i\sqrt{\frac{3}{2}} & 0 & 0 & 0 & i\sqrt{\frac{3}{2}} \\ 0 & -\frac{3}{2}i & 0 & \frac{1}{2}i & 0 \\ 0 & 0 & -i\sqrt{\frac{3}{2}} & 0 & i \end{pmatrix}, \tag{3.352}$$

$$Y\,X = \begin{pmatrix} i & 0 & 0i\sqrt{\frac{3}{2}} & 0 & 0 \\ 0 & \frac{1}{2}i & 0 & \frac{3}{2}i & 0 \\ -i\sqrt{\frac{3}{2}} & 0 & 0 & 0 & i\sqrt{\frac{3}{2}} \\ 0 & -\frac{3}{2}i & 0 & -\frac{1}{2}i & 0 \\ 0 & 0 & -i\sqrt{\frac{3}{2}} & 0 & -i \end{pmatrix}, \tag{3.353}$$

$$X\,Z = \begin{pmatrix} 0 & 1 & 0 & 0 & 0 \\ 2 & 0 & 0 & 0 & 0 \\ 0 & \sqrt{\frac{3}{2}} & 0 & -\sqrt{\frac{3}{2}} & 0 \\ 0 & 0 & 0 & 0 & -2 \\ 0 & 0 & 0 & -1 & 0 \end{pmatrix}, \tag{3.354}$$

$$Z\,X = \begin{pmatrix} 0 & 2 & 0 & 0 & 0 \\ 1 & 0 & \sqrt{\frac{3}{2}} & 0 & 0 \\ 0 & 0 & 0 & 0 & 0 \\ 0 & 0 & -\sqrt{\frac{3}{2}} & 0 & -1 \\ 0 & 0 & 0 & -2 & 0 \end{pmatrix}, \tag{3.355}$$

$$Y\,Z = \begin{pmatrix} 0 & i & 0 & 0 & 0 \\ -2i & 0 & 0 & 0 & 0 \\ 0 & -i\sqrt{\frac{3}{2}} & 0 & -i\sqrt{\frac{3}{2}} & 0 \\ 0 & 0 & 0 & 0 & -2i \\ 0 & 0 & 0 & i & 0 \end{pmatrix}, \tag{3.356}$$

$$Z\,Y = \begin{pmatrix} 0 & 2i & 0 & 0 & 0 \\ -i & 0 & i\sqrt{\frac{3}{2}} & 0 & 0 \\ 0 & 0 & 0 & 0 & 0 \\ 0 & 0 & i\sqrt{\frac{3}{2}} & 0 & -i \\ 0 & 0 & 0 & 2i & 0 \end{pmatrix}, \tag{3.357}$$

$$[X\,,Y] \,=\, i\,(X\,Y \,-\, Y\,X) \,=\, \begin{pmatrix} 2 & 0 & 0 & 0 & 0 \\ 0 & 1 & 0 & 0 & 0 \\ 0 & 0 & 0 & 0 & 0 \\ 0 & 0 & 0 & -1 & 0 \\ 0 & 0 & 0 & 0 & -2 \end{pmatrix} \,=\, Z, \tag{3.358}$$

$$[Y\,,Z] \,=\, i\,(Y\,Z \,-\, Z\,Y) \,=\, \begin{pmatrix} 0 & 1 & 0 & 0 & 0 \\ 1 & 0 & \sqrt{\frac{3}{2}} & 0 & 0 \\ 0 & \sqrt{\frac{3}{2}} & 0 & \sqrt{\frac{3}{2}} & 0 \\ 0 & 0 & \sqrt{\frac{3}{2}} & 0 & 1 \\ 0 & 0 & 0 & 1 & 0 \end{pmatrix} \,=\, X, \tag{3.359}$$

$$[Z\,,X] \,=\, i\,(Z\,X \,-\, X\,Z) \,=\, \begin{pmatrix} 0 & i & 0 & 0 & 0 \\ -i & 0 & i\sqrt{\frac{3}{2}} & 0 & 0 \\ 0 & -i\sqrt{\frac{3}{2}} & 0 & i\sqrt{\frac{3}{2}} & 0 \\ 0 & 0 & -i\sqrt{\frac{3}{2}} & 0 & i \\ 0 & 0 & 0 & -i & 0 \end{pmatrix}. \tag{3.360}$$

10. ELEMENTARY PHYSICAL QUANTITIES

We give the following quantities [9] in absolute units [10]. With a * we indicate the experimental quantities from which all the other ones can be derived. [11] [12]

---

[9]@ In the tables we report the updated value for each quantity. These slightly differ from the ones given by the author. The values for the last 10 quantities do not appear in the original manuscript.

[10]@ In particular, the author gives length in m, mass in g, time in s and electric charge in esu. Other units are derived from these.

[11]@ However, nowadays, fundamental quantities, which are most precisely known, do not coincide with the ones marked in the following tables.

[12]@ With reference to the 8th line of the following table, the present convention for the atomic mass unit is ^{12}C mass/12, rather than ^{16}O mass/16.

Quantity	Value	
e (electron charge)	4.80320×10^{-10}	*
m (electron rest mass)	0.91094×10^{-27}	
h (quantum of action)	6.62607×10^{-27}	
$h/2\pi$	1.05457×10^{-27}	
$k = R/N$ (Boltzmann constant)	1.38065×10^{-16}	
R (perfect gas constant)	8.31447×10^{7}	*
N (Avogadro number)	6.02214×10^{23}	
$M_H = 1/N$ (^{16}O mass/16)	1.66054×10^{-24}	
e/mc	1.75878×10^{7}	*
c (speed of light)	2.99792×10^{10}	*
$F = eN/c$ (Faraday constant)	9648.29	*
$R/c = (2\pi^2me^4)/(h^3c)$ (Rydberg wave number)	109734.564	*
$R = (2\pi^2me^4)/h^3$ (Rydberg frequency)	3.28984×10^{15}	
$Rh = (2\pi^2me^4)/h^2$ (Rydberg energy)	2.17987×10^{-11}	
$r = h^2/(4\pi^2me^2)$ (first Bohr radius)	0.52918×10^{-10}	
$\mu = (eh)/(4\pi mc)$ (Bohr magneton)	9.27378×10^{-21}	
$\nu = e/(4\pi mc)$ (Larmor frequency for a unitary field)	1.39959×10^{6}	
$e/(4\pi mc^2)$ (Larmor wave number for a unitary field)	4.66841×10^{-5}	
$(hc^2)/(10^4e)$ (volts corresponding to 1μ)	1.23990	
$(Rhc)/(10^8e)$ (volts corresponding to $1 Rydberg$)	13.60603	
$(mc^3)/(10^8e)$ (volts corresponding to m)	511037	
$(10^8e)/(ck)$ (temperature corresponding to $1V$)	11604.2	
$(10^4ch)/k$ (temperature corresponding to 1μ)	14388.1	

11. "CHASING THE DOG"

Let us consider a point Q in motion on the x axis with a constant velocity u such that its rectangular coordinates are $Q(ut, 0)$. Another point $P(x, y)$ is moving with constant velocity v towards Q; we have to determine the trajectory of P. The tangent in P to such trajectory intersects Q at an arbitrary time; the envelope of the lines PQ is, then, the path of pursuit. Let us introduce the parameter α, signifying the angle between PQ and the x-axis. The coordinates of P satisfy the equation of the straight line intersecting P and Q:

$$y = (ut - x) \tan \alpha; \tag{3.361}$$

and, since P also belongs to the envelope of such straight lines, x and y also satisfy the equation given by the derivative of Eq. (3.361) with respect to time [13]:

$$(ut - x) \left(1 + \tan^2 \alpha\right) \frac{d\alpha}{dt} + u \tan \alpha = 0; \tag{3.362}$$

from which we get

$$x = ut + u \frac{dt}{d\alpha} \sin \alpha \cos \alpha, \tag{3.363}$$

$$y = -u \frac{dt}{d\alpha} \sin^2 \alpha \tag{3.364}$$

and, on differentiation,

$$\dot{x} = 2u \cos^2 \alpha + u \frac{d^2 t}{d\alpha^2} \frac{d\alpha}{dt} \sin \alpha \cos \alpha, \tag{3.365}$$

$$\dot{y} = -2u \sin \alpha \cos \alpha - u \frac{d^2 t}{d\alpha^2} \frac{d\alpha}{dt} \sin^2 \alpha, \tag{3.366}$$

On the other hand, we have

$$\dot{x} = v \cos \alpha, \quad \dot{y} = -v \sin \alpha, \tag{3.367}$$

so that, comparison with Eq. (3.365) or (3.366) gives

$$2 \cos \alpha + \frac{d^2 t}{d\alpha^2} \frac{d\alpha}{dt} \sin \alpha = \frac{v}{u}, \tag{3.368}$$

[13]@ See also Eq. (3.367).

that is,

$$
\begin{aligned}
\frac{\mathrm{d}}{\mathrm{d}\alpha} \log \frac{\mathrm{d}t}{\mathrm{d}\alpha} &= \frac{v}{u \sin \alpha} - 2 \frac{\cos \alpha}{\sin \alpha}, \\
\log \frac{\mathrm{d}t}{\mathrm{d}\alpha} &= \frac{v}{u} \log \tan \frac{\alpha}{2} - 2 \log \sin \alpha + \text{const.}, \\
\frac{\mathrm{d}t}{\mathrm{d}\alpha} &= c_1 \frac{(\tan \alpha/2)^{v/u}}{\sin^2 \alpha}, \qquad (3.369) \\
t &= c_1 \int \frac{(\tan \alpha/2)^{v/u}}{\sin^2 \alpha} \mathrm{d}\alpha + c_2 . \qquad (3.370)
\end{aligned}
$$

Substitution into Eqs. (3.363) and (3.364), results in

$$x = u\, c_1 \left[\int \frac{(\tan \alpha/2)^{v/u}}{\sin^2 \alpha} \mathrm{d}\alpha + \frac{\cos \alpha}{\sin \alpha} (\tan \alpha/2)^{v/u} \right] + u\, c_2 \qquad (3.371)$$

$$y = -u\, c_1 (\tan \alpha/2)^{v/u} . \qquad (3.372)$$

As is natural to expect, the shape of the curves depends only on the ratio v/u. Let us assume, for example, $u = v$; we have

$$\int \frac{\tan \alpha/2}{\sin^2 \alpha} \mathrm{d}\alpha + \frac{1}{2} \log \tan \frac{\alpha}{2} + \frac{1}{4} \tan^2 \frac{\alpha}{2} - \frac{1}{4} \qquad (3.373)$$

and, setting

$$u = v = 1, \quad c_1 = -1, \quad c_2 = 0, \qquad (3.374)$$

we get

$$
\begin{aligned}
t &= -\frac{1}{2} \log \tan \frac{\alpha}{2} - \frac{1}{4} \tan^2 \frac{\alpha}{2} + \frac{1}{4}, \qquad (3.375) \\
x &= t - \frac{\tan \alpha/2}{\tan \alpha}, \qquad (3.376) \\
y &= \tan \frac{\alpha}{2}. \qquad (3.377)
\end{aligned}
$$

We use the last expression to eliminate α:

$$
\begin{aligned}
t &= -\frac{1}{2} \log y - \frac{1}{4} y^2 + \frac{1}{4}, \qquad (3.378) \\
x &= \frac{1}{2} \log y + \frac{1}{4} y^2 - \frac{1}{4}. \qquad (3.379)
\end{aligned}
$$

Since, for $t = 0$, one has $x = 0$, $y = 1$, t gives the length of the arc of the curve between the point $(0, 1)$ and the arbitrary point (x, y). From Eqs. (3.378) and (3.379), it follows that

$$t = x + \frac{1}{2} - \frac{1}{2} y^2. \qquad (3.380)$$

From Eq. (3.379) it is seen that the minimum value of x is 0 (for $y = 1$); the point $(0, 1)$, considered as the origin of the arcs, is thus the point in which the tangent to the curve is vertical.

12. STATISTICAL POTENTIAL IN MOLECULES

The potential between electrons in a gas satisfies, statistically, the differential equation

$$\nabla^2 V = -kV^{3/2}. \tag{3.381}$$

When the equipotential surfaces are approximately known, V can approximately be determined in the following way. Let

$$f(x, y, z) = p \tag{3.382}$$

be the approximate expression for the equipotential surfaces as function of a parameter p. Setting

$$V = V(p), \tag{3.383}$$

we have

$$\nabla V = \frac{\mathrm{d}V}{\mathrm{d}p} \nabla p \tag{3.384}$$

and, denoting with n the outward normal to the surface,

$$\int_\sigma \frac{\partial V}{\partial n} \mathrm{d}\sigma = \frac{\mathrm{d}V}{\mathrm{d}p} \int_\sigma \frac{\partial p}{\partial n} \mathrm{d}\sigma = y_1(p) \frac{\mathrm{d}V}{\mathrm{d}p}, \tag{3.385}$$

where $y_1(p)$ is a known function. While integrating Eq. (3.381) over the space between two equipotential surfaces corresponding to p and $p + \mathrm{d}p$, $V^{3/2}$ can be moved outside the integral:

$$\int_{\Delta S} \nabla^2 V \,\mathrm{d}S = -k\, V^{3/2} \int_\sigma \left(\frac{\partial p}{\partial n}\right)^{-1} \mathrm{d}p\, \mathrm{d}\sigma = -k\, V^{3/2}\, y_2(p)\, \mathrm{d}p, \tag{3.386}$$

where

$$y_2(p) = \int_\sigma \left(\frac{\partial p}{\partial n}\right)^{-1} \mathrm{d}\sigma \tag{3.387}$$

is again a known function of p. On the other hand, from the divergence theorem

$$\int_{\Delta S} \nabla^2 V \,\mathrm{d}S = \int_{\sigma(p+\mathrm{d}p)} \frac{\partial V}{\partial n} \mathrm{d}\sigma(p + \mathrm{d}p) - \int_{\sigma(p)} \frac{\partial V}{\partial n} \mathrm{d}\sigma(p)$$

$$
\begin{aligned}
&= y_1(p+\mathrm{d}p)\,V'(p+\mathrm{d}p) \;-\; y_1(p)\,V'(p) \\
&= (y_1\,V'' \;+\; y_1'\,V')\,\mathrm{d}p,
\end{aligned}
$$

so that, on comparison with Eq. (3.386),

$$y_1\,V'' \;+\; y_1'\,V' \;=\; -\,k\,V^{3/2}\,y_2. \tag{3.388}$$

This equation makes it possible to determine $V(p)$ when the boundary conditions are assigned.

Let us consider, for example, a diatomic molecule with identical nuclei and assume that it has the following approximate equipotential surfaces

$$p \;=\; \frac{r_1\,r_2}{r_1\,+\,r_2} \;=\; \left(\frac{1}{r_1}\,+\,\frac{1}{r_2}\right)^{-1}. \tag{3.389}$$

We then have

$$\boldsymbol{\nabla}\,p \;=\; -\,p^2\,\boldsymbol{\nabla}\,\frac{1}{p} \;=\; -\,p^2\,\boldsymbol{\nabla}\,\frac{1}{r_1} \;-\; p^2\,\boldsymbol{\nabla}\,\frac{1}{r_2}. \tag{3.390}$$

Thus denoting by $\mathbf{u}$ and $\mathbf{v}$ two unitary vectors in the directions of increasing r_1 and r_2, respectively,

$$\boldsymbol{\nabla}\,p \;=\; \frac{p^2}{r_1^2}\,\mathbf{u} + \frac{p^2}{r_2^2}\,\mathbf{v}, \tag{3.391}$$

$$\frac{\partial p}{\partial n} \;=\; |\boldsymbol{\nabla}\,p|\,, \tag{3.392}$$

from which we can calculate y_1 and y_2. However, it is better to perform the calculations using elliptic coordinates. Also note that

$$y_2 \;=\; \frac{\partial S}{\partial p}, \tag{3.393}$$

where S is the volume enclosed by the equipotential surface p. Moreover, y_1 is the outward flux of $\boldsymbol{\nabla}\,p = -p^2\boldsymbol{\nabla}\,(1/p)$; and, since $1/p$ is harmonic with singularities of the type $1/r_1$ and $1/r_2$ at the nuclei, the outward flux of $\boldsymbol{\nabla}\,(1/p)$ is -8π; it thus follows that

$$y_1(p) \;=\; 8\pi\,p^2. \tag{3.394}$$

Let us consider a meridian cross section of the volume enclosed by the surface p; in rectangular coordinates x and z, let the nuclei be situated on the x axis at the points $(a,0)$ and $(-a,0)$. Introducing the elliptic coordinates

$$u \;=\; (r_1+r_2)\,/2, \quad v \;=\; (r_1-r_2)\,/2, \tag{3.395}$$

we have

$$
\begin{aligned}
r_1 &= u+v, \quad r_2 = u-v, \quad x = \frac{u\,v}{a} \\
y^2 &= \frac{(u^2-a^2)(a^2-v^2)}{a^2}, \quad S = \pi \int_s y^2\,\mathrm{d}x,
\end{aligned}
\tag{3.396}
$$

where the integral is evaluated over the boundary of the meridian semi-cross section ($y > 0$). Equation (3.389), which must be satisfied on the boundary, becomes

$$
p = \left(u^2 - v^2\right) \Big/ 2u,
\tag{3.397}
$$

with

$$
v^2 = u^2 - 2\,u\,p, \quad v = \pm\sqrt{u^2 - 2\,u\,p}, \quad u = p + \sqrt{p^2+v^2}, \quad u > 0.
\tag{3.398}
$$

We have $y = 0$ at the points

$$
\begin{aligned}
u &= p + \sqrt{p^2+a^2}, & v &= a, \\
u &= p + \sqrt{p^2+a^2}, & v &= -\,a, \\
u &= a, & v &= \sqrt{a^2 - 2ap}, \\
u &= a, & v &= -\sqrt{a^2 - 2ap}.
\end{aligned}
$$

The first two points are always real; the last two are real and distinct only when $p < a/2$, while they coincide in $(u\,,\,v) = (a\,,\,0)$ when $p = a/2$. Let us introduce the variable

$$
t = \frac{v}{u},
\tag{3.399}
$$

wherein u and v can be expressed in terms of rational functions of t:

$$
u = \frac{2p}{1-t^2}, \quad v = \frac{2pt}{1-t^2};
\tag{3.400}
$$

and, since $x = uv/a$, it follows that

$$
\mathrm{d}x = \frac{1}{a}\,\mathrm{d}\,(uv) = \frac{2p}{a}\mathrm{d}\,\frac{t}{(1-t^2)^2} = \frac{2p}{a}\,\frac{1+3t^2}{(1-t^2)^3}\,\mathrm{d}t.
\tag{3.401}
$$

If we evaluate the integral only over the positive values allowed for t, we find

$$
S = \frac{4\pi p}{a^3} \int \left(\frac{4p^2}{(1-t^2)^2} - a^2 \right) \left(a^2 - \frac{4p^2t^2}{(1-t^2)^2} \right) \frac{1+3t^2}{(1-t^2)^3}\,\mathrm{d}t,
\tag{3.402}
$$

where the lower limit of the integral is zero when $p \geq a/2$, while it is $\sqrt{a^2-2ap}/a$ when $p < a/2$; the upper limit is, in any case, $a/(p + \sqrt{p^2+a^2})$.

13. THE GROUP OF PROPER UNITARY TRANSFORMATIONS IN TWO VARIABLES

Let us consider the group $U(2)$ of unitary transformations in two variables ξ and η, with determinant equal to 1. If

$$\begin{pmatrix} \alpha & \beta \\ \gamma & \delta \end{pmatrix}$$

is the matrix corresponding to a particular transformation, that is,

$$\xi' = \alpha\,\xi + \beta\,\eta, \quad \eta' = \gamma\,\xi + \delta\,\eta, \tag{3.403}$$

then the following relations must be true:

$$\begin{array}{rclrcl} \alpha^*\alpha + \beta^*\beta & = & 1, & \alpha^*\gamma + \beta^*\delta & = & 0, \\ \gamma^*\gamma + \delta^*\delta & = & 1, & \alpha\,\delta - \beta\,\gamma & = & 1. \end{array} \tag{3.404}$$

On making the substitutions

$$\begin{array}{rclrcl} \alpha & = & \alpha_1 + i\,\alpha_2, & \beta & = & \beta_1 + i\,\beta_2, \\ \gamma & = & \gamma_1 + i\,\gamma_2, & \delta & = & \delta_1 + i\,\delta_2, \end{array}$$

in Eq. (3.404), the following relations between real quantities must hold:

$$\begin{array}{rcl} \alpha_1^2 + \alpha_2^2 + \beta_1^2 + \beta_2^2 & = & 1, \\ \gamma_1^2 + \gamma_2^2 + \delta_1^2 + \delta_2^2 & = & 1, \\ \alpha_1\gamma_1 + \alpha_2\gamma_2 + \beta_1\delta_1 + \beta_2\delta_2 & = & 0, \\ \alpha_1\gamma_2 - \alpha_2\gamma_1 + \beta_1\delta_2 - \beta_2\delta_1 & = & 0, \\ \alpha_1\delta_1 - \alpha_2\delta_2 - \beta_1\gamma_1 + \beta_2\gamma_2 & = & 1, \\ \alpha_1\delta_2 + \alpha_2\delta_1 - \beta_1\gamma_2 - \beta_2\gamma_1 & = & 0. \end{array}$$

On multiplying the last four of these six equations by α_1, $-\alpha_2$, $-\beta_1$, $-\beta_2$, respectively, and summing the resulting expressions, from the first equation we infer

$$\gamma_1 = \beta_1.$$

Analogously, we find

$$\gamma_2 = \beta_2, \quad \delta_1 = \alpha_1, \quad \delta_2 = -\,\alpha_2.$$

We can thus arbitrarily choose α_1, α_2, β_1, β_2 so that they satisfy the first of the above set of six equations and then determine the other unknowns from the immediately forgoing relations; the remaining five equations, including the second one which has not explicitly been taken into account, will then be automatically satisfied. On setting

$$\alpha_1 = x, \quad \alpha_2 = \lambda, \quad \beta_1 = -\mu, \quad \beta_2 = \nu,$$

where x, λ, μ, ν are arbitrary real numbers that satisfy the equation

$$x^2 + \lambda^2 + \mu^2 + \nu^2 = 1, \tag{3.405}$$

the elements of the most general unitary matrix with determinant equal to 1 will read

$$\begin{array}{c} \alpha = x + i\lambda, \quad \beta = -\mu + i\nu, \\ \gamma = -\beta^* = \mu + i\nu, \quad \delta = \alpha^* = x - i\lambda. \end{array} \tag{3.406}$$

Every transformation of the group is defined by the 4 real numbers x, λ, μ, ν; it will accordingly simply be denoted by

$$(x, \lambda, \mu, \nu).$$

Let us consider two transformations of the group and their product:

$$A = \begin{pmatrix} x + i\lambda & -\mu + i\nu \\ \mu + i\nu & x - i\lambda \end{pmatrix}, \quad B = \begin{pmatrix} x' + i\lambda' & -\mu' + i\nu' \\ \mu' + i\nu' & x' - i\lambda' \end{pmatrix},$$
$$AB = \begin{pmatrix} x'' + i\lambda'' & -\mu'' + i\nu'' \\ \mu'' + i\nu'' & x'' - i\lambda'' \end{pmatrix},$$

where, per definition,

$$\begin{array}{rcl} x'' & = & xx' - \lambda\lambda' - \mu\mu' - \nu\nu', \\ \lambda'' & = & x\lambda' + \lambda x' - \mu\nu' + \nu\mu', \\ \mu'' & = & x\mu' + \lambda\nu' + \mu x' - \nu\lambda', \\ \nu'' & = & x\nu' - \lambda\mu' + \mu\lambda' + \nu x'; \end{array} \tag{3.407}$$

in short,

$$(x, \lambda, \mu, \nu)\,(x', \lambda', \mu', \nu') = (x'', \lambda'', \mu'', \nu''). \tag{3.408}$$

which coincides with the multiplication rule of quaternions.

In the space of $v + 1 = 2j + 1$ dimensions, let us consider the vector with components

$$\frac{\xi^r\,\eta^{v-r}}{f(v,r)}, \quad r = 0, 1, ..., v. \tag{3.409}$$

The present group transforms this into the vector with components

$$\frac{\xi'^r\,\eta'^{v-r}}{f(v,r)}, \quad r = 0, 1, ..., v. \tag{3.410}$$

From Eqs. (3.403), the components of the transformed vector can be obtained as a linear combination of the components of the original one, this combination being univocal, since the $v+1$ monomials $\xi^r \eta^{v-r}$ $(r=0,1,\ldots,v)$ are linearly independent. We thus have a $(2j+1)$-dimensional representation $\mathcal{D}_j$ of the group. Obviously, the same representation holds for all transformations of the kind in Eq. (3.403), that is, for all affine linear transformations in 2 dimensions, or for all members of the sub-group $O(2)$ of the linear transformations with unitary determinant, our group $U(2)$ being a sub group of it. The function $f(v,r)$ can be determined in such a way that the unitary transformations of the group $U(2)$ are represented with unitary transformations as well. To this end, it must be true that

$$\sum_r \frac{|\xi^r \eta^{v-r}|^2}{f^2(v,r)} \tag{3.411}$$

(we are assuming f to be real) depends only on $|\xi|^2+|\eta|^2$; that is, setting $|\xi|^2=a$ and $|\eta|^2=b$,

$$\sum_r \frac{a^r b^{v-r}}{f^2(v,r)} \tag{3.412}$$

is a function of $a+b$. For this to happen, it suffices to equate the first quantity to the v-th power of the second quantity, and thus [14]:

$$f(v,r) = \binom{v}{r}^{-1/2} = \sqrt{r!(v-r)!/v!}, \tag{3.413}$$

or, since $f(v,r)$ can always be multiplied by an r-independent quantity, more simply:

$$f(v,r) = \sqrt{r!(v-r)!}. \tag{3.414}$$

Thus, ξ and η define a vector in a space of $2j+1$ dimensions with components

$$\frac{\xi^r \eta^{v-r}}{\sqrt{r!(v-r)!}}, \qquad r=0,1,\ldots,v. \tag{3.415}$$

Let us consider the transformation

$$(x, \epsilon a, \epsilon b, \epsilon c)\,. \tag{3.416}$$

Once $\epsilon a, \epsilon b, \epsilon c$ are given, x is determined apart from its sign, which we may choose to be positive. Let us assume that ϵ is small; thus x will

[14]@ The author wanted to obtain the formula for the power of a binomial.

differ from unity by a second-order term, so that

$$\lim_{\epsilon\to 0} \frac{(x, \epsilon a, \epsilon b, \epsilon c) - (1,0,0,0)}{\epsilon} = (0, a, b, c)\,. \tag{3.417}$$

The transformation $(0, a, b, c)$, whose definition is given by Eqs. (3.406), is an infinitesimal transformation. In general, it does not belong to $U(2)$, but it is always a (real) multiple of a unitary transformation with determinant equal to 1. By contrast, the transformation

$$(x, \lambda, \mu, \nu) = \mathrm{e}^{(0,a,b,c)\,t} \tag{3.418}$$

does belong to $U(2)$, where t is an arbitrary real number; that is, we necessarily have $x^2 + \lambda^2 + \mu^2 + \nu^2 = 1$. Given a, b, c, the quantities x, λ, μ, ν are functions of t (from Eq. (3.418)), and we have

$$\begin{aligned} \left(\frac{\mathrm{d}x}{\mathrm{d}t}, \frac{\mathrm{d}\lambda}{\mathrm{d}t}, \frac{\mathrm{d}\mu}{\mathrm{d}t}, \frac{\mathrm{d}\nu}{\mathrm{d}t}\right) &= (x, \lambda, \mu, \nu)\,(0, a, b, c) \\ &= (0, a, b, c)\,(x, \lambda, \mu, \nu)\,, \end{aligned} \tag{3.419}$$

that is,

$$\begin{aligned} \frac{\mathrm{d}x}{\mathrm{d}t} &= -a\,\lambda - b\,\mu - c\,\nu, \\ \frac{\mathrm{d}\lambda}{\mathrm{d}t} &= a\,x - c\,\mu + b\,\nu = a\,x + c\,\mu - b\,\nu = a\,x, \\ \frac{\mathrm{d}\mu}{\mathrm{d}t} &= b\,x, \qquad \frac{\mathrm{d}\nu}{\mathrm{d}t} = c\,x. \end{aligned} \tag{3.420}$$

Differentiating the first equation with respect to t, one gets

$$\frac{\mathrm{d}^2x}{\mathrm{d}t^2} = -\left(a^2 + b^2 + c^2\right) x, \tag{3.421}$$

from which

$$x = \cos t\,\sqrt{a^2 + b^2 + c^2}, \tag{3.422}$$

$$\begin{aligned} \lambda &= \frac{a}{\sqrt{a^2 + b^2 + c^2}} \sin t\,\sqrt{a^2 + b^2 + c^2}, \\ \mu &= \frac{b}{\sqrt{a^2 + b^2 + c^2}} \sin t\,\sqrt{a^2 + b^2 + c^2}, \\ \nu &= \frac{c}{\sqrt{a^2 + b^2 + c^2}} \sin t\,\sqrt{a^2 + b^2 + c^2}. \end{aligned} \tag{3.423}$$

Choosing $t = 1$, it follows that the infinitesimal transformation $(0, a, b, c)$ can be deduced from the transformation

$$(x, \lambda, \mu, \nu) = \mathrm{e}^{(0,a,b,c)} \tag{3.424}$$

by setting

$$\begin{aligned}
x &= \cos\sqrt{a^2+b^2+c^2}, \\
\lambda &= \frac{a}{\sqrt{a^2+b^2+c^2}}\sin\sqrt{a^2+b^2+c^2}, \\
\mu &= \frac{b}{\sqrt{a^2+b^2+c^2}}\sin\sqrt{a^2+b^2+c^2}, \\
\nu &= \frac{c}{\sqrt{a^2+b^2+c^2}}\sin\sqrt{a^2+b^2+c^2}.
\end{aligned} \tag{3.425}$$

Equations (3.425) tells us that an arbitrary transformation of the group $U(2)$ can be cast in the form given by Eq. (3.424), where the constants a, b, c can be univocally determined from the conditions

$$a\,,b\,,c \geq 0, \quad 0 \leq \sqrt{a^2+b^2+c^2} \leq 2\pi. \tag{3.426}$$

Let us now consider an arbitrary representation of the group $U(2)$. We set

$$\lim_{\epsilon\to 0}\frac{\mathcal{U}(x,\epsilon a,\epsilon b,\epsilon c) - 1}{\epsilon} = a\,P_1 + b\,P_2 + c\,P_3. \tag{3.427}$$

As a consequence,

$$\begin{aligned}
e^{a\,P_1+b\,P_2+c\,P_3} &= \lim_{\epsilon\to 0}\left(1+\epsilon\,(a\,P_1+b\,P_2+c\,P_3)\right)^{1/\epsilon} \\
&= \lim_{\epsilon\to 0}\left(\mathcal{U}(x,\epsilon a,\epsilon b,\epsilon c)\right)^{1/\epsilon} = \lim_{\epsilon\to 0}\mathcal{U}(x,\epsilon a,\epsilon b,\epsilon c)^{1/\epsilon} \\
&= \lim_{\epsilon\to 0}\mathcal{U}\left((1,0,0,0)+\epsilon(0,a,b,c)\right)^{1/\epsilon} = \mathcal{U}\left(e^{(0,a,b,c)}\right)
\end{aligned}$$

and, from Eq. (3.424),

$$\mathcal{U}(x,\lambda,\mu,\nu) = \mathcal{U}\left(e^{(0,a,b,c)}\right) = e^{a\,P_1+b\,P_2+c\,P_3}, \tag{3.428}$$

which has to be true together with Eq. (3.425). It is then sufficient to know the matrices P_1, P_2, P_3 in order to have a representation of the group $U(2)$. However, the matrices P_1, P_2, P_3 cannot be chosen arbitrarily. This happen, first of all, because if one relaxes the constraints (3.426) for continuity reasons, an element of $U(2)$ can be represented by different sets $(a,b,c), (a',b',c'), \ldots$, while for the uniqueness of the representation we must have

$$e^{a\,P_1+b\,P_2+c\,P_3} = e^{a'\,P_1+b'\,P_2+c'\,P_3} = \ldots. \tag{3.429}$$

Secondly, there must be a correspondence between the product of two elements of the group and the product of the two related transformations. Let us suppose that the first condition is satisfied; then, for $t = 0$, we get

$$\mathcal{U}\left(\mathrm{e}^{(0,at,bt,ct)}\,\mathrm{e}^{(0,a't,b't,c't)}\right) = \mathrm{e}^{(a\,P_1 + b\,P_2 + c\,P_3)t}\,\mathrm{e}^{(a'\,P_1 + b'\,P_2 + c'\,P_3)t}. \tag{3.430}$$

Let us set

$$\mathrm{e}^{(0,at,bt,ct)}\,\mathrm{e}^{(0,a't,b't,c't)} = \mathrm{e}^{(0,x,y,z)}. \tag{3.431}$$

The quantities x, y, z will then be functions of t that can be determined in an infinite number of ways from Eqs. (3.424) and (3.425); however we will demand that they satisfy the continuity requirement and the conditions $x = y = z = 0$ for $t = 0$. From Eq. (3.428), Eq. (3.430) becomes

$$\mathrm{e}^{x\,P_1 + y\,P_2 + z\,P_3} = \mathrm{e}^{(a\,P_1 + b\,P_2 + c\,P_3)t}\,\mathrm{e}^{(a'\,P_1 + b'\,P_2 + c'\,P_3)t}. \tag{3.432}$$

Expanding Eq. (3.431) in a power series, we get

$$\begin{aligned}
&1 + xP_1 + yP_2 + zP_3 + \frac{1}{2}\left[x^2P_1^2 + y^2P_2^2 + z^2P_3^2\right.\\
&\left.+xy\,(P_1P_2 + P_2P_1) + yz\,(P_2P_3 + P_3P_2) + zx\,(P_1P_3 + P_3P_1)\right]\\
&+\frac{1}{6}\left[x^3P_1^3 + y^3P_2^3 + z^3P_3^3 + x^2y\left(P_1^2P_2 + P_1P_2P_1 + P_2P_1^2\right)\right.\\
&+xy^2\left(P_1P_2^2 + P_2P_1P_2 + P_2^2P_1\right) + y^2z\left(P_2^2P_2 + P_2P_3P_2 + P_3P_2^2\right)\\
&+yz^2\left(P_2P_3^2 + P_3P_2P_3 + P_3^2P_2\right) + z^2z\left(P_3^2P_1 + P_3P_1P_3 + P_1P_3^2\right)\\
&+x^2z\left(P_3P_1^2 + P_1P_3P_1 + P_1^2P_3\right) + xyz\,(P_1P_2P_3 + P_2P_3P_1\\
&\left.+P - 3P_1P_2 + P_1P_3P_2 + P_2P_1P_3 + P_3P_2P_1)\right] + \ldots\\
&= 1 + t\,(aP_1 + bP_2 + cP_3 + a'P_1 + b'P_2 + c'P_3)\\
&+\frac{t^2}{2}\left[a^2P_1^2 + b^2P_2^2 + c^2P_3^2 + ab\,(P_1P_2 + P_2P_1) + bc\,(P_2P_3 + P_3P_2)\right.\\
&+ca\,(P_1P_3 + P_3P_1) + a'^2P_1^2 + b'^2P_2^2 + c'^2P_3^2 + a'b'\,(P_1P_2 + P_2P_1)\\
&+b'c'\,(P_2P_3 + P_3P_2) + c'a'\,(P_1P_3 + P_3P_1) + 2aa'P_1^2 + 2bb'P_2^2\\
&+2cc'P_3^2 + 2ab'P_1P_2 + 2bc'P_2P_3 + 2ca'P_3P_1\\
&\left.+2ac'P_1P_3 + 2ba'P_2P_1 + 2cb'P_3P_2\right] + \ldots.
\end{aligned} \tag{3.433}$$

Since x, y, z are infinitesimals with respect to t and a, b, c, a', b', c' are arbitrary constants, by equating the terms of the same order in the two sides of Eq. (3.433), we find the relations that P_1, P_2, P_3 must satisfy.

Actually, we want to find the first terms in the expansion of x, y, z in t; to this end let us expand (3.431) in series up to second-order terms. We find

$$\begin{aligned}&1 + (0, at, bt, ct) + (0, a't, b't, c't) + \frac{1}{2}(0, at, bt, ct)^2 \\ &+\frac{1}{2}(0, a't, b't, c't)^2 + (0, at, bt, ct)(0, a't, b't, c't) + \ldots \\ &= (1, 0, 0, 0) + (0, x, y, z) + \frac{1}{2}(0, x, y, z)^2 + \ldots\end{aligned}$$

from which follows, on equating components of the quaternions corresponding

$$\begin{aligned}1 - \frac{1}{2}t^2 \left[(a + a')^2 + (b + b')^2 \right. &+ \left.(c + c')^2\right] + \ldots \\ &= 1 - \frac{1}{2}\left(x^2 + y^2 + z^2\right) + \ldots, \quad (3.434)\\ (a + a')t + (cb' - bc')t^2 + \ldots &= x + at + \ldots, \quad (3.435)\\ (b + b')t + (ac' - ca')t^2 + \ldots &= y + at + \ldots, \quad (3.436)\\ (c + c')t + (ba' - ab')t^2 + \ldots &= z + at + \ldots. \quad (3.437)\end{aligned}$$

From the last three equations (the first one is an obvious consequence of these), we can deduce the expansion to second-order of x, y, z. Substituting in Eq. (3.433), we find that, up to first-order terms, it is identically satisfied. Equating the second-order terms, we get

$$\begin{aligned}&(cb' - bc')P_1 + (ac' - ca')P_2 + (ba' - ab')P_3 \\ &+\frac{1}{2}\left[(a + a')^2 P_1^2 + (b + b')^2 P_2^2 + (c + c')^2 P_3^2\right. \\ &+(a + a')(b + b')(P_1P_2 + P_2P_1) + (b + b')(c + c')(P_2P_3 + P_3P_2) \\ &\left.+(c + c')(a + a')(P_3P_1 + P_1P_3)\right] \\ &= \frac{1}{2}\left[(a + a')^2 P_1^2 + (b + b')^2 P_2^2 + (c + c')^2 P_3^2\right. \\ &+(a + a')(b + b')(P_1P_2 + P_2P_1) + (b + b')(c + c')(P_2P_3 + P_3P_2) \\ &+(c + c')(a + a')(P_3P_1 + P_1P_3) + (ab' - ba')(P_1P_2 - P_2P_1) \\ &\left.+(bc' - cb')(P_2P_3 - P_3P_2) + (ca' - ac')(P_3P_1 - P_1P_3)\right]. \quad (3.438)\end{aligned}$$

Since these relations must be true for arbitrary values of the constants appearing in them, we get the following exchange relations:

$$\begin{aligned}P_1P_2 - P_2P_1 &= -2P_3, \\ P_2P_3 - P_3P_2 &= -2P_1, \\ P_3P_1 - P_1P_3 &= -2P_2.\end{aligned} \quad (3.439)$$

Let us consider the representations $\mathcal{D}_j$ of the group $U(2)$ composed of transformations acting on the vector (3.415). The vector with components

$$\frac{\xi^r\,\eta^{v-r}}{\sqrt{r!(v-r)!}}$$

is transformed by P_3 into the vector with components

$$\begin{aligned}\left[\frac{\mathrm{d}}{\mathrm{d}\epsilon}\,\frac{(\xi-i\epsilon\eta)^r\,(\eta+i\epsilon\xi)^{v-r}}{\sqrt{r!(v-r)!}}\right]_{\epsilon=0} &= \frac{r\,i\,\xi^{r-1}\,\eta^{v-r+1}}{\sqrt{r!(v-r)!}}\\ +\,\frac{(v-r)\xi^{r+1}\,\eta^{v-r-1}}{\sqrt{r!(v-r)!}} &= i\,\sqrt{r(v-r+1)}\,\frac{\xi^{r-1}\,\eta^{v-r+1}}{\sqrt{(r-1)!(v-r+1)!}}\\ &+i\,\sqrt{(r+1)(v-r)}\,\frac{\xi^{r+1}\,\eta^{v-r-1}}{\sqrt{(r+1)!(v-r-1)!}},\end{aligned}$$

so that, setting $m = v/2 - r = j - r$, we get the matrix P_3:

$$\begin{aligned}(P_3)_{m,m-1} &= i\,\sqrt{(j+m)(j-m+1)} = i\,\sqrt{j(j+1)-m(m-1)},\\ (P_3)_{m,m+1} &= i\,\sqrt{(j+m+1)(j-m)} = i\,\sqrt{j(j+1)-m(m+1)},\end{aligned} \tag{3.440}$$

with $m = -j, -j+1, \ldots$; and all the other components are zero. The matrix P_3 is thus emisymmetric, i.e., iP_3 is a Hermitian matrix. In other words, like all the infinitesimal unitary transformations, P_3 is a pure imaginary quantity.

The same vector (3.415) is transformed by P_2 into the vector with components

$$\begin{aligned}\left[\frac{\mathrm{d}}{\mathrm{d}\epsilon}\,\frac{(\xi-\epsilon\eta)^r\,(\eta+\epsilon\xi)^{v-r}}{\sqrt{r!(v-r)!}}\right]_{\epsilon=0} &= \frac{-\,r\,\xi^{r-1}\,\eta^{v-r+1}}{\sqrt{r!(v-r)!}}\\ +\,\frac{(v-r)\xi^{r+1}\,\eta^{v-r-1}}{\sqrt{r!(v-r)!}} &= -\,\sqrt{r(v-r+1)}\,\frac{\xi^{r-1}\,\eta^{v-r+1}}{\sqrt{(r-1)!(v-r+1)!}}\\ &+\sqrt{(r+1)(v-r)}\,\frac{\xi^{r+1}\,\eta^{v-r-1}}{\sqrt{(r+1)!(v-r-1)!}}.\end{aligned}$$

It follows that the only non-zero components of the matrix P_2 are

$$\begin{aligned}(P_2)_{m,m-1} &= -\,\sqrt{(j+m)(j-m+1)} = -\,\sqrt{j(j+1)-m(m-1)},\\ (P_2)_{m,m+1} &= \sqrt{(j+m+1)(j-m)} = \sqrt{j(j+1)-m(m+1)}.\end{aligned} \tag{3.441}$$

The matrix P_1 transforms the vector (3.415) into the vector with components

$$\left[\frac{\mathrm{d}}{\mathrm{d}\epsilon}\,\frac{(\xi+i\epsilon\xi)^r\,(\eta-i\epsilon\eta)^{v-r}}{\sqrt{r!(v-r)!}}\right]_{\epsilon=0}$$

$$= \frac{r\,i\,\xi^r\,\eta^{v-r}}{\sqrt{r!(v-r)!}} - \frac{(v-r)\,i\,\xi^r\,\eta^{v-r}}{\sqrt{r!(v-r)!}}.$$

Consequently the only non-zero elements of P_1 are the diagonal ones, with

$$(P_1)_{m,m} = 2\,m\,i. \tag{3.442}$$

We then have the following representations:

- $j = 0$

$$P_1 = 0, \quad P_2 = 0, \quad P_3 = 0.$$

- $j = \frac{1}{2}$

$$P_1 = \begin{pmatrix} i & 0 \\ 0 & -i \end{pmatrix}, \quad P_2 = \begin{pmatrix} 0 & -1 \\ 1 & 0 \end{pmatrix}, \quad P_3 = \begin{pmatrix} 0 & i \\ i & 0 \end{pmatrix}.$$

- $j = 1$

$$P_1 = \begin{pmatrix} 2i & 0 & 0 \\ 0 & 0 & 0 \\ 0 & 0 & -2i \end{pmatrix}, \quad P_2 = \begin{pmatrix} 0 & -\sqrt{2} & 0 \\ \sqrt{2} & 0 & -\sqrt{2} \\ 0 & \sqrt{2} & 0 \end{pmatrix},$$

$$P_3 = \begin{pmatrix} 0 & i\sqrt{2} & 0 \\ i\sqrt{2} & 0 & i\sqrt{2} \\ 0 & i\sqrt{2} & 0 \end{pmatrix}.$$

- $j = \frac{3}{2}$

$$P_1 = \begin{pmatrix} 3i & 0 & 0 & 0 \\ 0 & i & 0 & 0 \\ 0 & 0 & -1 & 0 \\ 0 & 0 & 0 & -3i \end{pmatrix}, \quad P_2 = \begin{pmatrix} 0 & -\sqrt{3} & 0 & 0 \\ \sqrt{3} & 0 & -2 & 0 \\ 0 & 2 & 0 & -\sqrt{3} \\ 0 & 0 & \sqrt{3} & 0 \end{pmatrix},$$

$$P_3 = \begin{pmatrix} 0 & i\sqrt{3} & 0 & 0 \\ i\sqrt{3} & 0 & 2i & 0 \\ 0 & 2i & 0 & i\sqrt{3} \\ 0 & 0 & i\sqrt{3} & 0 \end{pmatrix}.$$

- $j = 2$

$$P_1 = \begin{pmatrix} 4i & 0 & 0 & 0 & 0 \\ 0 & 2i & 0 & 0 & 0 \\ 0 & 0 & 0 & 0 & 0 \\ 0 & 0 & 0 & -2i & 0 \\ 0 & 0 & 0 & 0 & -4i \end{pmatrix}, \quad P_2 = \begin{pmatrix} 0 & -2 & 0 & 0 & 0 \\ 2 & 0 & -\sqrt{6} & 0 & 0 \\ 0 & \sqrt{6} & 0 & -\sqrt{6} & 0 \\ 0 & 0 & \sqrt{6} & 0 & -2 \\ 0 & 0 & 0 & 2 & 0 \end{pmatrix}$$

$$P_3 = \begin{pmatrix} 0 & 2i & 0 & 0 & 0 \\ 2i & 0 & i\sqrt{6} & 0 & 0 \\ 0 & i\sqrt{6} & 0 & i\sqrt{6} & 0 \\ 0 & 0 & i\sqrt{6} & 0 & 2i \\ 0 & 0 & 0 & 2i & 0 \end{pmatrix}.$$

It is easy to test that the exchange relations are satisfied by these matrices. In fact, if we also include the vanishing components, the matrices can be written in the form

$$\begin{aligned} (P_1)_{m,n} &= 2\,m\,i\,\delta_{m,n}, \\ (P_2)_{m,n} &= \sqrt{j(j+1) - mn}\,(\delta_{m,n-1} - \delta_{m,n+1}), \\ (P_3)_{m,n} &= \sqrt{j(j+1) - mn}\,(i\delta_{m,n-1} + i\delta_{m,n+1}). \end{aligned} \tag{3.443}$$

It follows that

$$\begin{aligned} (P_1P_2)_{m,n} &= 2\,m\,i\sqrt{j(j+1) - mn}\,(\delta_{m,n-1} - \delta_{m,n+1}), \\ (P_2P_1)_{m,n} &= 2\,n\,i\sqrt{j(j+1) - mn}\,(\delta_{m,n-1} - \delta_{m,n+1}), \\ (P_1P_2 - P_2P_1)_{m,n} &= \sqrt{j(j+1) - mn}\,(-2\,i\,\delta_{m,n-1} - 2\,i\,\delta_{m,n+1}) \\ &= -2\,(P_3)_{m,n}, \\ (P_2P_3)_{m,n} &= i\sqrt{j(j+1) - m(m+1)} \\ &\quad \times\sqrt{j(j+1) - (m+1)(m+2)}\,\delta_{m+2,n} \\ &\quad - 2\,m\,i\,\delta_{m,m} - i\sqrt{j(j+1) - m(m-1)} \\ &\quad \times\sqrt{j(j+1) - (m-1)(m-2)}\,\delta_{m-2,n}, \\ (P_3P_2)_{m,n} &= i\sqrt{j(j+1) - m(m+1)} \\ &\quad \times\sqrt{j(j+1) - (m+1)(m+2)}\,\delta_{m+2,n} \\ &\quad + 2\,m\,i\,\delta_{m,m} - i\sqrt{j(j+1) - m(m-1)} \end{aligned}$$

$$\begin{aligned}
& \times\sqrt{j(j+1)-(m-1)(m-2)}\,\delta_{m-2,n}, \\
(P_3P_2 \,-\, P_2P_3)_{m,n} &= -\,4\,m\,i\delta_{m,m} \;=\; -\,2\,(P_1))m,n, \\
(P_3P_1)_{m,n} &= 2\,n\,i\sqrt{j(j+1)-mn}\,(i\,\delta_{m,n-1} \,+\, i\,\delta_{m,n+1}), \\
(P_1P_3)_{m,n} &= 2\,m\,i\sqrt{j(j+1)-mn}\,(i\,\delta_{m,n-1} \,+\, i\,\delta_{m,n+1}), \\
(P_3P_1 \,-\, P_1P_3)_{m,n} &= \sqrt{j(j+1)-mn}\,(-\,2\,\delta_{m,n-1} \,+\, 2\,\delta_{m,n+1}) \\
&= -\,2\,(P_2)_{m,n}\,.
\end{aligned}$$

14. EXCHANGE RELATIONS FOR INFINITESIMAL TRANSFORMATIONS IN THE REPRESENTATIONS OF CONTINUOUS GROUPS

Let us consider a continuous group with n parameters

$$s \;=\; (q_1, q_2, \ldots, q_n)\,. \tag{3.444}$$

We choose the parameters in such a way that all the coordinates of the unit element are zero:

$$1 \;=\; (0, 0, \ldots, 0)\,. \tag{3.445}$$

Let us also consider a given representation of the group

$$\mathcal{U}(s) \;=\; \mathcal{U}\,(q_1, q_2, \ldots, q_n)\,. \tag{3.446}$$

An infinitesimal transformation is defined by

$$\lim_{\epsilon\to 0} \frac{\mathcal{U}(\epsilon a_1, \epsilon a_2, \ldots, \epsilon a_n) - 1}{\epsilon} \;=\; a_1P_1 \,+\, a_2P_2 \,+\, \ldots \,+\, a_nP_n, \tag{3.447}$$

i.e., the infinitesimal transformations can be expressed in terms of linear combinations of n particular transformations. The matrices $P_1, P_2, ..., P_n$ obey algebraic relations that do not depend on the number of dimensions and on the particular representation but only on the structure of the given group. The exchange relations are some of these. Let us consider the “commutator”

$$\begin{aligned}
c \;=\; (x_1, x_2, \ldots, x_n) \;&=\; (\alpha, 0, 0, \ldots, 0)\;(0, \beta, 0, \ldots, 0) \\
&\quad\times(\alpha, 0, 0, \ldots, 0)^{-1}\;(0, \beta, 0, \ldots, 0)^{-1},
\end{aligned}\tag{3.448}$$

i.e., setting

$$s = (\alpha, 0, 0, \ldots, 0), \quad t = (0, \beta, 0, \ldots, 0), \tag{3.449}$$
$$c = s\,t\,s^{-1}t^{-1}, \tag{3.450}$$

we have

$$s\,t = c\,t\,s, \tag{3.451}$$
$$\mathcal{U}(s)\mathcal{U}(t) = \mathcal{U}(c)\mathcal{U}(t)\mathcal{U}(s). \tag{3.452}$$

Take the derivative of this with respect to α,

$$\begin{aligned}\frac{\mathrm{d}\mathcal{U}(s)}{\mathrm{d}\alpha}\mathcal{U}(t) &= \sum_i \frac{\partial x_i}{\partial \alpha}\frac{\partial \mathcal{U}(s)}{\partial x_i}\mathcal{U}(t)\mathcal{U}(s) \\ &\quad + \mathcal{U}(c)\mathcal{U}(t)\frac{\mathrm{d}\mathcal{U}(s)}{\mathrm{d}\alpha},\end{aligned} \tag{3.453}$$

and then the derivative of the outcome with respect to β

$$\begin{aligned}\frac{\mathrm{d}\mathcal{U}(s)}{\mathrm{d}\alpha}\frac{\mathrm{d}\mathcal{U}(t)}{\mathrm{d}\beta} &= \sum_i \frac{\partial^2 x_i}{\partial\alpha\partial\beta}\frac{\partial \mathcal{U}(s)}{\partial x_i}\mathcal{U}(t)\mathcal{U}(s) \\ &+ \sum_{i,k}\frac{\partial x_i}{\partial \alpha}\frac{\partial x_k}{\partial \beta}\frac{\partial^2\mathcal{U}(c)}{\partial x_i \partial x_k}\mathcal{U}(t)\mathcal{U}(s) \\ &+ \sum_i \frac{\partial x_i}{\partial \alpha}\frac{\partial \mathcal{U}(c)}{\partial x_i}\frac{\partial \mathcal{U}(t)}{\partial \beta}\mathcal{U}(s) + \sum_i \frac{\partial x_i}{\partial \beta}\frac{\partial \mathcal{U}(c)}{\partial x_i}\mathcal{U}(t)\frac{\partial \mathcal{U}(s)}{\partial \alpha} \\ &+ \mathcal{U}(c)\frac{\partial \mathcal{U}(t)}{\partial \beta}\frac{\partial \mathcal{U}(s)}{\partial \alpha}.\end{aligned} \tag{3.454}$$

For vanishing α or β, the commutator reduces to the unit element. For $\alpha = \beta = 0$ we then have, for $i = 1, 2, \ldots, n$,

$$\mathcal{U}(c) = 1, \tag{3.455}$$
$$\frac{\partial x_i}{\partial \alpha} = \frac{\partial x_i}{\partial \beta} = 0, \tag{3.456}$$
$$\frac{\partial^2 x_i}{\partial\alpha\partial\beta} = a_i^{1,2}, \tag{3.457}$$
$$\frac{\partial \mathcal{U}(c)}{\partial x_i} = P_i, \tag{3.458}$$
$$\frac{\partial \mathcal{U}(s)}{\partial \alpha} = P_1, \tag{3.459}$$
$$\frac{\partial \mathcal{U}(t)}{\partial \beta} = P_2, \tag{3.460}$$
$$\mathcal{U}(s) = \mathcal{U}(t) = 1; \tag{3.461}$$

the superscripts 1 and 2 in Eq. (3.457) denote the first coordinate of s and the second one of t, respectively, the other coordinates all being zero; similar formulas hold for every pair r, s of coordinates and the $a_i^{r,s}$ are manifestly antisymmetric in the upper indices. The formula (3.454) then reads

$$P_1 P_2 = \sum_i a_i^{1,2} P_i + P_2 P_1, \tag{3.462}$$

that is,

$$P_1 P_2 - P_2 P_1 = \sum_i a_i^{1,2} P_i, \tag{3.463}$$

or, more generally,

$$P_r P_s - P_s P_r = \sum_i a_i^{r,s} P_i, \tag{3.464}$$

which are the so-called exchange relations

15. EMPIRICAL RELATIONS FOR A TWO-ELECTRON ATOM

Let us consider a two-electron atom with charge Z in its ground state. We denote by $a = <1/r_1> = <1/r_2>$ the mean value of the inverse of the distance of each electron from the nucleus, and with $b = < 1/r_{12} >$ the mean value of the inverse of the distance between the two electrons. Expressing the distances in electronic units and the energy in Ry, we have

$$E = -2\,a\,Z + b, \tag{3.465}$$

since the energy is equal to half the mean value of the potential energy. If we now consider an atom with atomic number $Z + dZ$, perturbation theory gives

$$\mathrm{d}E = -4\,a\,\mathrm{d}Z, \tag{3.466}$$

and thus we have two equations in the three unknown Z functions E, a, b. We now add another empirical relation between a and b, which is presumably a good approximation:

$$b = (2Z - 2a)\,(2a - Z). \tag{3.467}$$

This relation can be deduced from the following considerations. For sufficiently high values of Z, perturbation theory gives

$$E = -2\,Z^2 + \frac{5}{4}\,Z + \ldots; \tag{3.468}$$

but, on the other hand,

$$b = \frac{5}{8} Z + \ldots, \tag{3.469}$$

so that, from Eq. (3.465),

$$a = Z - \frac{5}{16} + \ldots, \tag{3.470}$$

which, in first approximation, satisfies Eq. (3.467). For very small values of Z we can consider that the first electron is next to the nucleus, while the other one is practically at an infinite distance; then we have $a \simeq Z/2$, $b \simeq 0$, and Eq. (3.467) is again satisfied. We finally assume that it is also a good approximation for intermediate values of Z.

Substituting Eq. (3.467) in Eq. (3.465), we find

$$\begin{aligned} E &= -2\,a\,Z + (2Z - 2a)\,(2a - Z) \\ &= -\,2\,Z^2 + 4\,a\,Z - 4\,a^2. \end{aligned} \tag{3.471}$$

On differentiating this relation, we get

$$\mathrm{d}E = -\,4\,Z\,\mathrm{d}Z + 4\,a\,\mathrm{d}Z + 4\,Z\,\mathrm{d}a - 8\,a\,\mathrm{d}a; \tag{3.472}$$

and, comparing this with Eq. (3.466), we find

$$\mathrm{d}Z = \mathrm{d}a, \tag{3.473}$$

from which, since we know the value of a for infinite Z, we deduce

$$a = Z - \frac{5}{16}. \tag{3.474}$$

Substitution of this in (3.471), yields

$$E = -2\,Z^2 + \frac{5}{4}\,Z - \frac{25}{64}. \tag{3.475}$$

This formula can be used only for $Z \geq 5/8$, since for $Z = 5/8$ we have $b = 0$. For the helium atom ($Z = 2$), we find $E = -5.89$, against the experimental value of 5.807, with an error in excess of 1.13 V for the ionization potential (25.59 V instead of 24.46 V). For the hydrogen atom ($Z = 1$) we find instead $E = -1.141$, from which the ionization potential would be 1.91 (electron affinity). [15] The procedure used here is not very

[15]@ Note that the actual experimental values of the ionization potential for the hydrogen atom, the (neutral) helium atom, and the once ionized helium atom are 13.5984 V, 24.5874 V, and 54.4178 V, respectively, corresponding to ionization energies of 0.9995, 1.8072, and 3.9998 (measured in Ry, as used in the text), respectively. The electron affinity of the hydrogen (i.e., the difference between the ground state energies of the neutral atom and the once-ionized atom) is 0.7542 eV.

satisfactory, since for very small Z the quantity b would vanish faster than a first-order term and it would become negative.

Instead of Eq. (3.467), let us now choose the approximation

$$b = \frac{5}{8}\sqrt{k^2+Z^2} - \frac{5}{8}k, \tag{3.476}$$

where k has to be determined. Substituting in (3.465), we obtain

$$E = -2\,a\,Z + \frac{5}{8}\sqrt{k^2+Z^2} - \frac{5}{8}k; \tag{3.477}$$

and, on differentiating,

$$\mathrm{d}E = -2\,a\,\mathrm{d}Z - 2\,Z\,\mathrm{d}a + \frac{5Z\,\mathrm{d}Z}{8\sqrt{k^2+Z^2}}. \tag{3.478}$$

Compare this with Eq. (3.466):

$$\begin{aligned}
-4\,a\,\mathrm{d}Z &= -2\,a\,\mathrm{d}Z - 2\,Z\,\mathrm{d}a + \frac{5Z\,\mathrm{d}Z}{8\sqrt{k^2+Z^2}},\\
2\,Z\,\mathrm{d}a &= 2\,a\,\mathrm{d}Z + \frac{5Z\,\mathrm{d}Z}{8\sqrt{k^2+Z^2}},\\
\frac{\mathrm{d}a}{\mathrm{d}Z} &= \frac{a}{Z} + \frac{5}{16\sqrt{k^2+Z^2}},
\end{aligned} \tag{3.479}$$

from which, noting that for $Z \to \infty$ the quantity a/Z tends to 1, we infer that

$$a = Z\left(1 + \int_0^\infty \frac{5\,\mathrm{d}Z}{16\,Z\,\sqrt{k^2+Z^2}}\right). \tag{3.480}$$

In this formula a becomes negative for sufficiently small values of Z; in order to eliminate this drawback, it is necessary that, for small Z, b vanishes as a term of order greater than 2.

Instead of (3.476), let us therefore choose

$$b = \frac{5}{8}\,Z\,\mathrm{e}^{-k/Z}, \tag{3.481}$$

so that Eq. (3.465) becomes

$$E = -2\,a\,Z + \frac{5}{8}\,Z\,\mathrm{e}^{-k/Z}. \tag{3.482}$$

Differentiation gives

$$\mathrm{d}E = -2\,a\,\mathrm{d}Z - 2\,Z\,\mathrm{d}a + \frac{5}{8}\,\mathrm{e}^{-k/Z}\,\mathrm{d}Z + \frac{5k}{8Z}\,\mathrm{e}^{-k/Z}\mathrm{d}Z; \tag{3.483}$$

comparing this with Eq. (3.466)

$$\frac{\mathrm{d}a}{\mathrm{d}Z} = \frac{a}{Z} + \frac{5}{16\,Z}\,\mathrm{e}^{-k/Z} + \frac{5k}{16\,Z^2}\,\mathrm{e}^{-k/Z}, \tag{3.484}$$

we get

$$a = Z\left(1 + \int_0^\infty \frac{5Z+5k}{16\,Z^3}\,\mathrm{e}^{-k/Z}\,\mathrm{d}Z\right). \tag{3.485}$$

Since for small Z we must have $a \simeq Z/2$, we can choose k such that

$$\int_0^\infty \frac{5Z+5k}{16\,Z^3}\,\mathrm{e}^{-k/Z}\,\mathrm{d}Z = \frac{1}{2}, \tag{3.486}$$

i.e., $k = 5/4$. However, in this way we obtain a bad approximation. Indeed, we would have

$$a = \frac{Z}{2} + \left(\frac{Z}{2} + \frac{5}{16}\right)\mathrm{e}^{-1.25/Z}, \tag{3.487}$$

$$b = \frac{5}{8}\,Z\,\mathrm{e}^{-1.25/Z}, \tag{3.488}$$

$$E = -Z^2 - Z^2\,\mathrm{e}^{-1.25/Z}, \tag{3.489}$$

and for helium ($Z = 2$) we would get $E = -5.14$, which is a value far from the experimental one.

Let us set

$$b = \frac{5}{8}\left(\sqrt[3]{k^3 + Z^3} - k\right). \tag{3.490}$$

Equation (3.465) now becomes

$$E = -2\,a\,Z + \frac{5}{8}\sqrt[3]{k^3+Z^3} - \frac{5}{8}\,k, \tag{3.491}$$

and thus

$$\mathrm{d}E = -2\,a\,\mathrm{d}Z - 2\,Z\,\mathrm{d}a - \frac{5Z^2 dZ}{8\,(k^3+Z^3)^{2/3}}. \tag{3.492}$$

On comparing with Eq. (3.466), we get

$$\frac{\mathrm{d}a}{\mathrm{d}Z} = \frac{a}{Z} + \frac{5Z}{16\,(k^3+Z^3)^{2/3}}, \tag{3.493}$$

from which we deduce

$$a = Z\left(1 + \int_0^\infty \frac{5\,dZ}{16\,(k^3+Z^3)^{2/3}}\right), \tag{3.494}$$

where the constant k can be determined in a way similar to that leading to Eq. (3.486). However, we note that relations analogous to Eq. (3.490) are arbitrary, and there is no *a priori* reason to prefer one or the other.

16. THE GROUP OF ROTATIONS $O(3)$ [16]

Let us consider a point on a unit sphere with coordinates (x, y, z) and its image on the equatorial plane $z = 0$ through the south pole, whose coordinates are $(0, 0, -1)$. The coordinates (x, y, z) are related to those of the image point (x', y', z') by the equations

$$x = \frac{2x'}{1 + x'^2 + y'^2}, \quad y = \frac{2y'}{1 + x'^2 + y'^2}, \quad z = \frac{1 - x'^2 - y'^2}{1 + x'^2 + y'^2}. \qquad (3.495)$$

On setting

$$x' + i\,y' = \eta/\xi, \qquad (3.496)$$

relations (3.495) transform into

$$x + i\,y = \frac{2\,\eta\xi^*}{\xi\xi^* + \eta\eta^*}, \quad x - i\,y = \frac{2\,\eta^*\xi}{\xi\xi^* + \eta\eta^*}, \quad z = \frac{\xi\xi^* - \eta\eta^*}{\xi\xi^* + \eta\eta^*}. \qquad (3.497)$$

Let us now consider a unitary transformation (with determinant equal to 1) of the group $SU(2)$ acting on ξ and η; the transformed variables are given by

$$\begin{aligned} \xi_1 &= x\,\xi + i\,\lambda\,\xi - \mu\,\eta + i\,\nu\,\eta, \\ \eta_1 &= \mu\,\xi + i\,\nu\,\xi + x\,\eta - i\,\lambda\,\eta \end{aligned}$$

(with $x^2 + \lambda^2 + \mu^2 + \nu^2$). As a consequence, the point (x, y, z) is transformed into the point with coordinates (x_1, y_1, z_1):

$$\begin{aligned} x_1 + iy_1 &= 2\,\frac{(x\mu + \lambda\nu + ix\nu - i\lambda\mu)(\xi\xi^* - \eta\eta^*)}{\xi\xi^* + \eta\eta^*} \\ &\quad + 2\,\frac{(x^2 - \lambda^2 - 2ix\lambda)\eta\xi^* + (-\mu^2 + \nu^2 - 2i\mu\nu)\xi\eta^*}{\xi\xi^* + \eta\eta^*}, \\ x_1 - iy_1 &= 2\,\frac{(x\mu + \lambda\nu - ix\nu + i\lambda\mu)(\xi\xi^* - \eta\eta^*)}{\xi\xi^* + \eta\eta^*} \\ &\quad + 2\,\frac{(x^2 - \lambda^2 + 2ix\lambda)\eta\xi^* + (-\mu^2 + \nu^2 + 2i\mu\nu)\xi\eta^*}{\xi\xi^* + \eta\eta^*}, \\ z_1 &= \frac{(x^2 + \lambda^2 - \mu^2 - \nu^2)(\xi\xi^* - \eta\eta^*)}{\xi\xi^* + \eta\eta^*} \\ &\quad + 2\,\frac{(-x\mu + \lambda\nu + ix\nu + i\lambda\mu)\eta\xi^* + (-x\mu + \lambda\nu - ix\nu - i\lambda\mu)\xi\eta^*}{\xi\xi^* + \eta\eta^*}, \end{aligned}$$

[16]@ In the original manuscript, this group is denoted by δ_3; however, here we use the modern notation $O(3)$. Note also that sometimes the author uses the same notation for a group and for its restriction to transformations with determinant equal to 1 (which, in modern notations, is denoted with an S preceding the name of the group; for example $O(3)$ and $SO(3)$).

that is,

$$\begin{aligned}
x_1 + iy_1 &= 2\,(x\mu + \lambda\nu + ix\nu - i\lambda\mu)\,z \\
&\quad + (x^2 - \lambda^2 - 2ix\lambda)(x + iy) + (-\mu^2 + \nu^2 - 2i\mu\nu)(x - iy), \\
x_1 - iy_1 &= 2\,(x\mu + \lambda\nu - ix\nu + i\lambda\mu)\,z \\
&\quad + (x^2 - \lambda^2 + 2ix\lambda)(x - iy) + (-\mu^2 + \nu^2 + 2i\mu\nu)(x + iy), \\
z_1 &= (x^2 + \lambda^2 - \mu^2 - \nu^2)\,z \,+\, 2\,(-x\mu + \lambda\nu + ix\nu + i\lambda\mu)(x + iy) \\
&\quad + (-x\mu + \lambda\nu - ix\nu - i\lambda\mu)(x - iy),
\end{aligned}$$

or

$$\begin{aligned}
x_1 &= (x^2 - \lambda^2 - \mu^2 + \nu^2)x \,+\, 2(x\lambda - \mu\nu)y \,+\, 2(x\mu + \lambda\nu)z, \\
y_1 &= -2(x\lambda + \mu\nu)x \,+\, (x^2 - \lambda^2 + \mu^2 - \nu^2)y \,+\, 2(x\nu - \lambda\mu)z, \\
z_1 &= 2(-x\mu + \lambda\nu)x \,+\, 2(-x\nu - \lambda\mu)y \,+\, (x^2 + \lambda^2 - \mu^2 - \nu^2)z.
\end{aligned} \tag{3.498}$$

This represents a rotation (the most general one) in three-dimensional space; for each rotation we can choose the constants x, λ, μ, ν in two ways, related by a sign change in the components of the quaternion. Note that Eqs. (3.498) correspond to the representation $\mathcal{D}_1$ of the group $SU(2)$. By inverting these relations (and losing, however, the uniqueness), we can consider $\mathcal{D}_j$ as representations of $O(3)$; the ones with non-integer j are twofold, while the ones with integer j are unique, since in these representations two equal and opposite quaternions correspond to the same transformation. In the representations $\mathcal{D}_j$ with non-integer j (for example $\mathcal{D}_{1/2}$), derived from the inversion of (3.498), to each rotation in the three-dimensional space there correspond two equal and opposite matrices.

An infinitesimal rotation (through an angle ϵ) about the z axis corresponds to the quaternion

$$\left(1, -\frac{1}{2}\epsilon, 0, 0\right)$$

(out of two possible ones, we have chosen the quaternion nearest to unity). Analogously, an infinitesimal rotation about the x axis corresponds to the quaternion

$$\left(1, 0, 0, -\frac{1}{2}\epsilon\right),$$

while an infinitesimal rotation about the y-axis corresponds to the quaternion

$$\left(1, 0, \frac{1}{2}\epsilon, 0\right).$$

It then follows that the infinitesimal rotations along the x,y,z axes can be expressed through the fundamental infinitesimal transformations P_1, P_2, and P_3 by the simple relations:

$$R_z = -\frac{1}{2}P_1, \quad R_x = -\frac{1}{2}P_3, \quad R_y = \frac{1}{2}P_2, \tag{3.499}$$

and this holds for an arbitrary representation of $U(2)$, as long as it is considered as a (unique or twofold) representation of $O(3)$. From Eqs. (3.439), the exchange relations

$$\begin{aligned} R_x R_y - R_y R_x &= R_z, \\ R_y R_z - R_z R_y &= R_x, \\ R_z R_x - R_x R_z &= R_y \end{aligned} \tag{3.500}$$

then follow. Moreover, from Eqs. (3.443) we deduce the following expressions for the matrices R_x, R_y, R_z in the $\mathcal{D}_j$ representations (changing the sign of m and n, i.e., setting $m = j - r$) :

$$\begin{aligned} \left(\frac{R_z}{i}\right)_{m,n} &= m\,\delta_{m,n}, \\ \left(\frac{R_x}{i}\right)_{m,n} &= -\frac{i}{2}\sqrt{j(j+1)-mn}\,\left(\delta_{m+1,n} + \delta_{m-1,n}\right), \\ \left(\frac{R_y}{i}\right)_{m,n} &= \frac{i}{2}\sqrt{j(j+1)-mn}\,\left(\delta_{m+1,n} - i\delta_{m-1,n}\right). \end{aligned} \tag{3.501}$$

We then have the following matrices:

- $j = 0$

$$\frac{R_z}{i} = 0, \quad \frac{R_x}{i} = 0, \quad \frac{R_y}{i} = 0.$$

- $j = \frac{1}{2}$

$$\frac{R_z}{i} = \begin{pmatrix} \frac{1}{2} & 0 \\ 0 & -\frac{1}{2} \end{pmatrix}, \quad \frac{R_x}{i} = \begin{pmatrix} 0 & -\frac{1}{2} \\ \frac{1}{2} & 0 \end{pmatrix}, \quad \frac{R_y}{i} = \begin{pmatrix} 0 & -\frac{i}{2} \\ \frac{i}{2} & 0 \end{pmatrix}.$$

- $j = 1$

$$\frac{R_z}{i} = \begin{pmatrix} 1 & 0 & 0 \\ 0 & 0 & 0 \\ 0 & 0 & -1 \end{pmatrix}, \quad \frac{R_x}{i} = \begin{pmatrix} 0 & -\frac{\sqrt{2}}{2} & 0 \\ -\frac{\sqrt{2}}{2} & 0 & -\frac{\sqrt{2}}{2} \\ 0 & -\frac{\sqrt{2}}{2} & 0 \end{pmatrix},$$

$$\frac{R_y}{i} = \begin{pmatrix} 0 & -i\frac{\sqrt{2}}{2} & 0 \\ i\frac{\sqrt{2}}{2} & 0 & -i\frac{\sqrt{2}}{2} \\ 0 & i\frac{\sqrt{2}}{2} & 0 \end{pmatrix}.$$

- $j = \frac{3}{2}$.

$$\frac{R_z}{i} = \begin{pmatrix} \frac{3}{2} & 0 & 0 & 0 \\ 0 & \frac{1}{2} & 0 & 0 \\ 0 & 0 & -\frac{1}{2} & 0 \\ 0 & 0 & 0 & -\frac{3}{2} \end{pmatrix}, \quad \frac{R_x}{i} = \begin{pmatrix} 0 & -\frac{\sqrt{3}}{2} & 0 & 0 \\ -\frac{\sqrt{3}}{2} & 0 & -1 & 0 \\ 0 & -1 & 0 & -\frac{\sqrt{3}}{2} \\ 0 & 0 & -\frac{\sqrt{3}}{2} & 0 \end{pmatrix},$$

$$\frac{R_y}{i} = \begin{pmatrix} 0 & -i\frac{\sqrt{3}}{2} & 0 & 0 \\ i\frac{\sqrt{3}}{2} & 0 & -i & 0 \\ 0 & i & 0 & -i\frac{\sqrt{3}}{2} \\ 0 & 0 & i\frac{\sqrt{3}}{2} & 0 \end{pmatrix}.$$

- $j = 2$

$$\frac{R_z}{i} = \begin{pmatrix} 2 & 0 & 0 & 0 & 0 \\ 0 & 1 & 0 & 0 & 0 \\ 0 & 0 & 0 & 0 & 0 \\ 0 & 0 & 0 & -1 & 0 \\ 0 & 0 & 0 & 0 & -2 \end{pmatrix}, \quad \frac{R_x}{i} = \begin{pmatrix} 0 & -1 & 0 & 0 & 0 \\ -1 & 0 & -\frac{\sqrt{6}}{2} & 0 & 0 \\ 0 & -\frac{\sqrt{6}}{2} & 0 & -\frac{\sqrt{6}}{2} & 0 \\ 0 & 0 & -\frac{\sqrt{6}}{2} & 0 & - \\ 0 & 0 & 0 & -1 & 0 \end{pmatrix}$$

$$\frac{R_y}{i} = \begin{pmatrix} 0 & -i & 0 & 0 & 0 \\ i & 0 & -i\frac{\sqrt{6}}{2} & 0 & 0 \\ 0 & i\frac{\sqrt{6}}{2} & 0 & -i\frac{\sqrt{6}}{2} & 0 \\ 0 & 0 & i\frac{\sqrt{6}}{2} & 0 & -i \\ 0 & 0 & 0 & i & 0 \end{pmatrix}.$$

- $j = \frac{5}{2}$

$$\frac{R_z}{i} = \begin{pmatrix} \frac{5}{2} & 0 & 0 & 0 & 0 & 0 \\ 0 & \frac{3}{2} & 0 & 0 & 0 & 0 \\ 0 & 0 & \frac{1}{2} & 0 & 0 & 0 \\ 0 & 0 & 0 & -\frac{1}{2} & 0 & 0 \\ 0 & 0 & 0 & 0 & -\frac{3}{2} & 0 \\ 0 & 0 & 0 & 0 & 0 & -\frac{5}{2} \end{pmatrix},$$

$$\frac{R_x}{i} = \begin{pmatrix} 0 & -\frac{\sqrt{5}}{2} & 0 & 0 & 0 & 0 \\ -\frac{\sqrt{5}}{2} & 0 & -\sqrt{2} & 0 & 0 & 0 \\ 0 & -\sqrt{2} & 0 & -\frac{3}{2} & 0 & 0 \\ 0 & 0 & -\frac{3}{2} & 0 & -\sqrt{2} & 0 \\ 0 & 0 & 0 & -\sqrt{2} & 0 & -\frac{\sqrt{5}}{2} \\ 0 & 0 & 0 & 0 & -\frac{\sqrt{5}}{2} & 0 \end{pmatrix},$$

$$\frac{R_y}{i} = \begin{pmatrix} 0 & -i\frac{\sqrt{5}}{2} & 0 & 0 & 0 & 0 \\ i\frac{\sqrt{5}}{2} & 0 & -i\sqrt{2} & 0 & 0 & 0 \\ 0 & i\sqrt{2} & 0 & -i\frac{3}{2} & 0 & 0 \\ 0 & 0 & i\frac{3}{2} & 0 & -i\sqrt{2} & 0 \\ 0 & 0 & 0 & i\sqrt{2} & 0 & -i\frac{\sqrt{5}}{2} \\ 0 & 0 & 0 & 0 & i\frac{\sqrt{5}}{2} & 0 \end{pmatrix}.$$

- $j = 3$

$$\frac{R_z}{i} = \begin{pmatrix} 3 & 0 & 0 & 0 & 0 & 0 & 0 \\ 0 & 2 & 0 & 0 & 0 & 0 & 0 \\ 0 & 0 & 1 & 0 & 0 & 0 & 0 \\ 0 & 0 & 0 & 0 & 0 & 0 & 0 \\ 0 & 0 & 0 & 0 & -1 & 0 & 0 \\ 0 & 0 & 0 & 0 & 0 & -2 & 0 \\ 0 & 0 & 0 & 0 & 0 & 0 & -3 \end{pmatrix},$$

$$\frac{R_x}{i} = \begin{pmatrix} 0 & -\frac{\sqrt{6}}{2} & 0 & 0 & 0 & 0 & 0 \\ -\frac{\sqrt{6}}{2} & 0 & -\frac{\sqrt{10}}{2} & 0 & 0 & 0 & 0 \\ 0 & -\frac{\sqrt{10}}{2} & 0 & -\sqrt{3} & 0 & 0 & 0 \\ 0 & 0 & -\sqrt{3} & 0 & -\sqrt{3} & 0 & 0 \\ 0 & 0 & 0 & -\sqrt{3} & 0 & -\frac{\sqrt{10}}{2} & 0 \\ 0 & 0 & 0 & 0 & -\frac{\sqrt{10}}{2} & 0 & -\frac{\sqrt{6}}{2} \\ 0 & 0 & 0 & 0 & 0 & -\frac{\sqrt{6}}{2} & 0 \end{pmatrix},$$

$$\frac{R_y}{i} = \begin{pmatrix} 0 & -i\frac{\sqrt{6}}{2} & 0 & 0 & 0 & 0 & 0 \\ i\frac{\sqrt{6}}{2} & 0 & -i\frac{\sqrt{10}}{2} & 0 & 0 & 0 & 0 \\ 0 & i\frac{\sqrt{10}}{2} & 0 & -i\sqrt{3} & 0 & 0 & 0 \\ 0 & 0 & i\sqrt{3} & 0 & -i\sqrt{3} & 0 & 0 \\ 0 & 0 & 0 & i\sqrt{3} & 0 & -i\frac{\sqrt{10}}{2} & 0 \\ 0 & 0 & 0 & 0 & i\frac{\sqrt{10}}{2} & 0 & -i\frac{\sqrt{6}}{2} \\ 0 & 0 & 0 & 0 & 0 & i\frac{\sqrt{6}}{2} & 0 \end{pmatrix}.$$

17. THE LORENTZ GROUP

This group is composed of the orthogonal transformations of the variables

$$ct,\ \frac{x}{i},\ \frac{y}{i},\ \frac{z}{i}.$$

Restricting our study to the orthogonal transformations with determinant equal to 1 (excluding, then, those with determinant equal to -1), we have the proper group of (real or complex) rotations $SO(4)$ in a four-dimensional space.

Let us consider the variables x_1, x_2, x_3, x_4 in a four-dimensional space and let $\xi_1, \xi_2, \xi_3, \xi_4$ be the variables of the dual space, which transform in the contragradient way. Thus, if the x variables linearly transform into the variables x'_1, x'_2, x'_3, x'_4, the variables ξ transform in such a way that the following relation must hold:

$$x_1 \xi_1 + x_2 \xi_2 + x_3 \xi_3 + x_4 \xi_4 = x'_1 \xi'_1 + x'_2 \xi'_2 + x'_3 \xi'_3 + x'_4 \xi'_4. \quad (3.502)$$

If

$$x'_i = \sum_k a_{ik}\, x_k, \quad (3.503)$$

by substituting in Eq. (3.502), we get

$$\sum_i x_i\, \xi_i = \sum_{i,k} a_{ik}\, \xi'_i\, x_k = \sum_k x_k \sum_i a_{ik}\, \xi'_i. \quad (3.504)$$

Since this relation must hold for arbitrary values of the x and ξ variables, we conclude

$$\xi_k = \sum_i a_{ik}\, \xi'_i, \quad (3.505)$$

which expresses the contragradient variation law for the ξ variables. A transformation acting only on some of the x variables, say x_1 and x_2, will act in the dual space only on the corresponding ξ variables (in the present case, ξ_1 and ξ_2) and vice-versa. This directly follows from Eqs. (3.503) and (3.505).

Consider a transformation σ_{12} belonging to the group $SL(2,\mathcal{C})$ [17] of the homogenous linear transformations in two variables with determinant equal to 1 that acts on the variables x_1 and x_2. Let also a transformation σ_{34} of the same group act on the variables x_3 and x_4. The transformations

$$\sigma = (\sigma_{12},\ \sigma_{34})\,, \quad (3.506)$$

---

[17] @In the original manuscript, this group is denoted by c_2; however, here we use the modern notation $SL(2,\mathcal{C})$.

acting on the 4 variables x_1, x_2, x_3, x_4, constitute a representation of the abstract group $(SL(2,\mathcal{C}))^2$ whose elements are pairs (σ, τ) of the elements of $SL(2,\mathcal{C})$ and satisfy the composition rule

$$(\sigma,\ \tau)\ (\sigma',\ \tau')\ =\ (\sigma\,\sigma',\ \tau\,\tau')\,. \tag{3.507}$$

Let us consider the expressions

$$z_1\ =\ x_1\,\xi_3, \quad z_2\ =\ x_2\,\xi_3, \quad z_3\ =\ x_1\,\xi_4, \quad z_4\ =\ x_2\,\xi_4, \tag{3.508}$$

which reduce the quadratic form

$$z_1\,z_4\ -\ z_2\,z_3 \tag{3.509}$$

to zero. Under a transformation σ, the x and ξ variables transform as follows:

$$\begin{array}{rclrcl} x'_1 & = & \alpha x_1 + \beta x_2, & x'_2 & = & \gamma x_1 + \delta x_2, \\ x'_3 & = & \alpha_1 x_3 + \beta_1 x_4, & x'_4 & = & \gamma_1 x_3 + \delta_1 x_4, \end{array} \tag{3.510}$$

with $\alpha\delta - \beta\gamma = \alpha_1\delta_1 - \beta_1\gamma_1 = 1$, and

$$\begin{array}{rclrcl} \xi'_1 & = & \delta\xi_1 - \gamma\xi_2, & \xi'_2 & = & -\beta\xi_1 + \alpha\xi_2, \\ \xi'_3 & = & \delta_1\xi_3 - \gamma_1\xi_4, & \xi'_4 & = & -\beta_1\xi_3 + \alpha_1\xi_4. \end{array} \tag{3.511}$$

On substituting in Eqs. (3.508), we get

$$\begin{array}{rcl} z'_1 & = & \alpha\delta_1 z_1 + \beta\delta_1 z_2 + \alpha\gamma_1 z_3 - \beta\gamma_1 z_4, \\ z'_2 & = & \gamma\delta_1 z_1 + \delta\delta_1 z_2 - \gamma\gamma_1 z_3 - \delta\gamma_1 z_4, \\ z'_3 & = & -\alpha\beta_1 z_1 - \beta\beta_1 z_2 + \alpha\alpha_1 z_3 + \beta\alpha_1 z_4, \\ z'_4 & = & -\gamma\beta_1 z_1 - \delta\beta_1 z_2 + \gamma\alpha_1 z_3 + \delta\alpha_1 z_4, \end{array} \tag{3.512}$$

from which we deduce

$$z'_1\,z'_4\ -\ z'_2\,z'_3\ =\ z_1\,z_4\ -\ z_2\,z_3, \tag{3.513}$$

i.e., the quadratic form (3.509) is invariant under the transformation (3.512). The matrix of the transformation (3.512) comes from the (commuting) product of matrices:

$$\begin{pmatrix} \alpha & \beta & 0 & 0 \\ \gamma & \delta & 0 & 0 \\ 0 & 0 & \alpha & \beta \\ 0 & 0 & \gamma & \delta \end{pmatrix} \cdot \begin{pmatrix} \delta_1 & 0 & -\gamma_1 & 0 \\ 0 & \delta_1 & 0 & -\gamma_1 \\ -\beta_1 & 0 & \alpha_1 & 0 \\ 0 & -\beta_1 & 0 & \alpha_1 \end{pmatrix}, \tag{3.514}$$

and its determinant is equal to 1, so that Eqs. (3.512) constitute a representation of $(SL(2,\mathcal{C}))^2$.

Every homogeneous transformation, with determinant equal to 1, that preserves the quadratic form (3.509) can be cast in the form (3.512) in two ways (the sign of the 8 constants $\alpha, \beta, \gamma, \delta, \alpha_1, \beta_1, \gamma_1, \delta_1$ is arbitrary). The quantities

$$z_1' = x_4\,\xi_2, \quad z_2' = -\,x_4\,\xi_1, \quad z_3' = -\,x_3\,\xi_2, \quad z_4' = x_3\,\xi_1 \tag{3.515}$$

transform as z_i since, from the unimodularity of σ_{12} and σ_{34}, the variables x_1 and x_2 transform as ξ_2 and $-\xi_1$, while ξ_3 and ξ_4 transform as x_4 and $-x_3$. Moreover, each linear combination of the vectors z and z' will transform in the same way; in particular this holds for the combination with components

$$z_1'' = z_1 + z_1', \quad z_2'' = z_2 + z_2', \quad z_3'' = z_3 + z_3', \quad z_4'' = z_4 + z_4'. \tag{3.516}$$

Let us introduce the quantities $ct,\ x/i,\ y/i,\ z/i$ and consider their transformation law as defined by

$$\begin{array}{rcccl} ct & \sim & z_1 + z_4 & \sim & z_1'' + z_4'', \\ x/i & \sim & (z_2 + z_3)/i & \sim & (z_2'' + z_3'')/i, \\ y/i & \sim & z_3 - z_2 & \sim & z_3'' - z_2'', \\ z/i & \sim & (z_1 - z_4)/i & \sim & (z_1'' - z_4'')/i. \end{array} \tag{3.517}$$

Then, we have

$$c^2t^2 - x^2 - y^2 - z^2 \sim 4\left(z_1''\,z_4'' - z_2''\,z_3''\right); \tag{3.518}$$

and since the r.h.s. is invariant, it follows that the transformation σ represents a Lorentz transformation of the space time variables x, y, z, t. Relations (3.517) can be written as

$$\begin{array}{rcl} ct & \sim & \xi_1 x_3 + \xi_2 x_4 + \xi_3 x_1 + \xi_4 x_2, \\ x/i & \sim & i\xi_1 x_4 + i\xi_2 x_3 - i\xi_3 x_2 - i\xi_4 x_1, \\ y/i & \sim & \xi_1 x_4 - \xi_2 x_3 - \xi_3 x_2 + \xi_4 x_1, \\ z/i & \sim & i\xi_1 x_3 - i\xi_2 x_4 - i\xi_3 x_1 + i\xi_4 x_2, \end{array} \tag{3.519}$$

and the expressions on the right are of the form

$$\sum_{ik} \gamma_{ik}^{\alpha}\,\xi_i\,x_k, \qquad \alpha = 1, 2, 3, 4. \tag{3.520}$$

The matrices γ_{ik}^{α} are Hermitian and satisfy the relation

$$\frac{1}{2}\left(\gamma_\alpha\,\gamma_\beta + \gamma_\beta\,\gamma_\alpha\right) = \delta_{\alpha\beta}. \tag{3.521}$$

Furthermore, if σ is a matrix defined by Eqs. (3.512), the transformed matrices $\sigma^{-1}\gamma_\alpha\sigma$ corresponding to $ct',\ x'/i,\ y'/i,\ z'/i$ are linear combinations of those corresponding to $ct,\ x/i,\ y/i,\ z/i$; see the next section.

18. DIRAC MATRICES AND THE LORENTZ GROUP

In an n-dimensional space, we have to construct the p Hermitian operators

$$\alpha_1, \ \alpha_2, \ \ldots, \ \alpha_p \tag{3.522}$$

obeying the relations

$$\frac{\alpha_i \alpha_k + \alpha_k \alpha_i}{2} = \delta_{ik}. \tag{3.523}$$

For arbitrary n and p, the problem might have no solutions or have only one fundamental solution (i.e., to which all the other possible sets of matrices $\alpha_1, \alpha_2, \ldots, \alpha_p$; $\alpha'_1, \alpha'_2, \ldots, \alpha'_p$; ... are related by means of a unitary transformation) or, even, have several different solutions that cannot be related by unitary transformations.

Let us consider the case $p = 1$. The only condition to be satisfied then is

$$\alpha_1^2 = 1, \tag{3.524}$$

so that all the eigenvalues of α_1 are either 1 or -1. Thus, the space R_n is divided into the sum of two subspaces $R'_r + R''_{n-r}$; the first one being r-dimensional ($0 \leq r \leq n$), corresponds to the positive eigenvalue $+1$ (which is degenerate $r - 1$ times), while the second one corresponds to the negative eigenvalue -1 (which is $n - r - 1$ times degenerate). If we assume that the first r fundamental vectors are r arbitrary unitary and orthogonal vectors of R'_r and that the last $n - r$ fundamental vectors are $n - r$ arbitrary unitary and orthogonal vectors of R''_{n-r}, then the α_1 matrix is diagonal, with the first r diagonal elements equal to 1 and the last $n - r$ ones equal to -1. Allowing r to assume values from n to 0, we then obtain the $n + 1$ fundamental solutions to the problem in the special case considered here.

Let us consider the case $p = 2$. The conditions to be satisfied now are

$$\alpha_1^2 = 1, \quad \alpha_2^2 = 1, \quad \alpha_1 \alpha_2 + \alpha_2 \alpha_1 = 0. \tag{3.525}$$

Let R'_r be the subspace corresponding to the eigenvalue $+1$ of α_1, and R''_{n-r} that corresponding to the eigenvalue -1. With a denoting a vector of R'_r; from the last equation of (3.525) we then have

$$(\alpha_1 \alpha_2 + \alpha_2 \alpha_1) \, a = 0, \tag{3.526}$$

or

$$(\alpha_1 + 1) \, \alpha_2 \, a = 0. \tag{3.527}$$

It follows that $\alpha_2 a$ belongs to R''_{n-r}, and since the determinant of α_2 is different from zero,

$$n - r \geq r \tag{3.528}$$

obtains. Now, let b be a vector of R''_{n-r}; from the last equation of (3.525), we get $(\alpha_1 - 1)\,\alpha_2 b = 0$, i.e., $\alpha_2 b$ belongs to R'_r and then $r \geq n - r$. By combining this relation with Eq. (3.528) we get

$$r = n/2. \tag{3.529}$$

It follows that in the case $p = 2$ solutions exist only if n is even. Assuming $n = 2r$ with integer r, we can choose r arbitrary unitary and orthogonal vectors of R'_r as the first r fundamental vectors. On the other hand, the last r fundamental vectors can be chosen to be the transformed vectors [18] with $\alpha_2 \leq 1$, $\alpha_2 \leq 2$, ..., $\alpha_2 \leq r$. These obviously are orthogonal to the first r fundamental vectors because they belong to R''_r (which is orthogonal to R'_r), but they are also unitary and mutually orthogonal since α_2 is Hermitian and coincides with its inverse. The matrices α_1 and α_2 then, have the forms

$$\alpha_1 = \begin{pmatrix} \begin{pmatrix} 1 & 0 & \dots & 0 \\ 0 & 1 & \dots & 0 \\ \dots & \dots & \dots & \dots \\ 0 & 0 & \dots & 1 \end{pmatrix} & \begin{pmatrix} 0 & 0 & \dots & 0 \\ 0 & 0 & \dots & 0 \\ \dots & \dots & \dots & \dots \\ 0 & 0 & \dots & 0 \end{pmatrix} \\ \begin{pmatrix} 0 & 0 & \dots & 0 \\ 0 & 0 & \dots & 0 \\ \dots & \dots & \dots & \dots \\ 0 & 0 & \dots & 0 \end{pmatrix} & \begin{pmatrix} -1 & 0 & \dots & 0 \\ 0 & -1 & \dots & 0 \\ \dots & \dots & \dots & \dots \\ 0 & 0 & \dots & -1 \end{pmatrix} \end{pmatrix},$$

$$\alpha_2 = \begin{pmatrix} \begin{pmatrix} 0 & 0 & \dots & 0 \\ 0 & 0 & \dots & 0 \\ \dots & \dots & \dots & \dots \\ 0 & 0 & \dots & 0 \end{pmatrix} & \begin{pmatrix} 1 & 0 & \dots & 0 \\ 0 & 1 & \dots & 0 \\ \dots & \dots & \dots & \dots \\ 0 & 0 & \dots & 1 \end{pmatrix} \\ \begin{pmatrix} 1 & 0 & \dots & 0 \\ 0 & 1 & \dots & 0 \\ \dots & \dots & \dots & \dots \\ 0 & 0 & \dots & 1 \end{pmatrix} & \begin{pmatrix} 0 & 0 & \dots & 0 \\ 0 & 0 & \dots & 0 \\ \dots & \dots & \dots & \dots \\ 0 & 0 & \dots & 0 \end{pmatrix} \end{pmatrix}. \tag{3.530}$$

Thus, for $p = 2$ and even n, the problem has only one fundamental solution, while for odd n there is no solution at all.

---

[18]@ That is, the vectors $\alpha_2 a$.

Let us now consider the case $p > 2$. We have the p matrices

$$\alpha_1,\ \alpha_2,\ \alpha_3,\ \ldots,\ \alpha_p. \tag{3.531}$$

As above, we choose the first $r = n/2$ fundamental vectors in the space R'_r corresponding to the positive eigenvalue $+1$ and the last $n-r$ vectors in the space R''_r obtained by applying the matrix α_2 to the first ones. The only difference with the previous case is that we don't choose arbitrarily the first r fundamental vectors in the space R'_r, but we choose them from a specific representation of $\alpha_3,\ \alpha_4, \ldots,\ \alpha_p$. To this end, we set

$$\begin{aligned} \alpha_2\alpha_3 &= i\beta_1, & \alpha_3 &= i\alpha_2\beta_1, \\ \alpha_2\alpha_4 &= i\beta_2, & \alpha_4 &= i\alpha_2\beta_2 \\ \ldots, & & & \\ \alpha_2\alpha_p &= i\beta_{p-2}, & \alpha_p &= i\alpha_2\beta_{p-2}. \end{aligned} \tag{3.532}$$

The operators $\beta_1, \beta_2, \ldots, \beta_{p-2}$ transform vectors of R'_r into vectors of R'_r, and vectors of R''_r into vectors of R''_r. Then, given

$$\alpha_{i+2} = \begin{pmatrix} \mathbf{0} & i\gamma_i \\ i\delta_i & \mathbf{0} \end{pmatrix}, \tag{3.533}$$

their matrix form is seen to have the structure

$$\beta_i = \begin{pmatrix} \delta_i & \mathbf{0} \\ \mathbf{0} & \gamma_i \end{pmatrix}, \qquad i = 1, 2, \ldots, p-2, \tag{3.534}$$

where γ_i and δ_i are matrices of dimension $n/2$. Moreover, from

$$\frac{\alpha_{i+2}\alpha_{k+2} + \alpha_{k+2}\alpha_{i+2}}{2} = \delta_{ik} \tag{3.535}$$

we deduce, with the help of Eqs. (3.532), that

$$\frac{1}{2}\left(\alpha_2\beta_i\alpha_2\beta_k + \alpha_2\beta_k\alpha_2\beta_i\right) = -\,\delta_{ik}; \tag{3.536}$$

and, since

$$\begin{aligned} i\beta_1\alpha_2 &= \alpha_2\alpha_{i+2}\alpha_2 = \alpha_2\left(\alpha_{i+2}\alpha_2\right) \\ &= -\,\alpha_2\left(\alpha_2\alpha_{i+2}\right) = -\,\alpha_{i+2} = -i\,\alpha_2\,\beta_i, \end{aligned}$$

it follows that

$$\beta_i\alpha_2 = -\,\alpha_2\beta_i, \quad \beta_k\alpha_2 = -\,\alpha_2\beta_k. \tag{3.537}$$

Thus, the relation (3.536) becomes

$$\begin{aligned} -\ & \alpha_2\beta_i\alpha_2\beta_k + \alpha_2\beta_k\alpha_2\beta_i = -\,\alpha_2\left(\beta_i\alpha_2\right)\beta_k + \alpha_2\left(\beta_k\alpha_2\right)\beta_i \\ =\ & \alpha_2\left(\alpha_2\beta_i\right)\beta_k + \alpha_2\left(\alpha_2\beta_k\right)\beta_i = \beta_1\beta_k + \beta_k\beta_i = 2\,\delta_{ik}, \end{aligned} \tag{3.538}$$

so that the matrices β satisfy relations analogous to Eqs. (3.523). However, we still have not used some of the relations in Eq. (3.523); in particular, we have not used the relations involving the matrices α_2 and α_{i+2} $(i+1,2,\ldots,p-2)$ or, from Eqs. (3.532), relation (3.537) from which relations (3.538) derive their validity. Now, given the form in Eq. (3.530) for α_2, Eq. (3.537) implies that the sub-matrices γ_i, δ_i of dimensions $n/2$ satisfy the relation

$$\gamma_i \ = \ - \delta_i, \tag{3.539}$$

so that Eq. (3.538) is satisfied when

$$\gamma_i \, \gamma_k \ + \ \gamma_k \, \gamma_i \ = \ 2 \, \delta_{ik}, \quad i,k = 1,2,\ldots,p-2. \tag{3.540}$$

Note that, by assuming the matrices γ_i to be Hermitian, the matrices α_{i+2} are automatically Hermitian; thus, our problem is fully analogous to the one with $n' = n/2$ and $p' = p-2$. If again $p' > 2$, we revert to one of the cases studied earlier: if n is odd, we have $n+1$ fundamental solutions when $p = 1$, otherwise there is no solution.If instead $p' \leq 2$, we return to one of the two cases studied early in this section.

We have solved the problem and obtained a procedure by which we can construct all the possible fundamental solutions.

Let us set p in the form $p = 2k$ or $p = 2k+1$; the problem has solutions only if n can be divided by 2^k; more precisely, there is only one solution if p is even, while there are $(n/2^k)+1$ solutions if p is odd. As particular cases we have the solutions for $p = 1$ and $p = 2$ discussed above.

We can also consider the special case in which p takes the maximum allowed value for fixed n. Decomposing n in first factors and denoting with t the exponent of the number 2 in such decomposition, we have $p_{\max} = 2t+1$ and the number of fundamental solutions is $(n/2^t)+1 \geq 2$, where the equal sign applies only when n is an integer power of 2.

Non-Hermitian operators. Let us now relax the assumption that the $\alpha_1, \alpha_2, \ldots, \alpha_p$ are Hermitian operators. In this case, all the solutions that are related via an *arbitrary* coordinate transformation can be regarded as the same fundamental solution. This means that, if α_i is a solution, then $\alpha_i' = S\alpha_i S^{-1}$ is also a solution if S is an arbitrary operator with determinant different from zero. For the representation of α_i we choose a *non-normal* coordinate system and proceed exactly as in the previous case, replacing the orthonormality condition with the condition of linear independence of the fundamental vectors. Thus we arrive exactly at the same *Hermitian* matrices obtained above but, since the coordinate system is non-normal, in general they will not represent Hermitian operators. Going back to normal coordinates, the matrices representing the Hermitian operators can be obtained from the fundamental ones by means of a unitary transformation, while the matrices representing non-Hermitian operators can be obtained from the same fundamental matrices by means of a non-unitary transformation; in general, these matrices will be non-Hermitian.

Examples. Let us now give some examples of fundamental matrices, when $n = 2^t$ and $p = 2t + 1$ takes their maximum allowed values. We always have two fundamental solutions which differ only in the sign of the last matrix.

- $n = 1,\ p = 1$

$$\alpha_i = \pm 1$$

- $n = 2,\ p = 3$

$$\alpha_1 = \begin{pmatrix} 1 & 0 \\ 0 & -1 \end{pmatrix}, \quad \alpha_2 = \begin{pmatrix} 0 & 1 \\ 1 & 0 \end{pmatrix}, \quad \alpha_3 = \begin{pmatrix} 0 & i \\ -i & 0 \end{pmatrix}.$$

- $n = 4,\ p = 5$

$$\alpha_1 = \begin{pmatrix} 1 & 0 & 0 & 0 \\ 0 & 1 & 0 & 0 \\ 0 & 0 & -1 & 0 \\ 0 & 0 & 0 & -1 \end{pmatrix}, \quad \alpha_2 = \begin{pmatrix} 0 & 0 & 1 & 0 \\ 0 & 0 & 0 & 1 \\ 1 & 0 & 0 & 0 \\ 0 & 1 & 0 & 0 \end{pmatrix},$$

$$\alpha_3 = \begin{pmatrix} 0 & 0 & i & 0 \\ 0 & 0 & 0 & -1 \\ -i & 0 & 0 & 0 \\ 0 & i & 0 & 0 \end{pmatrix}, \quad \alpha_4 = \begin{pmatrix} 0 & 0 & 0 & i \\ 0 & 0 & i & 0 \\ 0 & -i & 0 & 0 \\ -i & 0 & 0 & 0 \end{pmatrix},$$

$$\alpha_5 = \begin{pmatrix} 0 & 0 & 0 & -1 \\ 0 & 0 & 1 & 0 \\ 0 & 1 & 0 & 0 \\ -1 & 0 & 0 & 0 \end{pmatrix}.$$

• $n = 8,\ p = 7$

$$\alpha_1 = \begin{pmatrix} 1 & 0 & 0 & 0 & 0 & 0 & 0 & 0 \\ 0 & 1 & 0 & 0 & 0 & 0 & 0 & 0 \\ 0 & 0 & 1 & 0 & 0 & 0 & 0 & 0 \\ 0 & 0 & 0 & 1 & 0 & 0 & 0 & 0 \\ 0 & 0 & 0 & 0 & -1 & 0 & 0 & 0 \\ 0 & 0 & 0 & 0 & 0 & -1 & 0 & 0 \\ 0 & 0 & 0 & 0 & 0 & 0 & -1 & 0 \\ 0 & 0 & 0 & 0 & 0 & 0 & 0 & -1 \end{pmatrix},$$

$$\alpha_2 = \begin{pmatrix} 0 & 0 & 0 & 0 & 1 & 0 & 0 & 0 \\ 0 & 0 & 0 & 0 & 0 & 1 & 0 & 0 \\ 0 & 0 & 0 & 0 & 0 & 0 & 1 & 0 \\ 0 & 0 & 0 & 0 & 0 & 0 & 0 & 1 \\ 1 & 0 & 0 & 0 & 0 & 0 & 0 & 0 \\ 0 & 1 & 0 & 0 & 0 & 0 & 0 & 0 \\ 0 & 0 & 1 & 0 & 0 & 0 & 0 & 0 \\ 0 & 0 & 0 & 1 & 0 & 0 & 0 & 0 \end{pmatrix},$$

$$\alpha_3 = \begin{pmatrix} 0 & 0 & 0 & 0 & i & 0 & 0 & 0 \\ 0 & 0 & 0 & 0 & 0 & i & 0 & 0 \\ 0 & 0 & 0 & 0 & 0 & 0 & -i & 0 \\ 0 & 0 & 0 & 0 & 0 & 0 & 0 & -i \\ -i & 0 & 0 & 0 & 0 & 0 & 0 & 0 \\ 0 & -i & 0 & 0 & 0 & 0 & 0 & 0 \\ 0 & 0 & i & 0 & 0 & 0 & 0 & 0 \\ 0 & 0 & 0 & i & 0 & 0 & 0 & 0 \end{pmatrix},$$

$$\alpha_4 = \begin{pmatrix} 0 & 0 & 0 & 0 & 0 & 0 & i & 0 \\ 0 & 0 & 0 & 0 & 0 & 0 & 0 & i \\ 0 & 0 & 0 & 0 & i & 0 & 0 & 0 \\ 0 & 0 & 0 & 0 & 0 & i & 0 & 0 \\ 0 & 0 & -i & 0 & 0 & 0 & 0 & 0 \\ 0 & 0 & 0 & -i & 0 & 0 & 0 & 0 \\ -i & 0 & 0 & 0 & 0 & 0 & 0 & 0 \\ 0 & -i & 0 & 0 & 0 & 0 & 0 & 0 \end{pmatrix},$$

$$\alpha_5 = \begin{pmatrix} 0 & 0 & 0 & 0 & 0 & 0 & -1 & 0 \\ 0 & 0 & 0 & 0 & 0 & 0 & 0 & 1 \\ 0 & 0 & 0 & 0 & 1 & 0 & 0 & 0 \\ 0 & 0 & 0 & 0 & 0 & -1 & 0 & 0 \\ 0 & 0 & 1 & 0 & 0 & 0 & 0 & 0 \\ 0 & 0 & 0 & -1 & 0 & 0 & 0 & 0 \\ -1 & 0 & 0 & 0 & 0 & 0 & 0 & 0 \\ 0 & 1 & 0 & 0 & 0 & 0 & 0 & 0 \end{pmatrix},$$

$$\alpha_6 = \begin{pmatrix} 0 & 0 & 0 & 0 & 0 & 0 & 0 & -1 \\ 0 & 0 & 0 & 0 & 0 & 0 & -1 & 0 \\ 0 & 0 & 0 & 0 & 0 & 1 & 0 & 0 \\ 0 & 0 & 0 & 0 & 1 & 0 & 0 & 0 \\ 0 & 0 & 0 & 1 & 0 & 0 & 0 & 0 \\ 0 & 0 & 1 & 0 & 0 & 0 & 0 & 0 \\ 0 & -1 & 0 & 0 & 0 & 0 & 0 & 0 \\ -1 & 0 & 0 & 0 & 0 & 0 & 0 & 0 \end{pmatrix},$$

$$\alpha_7 = \begin{pmatrix} 0 & 0 & 0 & 0 & 0 & 0 & 0 & -i \\ 0 & 0 & 0 & 0 & 0 & 0 & i & 0 \\ 0 & 0 & 0 & 0 & 0 & i & 0 & 0 \\ 0 & 0 & 0 & 0 & -i & 0 & 0 & 0 \\ 0 & 0 & 0 & i & 0 & 0 & 0 & 0 \\ 0 & 0 & -i & 0 & 0 & 0 & 0 & 0 \\ 0 & -i & 0 & 0 & 0 & 0 & 0 & 0 \\ i & 0 & 0 & 0 & 0 & 0 & 0 & 0 \end{pmatrix}.$$

Interpretation according to group theory. Let us consider the operators

$$\alpha_1,\ \alpha_2,\ \ldots,\ \alpha_p, \tag{3.541}$$

obeying the usual relations

$$\alpha_i\alpha_k + \alpha_k\alpha_i = 2\,\delta_{ik}, \tag{3.542}$$

and the compound operators obtained by multiplying an arbitrary set of αs in arbitrary order. From the relations (3.542) it follows that they can always be cast in the form

$$g: \quad \pm\,\alpha_1^{\epsilon_1}\,\alpha_2^{\epsilon_2}\,\ldots\,\alpha_p^{\epsilon_p}, \tag{3.543}$$

and that they form a finite group. In the expression above, ϵ_i take the values 0 and 1. Operators (3.543) can be viewed as elements of a group containing 2^{p+1} elements. In order to have a representation of this group, it is sufficient to find p matrices that obey Eq. (3.542);

they correspond to p fundamental elements of the group (3.543), which are the operators α themselves and that can be obtained from group (3.543) by taking the $+$ sign and setting all the ϵ but one equal to zero. All the other elements of the group can be obtained by taking the product of these fundamental matrices. The problem of finding p matrices obeying relations (3.542) has already been solved for all the values of n, so that it can be solved in any possible way for a given n. We then have a corresponding number of representations for the group g. However, these are not all the possible representations. In fact, the composition rule for the elements of the group has been derived from Eq. (3.542), even though these cannot really be derived from the composition rule, except when opposite matrices correspond to elements of the group marked with the same ϵ but with opposite sign. Consider, in particular, irreducible representations. The element

$$-\,\alpha_1^0\,\alpha_2^0\,\ldots\,\alpha_p^0 \tag{3.544}$$

commutes with all the elements of the group; and, since its square is the identity, in the irreducible representations it will correspond to the unit matrix or to its opposite. Note that only in this second case the fundamental matrices satisfy the relations (3.542), and this happens for an arbitrary representation only when it can be decomposed into irreducible representations corresponding to the second case. Irreducible representations corresponding to the first case are, necessarily, one-dimensional ones since they are shortened representations of g, i.e., representations of the Abelian group g' that can be obtained from Eq. (3.543) by identifying with a unique element the ones with opposite signs. Group g is composed of equivalent elements with respect to the invariant subgroup formed with the unit element and the element in Eq. (3.544). Since the group g' contains 2^p elements, the one-dimensional irreducible representations, corresponding to the first case, are exactly 2^p. The irreducible elements obviously are

$$\eta_1^{\epsilon_1}\eta_2^{\epsilon_2}\ldots\eta_p^{\epsilon_p}. \tag{3.545}$$

Furthermore, let us assume that there exist s irreducible representations corresponding to the second case. For the "completeness" theorem we must have

$$n_1^2 + n_2^2 + \ldots + n_s^2 \;=\; 2^{p+1} - 2^p \;=\; 2^p. \tag{3.546}$$

Let us suppose that n_i are ordered in a non-decreasing way; in this case n_1 is the smallest value of n for which it is possible to find p matrices obeying Eq. (3.542). If $p = 2k$ is even, we know that such minimum value is $n = 2^k = 2^{p/2}$; then it follows from Eq. (3.546) that only one irreducible representation corresponding to the second case exists, with

$$n = 2^{p/2} = 2^k, \quad p = 2k. \tag{3.547}$$

Instead, if $p = 2k + 1$ is odd, we again have $n_1 = 2^k$, but a second irreducible representation of the same dimension must exist. Then, for odd p, we have two irreducible representations corresponding to the second case with

$$n_1 = n_2 = 2^{\frac{p-1}{2}} = 2^k, \quad p = 2k + 1. \tag{3.548}$$

Since a representation in which relations (3.542) are satisfied can be decomposed in irreducible representations corresponding to the second case, it is now easy to understand the theorem (see the paragraph *Non-Hermitian operators*), i.e., the problem of finding p matrices of dimension n obeying Eq. (3.542) can be solved only if n can be divided by 2^k. Moreover, we also understand that such solution is unique (apart from transformations) if p is even, since in this case the possible decomposition into irreducible matrices is unique. If p is odd, then there are $n/2^k + 1$ fundamental solutions, since in the decomposition of the representation of dimension n into irreducible representations of the second kind, with the last one having the same dimension 2^k, one of them can fit an integer number of times between 0 and $n/2^k$.
When n is a multiple of 2^k, we can adapt the coordinates to the decomposition into irreducible representations; we then obtain for the α matrices a form which is simpler than the one obtained in the direct way, since they are formed of partial matrices of dimension 2^k which, with a convenient choice of the coordinates, can be deduced from the ones already considered for the case $n = 2^k$. [19]

19. THE SPINNING ELECTRON

Let us consider the Dirac equations in the form

$$\mathcal{H}\psi \equiv \left[\frac{\alpha_1}{i}\left(mc + \frac{W}{c} + \frac{e}{c}\phi\right) + \alpha_2\left(p_x + \frac{e}{c}A_x\right)\right.$$
$$\left. + \alpha_3\left(p_y + \frac{e}{c}A_y\right) + \alpha_4\left(p_z + \frac{e}{c}A_z\right) + \frac{mc}{i}\right]\psi = 0, \tag{3.549}$$

---

[19]@ This section ends with: *For the connection between the Dirac matrices and the Lorentz group, see the section "Invariance of the Dirac equations."* However, in the five Volumetti there is no section that deals with this subject.

where the α matrices are the ones displayed in Sec. 3.18 with $n = 4$, $p = 5$. Let $\mathcal{H}_1$ be the operator obtained from $\mathcal{H}$ by changing its last term mc/i into $-mc/i$, and consider the quantity $\mathcal{H}_1 \mathcal{H} \psi$ [20]:

$$\begin{aligned}
&\left[-\left(mc + \frac{W}{c} + \frac{e}{c}\phi \right)^2 + \left(p_x + \frac{e}{c}A_x \right)^2 \right. \\
&+ \left(p_y + \frac{e}{c}A_y \right)^2 + \left(p_z + \frac{e}{c}A_z \right)^2 + m^2c^2 \\
&+ \alpha_1\alpha_2 \frac{e\hbar}{c} E_x + \alpha_1\alpha_3 \frac{e\hbar}{c} E_y + \alpha_1\alpha_4 \frac{e\hbar}{c} E_z \\
&\left. - \frac{e\hbar}{ci}\alpha_2\alpha_3 H_z - \frac{e\hbar}{ci}\alpha_3\alpha_4 H_x - \frac{e\hbar}{ci}\alpha_4\alpha_2 H_y \right] \psi = 0. \qquad (3.550)
\end{aligned}$$

The first five terms give the relativistic Hamiltonian for an electron without spin, while the others represent the corrections induced by spin. By noting that the square of the matrices $\alpha_1\alpha_2$, $\alpha_1\alpha_3$, $\alpha_1\alpha_4$, $\alpha_2\alpha_3$, $\alpha_3\alpha_4$, $\alpha_4\alpha_2$ is -1, so that their eigenvalues are $\pm i$, and that the classical Hamiltonian, in first approximation, is $\mathcal{H}_1\mathcal{H}/2m$, we deduce that the electron has a magnetic moment $e\hbar/2mc$ and an imaginary electric moment $e\hbar/2mci$.

Let us consider the equivalent, but more convenient, expressions for Eqs. (3.549):

$$\begin{aligned}
&\left[-\left(mc + \frac{W}{c} + \frac{e}{c}\phi \right) + \alpha_1\, mc + \alpha_2 \left(p_x + \frac{e}{c}A_x \right) \right. \\
&\left. + \alpha_3 \left(p_y + \frac{e}{c}A_y \right) + + \alpha_4 \left(p_z + \frac{e}{c}A_z \right) \right] \psi = 0, \qquad (3.551)
\end{aligned}$$

which can be cast in the form

$$\begin{aligned}
\mathcal{H}\psi \;\equiv\; &\left[(\alpha_1 - 1)\, mc^2 - \frac{e}{c}\phi + \alpha_2\, c \left(p_x + \frac{e}{c}A_x \right) \right. \\
&\left. + \; \alpha_3\, c \left(p_y + \frac{e}{c}A_y \right) + \alpha_4\, c \left(p_z + \frac{e}{c}A_z \right) \right] = W\,\psi. \qquad (3.552)
\end{aligned}$$

Let us assume that the magnetic field is constant, with intensity H, and that it is directed along the z axis. We then have

$$A_x = -\frac{1}{2}\, y\, H, \quad A_y = \frac{1}{2}\, x\, H, \quad A_z = 0, \qquad (3.553)$$

[20]@ In the original manuscript, the old notation $h/2\pi$ is used, while we here denote the same quantity with $\hbar$. Note also that ϕ and $\mathbf{A}$ are the scalar and vector electromagnetic potentials, respectively, while in the following $\mathbf{E}$ and $\mathbf{H}$ denote the electric and magnetic fields, respectively.

so that Eqs. (3.552) become

$$\mathcal{H}\psi \equiv \left[(\alpha_1 - 1)\, mc^2 - \frac{e}{c}\phi + \alpha_2\, c\left(p_x - \frac{e}{2c}yH\right) + \alpha_3\, c\left(p_y + \frac{e}{2c}xH\right) + \alpha_4\, c\, p_z\right] = W\psi. \tag{3.554}$$

Let us denote with ψ_n the scalar solutions of the Schrödinger equation

$$-\frac{\hbar^2}{2m}\nabla^2\psi^n - e\,\phi\,\psi^n - W_n\,\psi^n = 0 \tag{3.555}$$

and with

$$x_{nn'},\ y_{nn'},\ z_{nn'} \tag{3.556}$$

the polarization matrices. Then, Eqs. (3.554) have the explicit forms

$$-e\,\phi\,\psi_1 + c\left(p_x - \frac{e}{2c}yH\right)\psi_3 + c\,i\left(p_y + \frac{e}{2c}xH\right)\psi_3 + c\,i\,p_z\,\psi_4 = W\,\psi_1, \tag{3.557}$$

$$-e\,\phi\,\psi_2 + c\left(p_x - \frac{e}{2c}yH\right)\psi_4 - c\,i\left(p_y + \frac{e}{2c}xH\right)\psi_4 + c\,i\,p_z\,\psi_3 = W\,\psi_2, \tag{3.558}$$

$$-2mc^2\,\psi_3 - e\,\phi\,\psi_3 + c\left(p_x - \frac{e}{2c}yH\right)\psi_1 - c\,i\left(p_y + \frac{e}{2c}xH\right)\psi_1 - c\,i\,p_z\,\psi_2 = W\,\psi_3, \tag{3.559}$$

$$-2mc^2\,\psi_4 - e\,\phi\,\psi_4 + c\left(p_x - \frac{e}{2c}yH\right)\psi_2 + c\,i\left(p_y + \frac{e}{2c}xH\right)\psi_2 - c\,i\,p_z\,\psi_1 = W\,\psi_4. \tag{3.560}$$

Here we neglect the solutions corresponding to the positive electron (positron), i.e., in first approximation, the ones with large ψ_3 and ψ_4 and small ψ_1 and ψ_2. Thus, in first approximation, the Dirac equations are solved by the double system of vector functions ψ^{n1} and ψ^{n2} whose components are the following:

	1st comp.	2nd comp.	3rd comp.	4th comp.
ψ^{n1}	ψ^n	0	$(2mc)^{-1}(p_x - ip_y)\psi^n$	$-(2mc)^{-1}ip_z\psi^n$
ψ^{n2}	0	ψ^n	$-(2mc)^{-1}ip_z\psi^n$	$(2mc)^{-1}(p_x + ip_y)\psi^n$

(3.561)

These functions are mutually orthogonal and, in first approximation, normalized. To determine the eigenvalues in second approximation (i.e., when taking into account in first approximation relativistic, spin and magnetic field effects), we substitute Eqs. (3.557)-(3.560) into Eq. (3.555). For ψ^{n1} and ψ^{n2}, we obtain, respectively:

(a) For ψ^{n1} : (3.562)

$$
\begin{aligned}
& -e\,\phi\,\psi + c\left(p_x - \frac{e}{2c}\,y\,H\right)\psi_3 + c\,i\left(p_y + \frac{e}{2c}\,x\,H\right)\psi_3 \\
& + c\,i\,p_z\,\psi_4 - W_n\,\psi_1 \equiv (\delta\mathcal{H}\,\psi)_1 \\
& = \frac{e\,H}{4mc}\,(x\,p_y - y\,p_x)\,\psi^n + \frac{i\,e\,H}{4mc}\,(x\,p_x + y\,p_y)\,\psi^n;
\end{aligned}
$$

$$
\begin{aligned}
& -e\,\phi\,\psi_2 + c\left(p_x - \frac{e}{2c}yH\right)\psi_4 - c\,i\left(p_y - \frac{e}{2c}xH\right)\psi_4 \\
& + c\,i\,p_z\,\psi_3 - W_n\,\psi_2 \equiv (\delta\mathcal{H}\,\psi)_2 \\
& = -\frac{eH}{4mc}\,(x - i\,y)\,p_z\,\psi^n,
\end{aligned}
$$

$$
\begin{aligned}
& -2mc^2\,\psi_3 - e\,\phi\,\psi_3 + c\left(p_x - \frac{e}{2c}yH\right)\psi_1 \\
& - c\,i\left(p_y + \frac{e}{2c}xH\right)\psi_1 - c\,i\,p_z\,\psi_2 - W_n\,\psi_3 \equiv (\delta\mathcal{H}\,\psi)_3 \\
& = -\frac{1}{2mc}\,(W_n + e\,\phi)\,(p_x - i\,p_y)\,\psi^n - \frac{ieH}{2}\,(x - i\,y)\,\psi^n,
\end{aligned}
$$

$$
\begin{aligned}
& -2mc^2\,\psi_4 - e\,\phi\,\psi_4 + c\left(p_x - \frac{e}{2c}yH\right)\psi_2 \\
& + c\,i\left(p_y + \frac{e}{2c}xH\right)\psi_2 - c\,i\,p_z\,\psi_1 - W_n\,\psi_4 \equiv (\delta\mathcal{H}\,\psi)_4 \\
& = \frac{i}{2mc}\,(W_n + e\,\phi)\,p_z\,\psi^n.
\end{aligned}
$$

(b) For ψ^{n2} : (3.563)

$$
\begin{aligned}
& -e\,\phi\,\psi + c\left(p_x - \frac{e}{2c}\,y\,H\right)\psi_3 + c\,i\left(p_y + \frac{e}{2c}\,x\,H\right)\psi_3 \\
& + c\,i\,p_z\,\psi_4 - W_n\,\psi_1 \equiv (\delta\mathcal{H}\,\psi)_1 \\
& = \frac{eH}{4mc}\,(x + i\,y)\,p_z\,\psi^n,
\end{aligned}
$$

$$
\begin{aligned}
&-e\,\phi\,\psi_2 + c\left(p_x - \frac{e}{2c}yH\right)\psi_4 - c\,i\left(p_y - \frac{e}{2c}xH\right)\psi_4 \\
&+ c\,i\,p_z\,\psi_3 - W_n\,\psi_2 \;\equiv\; (\delta\mathcal{H}\,\psi)_2 \\
&= \frac{e\,H}{4mc}\,(x\,p_y - y\,p_x)\,\psi^n - \frac{i\,e\,H}{4mc}\,(x\,p_x + y\,p_y)\,\psi^n,
\end{aligned}
$$

$$
\begin{aligned}
&-2mc^2\,\psi_3 - e\,\phi\,\psi_3 + c\left(p_x - \frac{e}{2c}yH\right)\psi_1 \\
&- c\,i\left(p_y + \frac{e}{2c}xH\right)\psi_1 - c\,i\,p_z\,\psi_2 - W_n\,\psi_3 \;\equiv\; (\delta\mathcal{H}\,\psi)_3 \\
&= -\frac{i}{2mc}\,(W_n + e\,\phi)\,p_z\,\psi^n,
\end{aligned}
$$

$$
\begin{aligned}
&-2mc^2\,\psi_4 - e\,\phi\,\psi_4 + c\left(p_x - \frac{e}{2c}yH\right)\psi_2 + s \\
&+ c\,i\left(p_y + \frac{e}{2c}xH\right)\psi_2 - c\,i\,p_z\,\psi_1 - W_n\,\psi_4 \;\equiv\; (\delta\mathcal{H}\,\psi)_4 \\
&= -\frac{1}{2mc}\,(W_n + e\,\phi)\,(p_x + i\,p_y)\,\psi^n + \frac{ieH}{2}\,(x + i\,y)\,\psi^n.
\end{aligned}
$$

Assume that W_n has degeneracy q and let

$$
y^1,\; y^2,\; \ldots,\; y^q \tag{3.564}
$$

be the orthonormalized Schrödinger eigenfunctions associated with the eigenvalue W_n. Note that, because of the spin, we instead have $2q$ vector eigenfunctions with eigenvalue next to W_n. In first approximation, these will be obtained as linear combinations of the $2q$ approximate eigenfunctions $y^{11}, y^{21}, \ldots, y^{q1}, y^{12}, y^{22}, \ldots, y^{q2}$ given by Eqs. (3.561), with ψ^n replaced by $y^1, y^2, \ldots, y^q$, respectively. (We have set, generically, $y^{n1} = \psi^{n1}$, $y^{n2} = \psi^{n2}$.) The corrections to the eigenvalues will then be given, in first approximation, by the eigenvalues of the matrix $\delta\mathcal{H}$. We calculate this quantity in first approximation (a better knowledge is illusory) by setting

$$
\delta\mathcal{H}_{ri,sk} \;=\; \sum_{\gamma=1}^{4} \int y_\gamma^{ri*}\left(\delta\mathcal{H}\,y^{sk}\right)_\gamma \mathrm{d}\tau \tag{3.565}
$$

($i, k = 1, 2$ and $r, s = 1, 2, \ldots, q$). Note that the approximation consists in the fact that we use y^{r1} (or y^{sk}) as given by Eq. (3.561), which are normalized only in first approximation. We can divide the perturbation matrix $\delta\mathcal{H}_{ri,sk}$ into the sum of two quantities, the first one being independent of the magnetic field and the second one proportional to it:

$$
\delta\mathcal{H}_{ri,sk} \;=\; A_{ri,sk} + H\,B_{ri,sk}. \tag{3.566}
$$

We start the discussion with a particular case; let us assume that the magnetic field is zero and that W_n is a simple (non-degenerate) eigenvalue of the Schrödinger equation. [21] Since $q = 1$, there are only two basic functions, y^{11} and y^{12}. Neglecting the indices r and s, which are automatically equal to 1, and taking into account the expressions for $(\delta\mathcal{H}\, y^1)_\gamma$ and $(\delta\mathcal{H}\, y^2)_\gamma$ ($\gamma = 1,2,3,4$) in Eqs. (3.562) and (3.563), from Eq. (3.565) we have

$$
\begin{aligned}
\delta\mathcal{H}_{11} &= -\frac{1}{4m^2c^2}\left[\int \left(p_x^* + ip_y^*\right) y^{1*}\cdot (W_n + e\phi)\,(p_x - ip_y)\, y^1\, \mathrm{d}\tau \right.\\
&\qquad \left. + \int p_z^*\, y^{1*}\cdot (W_n + e\phi)\; p_z\, y^1\, \mathrm{d}\tau\right]\\
&= -\frac{1}{4m^2c^2}\left[\int y^{1*}\; (p_x + ip_y)\; (W_n + e\phi)\; (p_x - ip_y)\; y^1\, \mathrm{d}\tau \right.\\
&\qquad \left. + \int y^{1*}\, p_z\; (W_n + e\phi)\; p_z\, y^1\, \mathrm{d}\tau\right]\\
&= -\frac{1}{4m^2c^2}\int y^{1*}\; (W_n + e\phi)\; \left(p_x^2 + p_y^2\right)\; y^1\, \mathrm{d}\tau\\
&\quad -\frac{1}{4m^2c^2}\int -\frac{4e}{2\pi i}\, y^{1*}\left(\frac{\partial\phi}{\partial x} + i\frac{\partial\phi}{\partial y}\right)(p_x - ip_y)\; y^1\, \mathrm{d}\tau\\
&\quad -\frac{1}{4m^2c^2}\int y^{1*}\; (W_n + e\phi)\; p_z^2\, y^1\, \mathrm{d}\tau\\
&\quad -\frac{1}{4m^2c^2}\int -\frac{4e}{2\pi i}\, y^{1*}\,\frac{\partial\phi}{\partial z}\, p_z\, y^1\, \mathrm{d}\tau.
\end{aligned}
$$

On setting $V = -e\phi$ and noting that, from the Schrödinger equation,

$$
\left(p_x^2 + p_y^2 + p_z^2\right)\, y^1 \;=\; 2m\;(W_n \,-\, V)\; y^1, \tag{3.567}
$$

we get

$$
\begin{aligned}
\delta\mathcal{H}_{11} &= -\frac{1}{2mc^2}\int y^{1*}\; (W_n - V)^2\; y^1\, \mathrm{d}\tau\\
&\quad -\frac{\hbar^2}{4m^2c^2}\int y^{1*}\; (\boldsymbol{\nabla}\, V\, \boldsymbol{\nabla}\,)\; y^1\, \mathrm{d}\tau && (3.568)\\
&\quad +\frac{i\hbar^2}{4m^2c^2}\int y^{1*}\left(\frac{\partial V}{\partial x}\frac{\partial y^1}{\partial y} - \frac{\partial V}{\partial y}\frac{\partial y^1}{\partial x}\right)\mathrm{d}\tau. && (3.569)
\end{aligned}
$$

Using the assumption of no degeneracy, we obtain that y^1 is real, so that it coincides with y^{1*}; then the second integral in Eq. (3.568) can

[21]@ Note that the author considers degeneracy coming only from non-spinning properties; as discussed below (see discussion leading to Eq. (3.574)), the spin makes all the energy levels two-fold degenerate.

be simplified via an integration by parts, while the third one vanishes, so that we simply get

$$\delta\mathcal{H}_{11} = -\frac{1}{2mc^2}\int (W_n - V)^2\,(y^1)^2\,\mathrm{d}\tau + \frac{\hbar^2}{8m^2c^2}\int (y^1)^2\,\boldsymbol{\nabla}^2 V\,\mathrm{d}\tau. \tag{3.570}$$

Obviously we understand that when V has a singularity like $-k/r$, in an infinitesimal region $\Delta\tau$ around the singularity P_0 we find

$$\int_{\Delta\tau} (y^1)^2\,\boldsymbol{\nabla}^2 V\,\mathrm{d}\tau = 4\pi\,k\,(y^1)^2(P_0).$$

The expression for $\delta\mathcal{H}_{22}$ can be obtained from Eq. (3.568) by changing i into $-i$ in the third term; but, since the third integral is zero, it then coincides with the expression for $\delta\mathcal{H}_{11}$ in Eq. (3.570).

Now calculate $\delta\mathcal{H}_{12}$. We have

$$\begin{aligned}\delta\mathcal{H}_{12} &= \frac{i}{4m^2c^2}\left[\int \left(p_x^* + ip_y^*\right)\,y^{1*}\,(W_n - V)\,p_z\,y^1\,\mathrm{d}\tau\right.\\ &\quad \left. - \int p_z^*\,y^{1*}\,(W_n - V)\,(p_x + ip_y)\,y^1\,\mathrm{d}\tau\right]\\ &= \frac{i}{4m^2c^2}\left[\int y^{1*}\,(p_x + ip_y)\,(W_n - V)\,p_z\,y^1\,\mathrm{d}\tau\right.\\ &\quad \left. - \int y^{1*}\,p_z\,(W_n - V)\,(p_x + ip_y)\,y^1\,\mathrm{d}\tau\right]\\ &= \frac{i\,\hbar^2}{4m^2c^2}\int y^{1*}\left[\left(\frac{\partial V}{\partial x} + i\frac{\partial V}{\partial y}\right)\frac{\partial y^1}{\partial z}\right.\\ &\quad \left. - \frac{\partial V}{\partial z}\left(\frac{\partial y^1}{\partial x} + i\frac{\partial y^1}{\partial y}\right)\right]\mathrm{d}\tau. \end{aligned} \tag{3.571}$$

$\delta\mathcal{H}_{21}$ can be obtained from $\delta\mathcal{H}_{12}$ by changing i into $-i$ only in the expression under the integral sign, so that we obviously get

$$\delta\mathcal{H}_{21} = \delta\overline{\mathcal{H}}_{12}. \tag{3.572}$$

In the case considered, since y^1 is real, we get $\delta\mathcal{H}_{12} = \delta\mathcal{H}_{21} = 0$. We thus see that the eigenvalues of the perturbation matrix are all the same, and we simply have

$$\delta W_n = \delta\mathcal{H}_{11} = \delta\mathcal{H}_{22}. \tag{3.573}$$

Thus spin doesn't break the degeneracy of the energy level; however, the two degenerate levels are shifted by the magnetic field. Note that without a magnetic field all the energy levels are at least two-fold degenerate, and this holds not only in first approximation but is an exact property. Indeed from a solution of Eqs. (3.557)-(3.560) we can obtain another solution by setting

$$\psi_1' = -\psi_2^*, \quad \psi_2' = \psi_1^*, \quad \psi_3' = \psi_4^*, \quad \psi_4' = -\psi_3^*. \tag{3.574}$$

Since δW_n is equal to $\delta \mathcal{H}_{11}$, in absence of magnetic field and degeneracy, its expression given by Eq. (3.570) is made up of two terms: the first one represents the relativistic correction, while the second one gives the correction coming from the spin.

As an example, let us calculate the corrections (in second approximation) to the energy of the ground state of an atom with atomic number Z but with only one electron; we have

$$W_n = -Z^2 R h = -\frac{me^4 Z^2}{2\hbar^2}, \tag{3.575}$$

$$y^1 = c\,\mathrm{e}^{-Zr/a} = \left(e^3 m Z/\hbar^3\right)\sqrt{\pi m Z}\,\mathrm{e}^{-me^2 Zr/\hbar^2}, \tag{3.576}$$

$$\delta W_n = -\frac{5}{2}\frac{W_n^2}{mc^2} + 2\frac{W_n^2}{mc^2} = -\frac{1}{2}\frac{W_n^2}{mc^2}; \tag{3.577}$$

the first term in the sum for δW_n is the relativistic correction, while the second one is the correction for the spin. Note that relativistic effects are reduced by a factor 5 due to spin. We can deduce Eq. (3.577) from the known fine-structure formula

$$\frac{W}{mc^2} = \left[1 + \frac{\alpha^2 Z^2}{(n - j - 1/2 + \sqrt{(j+1/2)^2 - \alpha^2 Z^2})^2}\right]^{-1/2} - 1, \tag{3.578}$$

where n is the principal quantum number and $\alpha = e^2/\hbar c$ the fine structure constant. Expanding this formula up to second-order terms and denoting with $W_n = -R\,hZ^2/n^2$ the Balmer term, we find

$$W = W_n - \left(\frac{2n}{j+1/2} - \frac{3}{2}\right)\frac{W_n^2}{mc^2}. \tag{3.579}$$

Since in the case under consideration we have $n = 1$, $j = 1/2$, Eq. (3.577) immediately follows. Note that the (false) relativistic correction without a spin correction term could be obtained from Eq. (3.578) or from Eq. (3.579) by replacing j by the azimuthal quantum number k. In the case that we are considering we would have $k = 0$ and then, in first approximation, $\delta W_n = -(5/2)W_n/mc^2$, as already found. Writing down the Balmer term explicitly, Eq. (3.579) can be cast in the form

$$W = -\frac{R\,hZ^2}{n^2} - \frac{Z^2\alpha^2}{n^3}\left(\frac{1}{j+1/2} - \frac{3}{4n}\right)R\,h. \tag{3.580}$$

Consider now the case of a central field and let W_n be degenerate only for the presence of spin, i.e., with a $2k+1$ degeneracy, $k > 0$ being the azimuthal quantum number. In first approximation, the degenerate eigenfunctions will be

$$y^{11},\ y^{21}, \ldots,\ y^{(2k+1)1},\ y^{12},\ y^{22}, \ldots,\ y^{(2k+1)2},$$

or, introducing the equatorial quantum number m,

$$y^{m1},\ y^{m2}, \quad \text{with} \quad m = k, k-1, \ldots, -k+1, -k. \tag{3.581}$$

The perturbation matrix is the sum of two terms, the first of which, containing only the diagonal elements

$$\begin{aligned} \delta\mathcal{H}'_{m1,m1} &= \delta\mathcal{H}'_{m2,m2} = -\frac{1}{2mc^2}\int (W_n - V)^2\,\psi\,\psi^*\,\mathrm{d}\tau \\ &+ \frac{\hbar^2}{4m^2c^2}\int\left(\frac{1}{r}\frac{\mathrm{d}V}{\mathrm{d}r} + \frac{1}{2}\frac{\mathrm{d}^2V}{\mathrm{d}r^2}\right)\psi\,\psi^*\,\mathrm{d}\tau, \end{aligned} \tag{3.582}$$

depends on m and is an absolute constant to be added to the eigenvalues of the second matrix $\delta\mathcal{H}''$. The elements of this second matrix are

$$\delta\mathcal{H}''_{m1,n1} = \frac{\hbar}{4m^2c^2}\,u_{z\ mn}\int\frac{1}{r}\frac{\mathrm{d}V}{dr}\,\psi\,\psi^*\,\mathrm{d}\tau, \tag{3.583}$$

where u_z is the orbital (angular) momentum along the z axis. Analogously,

$$\begin{aligned} \delta\mathcal{H}''_{m1,n2} &= \frac{\hbar}{4m^2c^2}\left(-u_{y\ mn} + iu_{x\ mn}\right)\int\frac{1}{r}\frac{\mathrm{d}V}{\mathrm{d}r}\,\psi\,\psi^*\,\mathrm{d}\tau \qquad (3.584)\\ &= \frac{\hbar}{4m^2c^2}\left(-u^*_{y\ nm} + iu^*_{x\ nm}\right)\int\frac{1}{r}\frac{\mathrm{d}V}{\mathrm{d}r}\,\psi\,\psi^*\,\mathrm{d}\tau = \delta\mathcal{H}''_{n2,m1}, \\ \delta\mathcal{H}''_{m2,n1} &= \frac{\hbar}{4m^2c^2}\left(-u_{y\ mn} - iu_{x\ mn}\right)\int\frac{1}{r}\frac{\mathrm{d}V}{\mathrm{d}r}\,\psi\,\psi^*\,\mathrm{d}\tau \qquad (3.585)\\ &= \frac{\hbar}{4m^2c^2}\left(-u^*_{y\ nm} - iu^*_{x\ nm}\right)\int\frac{1}{r}\frac{\mathrm{d}V}{\mathrm{d}r}\,\psi\,\psi^*\,\mathrm{d}\tau = \delta\mathcal{H}''_{n1,m2}, \\ \delta\mathcal{H}''_{m2,n2} &= -\frac{\hbar}{4m^2c^2}\,u_{z\ mn}\int\frac{1}{r}\frac{\mathrm{d}V}{\mathrm{d}r}\,\psi\,\psi^*\,\mathrm{d}\tau. \qquad (3.586) \end{aligned}$$

We can assume that the $(2k+1)$-dimensional matrices u_z, u_x, u_y are given, up to a factor $\hbar$, by the matrices R_z/i, R_x/i, R_y/i in Eqs. (3.501), where j is replaced by k. However, to avoid imaginary quantities, we set

$$\begin{aligned} u_z &= \hbar\,\frac{R_z}{i} = \hbar\,T_z, \\ u_x &= -\hbar\,\frac{R_y}{i} = -\hbar\,T_y, \\ u_y &= \hbar\,\frac{R_x}{i} = \hbar\,T_x. \end{aligned} \tag{3.587}$$

We also write

$$\delta\mathcal{H}''_{mr,ns} = \frac{\hbar^2}{4m^2c^2}\, Q_{mr,ns} \int \frac{1}{r}\frac{\mathrm{d}V}{dr}\, \psi\, \psi^*\, \mathrm{d}\tau, \tag{3.588}$$

where the $(4k+2)$-dimensional matrix Q has the form

$$Q = \begin{pmatrix} T_z & -T_x - iT_y \\ -T_x + iT_y & -T_z \end{pmatrix}, \tag{3.589}$$

or explicitly, from Eqs. (3.501),

$$Q_{m1,n1} = m\, \delta_{m,n}, \tag{3.590}$$

$$Q_{m2,n2} = -\, m\, \delta_{m,n}, \tag{3.591}$$

$$Q_{m1,n2} = \sqrt{k(k+1) - mn}\, \delta_{m+1,n} = Q_{n2,m1}, \tag{3.592}$$

$$Q_{m2,n1} = \sqrt{k(k+1) - mn}\, \delta_{m-1,n} = Q_{n1,m2}. \tag{3.593}$$

It follows that the matrix Q is built from the $2k+1$ partial matrices composed of the following rows and columns:

$$\begin{array}{l} k,1; \quad k-1,1; \quad \text{and} \quad k,2; \quad k-2,1; \quad \text{and} \quad k-1,2; \quad \ldots; \\ \qquad k-r,1; \quad \text{and} \quad k-r+1,2; \quad \ldots; \\ \qquad -k,1; \quad \text{and} \quad -k+1,2; \quad -k,2. \end{array} \tag{3.594}$$

The first and the last partial matrices have only one element, and their eigenvalue is k. The $2k$ intermediate matrices ($r = 1, 2, \ldots, 2k$) have the form

$$\begin{pmatrix} k-r & \sqrt{k(k+1)-(k-r)(k-r+1)} \\ \sqrt{k(k+1)-(k-r)(k-r+1)} & -k+r-1 \end{pmatrix}, \tag{3.595}$$

and their eigenvalues are k and $-(k+1)$. We then have $2k+2$ eigenfunctions corresponding to the eigenvalue k of Q and $2k$ eigenfunctions corresponding to the eigenvalue $-(k+1)$ of Q. The first group of eigenfunctions is represented by the internal quantum number $j = k + 1/2$ and the second group by the internal quantum number $j = k - 1/2$. The spin thus splits the energy level in two parts, but the degeneracy remains for both levels since, even in the best case for which $k = 1$, the highest energy level has degeneracy 4 while the lowest has degeneracy 2. This is consistent with Eqs. (3.574): All the energy levels are at least two-fold degenerate in the absence of a magnetic field.

If in Eq. (3.595) we set

$$\ell = k + \frac{1}{2} - r, \tag{3.596}$$

then the matrix takes the form

$$\begin{pmatrix} \ell - 1/2 & \sqrt{(k+1/2)^2 - \ell^2} \\ \sqrt{(k+1/2)^2 - \ell^2} & -(\ell + 1/2) \end{pmatrix}. \tag{3.597}$$

The quantity ℓ represents the total (angular) momentum along the z axis, which is common to the two solutions corresponding to Eq. (3.597). The eigenfunctions corresponding to $j = k + 1/2$ are given, in first approximation, by

$$\psi'^{\ell} = \frac{1}{\sqrt{2k+1}} \left(\sqrt{k + \ell + 1/2}\; y^{\ell-1/2,1} - \sqrt{k - \ell + 1/2}\; y^{\ell+1/2,2} \right). \tag{3.598}$$

On varying ℓ between j $(= k + 1/2)$ and $-j$ $(= -k - 1/2)$, we obtain not only the solutions deriving from matrix (3.595) but also the ones with $l = \pm(k + 1/2)$ deriving from the first and the last matrices in Eq. (3.594). These extreme solutions, corresponding to an (angular) momentum along the z axis equal to $\pm(k+1/2)$, are not included in the solutions with $j_2 = k - 1/2$. These $2j_2 + 1 = 2k$ solutions are, instead, given in first approximation by

$$\psi''^{\ell} = \frac{1}{\sqrt{2k+1}} \left(\sqrt{k - \ell + 1/2}\; y^{\ell-1/2,1} + \sqrt{k + \ell + 1/2}\; y^{\ell+1/2,2} \right), \tag{3.599}$$

where ℓ is an integer varying from j_2 to $-j_2$. Notwithstanding the apparent symmetry between Eqs. (3.598) and (3.599), we have $2k+2$ solutions of the first kind and $2k$ of the second. Since y^{s1} and y^{s2} are defined only for $|s| < k$, note that, if we set $\ell = \pm(k+1/2)$ in Eq. (3.599), these relations do not make sense any more; this does not happen for Eq. (3.598) because in these relations the Schrödinger eigenfunctions are multiplied by 0.

From Eqs. (3.582) and (3.588), the corrections to the eigenvalue due to relativistic effects and to the spin are given, in first approximation, by:

For $j = k + 1/2$:

$$\begin{aligned} \delta W_n' &= -\frac{1}{2mc^2} \int (W_n - V)^2\, \psi\, \psi^*\, \mathrm{d}\tau \\ &\quad + \frac{\hbar^2}{4m^2c^2} \int \left(\frac{k+1}{r} \frac{\mathrm{d}V}{\mathrm{d}r} + \frac{1}{2} \frac{\mathrm{d}^2 V}{\mathrm{d}r^2} \right) \psi\, \psi^*\, \mathrm{d}\tau; \end{aligned} \tag{3.600}$$

For $j = k - 1/2$:

$$\delta W_n'' = -\frac{1}{2mc^2} \int (W_n - V)^2\, \psi\, \psi^*\, \mathrm{d}\tau$$

$$+\frac{\hbar^2}{4m^2c^2}\int\left(\frac{-k}{r}\frac{\mathrm{d}V}{\mathrm{d}r}+\frac{1}{2}\frac{\mathrm{d}^2V}{\mathrm{d}r^2}\right)\psi\,\psi^*\,\mathrm{d}\tau. \tag{3.601}$$

The splitting of the energy level is, in first approximation:

$$\delta W_n' \; - \; \delta W_n'' \;=\; \frac{(2k+1)\hbar^2}{4m^2c^2}\int\frac{1}{r}\frac{\mathrm{d}V}{\mathrm{d}r}\,\psi\,\psi^*\,\mathrm{d}\tau, \tag{3.602}$$

or, in terms of the wave number,

$$\delta n \;=\; \frac{(2k+1)\hbar}{4m^2c^3}\int\frac{1}{r}\frac{\mathrm{d}V}{\mathrm{d}r}\,\psi\,\psi^*\,\mathrm{d}\tau. \tag{3.603}$$

20. CHARACTERS OF $\mathcal{D}_j$ AND REDUCTION OF $\mathcal{D}_j \times \mathcal{D}_j'$ [22]

The representations $\mathcal{D}_j$ of $O(3)$, both unique and double, can be viewed as (unique) irreducible representations of the group $SU(2)$ of unitary transformations in two dimensions with determinant equal to 1. In particular, $O(3)$ itself, considered as equivalent to $\mathcal{D}_j$, is an irreducible representation of $SU(2)$. The law of this representation is expressed in Eqs. (3.498).

Every element of $SU(2)$ can be reduced, by a unitary transformation, to diagonal form

$$\left(\begin{array}{cc}\epsilon & 0\\ 0 & \epsilon^{-1}\end{array}\right), \tag{3.604}$$

with $|\epsilon| = 1$. The matrix in Eq. (3.604) is an element of $SU(2)$, and since we can always require the unitary transformation to have determinant equal to 1, so that it belongs to $SU(2)$, the considered element of our group is conjugate to the main element (3.604). Now, all the elements conjugate to the form (3.604) form a class; more precisely, they form the most general class of conjugate elements with ϵ varying under the constraint $|\epsilon| = 1$. Each class is thus characterized by the eigenvalues ϵ and $1/\epsilon$ of an arbitrary element belonging to it; the order of these eigenvalues is not determined. The angle ω in

$$\epsilon \;=\; \mathrm{e}^{i\omega}, \quad \frac{1}{\epsilon} \;=\; \mathrm{e}^{-i\omega}, \tag{3.605}$$

[22]@ In modern terminology, the word "character" usually corresponds to "trace."

defined apart for its sign, defines a class.

Since the character is a function of the class, we can limit our discussion to the characters of the main elements in the form (3.604).

In the representation $\mathcal{D}_j$ (with dimension $2j+1 = v+1$ of $SU(2)$) the matrix associated with the element (3.604) transforms the vector with components

$$\frac{\xi^r\,\eta^{v-r}}{\sqrt{r!(v-r)!}}, \qquad v = 2j, \quad r = 0, 1, \ldots, v, \tag{3.606}$$

into the vectors with components

$$\frac{\xi'^r\,\eta'^{v-r}}{\sqrt{r!(v-r)!}} = \frac{\xi^r\,\eta^{v-r}}{\sqrt{r!(v-r)!}}\,\epsilon^{2r-v}, \quad r = 0, 1, \ldots, v. \tag{3.607}$$

This matrix is thus diagonal, with elements

$$\epsilon^{2r-2j}, \quad r = v, v-1, \ldots, 0 \tag{3.608}$$

that is

$$\epsilon^{2j},\ \epsilon^{2j-2},\ \ldots,\ \epsilon^{-2j}. \tag{3.609}$$

The character is then given by

$$\chi_i = \epsilon^{2j} + \epsilon^{2j-2} + \ldots + \epsilon^{-2j} = \frac{\epsilon^{2j+1} - \epsilon^{-(2j+1)}}{\epsilon - \epsilon^{-1}}. \tag{3.610}$$

Let us consider an abstract group and two representations $\mathcal{G}$ and $\mathcal{G}'$ of it, the first one of dimension n and the second of dimension n'. To each element σ of the group, there corresponds a matrix S in $\mathcal{G}$ acting on the variables x:

$$x'_i = \sum_k S_{ik}\,x_k, \quad i, k = 1, 2, \ldots, n, \tag{3.611}$$

and a matrix S' in $\mathcal{G}'$ acting on the variables y:

$$y'_r = \sum_s S'_{rs}\,x_s, \quad r, s = 1, 2, \ldots, n'. \tag{3.612}$$

The matrices $S\times S'$ with dimension nn' are defined as the matrices which transform the products $x_i y_r$ into the products $x'_i y'_r$. They obviously form a representation, denoted by $S\times S'$, of the same abstract group. From Eqs. (3.611) and (3.612), we have

$$x'_i\,y'_r = \sum_{k,s} S_{ik}\,S'_{rs}\,x_k\,y_s, \tag{3.613}$$

from which we get the explicit definition of the matrices $S\times S'$:

$$(S\times S')_{ir,ks} = S_{ik}\, S'_{rs}. \tag{3.614}$$

On setting $k = i$ and $s = r$, we obtain the diagonal elements of $S\times S'$:

$$(S\times S')_{ir,ir} = S_{ii}\, S'_{rr}, \quad i = 1, 2, ;n; \quad r = 1, 2, \ldots, n', \tag{3.615}$$

from which follows that

$$\chi_{(S\times S')} = \chi_S\, \chi_{S'}. \tag{3.616}$$

Let us consider the representations $\mathcal{D}_j\times\mathcal{D}'_j$ of the group $SU(2)$, the character of which is given by $\chi_j\chi_{j'}$. Decomposing $\mathcal{D}_j\times\mathcal{D}'_j$ into irreducible representations $\mathcal{D}_\tau$, we find

$$\chi_j\, \chi_{j'} = \sum \chi_\tau. \tag{3.617}$$

From Eq. (3.610), on multiplying by $\epsilon - \epsilon^{-1}$, we get

$$\begin{aligned}&\left(\epsilon^{2j} + \epsilon^{2j-2} + \ldots + \epsilon^{-2j}\right)\left(\epsilon^{2j'+1} - \epsilon^{-(2j'+1)}\right)\\ &\quad = \sum \left(\epsilon^{2\tau+1} - \epsilon^{-(2\tau+1)}\right).\end{aligned} \tag{3.618}$$

Since the first term in Eq. (3.618) can be rewritten as [23]

$$\begin{aligned}&\epsilon^{1+2j'+2j} - \epsilon^{-(1+2j'+2j)} + \epsilon^{1+2j'+2j-2} - \epsilon^{-(1+2j'+2j-2)}\\ &\quad + \ldots + \epsilon^{1+2j'-2j} - \epsilon^{-(1+2j'-2j)},\end{aligned} \tag{3.619}$$

it follows that Eq. (3.618) can be identically satisfied only if the values of τ are univocally (i.e., each value can be realized only once) given by

$$\begin{array}{ll} j'+j,\ j'+j-1,\ \ldots,\ j'-j, & \text{if } j' \geq j,\\ j+j',\ j+j'-1,\ \ldots,\ j-j' & \text{if } j \geq j'. \end{array} \tag{3.620}$$

(The second part of rules (3.620) can be derived from obvious symmetry arguments, since, in Eq. (3.618) and in Eq. (3.619), j and j' can be exchanged.)

Note that the main element (3.604) represents a rotation in ordinary space. On setting

$$x = \cos\omega, \quad \lambda = \sin\omega, \quad \mu = \nu = 0 \tag{3.621}$$

[23]@ Equation (3.619) can be obtained from Eq. (3.618) by multiplying the first term in the first bracket by the first term in the second bracket and the last one in the first bracket by the second one in the second bracket, and so on.

in Eqs. (3.498), we see that such a rotation is given by

$$\begin{aligned} x' &= x \cos 2\omega + y \sin 2\omega, \\ y' &= -x \sin 2\omega + y \cos 2\omega, \\ z' &= z, \end{aligned} \tag{3.622}$$

i.e., we have a rotation along the z axis through an angle -2ω.

21. INTENSITY AND SELECTION RULES FOR A CENTRAL FIELD

Consider an energy level with internal quantum number j, so that it has spin degeneracy $2j+1$, and let us assume that it has no further degeneracy. Let us furthermore suppose that a perturbation which is symmetric along the z axis is acting on the system. Introducing the magnetic quantum number m $(= j, j-1, \ldots, -j)$ to label the $2j+1$ independent quantum states, the perturbation matrix $W(m, m')$ is necessarily diagonal, since the Hermitian form

$$\sum W(m, m')\, x_m^*\, x'_{m'} \tag{3.623}$$

has to be invariant under rotations around the z-axis (see the previous section), that is, when we change x_m into

$$y_m = \epsilon^{2m}\, x^m. \tag{3.624}$$

It follows that, in general, this perturbation breaks the degenerate level into $2j+1$ adjacent levels labeled by the magnetic quantum number.
There also exists another level, with internal quantum number j', which is broken by the perturbation into $2j'+1$ levels labeled by the magnetic quantum number m'. Let q be the electric moment of the atom with components q_x, q_y, q_z:

$$q_x = -e\,(x_1 + x_2 + \ldots), \quad \text{etc.} \tag{3.625}$$

The intensity of the spectral line $jm - j'm'$ is proportional to the square of the element (m, m') of the part of the q matrix

$$q(m, m') \tag{3.626}$$

that corresponds to the transformation $R_j - R_{j'}$. Let us consider a rotation s acting on the system; a transformation corresponding to s in the representation $\overline{\mathcal{D}}_j \times \mathcal{D}'_j$ acts on the Hermitian form

$$\sum q(m, m')\, x_m^*\, x'_{m'}. \tag{3.627}$$

On the other hand, under the action of such transformation, the components q_x, q_y, q_z from (3.627) must transform as x, y, z under the action of s; this follows from Eq. (3.625) and expresses the fact that q is a vector. The quantity in (3.627) is said to be a vector quantity in the representative space of $\overline{\mathcal{D}}_j \times \mathcal{D}'_j$ and of $\mathcal{D}_j \times \mathcal{D}'_j$; (since the $\mathcal{D}_j$ representations are defined except for a (unitary) transformation, $\overline{\mathcal{D}}$ and $\mathcal{D}_j$ are equivalent (in a narrow sense)). In turn, the transformation s acting on q_x, q_y, q_z is equivalent to $\mathcal{D}_j$. The problem of determining the maximum number of linear independent combinations of such vector quantities can be answered in terms of a general rule. Let d be a vector quantity, i.e., defined by r components:

$$\begin{aligned} \mathrm{d}_1 &= a_{11}x_1 + a_{12}x_2 + \ldots + a_{1r}x_r, \\ \mathrm{d}_2 &= a_{21}x_1 + a_{22}x_2 + \ldots + a_{2r}x_r, \\ \ldots, & \\ \mathrm{d}_r &= a_{r1}x_1 + a_{r2}x_2 + \ldots + a_{rr}x_r, \end{aligned} \tag{3.628}$$

which are linear combinations of m $(\geq r)$ variables x_j and let us consider two representations of a group g, the first one h being r-dimensional and irreducible and the second one $\mathcal{H}$ being n-dimensional. To each element σ of the group there correspond the matrices

$$s \text{ in } h \quad \text{and} \quad S \text{ in } \mathcal{H}. \tag{3.629}$$

If the transformation S acts on the x_i, and the d_i given by Eq. (3.628) transform exactly as they do under the action of s, then the vector quantity d is said to be h-covariant. Consider the problem of determining the maximum number of such covariant quantities that are linearly independent. To solve this problem, let us adapt the coordinates in the representative space of $\mathcal{H}$ to the decomposition into irreducible representations of y and let the irreducible representation h be present k times. A part $(k\,r)$ of the new variables forms a basis for the irreducible representations of h:

$$x_1^1,\ x_2^1,\ \ldots,\ x_r^1;\ x_1^2,\ x_2^2,\ \ldots,\ x_r^2;\ \ldots;\ x_1^k,\ x_2^k,\ \ldots,\ x_r^k \tag{3.630}$$

and, eventually, other variables remain on which the remaining irreducible representations act. The components y of a covariant quantity can be expressed by the formula

$$y = A^1 x^1 + A^2 x^2 + \ldots + A^k x^k + \ldots + A^{k+l} x^{x+l} + \ldots, \tag{3.631}$$

where A^1, A^2, A^k are square matrices of dimension k^r, while the A^{k+l} are matrices with r rows and p_l columns, p_l being the number of x^{k+l}

variables $(x_1^{k+l}, x_2^{k+l}, \ldots, x_{p_l}^{k+l})$. On these variables an arbitrary inequivalent irreducible representation of h, present in the decomposition of $\mathcal{H}$, acts. From the definition of a h covariant quantity, we must have

$$\begin{aligned} & A^1 s x^1 + A^2 s x^2 + \ldots + A^k s x^k + \ldots + A^{k+l} s^l x^{x+l} + \ldots = sy \\ & = sA^1 x^1 + sA^2 x^2 + \ldots + sA^k x^k + \ldots + A^{k+l} s x^{x+l} + \ldots, \end{aligned} \tag{3.632}$$

from which, since the x_j are arbitrary,

$$\begin{aligned} sA^1 &= A^1 s; \quad sA^2 = A^2 s; \quad \ldots; \quad sA^k = A^k s; \quad \ldots; \\ sA^{k+l} &= A^{k+l} s^l; \quad \ldots . \end{aligned} \tag{3.633}$$

From the fundamental theorem on irreducible representations, [24] assuming that s and s^l are inequivalent irreducible representations of the group g, we deduce that

$$\begin{array}{ll} A^1,\ A^2,\ \ldots,\ A^k & \text{are multiples of the unit matrix,} \\ A^{k+l},\ \ldots & \text{are zero.} \end{array} \tag{3.634}$$

It follows that all the h-covariant quantities are linear combinations, k of which are independent; indeed, since

$$d_i = a_1 x_i^1 + a_2 x_i^2 + a_k x_i^k \tag{3.635}$$

(with constant a), we have

$$d = \alpha_1 d^1 + \alpha_2 d^2 + \ldots + \alpha_k \mathrm{d}^k, \tag{3.636}$$

where d^γ $(\gamma = 1, 2, \ldots, k)$ are the components

$$d_i^\gamma = x_i^\gamma, \quad \gamma = 1, 2, \ldots, k; \ i = 1, 2, \ldots, r, \tag{3.637}$$

so that they are linearly independent. The number of d^γ's is equal to the number of times the irreducible representation h is contained in $\mathcal{H}$, and this evidently solves the posed problem.

Going back to the quantity (3.627), which is $\mathcal{D}_j$-covariant in the space of the representation $\mathcal{D}_j \times \mathcal{D}_{j'}$ of the group $SU(2)$, the problem of determining the maximum number of such covariant quantities that are linearly independent is translated into the problem of determining how

[24]@ In the original manuscript, there appears a reference: *see W. page 124.* Most probably the author refers to p. 124 of *Gruppentheorie und Quantenmechanik* by H. Weyl (Hirzel, Leipzig, 1928). For the English version, see p.153 of H. Weyl, *The Theory of Groups and Quantum Mechanics* (translation from the 2nd revised German edition) (Dover, New York, 1931).

many times $\mathcal{D}_j$ is contained in $\mathcal{D}_j \times \mathcal{D}_{j'}$. According to the rule established previously, we have a non-vanishing value for quantities such as those appearing in Eq. (3.627) only in the following three cases:

$$j' = j - 1, \quad j' = j \neq 0, \quad j' = j + 1, \tag{3.638}$$

which expresses the selection rules for the internal quantum number. Note that, when conditions (3.638) are met, the form (3.627) is determined, except for a constant factor, by arguments based on group theory.[25] The selection rules for the magnetic quantum number is simple as well. The component q_z has to remain unchanged when a rotation along the z axis is performed, so that $q_z(m, m')$ must be diagonal. On the other hand, the quantity $q_x + iq_y$ acquires a factor ϵ^{-2} when a rotation is performed, while $q_x - iq_y$ acquires a factor ϵ^2. In turn, the product $x^*_m x'_{m'}$ acquires a factor $\epsilon^{-2(m'-m)}$ under a rotation, so that the allowed transitions are (for the convention on the sign of m, see its definition in terms of j in Sec. 3.16):

$$\begin{array}{ll} \text{for } q_z, & m \rightarrow m, \\ \text{for } q_x + iq_y, & m \rightarrow m + 1, \\ \text{for } q_x - iq_y, & m \rightarrow m - 1. \end{array} \tag{3.639}$$

Let us now determine [26] the vector form (3.627), except for a constant factor, when it is different from zero (according to form (3.627)). To this end, let us consider the expression

$$\frac{1}{k!} \left(\xi^* \xi' + \eta^* \eta'\right)^k, \tag{3.640}$$

which is invariant if a unitary transformation of the group $SU(2)$ acts on ξ,η and ξ',η'. From Eqs. (3.497), it follows that $x + iy$, $x - iy$ and z transform as $\eta\xi^*$, $\eta^*\xi$, and $\xi\xi^* - \eta\eta^*$, respectively. On the other hand, a transformation of the group $SU(2)$ can be written as

$$\xi_1 = \alpha \xi + \beta \eta, \quad \eta_1 = -\beta^* \xi + \alpha^* \eta, \tag{3.641}$$

from which

$$\xi_1^* = \alpha^* \xi^* + \beta^* \eta^*, \quad \eta_1^* = -\beta \xi^* + \alpha \eta^*, \tag{3.642}$$

---

[25]@ In the original manuscript, there is here a reference: *see W. page 158.* Most probably the author refers again to Weyl's book, p.158 (in German) or p.199 (in the Dover edn.).

[26]@ In the original manuscript, there is here a reference: *see W. p. 154.* Most probably the author refers again to Weyl's book, p.154 (in German) or p.197 et seq. (in the Dover edn.)

or

$$\eta_1^* = \alpha\,\eta^* + \beta\,(-\,\xi^*)\,, \quad -\,\xi_1^* = -\,\beta^*\,\eta^* + \alpha^*\,(-\,\xi^*)\,, \qquad (3.643)$$

so that (ξ, η) transform as $(\eta^*, -\xi^*)$. On changing the sign in the first one of Eqs. (3.641), we get

$$\eta_1 = \alpha^*\,\eta + \beta^*\,(-\,\xi)\,, \quad -\,\xi_1^* = -\,\beta\,\eta + \alpha\,(-\,\xi)\,, \qquad (3.644)$$

so that, inversely, from Eqs. (3.642), (ξ^*, η^*) transform as $(\eta, -\xi)$. It follows that $(x+iy, x-iy, z)$ transform as $(\eta^2, -\xi^2, \xi\eta)$ or $(-\xi^{*2}, \eta^{*2}, \xi^*\eta^*)$:

$$x + i\,y \sim \eta^2, \quad x - i\,y \sim -\,\xi^2, \quad z \sim \xi\eta, \qquad (3.645)$$

$$x + i\,y \sim -\,\xi^{*2}, \quad x - i\,y \sim -\,\eta^{*2}, \quad z \sim \xi^*\eta^*. \qquad (3.646)$$

Allowing (ξ', η') to transform as (ξ, η), we also have

$$x + i\,y \sim 2\eta'\xi^*, \quad x - i\,y \sim 2\xi'\eta^*, \quad z \sim \xi'\xi^* - \eta'\eta^*, \qquad (3.647)$$

$$x + i\,y \sim \eta'^2, \quad x - i\,y \sim -\,\xi'^2, \quad z \sim \xi'\eta'. \qquad (3.648)$$

On multiplying the invariant (3.630) by the right-hand sides of Eqs. (3.646), (3.647), or (3.648), we always obtain the components of a vector quantity, which transform as $x + iy$, $x - iy$, z.

Let us first consider in Eqs. (3.640) the case where $k = 2j - 2 = 2j'$, corresponding to the first case in (3.638). Let us multiply the invariant by Eqs. (3.646); we then obtain the Hermitian forms

$$\begin{array}{c} (q_x + i\,q_y)\,(m, m')\,x_m^*\,x'_{m'}, \\ (q_x - i\,q_y)\,(m, m')\,x_m^*\,x'_{m'}, \\ q_z(m, m'), x_m^*\,x'_{m'}, \end{array} \qquad (3.649)$$

where

$$x_m = \frac{\xi^{j-m}\eta^{j+m}}{\sqrt{(j-m)!(j+m)!}}, \qquad m = j, j-1, \ldots, -j, \qquad (3.650)$$

$$x'_{m'} = \frac{\xi'^{j'-m'}\eta'^{j'+m'}}{\sqrt{(j'-m')!(j'+m')!}}, \qquad m' = j', j'-1, \ldots, -j', \qquad (3.651)$$

so that the monomials $x_m^* x'_{m'}$ transform according to $\overline{\mathcal{D}}_j \times \mathcal{D}_{j'}$. It follows that the expressions in Eq. (3.649) are, except for a constant factor, the components of the vector in Eq. (3.627) for the transition $j \to j' = j-1$. The product of Eq. (3.640) with the r.h.s.s of Eqs. (3.646) yields the polarization matrices for the transition $j \to j' = j - 1$:

$$\begin{aligned} (q_x + i\,q_y)\,(m, m') &= \sqrt{(j-m)(j-m')}\ \delta_{m+1,m'}, \\ (q_x - i\,q_y)\,(m, m') &= \sqrt{(j+m)(j+m')}\ \delta_{m-1,m'}, \qquad (3.652) \\ q_z(m, m') &= \sqrt{(j+m)(j-m)}\ \delta_{m,m'}. \end{aligned}$$

For the second transition in Eq. (3.638), on multiplying Eq. (3.640), with $k = 2j - 1 = 2j' - 1$, by the r.h.s. of (3.647), we obtain, for $j \to j' = j \neq 0$,

$$\begin{aligned}(q_x + i\,q_y)\,(m,m') &= \sqrt{(j-m)(j+m')}\;\delta_{m+1,m'},\\ (q_x - i\,q_y)\,(m,m') &= \sqrt{(j+m)(j-m')}\;\delta_{m-1,m'},\\ q_z(m,m') &= -m\;\delta_{m,m'}.\end{aligned}$$

These relations coincide with the ones in Eq. (3.501) for the elementary rotations $-(R_x+iR_y)/i$, $-(R_x-iR_y)/i$, $-R_z/i$ in the representation $\mathcal{D}_j$. This is not surprising, since such elementary rotations can be viewed as corresponding to a vector-type quantity in the space of representations $\overline{\mathcal{D}}_j \times \mathcal{D}_j$.

For the last transition in Eq. (3.638), we have to multiply the expression (3.640) with $k = 2j = 2j' - 2$ by the right-hand sides of Eqs. (3.648); we find, for $j \to j' = j + 1$:

$$\begin{aligned}(q_x + i\,q_y)\,(m,m') &= \sqrt{(j+m+1)(j+m'+1)}\;\delta_{m+1,m'},\\ (q_x - i\,q_y)\,(m,m') &= -\sqrt{(j-m+1)(j-m'+1)}\;\delta_{m-1,m'},\\ q_z(m,m') &= \sqrt{(j+m+1)(j-m+1)}\;\delta_{m,m'}.\end{aligned}$$

We notice that the selection rules (3.639) for the magnetic quantum number are satisfied. Extending $SO(3)$ to $O(3)$ with the inclusion of the improper rotations, we have two kinds of irreducible representations, $\mathcal{D}_j^+$ and $\mathcal{D}_j^-$. A polar vector, such as the electric moment, is $\mathcal{D}_j^-$-covariant and, in its matrix, the components corresponding to the intersection of two irreducible spaces, $\mathcal{R}_j^+$ and $\mathcal{R}_{j'}^+$ or $\mathcal{R}_j^-$ and $\mathcal{R}_{j'}^-$, are missing. We thus have the selection rule for the signature: Only the transitions $j \to -j$ and $+j$ are allowed. The scalar wave theory applied to the electron gives only the unique irreducible representations (with integer j) for the group $O(3)$ and, from symmetry properties of the spherical functions, the signature is $+1$ for even j and -1 for odd j; then the selection rule for the signature excludes the transition $j \to j' = j$. However, in the presence of spin this restriction is relaxed; to be precise, this restriction remains (although it is an approximation) but it applies to the orbital (angular) momentum k rather than to the internal quantum number. Thus the (approximate) selection rules to be added to Eq. (3.638) are the following:

$$k \to k'\, k+1, \quad k \to k'\, k-1. \tag{3.653}$$

22. THE ANOMALOUS ZEEMAN EFFECT (ACCORDING TO THE DIRAC THEORY)[27]

Let us turn to the Dirac equations (3.557)-(3.560) and their approximate solutions for a given eigenvalue of the Schrödinger equation (3.562), (3.563). We again consider the case of a central field and furthermore assume a constant magnetic field along the z axis. With reference to Eq. (3.566), we calculated only the field-independent part of the perturbation matrix, which is the sum of the constant diagonal term $\delta\mathcal{H}'$ in Eq. (3.582) and of the matrix

$$\delta\mathcal{H}'' = \frac{\hbar^2}{4m^2c^2}\, Q \int r^{-1}\frac{\mathrm{d}V}{\mathrm{d}r}\, \psi\psi^*\, \mathrm{d}\tau,$$

where Q is given by Eqs. (3.590)-(3.593). Let us evaluate the matrix $B_{ri,sk}$ in Eq. (3.566); we find

$$B_{r1,s1} = \frac{e\hbar}{2mc}\,(r+1)\,\delta_{rs}, \tag{3.654}$$

$$B_{r2,s2} = \frac{e\hbar}{2mc}\,(r-1)\,\delta_{rs}, \tag{3.655}$$

$$B_{r1,s2} = B_{r2,s1} = 0. \tag{3.656}$$

The problem is then to diagonalize $\delta\mathcal{H}'' + HB$ or, setting

$$\epsilon = \frac{2emcH}{\hbar \int r^{-1}\,(\mathrm{d}V/\mathrm{d}r)\,\psi\psi^*\,\mathrm{d}\tau}, \tag{3.657}$$

$$B = \frac{e\hbar}{2mc}\,T, \tag{3.658}$$

so that, from Eqs. (3.654)-(3.656),

$$\begin{aligned} T_{r1,s1} &= (r+1)\,\delta_{rs}, \\ T_{r2,s2} &= (r-1)\,\delta_{rs}, \\ T_{r1,s2} &= T_{r2,s1} = 0, \end{aligned} \tag{3.659}$$

to diagonalize the matrix

$$\frac{\hbar^2}{4m^2c^2}\int \frac{1}{r}\,\frac{\mathrm{d}V}{\mathrm{d}r}\,\psi\psi^*\,\mathrm{d}\tau\;(Q+\epsilon T) \tag{3.660}$$

---

[27] See Sec. 3.19.

or, simply,

$$Q + \epsilon T = S. \tag{3.661}$$

From Eq. (3.590)-(3.593) and (3.659) we get

$$S_{m1,m'1} = (m + \epsilon m + \epsilon)\, \delta_{mm'}, \tag{3.662}$$
$$S_{m2,m'2} = (-m + \epsilon m - \epsilon)\, \delta_{mm'}, \tag{3.663}$$
$$S_{m1,m'2} = \sqrt{k(k+1) - mm'}\, \delta_{m+1,m'}, \tag{3.664}$$
$$S_{m2,m'1} = \sqrt{k(k+1) - mm'}\, \delta_{m-1,m'}. \tag{3.665}$$

The matrix S can be decomposed into $2k+2$ submatrices, the first and the last of which consist of only one element, while the other matrices are 2×2 matrices, precisely as the matrix Q in Eq. (3.594). In fact, the term ϵT does not alter the decomposition property of Q, since it is diagonal with a suitable choice of the coordinates. The first sub-matrix, possessing only the $k1, k1$ element, has the eigenvalue

$$k1, k1: \quad k + \epsilon\,(k + 1)\,, \quad \ell = k + \frac{1}{2}. \tag{3.666}$$

The last sub-matrix, with the single element $-k2, -k2$, has the eigenvalue

$$-k2, -k2: \quad k - \epsilon\,(k + 1)\,, \quad \ell = -k - \frac{1}{2}. \tag{3.667}$$

The other $2k$ square sub-matrices can be labeled, as done for Q (see Eqs. (3.596) and (3.597)), by the total (angular) momentum ℓ along the z-axis ($\ell = k - 1/2, k - 3/2, \dots, -k + 1/2$). They have the form

$$\begin{pmatrix} \ell - 1/2 + \epsilon(\ell + 1/2) & \sqrt{(k+1/2)^2 - \ell^2} \\ \sqrt{(k+1/2)^2 - \ell^2} & -(\ell + 1/2) + \epsilon(\ell - 1/2) \end{pmatrix}. \tag{3.668}$$

The eigenvalues of Eq. (3.668) are

$$-\frac{1}{2} + \ell\epsilon \pm \sqrt{\left(k + \frac{1}{2}\right)^2 + \epsilon\ell + \frac{1}{4}\epsilon^2}. \tag{3.669}$$

Taking the factor of $Q + \epsilon T = S$ in Eq. (3.660) as the energy unit, and the corresponding frequency as defined by the Einstein law [28] as the frequency unit, the separation of the doublet (which is present when

[28] This is more widely known as the Planck law $E = h\nu$.

the field is turned off) is $2k+1$ and the Larmor frequency is ϵ. Then, for $\epsilon << 2k+1$, the formulae for the weak-field limit apply, while for $\epsilon >> 2k+1$ we have the strong-field limit (Paschen-Back effect). Let us assume that ϵ is small and expand the eigenvalues to first-order in ϵ. The eigenvalues given by Eqs. (3.666) and (3.667) are already (and exactly) of the first-order in ϵ; for the eigenvalues (3.669) we instead have, depending on the sign in front of the square root,

$$k + \epsilon\ell\,\frac{2k+2}{2k+1} = k + \epsilon\, g'\,\ell, \quad g' = \frac{2k+2}{2k+1}, \tag{3.670}$$

$$-k - 1 + \epsilon\ell\,\frac{2k}{2k+1} = k + \epsilon\, g''\,\ell, \quad g'' = \frac{2k}{2k+1}. \tag{3.671}$$

The eigenvalue of the first kind, as well as the eigenvalues (3.666) and (3.667), correspond to the internal quantum number $j = k + 1/2$ for $H \to 0$. In fact, the eigenvalues (3.666) and (3.667) can be cast in the form (3.670) with $\ell = k + 1/2$ and $\ell = -k - 1/2$ (with the same value of g'), respectively. We can thus write the following general formulae for the weak-field limit:

$$j = k + \frac{1}{2}$$

$$\text{eigenvalues}: \quad k + \epsilon\, g'\,\ell \tag{3.672}$$
$$\left(\ell = k + \frac{1}{2},\; k - \frac{1}{2},\; \ldots,\; -k - \frac{1}{2}\right),$$

$$j = k - \frac{1}{2}$$

$$\text{eigenvalues}: \quad -k - 1 + \epsilon\, g''\,\ell \tag{3.673}$$
$$\left(\ell = k - \frac{1}{2},\; k - \frac{3}{2},\; \ldots,\; -k + \frac{1}{2}\right).$$

The separation constants g' and g'' given in Eqs. (3.670) and (3.671) can then be expressed by the single formula

$$g = \frac{2j+1}{2k+1}; \tag{3.674}$$

and we always have

$$g' > 1, \quad g'' < 1, \quad g' + g'' = 2. \tag{3.675}$$

For example, for $k = 1$ we get $g' = 4/3$, $g'' = 2/3$. We also have a general expression for g' and g'' holding for an arbitrary number of electrons:

$$g = 1 + \frac{j(j+1) + s(s+1) - k(k+1)}{2j(j+1)}. \tag{3.676}$$

In fact, from (3.676),

$$\begin{aligned} g' &= 1 + \frac{(k+1/2)(k+3/2) + 3/4 - k(k+1)}{(2k+1)(k+3/2)} \\ &= \frac{2k+2}{2k+1} \end{aligned} \tag{3.677}$$

$$\begin{aligned} g'' &= 1 + \frac{(k-1/2)(k+1/2) + 3/4 - k(k+1)}{2(k-1/2)(k+1/2)} \\ &= \frac{2k}{2k+1}, \end{aligned} \tag{3.678}$$

which coincide with the expressions in Eqs. (3.670) and (3.671).

Let us now consider the other limiting case $\epsilon \to \infty$. The eigenvalues of S are infinities of the first-order; we will expand them up to terms which do not depend on ϵ. Even now the two eigenvalues (3.666) and (3.667) are exactly expressed by their expansion up to the second term. For the eigenvalues in Eq. (3.669), we have, depending on the sign in front of the square root:

$$\left(\ell + \frac{1}{2}\right)\epsilon + \ell - \frac{1}{2}, \tag{3.679}$$

$$\left(\ell - \frac{1}{2}\right)\epsilon - \left(\ell + \frac{1}{2}\right). \tag{3.680}$$

The eigenvalue (3.666) belongs to the class of Eq. (3.679), while the one in Eq. (3.667) belongs to that of Eq. (3.680), so that, summing up, we have two sets of eigenvalues, each with $2k+1$ elements:

$$\left(\ell + \frac{1}{2}\right)\epsilon + \ell - \frac{1}{2} \tag{3.681}$$

(for $\ell = k + 1/2,\ k - 1/2,\ \ldots,\ -k + 1/2$),

$$\left(\ell - \frac{1}{2}\right)\epsilon - \left(\ell + \frac{1}{2}\right) \tag{3.682}$$

(for $\ell = k - 1/2,\ k - 3/2,\ \ldots,\ -k - 1/2$). In the limiting case we have approximately no transition between different eigenvalues, since the first kind of eigenvalue corresponds to the electron "spin" oriented along the field, while the second kind correspond to the "spin" oriented along the opposite direction of the field. From this, the Zeeman effect (Paschen-Back effect) follows. Since for large ϵ the second term in Eq. (3.661) is the largest one and T is diagonal as well as (although approximately) the orbital (angular) momentum along the z-axis, it follows that in first approximation both ℓ and the orbital (angular) momentum m are constant. Labelling the eigenvalues according to m, in place of Eqs. (3.681) and (3.682) we now have:

$$(m + 1)\,\epsilon + m, \qquad m = k,\ k-1,\ \ldots,\ -k; \tag{3.683}$$

$$(m - 1)\,\epsilon - m, \qquad m = k,\ k-1,\ \ldots,\ -k. \tag{3.684}$$

The sum of the eigenvalues is equal to the sum of the diagonal terms of S, so that this quantity is always vanishing.

In Fig. 3.2 we show the transition from the anomalous Zeeman effect to the Paschen-Back effect for $k = 1$. [29] In the limiting case, for strong fields, the distance between the energy levels corresponding to the first kind of eigenvalues is $\epsilon + 1$, ϵ being the Larmor frequency in units such that $2k + 1$ is the separation of the doublet, while the distance between the energy levels of the second kind is $\epsilon - 1$.

23. COMPLETE SETS OF FIRST-ORDER DIFFERENTIAL EQUATIONS[30]

Let $A_1, A_2, \ldots, A_j, \ldots, A_r$ be linear homogeneous differential operators of $2n$ variables:

$$A_j = \sum_{k=1}^{2n} a_j^k \frac{\partial}{\partial x_k}, \tag{3.685}$$

where a_j^k are functions of $x_1, x_2, \ldots, x_{2n}$. The problem is to find the common solutions of the following set of equations:

$$A_j\, y = 0, \quad j = 1, 2, \ldots, r. \tag{3.686}$$

---

[29]@ Figure 3.2 reproduces qualitatively the scheme reported in the original manuscript. We remark the fact that the analysis made in the text apply only to the weak field limit ($\epsilon < 1$) and to the strong field limit ($\epsilon \gg 1$). The intermediate region has to be studied by solving the Dirac equation without approximations (which can only be done numerically). It is interesting that the author also reports the spectra for the intermediate region in the figure.

[30]@ In the original manuscript, there is here a reference: *see Franck, Physikal. 15, April 1929.* Most probably the author refers to the following paper (in German): Philipp Franck, *Phys. Z.* **30** (8), 209 (15 April 1929).

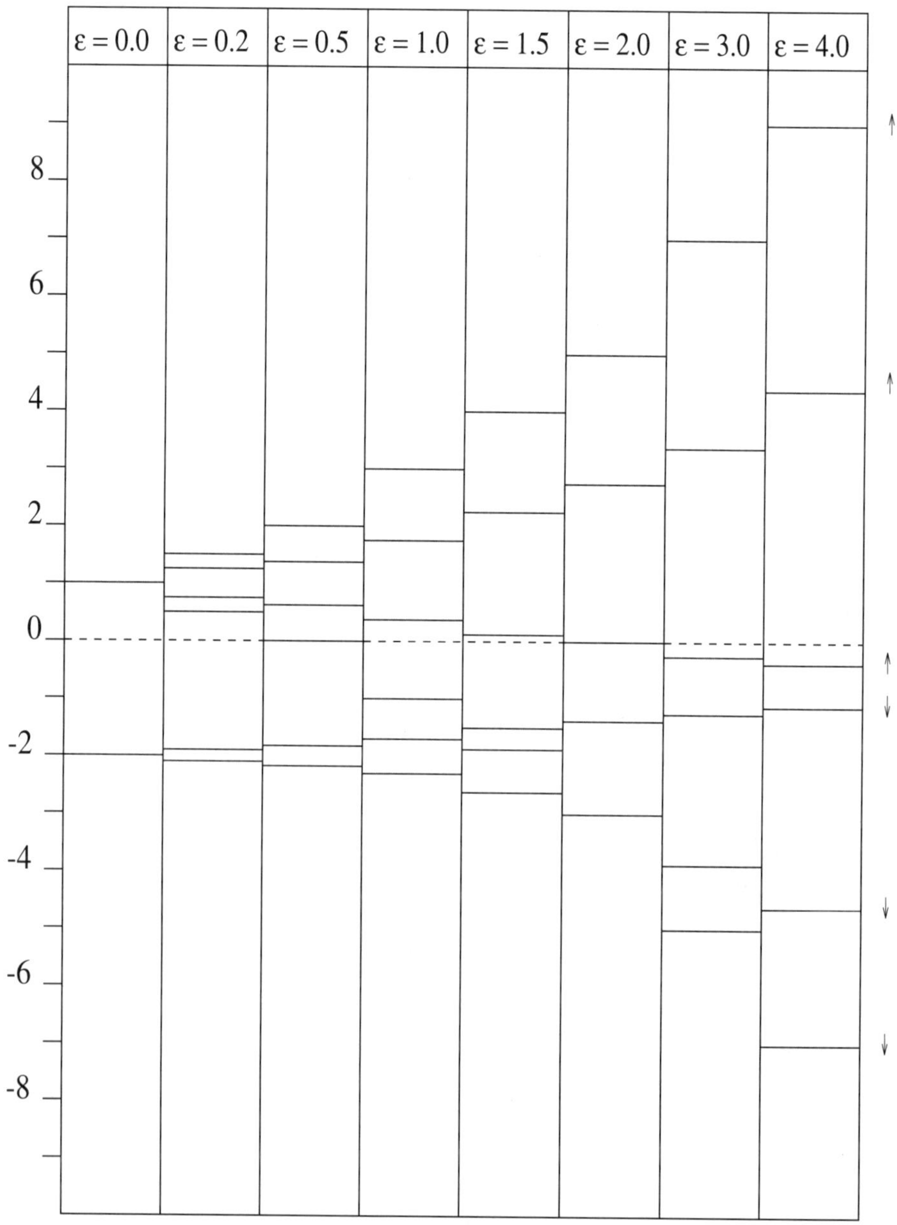

Fig. 3.2. Transition from the anomalous Zeeman effect to the Paschen-Back effect for $k = 1$; note that for $\epsilon = 3$ we have $\epsilon = 2k + 1$.

Let us assume that A_j are linearly independent (but, in general, the coefficients of the linear combinations are x-dependent), so that we necessarily have $r \leq 2n$. From Eq. (3.686), we can deduce other linear homogeneous differential equations that y must obey:

Let us first apply the operator A_j and then the operator $A_{j'}$ to y and vice-versa; from Eq. (3.686) we obtain:

$$(A_j\,A_{j'} - A_{j'}\,A_j)\,y = 0. \tag{3.687}$$

On setting

$$B_{jj'} = A_j\,A_{j'} - A_{j'}\,A_j; \quad B_{jj'}\,y = 0, \tag{3.688}$$

from Eq. (3.685) we get

$$\begin{aligned} B_{jj'} &= \left(\sum_{k=1}^{2n} a_j^k\,\frac{\partial}{\partial x_k}\right)\left(\sum_{k'=1}^{2n} a_{j'}^{k'}\,\frac{\partial}{\partial x_{k'}}\right) \\ &\quad - \left(\sum_{k'=1}^{2n} a_{j'}^{k'}\,\frac{\partial}{\partial x_{k'}}\right)\left(\sum_{k=1}^{2n} a_j^k\,\frac{\partial}{\partial x_k}\right) \\ &= \sum_{k,k'} a_j^k\,a_{j'}^{k'}\,\frac{\partial^2}{\partial x_k \partial x_{k'}} + \sum_{k,k'} a_j^k\,\frac{\partial a_{j'}^{k'}}{\partial x_k}\,\frac{\partial}{\partial x_{k'}} \\ &\quad - \sum_{k,k'} a_{j'}^{k'}\,a_j^k\,\frac{\partial^2}{\partial x_{k'} \partial x_k} - \sum_{k,k'} a_{j'}^{k'}\,\frac{\partial a_j^k}{\partial x_{k'}}\,\frac{\partial}{\partial x_k} \\ &= \sum_k \sum_{k'} \left(a_j^{k'}\,\frac{\partial a_{j'}^k}{\partial x_{k'}} - a_{j'}^{k'}\,\frac{\partial a_j^k}{\partial x_{k'}}\right)\frac{\partial}{\partial x_k} \\ &= \sum_k \left(A_j\,a_{j'}^k - A_{j'}\,a_j^k\right)\frac{\partial}{\partial x_k} \equiv \sum_k b_{jj'}^k\,\frac{\partial}{\partial x_k}, \end{aligned} \tag{3.689}$$

where

$$b_{jj'}^k = A_j\,a_{j'}^k - A_{j'}\,a_j^k. \tag{3.690}$$

It follows that Eq. (3.687) can be cast in the form

$$B_{jj'}\,y = \sum_k b_{jj'}^k\,\frac{\partial y}{\partial x_k}\,y = 0, \tag{3.691}$$

which is again an equation of the kind (3.686), that y must obey. By using the same procedure for each pair of operators A_j and $A_{j'}$ and then for each pair of operators A, B and B, B', we obtain new linear differential equations that y must obey. However, taking into account only the linearly independent equations, the number of which must be $\leq 2n$, we see that there comes a point where the procedure does not give new equations anymore. The set of linear differential equations thus obtained is then said to be complete.

Let us assume, for simplicity, that the set (3.686) is already complete; the conditions for this to happen are

$$A_j\, A_{j'} \;-\; A_{j'}\, A_j \;=\; \sum_r c^r_{jj'}\, A_r, \tag{3.692}$$

where c are x-dependent functions. There is a theorem [31] according to which the complete set (3.686) has exactly $2n - r$ independent solutions. All the possible solutions are thus arbitrary functions of the above $2n - r$ *Poisson brackets.* Let us divide the independent variables $x_1, x_2, \ldots$ into two groups denoted by

$$q_1,\; q_2,\; \ldots,\; q_n,\quad p_1,\; p_2,\; \ldots,\; p_n. \tag{3.693}$$

Let F and G be two arbitrary functions of q and p; we then define the Poisson bracket of F and G to be the expression

$$[F,\; G] \;=\; \sum_{i=1}^{n} \left(\frac{\partial F}{\partial q_i}\frac{\partial G}{\partial p_i} \;-\; \frac{\partial F}{\partial p_i}\frac{\partial G}{\partial q_i} \right). \tag{3.694}$$

The following properties hold:

$$[F,\; G] \;=\; -\; [G,\; F], \quad [F,\; F] \;=\; 0, \tag{3.695}$$

$$[q_i,\; q_k] \;=\; 0, \quad [p_i,\; p_k] \;=\; 0, \quad [q_i,\; p_k] \;=\; \delta_{ik}. \tag{3.696}$$

From Eqs. (3.696) it follows that Eq. (3.694) can be also written as

$$[F, G] \;=\; \sum_{s>r} \frac{\partial(F, G)}{\partial(x_r, x_s)}\, [x_r,\; x_s]\,, \tag{3.697}$$

where

$$\begin{array}{l} x_1 \;=\; q_1, \quad x_2 \;=\; q_2, \quad \ldots, \quad x_n \;=\; q_n, \\ x_{n+1} \;=\; p_1, \quad \ldots, \quad x_{2n} \;=\; p_n. \end{array} \tag{3.698}$$

In Eq. (3.697) the sum can be extended over each pair of indices for which $s < r$, since the result is the same; the important thing is, evidently, that each pair of variables x_r and x_s appears only once in the sum, albeit with an arbitrary order. Let us assume that G is given; then $[F, G]$ can be regarded as the result of an operation performed on F.

---

[31]@ In the original manuscript, there is here a reference: *see for example Goursat, Vorlesungen über die Integration....* However, it is not clear to what paper the author refers.

This operation has some properties similar to those of the derivation; in fact, if F is a function of f, then the following rule holds:

$$[F, G] = \frac{\mathrm{d}F}{\mathrm{d}f}\,[f, G]\,; \tag{3.699}$$

and, more generally, if F is a function of a functions $f_1, f_2, \ldots, f_a$, the rule

$$[F, G] = \sum_{i=1}^{a} \frac{\mathrm{d}F}{\mathrm{d}f_i}\,[f_i, G] \tag{3.700}$$

obtains, which is the analog of the derivation rule for the composite functions.

Equation (3.697) has several generalizations, one of which leads to Eq. (3.700) as a particular case. Let us assume that F and G are functions of b functions depending on the coordinates $g_1, g_2, \ldots, g_b$. From Eq. (3.694) it follows that

$$[F, G] = \sum_{s>r} \frac{\partial(F, G)}{\partial(g_r, g_s)}\,[g_r,\ g_s]\,, \tag{3.701}$$

from which Eqs. (3.697) and (3.700) can be deduced as particular cases. For example, to obtain Eq. (3.700), it is sufficient to set $b = a + 1$, $f_1 = g_1, f_2 = g_2, \ldots, f_a = g_a$, $F = F(g_1, g_2, \ldots, g_a)$, $G = f_b$.

Given three functions F, G, H depending on the coordinates, the following Jacobi identity holds:

$$[F, [G, H]] + [G, [H, F]] + [H, [F, G]] = 0. \tag{3.702}$$

We also note the following rule that stems from Eq. (3.700):

$$[a\,b, F] = a\,[b, F] + b\,[a, F]\,. \tag{3.703}$$

Two functions f and g are said to be "involute" if $[f, g] = 0$; more in general, several functions are called involute when each pair of functions is involute. From this definition, it then follows that the q coordinates are involuted between them, and the same applies to the p coordinates. By contrast, f and g are "conjugate" if $[f, g] = 1$; thus, in particular, q_i and p_i are conjugate.

Let $F_1, F_2, \ldots, F_r$ be r functions that depend on the coordinates; the problem is to find the functions that are involuted with all the given F. Let g be a solution (if it exists) of our problem; the following equations then hold:

$$[g, F_1] = [g, F_2] = \ldots = [g, F_r] = 0. \tag{3.704}$$

In the particular case $r = 1$ and $F_1 = H$, our problem reduces to the general problem of classical mechanics, consisting of the search of the integrals of motion for a mechanical system described by the Hamiltonian H. In fact, the relation $[g, H] = 0$ expresses exactly the fact that g is a constant of motion when the Hamilton equations are satisfied. In this particular case, we have $2n$ functions (q and p) of time that satisfy $2n$ ordinary first-order differential equations, and thus there are $2n$ arbitrary constants to fix the initial values of the q and the p coordinates; then there will be $2n$ independent functions satisfying the relation $[g, H] = 0$.

Let us return to the general case (3.704). The jth equation in Eq. (3.704) is

$$\sum_i \frac{\partial F_j}{\partial p_i}\frac{\partial g}{\partial q_i} - \sum_i \frac{\partial F_j}{\partial q_i}\frac{\partial g}{\partial p_i} = A_j\, g = 0, \tag{3.705}$$

where A_j is a first-order homogeneous linear differential operator. By varying j from 1 to r, we obtain a set of r equations. The question is: What conditions must F satisfy for the set to be complete? It is sufficient to substitute the expression for A_j in Eq. (3.692). From Eqs. (3.688) and (3.689) it follows that

$$\begin{aligned} \frac{\partial}{\partial p_s}\,[F_j, F_{j'}] &= \sum_r c^r_{jj'}\,\frac{\partial F_r}{\partial p_s}, \\ \frac{\partial}{\partial q_s}\,[F_j, F_{j'}] &= -\sum_r c^r_{jj'}\,\frac{\partial F_r}{\partial q_s}, \end{aligned} \tag{3.706}$$

or, written as a single vector equation:

$$\nabla\,[F_j, F_{j'}] = -\sum_r c^r_{jj'}\,\nabla F_r. \tag{3.707}$$

Since $c^r_{jj'}$ are arbitrary functions of the coordinates, from Eq. (3.707) it follows that, considering a subspace with $2n - r$ dimensions in which all the functions F are constants, all the Poisson brackets $[F_i, F_j]$ are constant as well, so that they are functions of F themselves:

$$[F_j, F_{j'}] = f_{jj'}\,(F_1, F_2, \ldots F_r)\,. \tag{3.708}$$

The set of differential equations (3.704) is then complete if Eq. (3.708) are satisfied. It then has exactly $2n - r$ independent solutions. Then, the r functions $F_1, F_2, \ldots, F_r$, satisfying Eq. (3.708), form a basis in the group of all the functions of F. From Eqs. (3.701) and (3.708) it follows that the Poisson brackets of two functions belonging to this group belongs itself to the group.

Note that also the $2n - r$ solutions of Eqs. (3.704) with (3.708) form a group, that is they satisfy equations similar to Eqs. (3.708). As a particular case, we then have the known theorem according to which, for a given mechanical problem, the Poisson bracket of two integrals of motion is itself an integral of motion.

4

VOLUMETTO IV: 24 APRIL 1930

1. CONNECTION BETWEEN THE SUSCEPTIBILITY AND THE ELECTRIC MOMENT OF AN ATOM IN ITS GROUND STATE

Let us consider an atom with n electrons in the ground state, which we will assume is an s level described by the eigenfunction ψ_0 corresponding to the eigenvalue E_0. The component of the electric moment along the z axis is given by

$$M \;=\; -\,e\,(z_1 \,+\, z_2 \,+\, \ldots \,+\, z_n) \;=\; -\,e\,z, \tag{4.1}$$

with $z = z_1+z_2+\ldots+z_n$. An electric field of intensity E acting along the z axis induces a perturbation of the atom that depends on the potential $EM = H$. Assuming that the ground state is not degenerate or, more precisely, that no p levels correspond to the eigenvalue E_0, the element M_{00} of the perturbation matrix certainly vanishes. Consequently, for weak fields the variation of the eigenvalue is given by the second-order formula

$$\delta E_0 \;=\; \sum_1^\infty \frac{|M_{0k}|^2}{E_0 - E_k} \;=\; \sum_1^\infty e^2\,E^2\,\frac{|z_k|^2}{E_0 - E_k}, \tag{4.2}$$

where

$$z\,\psi_0 \;=\; \sum_k z_k\,\psi_k. \tag{4.3}$$

However, if α is the atomic electric susceptibility, the variation of the energy is given, to first approximation, by

$$\delta E_0 \;=\; -\,\frac{1}{2}\,E^2\,\alpha, \tag{4.4}$$

from which, on comparison with Eq. (4.2),

$$\alpha \;=\; 2\,e^2\sum_k \frac{|z_k|^2}{E_k - E_0}. \tag{4.5}$$

Moreover, from (4.3), we deduce

$$\sum_k |z_k|^2 = \int z^2\, \psi_0^2\, \mathrm{d}\tau = \overline{z^2}. \tag{4.6}$$

The number of dispersion electrons $f = n$ (from a known theorem) is given by [1]

$$n = \sum_1^\infty (2m/\hbar^2)(E_k - E_0)\, |z_k|^2. \tag{4.7}$$

Let us consider the expressions

$$A = \sum_k (E_k - E_0)\, |z_k|^2 = \frac{n\hbar^2}{2m},$$

$$B = \sum_k |z_k|^2 = \overline{z^2}, \tag{4.8}$$

$$C = \sum_k \frac{|z_k|^2}{E_k - E_0} = \frac{\alpha}{2e^2}. \tag{4.9}$$

We necessarily have

$$B \leq \sqrt{AC} \tag{4.10}$$

(the equality sign holding only in the not realistic case in which z_k is different from zero only for one given value of $E_k - E_0$), that is,

$$\overline{z^2} < \sqrt{\frac{n\alpha\hbar^2}{4me^2}} = \frac{\sqrt{n\alpha a_0}}{2}, \tag{4.11}$$

where $a_0 = \hbar^2/me^2$ is the Bohr radius of the ground state orbit of the hydrogen atom. As long as we assume that the differences $E_k - E_0$ in the most relevant terms of the expressions A, B, C take nearly the same value (as happens for H- and He-like atoms but not for Li-like atoms), the relation (4.11) can be cast in the form

$$\overline{z^2} \lesssim \frac{\sqrt{n\alpha a_0}}{2}. \tag{4.12}$$

For hydrogen, we have $n = 1$; from the Stark effect we deduce that α is approximately given by $4.5a_0^3$, so that

$$\overline{z^2} < 1.06\, a_0^2 \tag{4.13}$$

[1] @ In the original manuscript, the old notation $h/2\pi$ instead of $\hbar$ is used..

(however, a direct calculation gives $\overline{z^2} = a_0^2$).

For helium, we have $n = 2$ and $\alpha \simeq 1.44 a_0^3$ (this can be deduced from the value of the dielectric constant), so that

$$\overline{z^2} < 0.85\, a_0^2. \tag{4.14}$$

Since the ground state of the helium atom is known with a good accuracy, we could evaluate $\overline{z^2}$ directly. Here we shall only make a rough estimate of this quantity from the known eigenfunction

$$c \exp\left\{-\frac{27}{16}\frac{z_1+z_2}{a_0}\right\}.$$

This corresponds to the independent motion of the two electrons, so that $\overline{z^2} = \overline{z_1^2} + \overline{z_2^2}$, while we would have $\overline{z^2} < \overline{z_1^2} + \overline{z_2^2}$; however, this eigenfunction also gives values for $\overline{z_1^2}$ and $\overline{z_2^2}$ that certainly are lower that the true values; so, since the two errors of approximation go in opposite directions, we can presume a good accuracy in our estimation of $\overline{z^2}$. We find

$$\overline{z^2} = 2\left(\frac{16}{27}\right)^2 a_0^2 = 0.70\, a_0^2, \tag{4.15}$$

confirming (4.12). Obviously, this approximation is not as accurate as it is for the hydrogen atom.

Finally, let us consider an *He*-like atom with infinite z. We have

$$\alpha = 2 \cdot 4.5\, a_0^3/Z^4 = 9a_0^3/Z^4$$

(the value $\alpha = 1.44 a_0^3$ for helium follows from this formula if we put $Z^4 = 1.58$) and, from (4.12),

$$\overline{z^2} < \frac{3}{\sqrt{2}}\frac{a_0^2}{Z^2} = 1.06\,\frac{2a_0^2}{Z_2}, \tag{4.16}$$

while the direct calculation gives $\overline{z^2} = 2a_0^2/Z^2$.
The error of (4.12) for helium thus turns out to be equal to that for hydrogen.[2]

[2]@ In the original manuscript, this paragraph ends with the following remark: "A more accurate estimate of $\overline{z^2}$ would probably reduce the error for helium which, however, must be appreciably greater than the one for the limiting case $z = \infty$. This happens because in helium the two electrons must jump jointly from one level to another due to their independent motions; this makes the range of values for $E_k - E_0$ wider." This note is followed by a question mark, reflecting the unclear meaning of his assertion.

2. IONIZATION PROBABILITY FOR A HYDROGEN ATOM IN AN ELECTRIC FIELD

Let us consider a hydrogen-like atom of charge Z. Using electronic units (i.e. $e = 1$, $\hbar = 1$, first Bohr radius $a_0 = 1$; and energy unit $e^2/a_0 = 2Ry$), the electron eigenfunction satisfies the differential equation

$$\nabla^2 \psi + 2\left(E + \frac{Z}{r}\right)\psi = 0. \tag{4.17}$$

On introducing an electric field F in the direction of the x axis, Eq. (4.17) becomes

$$\nabla^2 \psi + 2\left(E + \frac{Z}{r} - F\,x\right)\psi = 0. \tag{4.18}$$

Let us adopt the parabolic coordinates

$$\begin{aligned} &\xi = r + x, \quad \eta = r - x, \quad \tan\phi = \frac{x}{y}; \\ &x = \frac{1}{2}(\xi - \eta), \quad y = \sqrt{\xi\eta}\cos\phi, \quad z = \sqrt{\xi\eta}\sin\phi. \end{aligned} \tag{4.19}$$

We find

$$\begin{aligned} \nabla^2 \psi = {} & \frac{\partial\psi}{\partial\xi}\nabla^2\xi + \frac{\partial\psi}{\partial\eta}\nabla^2\eta + \frac{\partial\psi}{\partial\phi}\nabla^2\phi \\ & + \frac{\partial^2\psi}{\partial\xi^2}|\nabla\xi|^2 + \frac{\partial^2\psi}{\partial\eta^2}|\nabla\eta|^2 + \frac{\partial^2\psi}{\partial\phi^2}|\nabla\phi|^2 \\ & + 2\frac{\partial^2\psi}{\partial\xi\partial\eta}\nabla\xi\cdot\nabla\eta + 2\frac{\partial^2\psi}{\partial\xi\partial\phi}\nabla\xi\cdot\nabla\phi + 2\frac{\partial^2\psi}{\partial\eta\partial\phi}\nabla\eta\cdot\nabla\phi; \end{aligned} \tag{4.20}$$

and, since

$$\nabla^2\xi = \frac{3}{2}, \quad \nabla^2\eta = \frac{3}{2}, \quad \nabla^2\phi = 0,$$

$$|\nabla\xi|^2 = \frac{2\xi}{r}, \quad |\nabla\eta|^2 = \frac{2\eta}{r}, \quad |\nabla\phi|^2 = \frac{1}{r^2 - x^2},$$

$$\nabla\xi\cdot\nabla\eta = 0, \quad \nabla\xi\cdot\nabla\phi = 0, \quad \nabla\eta\cdot\nabla\phi = 0,$$

we get

$$\nabla^2\psi = \frac{2}{r}\left(\frac{\partial\psi}{\partial\xi} + \frac{\partial\psi}{\partial\eta} + \xi\frac{\partial^2\psi}{\partial\xi^2} + \eta\frac{\partial^2\psi}{\partial\eta^2} + \frac{r}{2(r^2 - x^2)}\frac{\partial^2\psi}{\partial\phi^2}\right). \tag{4.21}$$

On setting

$$\psi = e^{im\phi}\, y(\xi,\eta) \tag{4.22}$$

and substituting in Eq. (4.17), we find

$$\xi\frac{\partial^2 y}{\partial \xi^2} + \frac{\partial y}{\partial \xi} + \eta\frac{\partial^2 y}{\partial \eta^2} + \frac{\partial y}{\partial \eta} - \frac{m^2}{4\xi\eta}(\xi+\eta)\, y + \frac{1}{2}(\xi+\eta)\, E\, y$$
$$+ Z\, y - \frac{F}{4}\left(\xi^2 - \eta^2\right) y = 0. \tag{4.23}$$

If we then use the separation of variables

$$y = P(\xi)\, Q(\eta), \tag{4.24}$$

we obtain [3]

$$\begin{aligned}
\xi\, P'' + P' - \frac{m^2}{4\xi} P + \frac{1}{2}\xi\, E\, P + \frac{1}{2}(Z+\lambda)\, P - \frac{F}{4}\xi^2 P &= 0,\\
\eta\, Q'' + Q' - \frac{m^2}{4\eta} Q + \frac{1}{2}\eta\, E\, Q + \frac{1}{2}(Z-\lambda)\, Q + \frac{F}{4}\eta^2 Q &= 0.
\end{aligned} \tag{4.25}$$

Equations (4.25) are self-adjoint; they can be written in the simple form:

$$\begin{aligned}
P'' + \frac{1}{\xi} P' + \left(-\frac{m^2}{4\xi^2} + \frac{1}{2}E + \frac{Z+\lambda}{2\xi} - \frac{F}{4}\xi\right) P &= 0,\\
Q'' + \frac{1}{\eta} Q' + \left(-\frac{m^2}{4\eta^2} + \frac{1}{2}E + \frac{Z-\lambda}{2\eta} + \frac{F}{4}\eta\right) Q &= 0.
\end{aligned} \tag{4.26}$$

The ground state is characterized by $m = 0$ and by the absence of nodal points in P and Q. On setting

$$F = 2\,\epsilon \tag{4.27}$$

for weak fields (ϵ denotes a small quantity), we thus have

$$\begin{aligned}
P'' + \frac{1}{\xi} P' + \frac{1}{2}\left(E + \frac{Z+\lambda}{\xi} - \epsilon\,\xi\right) P &= 0,\\
Q'' + \frac{1}{\eta} Q' + \frac{1}{2}\left(E + \frac{Z-\lambda}{\eta} + \epsilon\,\eta\right) Q &= 0.
\end{aligned} \tag{4.28}$$

Let us put:

$$P = u\, e^{-\sqrt{-E/2}\,\xi}, \quad Q = v\, e^{-\sqrt{-E/2}\,\eta}. \tag{4.29}$$

[3]@ Note that λ is an arbitrary parameter.

The first equation in (4.28) becomes

$$u'' + u' \left(\frac{1}{\xi} - 2\sqrt{-\frac{E}{2}} \right) + u \left(\frac{Z + \lambda - 2\sqrt{\frac{E}{2}}}{2\xi} - \frac{\epsilon}{2}\,\xi \right) = 0, \quad (4.30)$$

while an analogous equation, derived from the second relation in Eq. (4.28), is obtained by replacing u and ξ with v and η, respectively, and by changing the signs of λ and ϵ. In the following we will focus only on the above equation; similar considerations will be true if we perform the mentioned replacements. For the ground state, if we switch off the field ($\epsilon = 0$), we have $\lambda = 0$, $E = -Z^2/2$, and $u = 1$ (except for a normalization factor). When the field is non-zero, we set

$$u = 1 + a_1\,\xi + a_2\,\xi^2 + a_3\,\xi^3 + \ldots; \quad (4.31)$$

and, from (4.30), the coefficients will be determined from the relation

$$a_n = \frac{1}{2n^2}\left[(2n-1)\sqrt{-2E} - (Z+\lambda)\right] a_{n-1} + \frac{\epsilon}{2n^2}\, a_{n-3}. \quad (4.32)$$

However, we can no longer require that u be a finite polynomial; in the presence of an electric field, strictly stationary discrete states do not exist. Thus, we have to use a method of iterative approximations in which u is a finite polynomial except for terms that go to zero with ϵ more rapidly than ϵ^n. This means that we neglect quantities that become appreciable at larger distances from the nucleus as long as ϵ decreases. We then set

$$a_n = a_n^{(0)} + \epsilon\, a_n^{(1)} + \epsilon^2\, a_n^{(2)} + \epsilon^3\, a_n^{(3)} + \ldots, \quad (4.33)$$

and the order at which each series i for the constants $a_n^{(i)}$ terminates, depends on n. Thus for the constants $a^{(0)}$ we have

$$a_0^{(0)} = 1, \quad a_1^{(0)} = a_2^{(0)} = \ldots = a_n^{(0)} = 0. \quad (4.34)$$

We also put

$$\lambda = \epsilon\,\lambda_1 + \epsilon^2\,\lambda_2 + \epsilon^3\,\lambda_3 + \ldots \quad (4.35)$$

$$\sqrt{-2E} = Z + \epsilon\, k_1 + \epsilon^2\, k_2 + \epsilon^3\, k_3 + \ldots. \quad (4.36)$$

Note that, since $a_0 = a_0^{(0)} = 1$, for $r > 1$ we would have

$$a_0^{(r)} = 0, \quad r > 1. \quad (4.37)$$

By substituting the above relations in (4.32) for the ϵ-independent part, we find

$$a_n^{(0)} = \frac{n-1}{n^2} Z a_{n-1}^{(0)}, \tag{4.38}$$

which satisfies Eq. (4.34). Furthermore, keeping first-order terms, we get the relation

$$a_n^{(1)} = \frac{n-1}{n^2} Z a_{n-1}^{(1)} + \frac{1}{2n^2} [(2n-1) k_1 - \lambda_1] a_{n-1}^{(0)} + \frac{1}{2n^2} a_{n-3}^{(0)}. \tag{4.39}$$

Requiring that the series of constant $a^{(1)}$ terminates at a certain point, we deduce that $a_3^{(1)} = 0$; we then obtain identically $a_{3+r}^{(1)} = 0$. On the other hand, we have

$$a_0^{(1)} = 0, \tag{4.40}$$

$$a_1^{(1)} = \frac{1}{2}(k_1 - \lambda_1), \tag{4.41}$$

$$a_2^{(1)} = \frac{1}{8}(k_1 - \lambda_1) Z, \tag{4.42}$$

$$a_3^{(1)} = \frac{k_1 - \lambda_1}{36} Z^2 + \frac{1}{18} = 0, \tag{4.43}$$

from which we get

$$k_1 - \lambda_1 + 2/Z^2 = 0. \tag{4.44}$$

In order to deduce the analogous relations for the function v in (4.29), λ and ϵ have to change sign (as discussed above), while $\sqrt{-2E}$ stays constant. This means that we have to change the sign of k_i (or a_n) if i is odd while we have to preserve its sign if i is even; furthermore, we have to change the sign of λ_i if i is even while preserving it if i is odd. In addition to (4.44), we then have the relation

$$-k_1 - \lambda_1 + \frac{2}{Z^2} = 0, \tag{4.45}$$

from which we infer that

$$K_1 = 0, \quad \lambda_1 = \frac{2}{Z^2}, \quad a_1^{(1)} = -\frac{1}{Z^2}, \quad a_2^{(1)} = -\frac{1}{4Z}. \tag{4.46}$$

Thus, to first approximation, we find

$$\begin{aligned} u &= 1 - \epsilon \left(\frac{1}{Z^2}\, \xi + \frac{1}{4Z}\, \xi^2 \right), \\ v &= 1 - \epsilon \left(\frac{1}{Z^2}\, \xi + \frac{1}{4Z}\, \xi^2 \right), \\ \sqrt{-2E} &= Z + \epsilon \cdot 0, \\ \lambda &= \frac{2}{Z^2}\, \epsilon, \qquad F = 2\, \epsilon. \end{aligned} \tag{4.47}$$

Note that the third equation (4.47) expresses the fact that, to first-order, there is no Stark effect for the ground state. Let us now consider second-order terms; by equating the terms with ϵ^2 in Eq. (4.32), we get

$$a_n^{(2)} = \frac{n-1}{n^2}\, Z\, a_{n-1}^{(2)} + \frac{1}{2n^2}\, [(2n-1)\, k_1 - \lambda_1]\, a_{n-1}^{(1)} \tag{4.48}$$

$$+ \frac{1}{2n^2}\, [(2n-1)\, k_2 - \lambda_2]\, a_{n-1}^{(0)} + \frac{1}{2n^2}\, a_{n-3}^{(1)}. \tag{4.49}$$

For the series $a^{(2)}$ to be finite, it is simple to show that we must have $a_5^{(2)} = 0$, from which it follows that $a_{5+r}^{(2)} = 0$. From Eqs. (4.34), (4.37), and (4.47), we have

$$a_0^{(2)} = 0, \tag{4.50}$$

$$a_1^{(2)} = \frac{1}{2}(k_2 - \lambda_2), \tag{4.51}$$

$$a_2^{(2)} = \frac{k_2 - \lambda_2}{8}\, Z + \frac{1}{4Z^4}, \tag{4.52}$$

$$\begin{aligned} a_3^{(2)} &= \frac{k_2 - \lambda_2}{36}\, Z^2 + \frac{1}{18Z^3} + \frac{1}{36Z^3} \\ &= \frac{k_2 - \lambda_2}{36}\, Z^2 + \frac{1}{12Z^3}, \end{aligned} \tag{4.53}$$

$$\begin{aligned} a_4^{(2)} &= \frac{k_2 - \lambda_2}{192}\, Z^3 + \frac{1}{64Z^2} - \frac{1}{32Z^2} \\ &= \frac{k_2 - \lambda_2}{192}\, Z^3 - \frac{1}{64Z^2}, \end{aligned} \tag{4.54}$$

$$\begin{aligned} a_5^{(2)} &= \frac{k_2 - \lambda_2}{1200}\, Z^4 - \frac{1}{400Z} + \frac{1}{200Z} \\ &= \frac{k_2 - \lambda_2}{1200}\, Z^4 - \frac{3}{400Z} = 0, \end{aligned} \tag{4.55}$$

from which

$$k_2 - \lambda_2 - \frac{9}{25} = 0. \tag{4.56}$$

However, a relation similar to (4.56), in which the sign of k_2 is preserved and that of λ_2 is reversed, holds:

$$k_2 + \lambda_2 - 9/25 = 0, \tag{4.57}$$

so that

$$\lambda_2 = 0, \quad k_2 = 9/25, \tag{4.58}$$

and

$$a_1^{(2)} = \frac{9}{2Z^5}, \quad a_2^{(2)} = \frac{11}{8Z^4}, \quad a_3^{(2)} = \frac{1}{3Z^3}, \quad a_1^{(2)} = \frac{1}{32Z^2}. \tag{4.59}$$

The results obtained to second-order can then be summarized in the following equations:

$$\begin{aligned} u &= 1 - F\left(\frac{1}{2Z^2}\,\xi + \frac{1}{8Z}\,\xi^2\right) \\ &\quad + F^2\left(\frac{9}{8Z^5}\,\xi + \frac{11}{32Z^4}\,\xi^2 + \frac{1}{12Z^3}\,\xi^3 + \frac{1}{128Z^2}\,\xi^4\right), \\ v &= 1 + F\left(\frac{1}{2Z^2}\,\eta + \frac{1}{8Z}\,\eta^2\right) \\ &\quad + F^2\left(\frac{9}{8Z^5}\,\eta + \frac{11}{32Z^4}\,\eta^2 + \frac{1}{12Z^3}\,\eta^3 + \frac{1}{128Z^2}\,\eta^4\right), \quad (4.60) \\ \sqrt{-2E} &= Z + F^2\,\frac{9}{4Z^5}, \\ E &= -\frac{1}{2}\,Z^2 - \frac{9}{4Z^4}\,F^2, \\ \lambda &= \frac{1}{Z^2}\,F + 0\,F^2. \end{aligned}$$

From Eqs. (4.24) and (4.29), to second order, the complete eigenfunction is

$$\psi = \exp\left\{-\left(Z + \frac{9}{4Z^5}\,F^2\right)\frac{\xi+\eta}{r}\right\}\,u(\xi)\,v(\eta), \tag{4.61}$$

where u and v are given by Eqs. (4.60). Using rectangular coordinates and neglecting third-order terms in F, we have

$$\begin{aligned} \psi &= \exp\left\{-\left(Z + \frac{9}{4Z^5}\,F^2\right) r\right\}\left[1 - F\left(\frac{1}{Z^2}\,x + \frac{1}{2Z}\,rx\right)\right. \\ &\quad + F^2\left(\left(\frac{9}{4Z^5}\,r + \frac{1}{16Z^4}\right)\left(7r^2 + 15x^2\right) + \frac{1}{24Z^3}\left(r^3 + 15rx^2\right)\right. \\ &\quad \left.\left. + \frac{1}{8Z^2}\,r^2x^2\right)\right]. \end{aligned} \tag{4.62}$$

It is useful, also, to expand ψ in spherical functions. Denoting by θ the angle between the position vector $\mathbf{r}$ and the x axis, we find (see the next section)

$$\begin{aligned}\psi &= \exp\left\{-\left(Z+\frac{9}{4Z^5}F^2\right)r\right\}\\ &\times\left[1+F^2\left(\frac{9r}{4Z^5}+\frac{3r^2}{4Z^4}+\frac{r^3}{4Z^3}+\frac{r^4}{24Z^2}\right)\right.\\ &-F\left(\frac{r}{Z^2}+\frac{r^2}{2Z}\right)P_1(\cos\theta)\\ &\left.+F^2\left(\frac{5r^2}{8Z^4}+\frac{5r^3}{12Z^3}+\frac{r^4}{12Z^2}\right)P_2(\cos\theta)\right], \end{aligned} \tag{4.63}$$

or, neglecting third-order terms,

$$\begin{aligned}\psi &= \mathrm{e}^{-Zr}\left[1+F^2\left(\frac{3r^2}{4Z^4}+\frac{r^3}{4Z^3}+\frac{r^4}{24Z^2}\right)\right.\\ &-F\left(\frac{r}{Z^2}+\frac{r^2}{2Z}\right)P_1(\cos\theta)\\ &\left.+F^2\left(\frac{5r^2}{8Z^4}+\frac{5r^3}{12Z^3}+\frac{r^4}{12Z^2}\right)P_2(\cos\theta)\right]. \end{aligned} \tag{4.64}$$

It is also simple (see the next section) to obtain an expression for ψ^2 to second-order in F:

$$\begin{aligned}\psi^2 &= \mathrm{e}^{-2Zr}\left[1+F^2\left(\frac{11r^2}{6Z^4}+\frac{5r^3}{6Z^3}+\frac{r^4}{6Z^2}\right)\right.\\ &-F\left(\frac{2r}{Z^2}+\frac{r^2}{Z}\right)P_1(\cos\theta)\\ &\left.+F^2\left(\frac{23r^2}{12Z^4}+\frac{3r^3}{2Z^3}+\frac{r^4}{3Z^2}\right)P_2(\cos\theta)\right]. \end{aligned} \tag{4.65}$$

3. EXPANSION OF LEGENDRE POLYNOMIALS IN THE INTERVAL $-1 < x \leq 1$ [4]

$$
\begin{aligned}
1 &= P_0, \\
x &= P_1, \\
x^2 &= \frac{2}{3}\,P_2 + \frac{1}{3}\,P_0, \\
x^3 &= \frac{2}{5}\,P_3 + \frac{3}{5}\,P_1, \\
x^4 &= \frac{8}{35}\,P_4 + \frac{4}{7}\,P_2 + \frac{1}{5}\,P_0, \\
&\cdots \\
x^n &= \sum_{2\alpha\leq n} 2^{n-2\alpha}\,(2n-4\alpha+1)\,\frac{(n-\alpha)!\,n!}{\alpha!\,(2n-2\alpha+1)!}\,P_{n-2\alpha}(x).
\end{aligned}
$$

4. MULTIPLICATION RULES FOR LEGENDRE POLYNOMIALS

We have [5]

[4]See Sec. 1.42.

[5]@ In the original manuscript, the explicit form of the products P_2P_3, P_2P_4, P_3P_1, P_3P_2, P_3P_3, P_3P_4, P_4P_1, P_4P_2, P_4P_3, P_4P_4 are not reported. They are

$$
\begin{aligned}
P_2P_3 &= P_3P_2 = \frac{9}{35}P_1 + \frac{4}{15}P_3 + \frac{10}{21}P_5, \\
P_2P_4 &= P_4P_2 = \frac{2}{7}P_2 + \frac{20}{77}P_4 + \frac{5}{11}P_6, \\
P_3P_1 &= P_1P_3, \\
P_3P_3 &= \frac{1}{7}P_0 + \frac{4}{21}P_2 + \frac{18}{77}P_4 + \frac{100}{231}P_6, \\
P_3P_4 &= P_4P_3 = \frac{4}{21}P_1 + \frac{2}{11}P_3 + \frac{20}{91}P_5\,\frac{175}{429}P_7, \\
P_4P_4 &= \frac{1}{9}P_0 + \frac{100}{693}P_2 + \frac{162}{1081}P_4 + \frac{20}{99}P_6 + \frac{490}{1287}P_8.
\end{aligned}
$$

	P_0	P_1	P_2	P_3	P_4
P_0	P_0	P_1	P_2	P_3	P_4
P_1	P_1	$\frac{1}{3}P_0 + \frac{2}{3}P_2$	$\frac{2}{5}P_1 + \frac{3}{5}P_3$	$\frac{3}{7}P_2 + \frac{4}{7}P_4$	$\frac{4}{9}P_3 + \frac{5}{9}P_5$
P_2	P_2	$\frac{2}{5}P_1 + \frac{3}{5}P_3$	$\frac{1}{5}P_0 + \frac{2}{7}P_2 + \frac{18}{35}P_4$	P_2P_3	P_2P_4
P_3	P_3	P_3P_1	P_3P_2	P_3P_3	P_3P_4
P_4	P_4	P_4P_1	P_4P_2	P_4P_3	P_4P_4

5. GREEN FUNCTIONS FOR THE DIFFERENTIAL EQUATION $y'' + (2/x - 1)\,y + \phi(x) = 0$

The differential equation

$$y'' + \left(\frac{2}{x} - 1\right) y = -\phi(x), \tag{4.66}$$

with boundary conditions

$$y(0) = y(\infty) = 0, \tag{4.67}$$

has solutions in the interval $[0, \infty)$ only if $-\phi(x)$ is orthogonal to the solution χ of the homogeneous equation

$$\chi'' + \left(\frac{2}{x} - 1\right) \chi = 0, \tag{4.68}$$

with boundary conditions analogous to (4.67). A non-vanishing (normalized) solution of (4.68) is

$$\chi = 2x\,\mathrm{e}^{-x}. \tag{4.69}$$

Then it follows that, if $\phi(x)$ is an arbitrary continuous function, the differential equation

$$y'' + \left(\frac{2}{x} - 1\right) y = -\phi(x) + 4x\,\mathrm{e}^{-x} \int_0^\infty \phi(x)\,x\,\mathrm{e}^{-x}\,\mathrm{d}x \tag{4.70}$$

has solutions satisfying the boundary conditions (4.67). We can set

$$y(x) = \int G(x, \xi)\,\phi(\xi)\,\mathrm{d}\xi. \tag{4.71}$$

Both y and $G(x, \xi)$ (viewed as functions of x for each value of ξ) are defined apart from a function proportional to χ. We resolve this ambiguity by choosing G (and then y) to be orthogonal to χ. In this way, the Green function $G(x, \xi)$ is necessarily symmetric in x and ξ. If we define

$$L = \frac{\mathrm{d}^2}{\mathrm{d}x^2} + \left(\frac{2}{x} - 1\right), \tag{4.72}$$

the Green function is seen to satisfy the differential equation

$$L\, G(x, \xi) = 4x\, \mathrm{e}^{-x}\, \xi\, \mathrm{e}^{-\xi}, \tag{4.73}$$

and its first derivative must have a singularity for $x = \xi$ in such a way that

$$\left[\frac{\mathrm{d}}{\mathrm{d}x} G(x, \xi)\right]_{x=\xi+0} - \left[\frac{\mathrm{d}}{\mathrm{d}x} G(x, \xi)\right]_{x=\xi-0} = -1. \tag{4.74}$$

Let us set

$$G(x, \xi) = 4\xi\, \mathrm{e}^{-\xi}\, p(x, \xi) \tag{4.75}$$

and consider, for the moment, p as a function of x for constant ξ. From Eq. (4.73), we have

$$L\, p = x\, \mathrm{e}^{-x}. \tag{4.76}$$

The general solution of Eq. (4.76) is obtained by adding a particular solution of it to the general solution of the associated homogeneous equation

$$L\, p = 0. \tag{4.77}$$

On the other hand, the general solution of Eq. (4.76) is a linear combination of two independent solutions, one of which is χ defined in Eq. (4.69); the other is, as well known,

$$\chi_1 = \chi \int \frac{\mathrm{d}x}{\chi^2} = -\mathrm{e}^x + 2x\, \mathrm{e}^{-x} \int \frac{\mathrm{e}^{2x}}{x}\, \mathrm{d}x \tag{4.78}$$

or, since we can arbitrarily fix the lower limit of the integral,

$$\chi_1 = 2x\, \mathrm{e}^{-x} \int_0^x \frac{\mathrm{e}^{2x} - 1}{x}\, \mathrm{d}x + 2x\, \mathrm{e}^{-x} \log x - \mathrm{e}^x. \tag{4.79}$$

A particular solution of Eq. (4.76), the one that vanishes for $x = 0$ along with its first derivative, is

$$p_0 = \frac{1}{2} x\, \mathrm{e}^{-x} \int_0^x \frac{\mathrm{e}^{2x} - 1}{x}\, \mathrm{d}x - \frac{1}{4}\mathrm{e}^x + \left(\frac{1}{4} + \frac{1}{2}x - \frac{1}{2}x^2\right) \mathrm{e}^{-x}. \tag{4.80}$$

It follows from Eq. (4.75) that the Green function can be cast in the form

$$G(x,\xi) \;=\; 4\xi\,\mathrm{e}^{-\xi}\,p_0(x) \;+\; a_i(\xi)\,\chi(x) \;+\; b_i(\xi)\chi_1(x), \tag{4.81}$$

where the index i takes the value 1 for $x<\xi$ and the value 2 for $x>\xi$, so that the problem is reduced to determining the quantities $a_1(\xi)$, $b_1(\xi)$, $a_2(\xi)$, $b_2(\xi)$ which are constant with respect to x. These are determined by the boundary conditions $G=0$ for $x=0$ and $x=\infty$, by the discontinuity condition (4.74) for $x=\xi$, and by the orthogonality condition between the Green function and the solution χ of the homogeneous equation satisfying the boundary conditions. The condition $G(0,\xi)=0$ implies that

$$b_1 \;=\; 0. \tag{4.82}$$

The condition $G(\infty,\xi)=0$ is satisfied when

$$b_2 \;=\; -\,\xi\,\mathrm{e}^{-\xi}. \tag{4.83}$$

From the condition (4.74), it follows that

$$(a_1 \;-\; a_2)\;\chi'(\xi) \;+\; (b_1 \;-\; b_2)\;\chi_1'(\xi) \;=\; 1, \tag{4.84}$$

that is, taking into account Eqs. (4.82) and (4.83),

$$a_2 \;=\; a_1 \;+\; \xi\,\mathrm{e}^{-\xi}\,\frac{\chi_1'(\xi)}{\chi'(\xi)} \;-\; \frac{1}{\chi'(\xi)}; \tag{4.85}$$

and, performing the calculation, we get

$$a_2 \;=\; a_1 \;+\; \xi\,\mathrm{e}^{-\xi}\int_0^{\xi}\frac{\mathrm{e}^{2\xi}-1}{\xi}\,\mathrm{d}\xi \;+\; \xi\,\mathrm{e}^{-\xi}\log\xi \;-\; \frac{\mathrm{e}^{\xi}}{2}. \tag{4.86}$$

In terms of the function a_1 - yet to be determined - we have the following expressions for the Green function for $x<\xi$ and $x>\xi$, respectively:

$$\begin{aligned} G(x,\xi) \;&=\; \xi\,\mathrm{e}^{-\xi}\left[2x\,\mathrm{e}^{-x}\int_0^{x}\frac{\mathrm{e}^{2x}-1}{x}\,\mathrm{d}x \;-\; \mathrm{e}^{x}\right.\\ &\quad\left. +\;\left(1 \;+\; 2x \;-\; 2x^2\right)\,\mathrm{e}^{-x}\right]\\ &\quad +\; 2\,a_1(\xi)\,x\,\mathrm{e}^{-x}, \qquad x<\xi; \end{aligned} \tag{4.87}$$

$$\begin{aligned} G(x,\xi) \;&=\; x\,\mathrm{e}^{-x}\left(2\xi\,\mathrm{e}^{-\xi}\int_0^{\xi}\frac{\mathrm{e}^{2\xi}-1}{\xi}\,\mathrm{d}\xi \;-\; \mathrm{e}^{\xi}\right)\\ &\quad +\; \xi\,\mathrm{e}^{-\xi}\left(1 \;+\; 2x \;-\; 2x^2\right) \;+ 2\xi\,\mathrm{e}^{-\xi}\,x\,\mathrm{e}^{-x}\,(\log\xi \;-\; \log x)\\ &\quad +\; 2\,a_1(\xi)\,x\,\mathrm{e}^{-x}, \qquad x>\xi. \end{aligned} \tag{4.88}$$

We can now determine a_1 through the orthogonality condition

$$\int \chi(x)\, G(x,\xi)\, \mathrm{d}x \;=\; 0.$$

We find

$$a_1 \;=\; \left(\frac{1}{2} + \frac{5}{2}\xi - \xi^2\right) \mathrm{e}^{-\xi} \;-\; C\,\xi\,\mathrm{e}^{-\xi} \;-\; \xi\,\mathrm{e}^{-\xi}\log 2\xi, \qquad (4.89)$$

where C is the Euler constant. We find the final expressions for the Green function by substitution in Eqs. (4.87) and (4.88):

$$\begin{aligned} G(x,\xi) \;&=\; \mathrm{e}^{-\xi}\,\mathrm{e}^{-x}\left(\xi + x + (7-2C)\,\xi\,x \;-\; 2\,\xi^2\,x \;-\; 2\,\xi\,x^2\right) \\ &\quad+\; 2\xi\,\mathrm{e}^{-\xi}\,x\mathrm{e}^{-x}\int_0^x \frac{\mathrm{e}^{2x}-1}{x}\,\mathrm{d}x \\ &\quad-\; \xi\,\mathrm{e}^{-\xi}\,\mathrm{e}^{x} \;-\; 2\xi\,\mathrm{e}^{-\xi}\,x\,\mathrm{e}^{-x}\log 2\xi, \qquad x < \xi; \qquad (4.90) \\ G(x,\xi) \;&=\; \mathrm{e}^{-\xi}\,\mathrm{e}^{-x}\left(\xi + x + (7-2C)\,\xi\,x \;-\; 2\,\xi^2\,x \;-\; 2\,\xi\,x^2\right) \\ &\quad+\; 2\xi\,\mathrm{e}^{-\xi}\,x\mathrm{e}^{-x}\int_0^\xi \frac{\mathrm{e}^{2\xi}-1}{\xi}\,\mathrm{d}\xi \\ &\quad-\; x\,\mathrm{e}^{-x}\,\mathrm{e}^{\xi} \;-\; 2\xi\,\mathrm{e}^{-\xi}\,x\,\mathrm{e}^{-x}\log 2x, \qquad x < \xi. \qquad (4.91) \end{aligned}$$

Note that, as we expected, $G(x,\xi)$ is symmetric in x and ξ, since Eq. (4.91) can be obtained from Eq.(4.90) by the interchange $x \leftrightarrow \xi$.

6. ON THE SERIES EXPANSION OF THE INTEGRAL LOGARITHM FUNCTION

The integral logarithm function is defined by the relation

$$Ei(-x) \;=\; -\,A(x), \qquad (4.92)$$

where [6]

$$A(x) \;=\; \int_x^\infty \frac{\mathrm{e}^{-\xi}}{\xi}\,\mathrm{d}\xi. \qquad (4.93)$$

For $\xi > x$ we can expand the term $1/\xi$ in a power series using the methods of finite difference calculus, by requiring that the first n terms

[6] This function is also known as the incomplete gamma function $\Gamma(0,x)$.

of the expansion give the exact result for $1/\xi$, where $\xi = x, x+1, \dots, x+n-1$. On setting

$$\xi = x + y, \tag{4.94}$$

we find the formula

$$\frac{1}{\xi} = \frac{1}{x} - \frac{y}{x(x+1)} + \frac{y(y-1)}{x(x+1)(x+2)} - \dots, \tag{4.95}$$

and Eq. (4.93) becomes

$$A(x) = \frac{e^{-x}}{x} \left(1 - \frac{\int_0^\infty y\, e^{-y}\, dy}{x+1} + \frac{\int_0^\infty y\,(y-1)\, e^{-y}\, dy}{(x+1)(x+2)} - \dots \right.$$
$$\left. \dots \pm \frac{I_n}{(x+1)(x+2)\cdots(x+n)} \mp \dots \right), \tag{4.96}$$

where

$$I_n = \int_0^\infty y\,(y-1)\cdots(y-n+1)\, e^{-y}\, dy. \tag{4.97}$$

We find

$$I_1 = 1, \quad I_2 = 1, \quad I_3 = 2, \quad I_4 = 4, \quad I_5 = 14,$$
$$I_6 = 38, \quad I_7 = 216, \quad I_8 = 600, \dots. \tag{4.98}$$

On substituting in Eq. (4.96), we get

$$\int_x^\infty \frac{e^{-\xi}}{\xi}\, d\xi = \frac{e^{-x}}{x} \left(1 - \frac{1}{x+1} + \frac{1}{(x+1)(x+2)} \right.$$
$$- \frac{2}{(x+1)(x+2)(x+3)} + \frac{4}{(x+1)(x+2)(x+3)(x+4)}$$
$$- \frac{14}{(x+1)(x+2)\cdots(x+5)} + \frac{38}{(x+1)(x+2)\cdots(x+6)}$$
$$\left. - \frac{216}{(x+1)(x+2)\cdots(x+7)} + \frac{600}{(x+1)(x+2)\cdots(x+8)} - \dots \right). \tag{4.99}$$

From this we can deduce the expansion of $\log(1+y)$:

$$\log(1+y) = y - \frac{y^2}{4} + \frac{2y^3 - 3y^2}{36} - \frac{y^4 - 4y^3 + 4y^2}{96} + \dots. \tag{4.100}$$

More generally, from Eq. (4.94), we obtain

$$\log\left(1+\frac{y}{x}\right) = \frac{y}{x} - \frac{y^2}{2x(x+1)} + \frac{2y^3-3y^2}{6x(x+1)(x+2)} - \frac{y^4-4y^3+4y^2}{4x(x+1)(x+2)(x+3)} + \dots . \quad (4.101)$$

Let us put $t = y/x$ and consider only n terms in the expansion (4.101) by setting $y = n-1$, so that $x = (n-1)/t$ except for the case $n = 1$, for which we keep y arbitrary. We thus obtain the following formulae for $\log(1+t)$ with a decreasing degree of approximation:

$$n = 1: \quad \log(1+t) = t, \quad (4.102)$$

$$n = 2: \quad \log(1+t) = t - \frac{t^2}{2(1+t)}, \quad (4.103)$$

$$n = 3: \quad \log(1+t) = t - \frac{t^2}{2+t} + \frac{t^3}{3(2+t)(2+2t)}, \quad (4.104)$$

$$n = 4: \quad \log(1+t) = t - \frac{3t^2}{2(3+t)} + \frac{3t^3}{2(3+t)(3+2t)} - \frac{3t^4}{4(3+t)(3+2t)(3+3t)}. \quad (4.105)$$

For example, the above expressions give for log 2 ($t = 1$) and log 10 ($t = 9$), respectively:

n	log 2	log 10
1	1.0000	9.00
2	0.7500	4.95
3	0.6944	2.74
4	0.6937	2.56

7. FUNDAMENTAL CHARACTERS OF THE GROUP OF PERMUTATIONS OF *F* OBJECTS

We have[7]

[7]@ In the following,P.N. denotes the number of "Partitio Numerorum", that is, the number of ways one can collect f objects. In the tables, the given "partitio" is stated in the first row

f = 1 (P.N. = 1)

n_k	Partitio → Class ↓	1
1 +	(1)	1

f = 2 (P.N. = 2)

	Partitio →		1+
n_k	Class ↓	2	1
1+	(1)(2)	1	1
1−	(12)	1	-1

f = 3 (P.N. = 3)

				1+
	Partitio →		2+	1+
n_k	Class ↓	3	1	1
1+	(1)(2)(3)	1	2	1
3−	(12)(3)	1	0	-1
2+	(123)	1	-1	1

f = 4 (P.N. = 5)

						1+
					2+	1+
	Partitio →		3+	2+	1+	1+
n_k	Class ↓	4	1	2	1	1
1+	(1) ...	1	3	2	3	1
6−	(12) ...	1	1	0	-1	-1
3+	(12)(34)	1	-1	2	-1	1
8+	(123) ...	1	0	-1	0	1
6−	(1234)	1	-1	0	1	-1

from the third column onwards. Instead, in the second column the classes of the cycles of permutations of f objects are stated. Finally, in the first column the number of cycles of the considered class are indicated. In each table, the characters corresponding to a given class and "partitio" are shown from the third column on and from the second row downwards.

f = 5 (P.N. = 7)

n_k	Partitio → / Class ↓	5	4+1	3+2	3+1+1	2+2+1	2+1+1+1	1+1+1+1+1
1+	(1) ...	1	4	5	6	5	4	1
10−	(12) ...	1	2	1	0	-1	-2	-1
15+	(12)(34) ...	1	0	1	-2	1	0	1
20+	(123) ...	1	1	-1	0	-1	1	1
20−	(123)(45)	1	-1	1	0	-1	1	-1
30−	(1234) ...	1	0	-1	0	1	0	-1
24+	(12345)	1	-1	0	1	0	-1	1

f = 6 (P.N. = 11)

n_k	Partitio → / Class ↓	6	5+1	4+2	4+1+1	3+3	3+2+1	3+1+1+1	2+2+2	2+2+1+1	2+1+1+1+1	1+1+1+1+1+1
1+	(1) ...	1	5	9	10	5	16	10	5	9	5	1
15−	(12) ...	1	3	3	2	1	0	-2	-1	-3	-3	-1
45+	(12)(34) ...	1	1	1	-2	1	0	-2	1	1	1	1
15−	(12)(34)(56)	1	-1	3	-2	-3	0	2	3	-3	1	-1
40+	(123) ...	1	2	0	1	-1	-2	1	-1	0	2	1
120−	(123)(45) ...	1	0	0	-1	1	0	1	-1	0	0	-1
40+	(123)(456)	1	-1	0	1	2	-2	1	2	0	-1	1
90−	(1234) ...	1	1	-1	0	-1	0	0	1	1	-1	-1
90+	(1234)(56)	1	-1	1	0	-1	0	0	-1	1	-1	1
144+	(12345) ...	1	0	-1	0	0	1	0	0	-1	0	1
120−	(123456)	1	-1	0	1	0	0	-1	0	0	1	-1

Degrees of the irreducible representations and reciprocal systems: [8]

f = 2

2	1+1
1+1	2
1	1

[8]@ In the following tables the first row lists the irreducible representation considered, the second row states the corresponding reciprocal system while the third one lists the degree of the considered representation.

f = 3

3	2+1	1+1+1
1+1+1	2+1	3
1	2	1

f = 4

4	3+1	2+2	2+1+1	1+1+1+1
1+1+1+1	2+1+1	2+2	3+1	4
1	3	2	3	1

f = 5

5	4+1	3+2	3+1+1
1+1+1+1+1	2+1+1+1	2+2+1	3+1+1
1	4	5	6

2+2+1	2+1+1+1	1+1+1+1+1
3+2	4+1	5
5	4	1

f = 6

6	5+1	4+2	4+1+1	3+3	3+2+1
1+1+1+1+1+1	2+1+1+1+1	2+2+1+1	3+1+1+1	2+2+2	3+2+1
1	5	9	10	5	16

3+1+1+1	2+2+2	2+2+1+1	2+1+1+1+1	1+1+1+1+1+1
4+1+1	3+3	4+2	5+1	6
10	5	9	5	1

f = 7

7	6+1	5+2	5+1+1
1+1+1+1+1+1+1	2+1+1+1+1+1	2+2+1+1+1	3+1+1+1+1
1	6	14	15

4+3	4+2+1	4+1+1+1	3+3+1	3+2+2	3+2+1+1
2+2+1	3+2+1+1	4+1+1+1	3+2+2	3+3+1	4+2+1
14	35	20	21	21	35

3+1+1+1+1	2+2+2+1	2+2+1+1+1
5+1+1	4+3	5+2
15	14	14

2+1+1+1+1+1	1+1+1+1+1+1+1
6+1	7
6	1

8. EXPANSION OF A PLANE WAVE IN SPHERICAL HARMONICS

The plane wave

$$u = e^{ikz} = e^{ikr\cos\theta} \tag{4.106}$$

satisfies the differential equation

$$\nabla^2 u + k^2 u = 0. \tag{4.107}$$

Each solution of Eq. (4.107) can be expressed in terms of linear combinations of particular solutions

$$\frac{1}{\sqrt{\rho}} I_{n+1/2}(\rho)\, \varphi_n^i(\theta, \phi) \tag{4.108}$$

($\rho = kr$; $n = 0, 1, 2, \ldots$; $i = -n, -n+1, \ldots, n$), where $I_{n+1/2}$ denote the Bessel functions of order $n + 1/2$ and φ_n^i a generic surface spherical function of order n. The function u in Eq. (4.106) is symmetric along the z axis, and only terms that depend on ρ and $\cos\theta$ will appear in its expansion:

$$u = e^{ikz} = e^{i\rho\cos\theta} = \sum_{n=0}^{\infty} \frac{a_n}{\sqrt{\rho}} I_{n+1/2}(\rho)\, P_n(\cos\theta), \tag{4.109}$$

where P_n are the Legendre polynomials. In order to determine the constants a_n, we can multiply each side of Eq. (4.109) by $P_n(\cos\theta)$ and integrate over a sphere of radius $r = \rho/k$; dividing the result by $2\pi r^2$ we get

$$\int_{-1}^{1} \mathrm{e}^{i\rho t} \, P_n(t) \, \mathrm{d}t = \frac{2}{2n+1} \frac{a_n}{\sqrt{\rho}} \, I_{n+1/2}(\rho). \tag{4.110}$$

We consider ρ to be a small quantity and expand the above expression in powers of ρ. The first non-vanishing term on the l.h.s. is (see Sec. 4.3)

$$\begin{aligned} \frac{i^n \rho^n}{n!} \int_{-1}^{1} t^n \, P_n(t) \, \mathrm{d}t &= \frac{i^n \rho^n}{n!} \frac{2}{3} \cdot \frac{3}{5} \cdots \frac{n}{2n-1} \cdot \frac{2}{2n+1} \\ &= \frac{2^{n+1} n!}{(2n+1)!} \, i^n \, \rho^n, \end{aligned} \tag{4.111}$$

while on the r.h.s. we obtain

$$\begin{aligned} \frac{2}{2n+1} \frac{a_n}{\sqrt{\rho}} \frac{1}{(n+1/2)!} \left(\frac{\rho}{r}\right)^{n+1/2} &= \frac{2}{2n+1} \sqrt{\frac{2}{\pi}} \frac{a_n}{1 \cdot 3 \cdots (2n+1)} \, \rho^n \\ &= \frac{a_n}{2n+1} \sqrt{\frac{2}{\pi}} \frac{2^{n+1} \, n!}{(2n+1)!} \, \rho^n. \end{aligned} \tag{4.112}$$

Comparison with the previous expression gives

$$a_n = (2n+1) \sqrt{\frac{\pi}{2}} \, i^n. \tag{4.113}$$

On substituting this result into Eq. (4.110), we obtain the following remarkable relation:

$$I_{n+1/2}(\rho) = \sqrt{\frac{\rho}{2\,\pi}} \, (-i)^n \int_{-1}^{1} \mathrm{e}^{i\rho t} \, P_n(t) \, \mathrm{d}t. \tag{4.114}$$

Examples:

$$I_{1/2}(\rho) = \sqrt{\frac{2}{\pi\,\rho}} \, \sin \rho, \tag{4.115}$$

$$I_{3/2}(\rho) = \sqrt{\frac{2}{\pi\,\rho}} \left(- \cos \rho + \frac{1}{\rho} \sin \rho \right). \tag{4.116}$$

On substituting Eq. (4.113) in Eq.(4.109), we find the expansion of the plane wave:

$$\mathrm{e}^{ikz} = \sum_{n=0}^{\infty} \frac{2n+1}{\rho} \sqrt{\frac{\pi\rho}{2}} \, i^n \, I_{n+1/2}(\rho) \, P_n(\cos \theta). \tag{4.117}$$

Let us write Eq. (4.114) in a form which is similar to that of Eq. (4.110):

$$\int_{-1}^{1} e^{i\rho t} P_n(t)\, dt = \sqrt{\frac{2\pi}{\rho}}\, i^n\, I_{n+1/2}(\rho) \tag{4.118}$$

and expand in powers of ρ. It is simple to see that only the terms with $\rho^{n+2\alpha}$ ($\alpha = 0, 1, 2, \ldots$) are different from zero. Equating the coefficients of $\rho^{n+2\alpha}$ on both sides, we find

$$\frac{i^{n+2\alpha}}{(n+2\alpha)!} \int_{-1}^{1} t^{n+2\alpha} P_n(t)\, dt = \frac{\sqrt{2\pi}\, i^n\, (-1)^\alpha}{\alpha!\, (n+\alpha+1/2)!\, 2^{n+2\alpha+1/2}}. \tag{4.119}$$

Simplifying this expression

$$\frac{1}{(n+2\alpha)!} \int_{-1}^{1} t^{n+2\alpha} P_n(t)\, dt = \frac{\sqrt{\pi}}{\alpha!\, (n+\alpha+1/2)!\, 2^{n+2\alpha}} \tag{4.120}$$

and noting that

$$\begin{aligned} \left(n + \alpha + \frac{1}{2}\right)! &= \frac{\sqrt{\pi}}{2} \cdot \frac{3}{2} \cdot \frac{5}{2} \cdots \left(n + \alpha + \frac{1}{2}\right) \\ &= \frac{\sqrt{\pi}}{2}\, \frac{1}{2^{n+\alpha}} \cdot 3 \cdot 5 \cdot 7 \cdots (2n + 2\alpha + 1) \\ &= \frac{\sqrt{\pi}}{2}\, \frac{(2n + 2\alpha + 1)!}{(n+\alpha)!\, 2^{2n+2\alpha}}, \end{aligned} \tag{4.121}$$

we get

$$\int_{-1}^{1} t^{n+2\alpha} P_n(t)\, dt = 2^{n+1}\, \frac{(n+\alpha)!\, (n+2\alpha)!}{\alpha!\, (2n+2\alpha+1)!}. \tag{4.122}$$

Replacing n by $n - 2\alpha$ in Eq. (4.122), we obtain

$$\int_{-1}^{1} t^n P_{n-2\alpha}(t)\, dt = 2^{n-2\alpha+1}\, \frac{(n-\alpha)!\, n!}{\alpha!\, (2n-2\alpha+1)!} \tag{4.123}$$

(with $2\alpha \leq n$); and, using the normalization condition for the Legendre polynomials,

$$\int_{-1}^{1} P_n^2(t)\, dt = \frac{2}{2n+1},$$

we deduce the expansion of t^n ($n - 1 \leq t \leq 1$) in terms of the Legendre polynomials:

$$t^n = \sum_{2\alpha \leq n} 2^{n-2\alpha}\, (2n - 4\alpha + 1)\, \frac{(n-\alpha)!\, n!}{\alpha!\, (2n-2\alpha+1)!}\, P_{n-2\alpha}(t). \tag{4.124}$$

9. THE RUTHERFORD FORMULA DEDUCED FROM CLASSICAL MECHANICS

Let us consider a uniform beam of particles with charge $Z'e$ and mass m moving along the z axis with a speed v. Let i_o/v be the number of particles per unit volume, so that i_o is the flux per unit surface (normal to the z axis) and per unit time. Moreover, let us consider a scattering body of charge Ze placed at the origin of the coordinate system; the problem is to determine the number of scattered particles at the angle θ per unit time and solid angle. This number can be written as $f(\theta)\, i_o$, where $f(\theta)$ has the dimension of a surface (cross section). To solve the problem, let us note that each particle moves in a plane containing the z axis. Using polar coordinates in this plane, we have

$$\begin{aligned} \ddot{\rho} - \rho\,\dot{\theta}^2 &= \frac{k}{\rho^2}, \quad k = \frac{ZZ'e^2}{m}; \\ \rho^2\dot{\theta} &= c. \end{aligned} \tag{4.125}$$

Eliminating θ from these expressions, we get

$$\ddot{\rho} = \frac{k}{\rho^2} + \frac{c^2}{\rho^3}. \tag{4.126}$$

Introducing the new variable $y = 1/\rho$, we find

$$\rho = \frac{1}{y}, \tag{4.127}$$

$$\dot{\rho} = -\frac{1}{y^2}\,\dot{y} = -\frac{\mathrm{d}y}{\mathrm{d}\theta}\,\rho^2\,\dot{\theta} = -\,c\,\frac{\mathrm{d}y}{\mathrm{d}\theta}, \tag{4.128}$$

$$\ddot{\rho} = -\,c\,\frac{\mathrm{d}^2y}{\mathrm{d}\theta^2}\,\dot{\theta} = -\,c^2\,y^2\,\frac{\mathrm{d}^2y}{\mathrm{d}\theta^2}. \tag{4.129}$$

On substituting into Eq. (4.126), we obtain the equation

$$\frac{\mathrm{d}^2y}{\mathrm{d}\theta^2} + y + \frac{k}{c^2} = 0, \tag{4.130}$$

whose general solution is

$$\frac{1}{\rho} = y = -\,\frac{k}{c^2} + a\,\cos\,\theta + b\,\sin\,\theta. \tag{4.131}$$

The arbitrary constants will be determined by the initial conditions. For $\theta = \pi$ we have $\rho = \infty$, since we consider the case in which the particles come from $-\infty$ along the z axis. This implies the condition

$$a = -k/c^2. \tag{4.132}$$

Moreover, according to the previous hypothesis, for $\theta = \pi$ we will also have $\dot{\rho} = -v$ and, since from Eq. (4.128) we have $\dot{\rho} = -c\,\mathrm{d}y/\mathrm{d}\theta$, it follows that

$$c\,b = -v \quad \text{or} \quad b = -v/c, \tag{4.133}$$

so that Eq. (4.131) becomes

$$\frac{1}{\rho} = -\frac{k}{c^2} - \frac{k}{c^2}\cos\theta - \frac{v}{c}\sin\theta. \tag{4.134}$$

This is the equation of an hyperbole with asymptotes along the directions $\theta_1 = \pi$ and $\theta_2 = -2\arctan(k/vc)$. We still have to determine the geometric meaning of c, which can be deduced from the second equation (4.125); however, we prefer to start from Eq. (4.134) and introduce the rectangular coordinates $z = \rho\cos\theta$ and $\xi = \rho\sin\theta$ in the orbital plane, so that Eq. (4.134) becomes

$$1 - \frac{k}{c^2}\sqrt{z^2+\xi^2} + \frac{k}{c^2}\,z + \frac{v}{c}\,\xi = 0. \tag{4.135}$$

From this we find

$$z^2 + \xi^2 = \left(z + \frac{vc}{k}\,\xi + \frac{c^2}{k}\right)^2, \tag{4.136}$$

that is,

$$\left(1 - \frac{v^2c^2}{k^2}\right)\xi^2 - 2\,\frac{vc}{k}\,\xi\,z - 2\,\frac{c^2}{k}\,z - 2\,\frac{vc^3}{k^2}\,\xi - \frac{c^4}{k^2} = 0, \tag{4.137}$$

from which we get the equation of the first asymptote:

$$\xi = -c/v. \tag{4.138}$$

The absolute value of ξ equals the initial value of the distance of the particle from the z axis (along which the particle moves). If we choose the direction of the ξ axis in such a way that ξ is initially positive (and thus, assuming $ZZ' > 0$, during the entire motion), we have

$$c = -v\,\delta. \tag{4.139}$$

From the above remark on the direction of the second asymptote, the angular deflection of the particle will be

$$\theta = 2\arctan(k/v^2\,\delta). \tag{4.140}$$

The scattering angle θ increases with decreasing δ, and the particles scattered at angles greater than θ are those coming, for high negative values of z, from a circle of radius δ normal to the z axis. The number of such particles per unit time is

$$n = \pi \delta^2 i_o, \tag{4.141}$$

or, after substituting δ from (4.140),

$$n = \frac{\pi k^2 i_o}{v^4 \tan^2 \theta/2}. \tag{4.142}$$

The number of particles scattered per unit solid angle will be

$$\frac{dn}{d\omega} = \frac{dn}{-2\pi \sin\theta \, d\theta}$$

and, since $dn/d\omega = f(\theta) \, i_o$, differentiations of Eq. (4.142) and division by $-2\pi i_o \sin\theta d\theta$, give

$$f(\theta) = \frac{Z^2 Z'^2 e^4}{4m^2 v^4 \sin^4 \theta/2} = \frac{Z^2 Z'^2 e^4}{16 W^2 \sin^4 \theta/2}, \tag{4.143}$$

where W is the kinetic energy of the free particle. On setting

$$W = \frac{Z Z' e^2}{l}, \tag{4.144}$$

l being a length (positive or negative according to the sign of ZZ'), and substituting in Eq. (4.143), we obtain the following remarkable formula for the cross section:

$$f(\theta) = \frac{l^2}{16 \sin^4 \theta/2}. \tag{4.145}$$

We can also define a different cross section as the ratio between the number n of particles scattered at angles *greater* than θ in unit time and i_o. From Eq. (4.142), we have

$$n = \frac{\pi Z^2 Z'^2 e^4 i_o}{m^2 v^4 \tan^2 \theta/2} = \frac{\pi Z^2 Z'^2 e^4 i_o}{4W^2 \tan^2 \theta/2} = \frac{\pi l^2 i_o}{4 \tan^2 \theta/2}, \tag{4.146}$$

from which it follows that

$$F(\theta) = \frac{\pi l^2}{4 \tan^2 \theta/2}. \tag{4.147}$$

The relation between the two cross sections is obviously the following:

$$F'(\theta) = -2\pi \sin\theta \, f(\theta). \tag{4.148}$$

Table 4.1. The scattering angle θ as a function of the parameter ϵ (see text).

ϵ	θ
0	0
1	arctan 4/3
2	$\pi/2$
3	π − arctan 12/5
4	π − arctan 4/3
5	π − arctan 21/20

The relation between θ and δ expressed by Eq. (4.140) can be cast in the form

$$\tan\frac{\theta}{2} = \frac{\epsilon}{2}, \tag{4.149}$$

where $\epsilon = l/\delta$.

10. THE RUTHERFORD FORMULA DEDUCED AS A FIRST APPROXIMATION TO THE BORN METHOD

Let us consider the plane wave [9]

$$\psi_0 = \mathrm{e}^{+i\gamma z}, \tag{4.150}$$

representing a uniform flux of particles along the direction of the z axis. Denoting with m their mass, each particle has the kinetic energy

$$W = \frac{\hbar^2}{2m}\gamma^2. \tag{4.151}$$

Let us consider the scattering center of charge Ze at the origin of the coordinate system, and let $Z'e$ be the charge of the scattered particles. If the energy has the form given in Eq.(4.151), the wavefunction obeys the differential equation

$$\boldsymbol{\nabla}^2\psi + \left(\gamma^2 - \frac{2m^2 k}{\hbar^2 r}\right)\psi = 0, \quad k = \frac{ZZ'e^2}{m}. \tag{4.152}$$

[9]@ In the original manuscript, the fundamental commutation relation $pq - qp = \hbar/i$ is written near Eq.(4.150). As already stated, here and in the following we prefer to use the modern notation $\hbar$ in place of $h/2\pi$.

If one assumes that the scattering potential is small, one can consider (4.150) as the unperturbed eigenfunction and set

$$\psi = \psi_0 + \psi_1, \tag{4.153}$$

ψ_1 being a small correction term. On substituting into Eq. (4.152) and neglecting second-order terms, to first approximation we get

$$\nabla^2 \psi_1 + \gamma^2 \psi_1 = \frac{2m^2 k}{\hbar^2 r} \, e^{i\gamma z}. \tag{4.154}$$

In order to make the solution of Eq. (4.154) unique, we have to require: (1) that ψ_1 vanishes at infinity, i.e., ψ must coincide with the unperturbed ψ_0 for large distances from the scattering center; (2) for its phenomenological meaning, ψ_1 must represent a diverging spherical wave. The desired solution can be obtained through the Green method, using $-e^{i\gamma r}/4\pi r$ as characteristic function. However, in such a way, convergence problems arise. In order to avoid these, we will assume that the scattering force acts only for distances less than R, and let $R \to \infty$ at the end of calculations. Then, Eq. (4.154) must be rewritten as follows:

$$\begin{aligned} \nabla^2 \psi_1 + \gamma^2 \psi_1 &= \frac{2m^2 k}{\hbar^2} \left(\frac{1}{r} - \frac{1}{R} \right) e^{i\gamma z}, \quad \text{for } r < R, \\ \nabla^2 \psi_1 + \gamma^2 \psi_1 &= 0, \quad \text{for } r > R. \end{aligned} \tag{4.155}$$

Let us write (4.155) in a slightly different form using the velocity of the free particles

$$v = \gamma \frac{\hbar}{m}. \tag{4.156}$$

We then have

$$\begin{aligned} \nabla^2 \psi_1 + \gamma^2 \psi_1 &= \frac{2\gamma^2 k}{v^2} \left(\frac{1}{r} - \frac{1}{R} \right) e^{i\gamma z}, \quad \text{for } r < R, \\ \nabla^2 \psi_1 + \gamma^2 \psi_1 &= 0, \quad \text{for } r > R; \end{aligned} \tag{4.157}$$

and, by using the Green method,

$$\psi_i(P_1) = \frac{1}{4\pi} \int_S \frac{2\gamma^2 k}{v^2} \left(\frac{1}{r} - \frac{1}{R} \right) \frac{1}{|\mathbf{r_1} - \mathbf{r}|} \, e^{i\gamma(|\mathbf{r_1} - \mathbf{r}| + z)} \, d\tau, \tag{4.158}$$

where the integral has to be calculated inside a sphere of radius R. (Let r_1, θ_1, ϕ_1 be the coordinates of P_1 and r, θ, ϕ those of an arbitrary point of the integration domain.) Let us assume that $r_1 \gg R$ and neglect in

ψ_1 terms of order $1/r^2$; in this way we can approximate the quantity $1/|\mathbf{r_1} - \mathbf{r}|$ by $1/r_1$ and Eq. (4.158) becomes

$$\psi_1(P_1) = \frac{\gamma^2 k}{2\pi v^2 r_1} \int_S \left(\frac{1}{r} - \frac{1}{R}\right) e^{i\gamma(|\mathbf{r_1} - \mathbf{r}| + z)} \, d\tau. \tag{4.159}$$

Note that we can neglect terms of order $1/r$ inside the integral; thus we can set

$$|\mathbf{r_1} - \mathbf{r}| \simeq r_1 - r\,(\cos\theta_1 \cos\theta + \sin\theta_1 \sin\theta \cos(\phi_1 - \phi)) \tag{4.160}$$

and, since $z = r\cos\theta$, the integral appearing in (4.159) becomes

$$e^{i\gamma r_1} \int_s \left(\frac{1}{r} - \frac{1}{R}\right) e^{i\gamma r\,((1-\cos\theta_1)\cos\theta - \sin\theta_1 \sin\theta \cos(\phi_1 - \phi))} \, d\tau. \tag{4.161}$$

We can set $\phi_1 = 0$ without restrictions, but the integral is not really simplified. It is useful to choose a different system of polar coordinates, using the outward bisector of the angle between r and r_1 as the polar axis. The plane containing the z axis and the point P_1 is the meridian plane ($\Phi = 0$); the polar coordinates in this plane will be

$$r_1, \quad \Theta_1 = \frac{\pi}{2} - \frac{\theta_1}{2}, \quad \Phi_1 = 0, \tag{4.162}$$

and, moreover,

$$\cos\theta = -\sin(\theta_1/2)\cos\Theta + \cos(\theta_1/2)\cos\Phi\sin\Theta, \tag{4.163}$$

so that

$$z = -r\sin(\theta_1/2)\cos\Theta + r\cos(\theta_1/2)\cos\Phi\sin\Theta, \tag{4.164}$$

$$|\mathbf{r_1} - \mathbf{r}| \simeq r_1 - r\,(\sin(\theta_1/2)\cos\Theta + \cos(\theta_1/2)\cos\Phi\sin\Theta), \tag{4.165}$$

$$z + |\mathbf{r_1} - \mathbf{r}| \simeq r_1 - 2\sin(\theta_1/2)\,r\cos\Theta. \tag{4.166}$$

Substitution into the integral in Eq. (4.159) gives

$$e^{i\gamma r_1} \int_S \left(\frac{1}{r} - \frac{1}{R}\right) e^{-2i\gamma r \sin(\theta_1/2)\cos\theta} \, d\tau. \tag{4.167}$$

Using $d\tau = r^2 dr\, d\cos\Theta\, d\Phi$ and integrating over Φ, we get

$$\begin{aligned}
&2\pi\, e^{i\gamma r_1} \int_0^R r^2 \left(\frac{1}{r} - \frac{1}{R}\right) dr \int_{-1}^{1} e^{-2i\gamma r \sin(\theta_1/2)\cos\theta} \, d\cos\Theta \\
&= \frac{2\pi\, e^{i\gamma r_1}}{\gamma\sin(\theta_1/2)} \int_0^R r \left(\frac{1}{r} - \frac{1}{R}\right) \sin(2\gamma r \sin(\theta_1/2)) \, dr \\
&= \frac{2\pi\, e^{i\gamma r_1}}{\gamma\sin(\theta_1/2)} \left[\frac{1}{2\gamma\sin(\theta_1/2)} \left(1 - \frac{\sin(2\gamma R\sin(\theta_1/2))}{2\gamma R\sin(\theta_1/2)}\right)\right].
\end{aligned} \tag{4.168}$$

On taking the limit $R \to \infty$, the term with R in the denominator vanishes (assuming $\theta_1 \neq 0$); substituting in Eq. (4.159) and replacing θ_1 with θ, we then have

$$\psi_1(P_1) = \frac{k\, e^{i\gamma r_1}}{2\, v^2\, r_1\, \sin^2(\theta/2)}. \qquad (4.169)$$

The relevant quantity is the ratio

$$\frac{i_1}{i_0} = \frac{|\psi_1|^2}{|\psi_0|^2}$$

between the scattered wave and the incident wave. Using the expression (4.150) for ψ_0 and introducing the energy of the particle instead of its speed, we find

$$i_1 = \frac{Z^2\, Z'^2\, e^4\, i_0}{16\, W^2\, r_1^2\, \sin^4(\theta/2)}. \qquad (4.170)$$

This formula coincides with that in Eq. (4.143), considering that the cross section introduced in the previous section by definition is

$$f(\theta) = r_1^2\, \frac{i_1}{i_0}. \qquad (4.171)$$

11. THE LAPLACE EQUATION

Let us consider the differential equation

$$u'' + \left(\delta_0 + \frac{\delta_1}{r}\right) u' + \left(\epsilon_0 + \frac{\epsilon_1}{r}\right) u = 0 \qquad (4.172)$$

and apply the Laplace transformation

$$u = \int_L f(z)\, e^{zr}\, dz. \qquad (4.173)$$

We have

$$u' = \int_L z\, f(z)\, e^{zr}\, dz, \qquad (4.174)$$

$$u'' = \int_L z^2\, f(z)\, e^{zr}\, dz. \qquad (4.175)$$

On substituting into Eq.(4.172), we deduce

$$0 = \int_L \left[z^2 f(z) e^{zr} + \left(\delta_0 + \frac{\delta_1}{r}\right) z f(z) e^{zr} + \left(\epsilon_0 + \frac{\epsilon_1}{r}\right) f(z) e^{zr} \right] dz; \tag{4.176}$$

and, multiplying the previous equation by r and noting that

$$r e^{zr} = \frac{d}{dz} e^{zr},$$

we find

$$\begin{aligned} 0 &= \int_L \left[\delta_1 z f(z) e^{zr} + \epsilon_1 f(z) e^{zr} + z^2 f(z) \frac{d}{dz} e^{zr} \right. \\ &\quad \left. + \delta_0 z f(z) \frac{d}{dz} e^{zr} + \epsilon_0 f(z) \frac{d}{dz} e^{zr} \right] dz \\ &= \int_L \left[\delta_1 z f(z) + \epsilon_1 f(z) \right. \\ &\quad \left. - \frac{d}{dz} \left(z^2 f(z) + \delta_0 z f(z) + \epsilon_0 f(z) \right) \right] e^{zr} dz \\ &\quad + \int_L \frac{d}{dz} \left(z^2 f(z) e^{zr} + \delta_0 z f(z) e^{zr} + \epsilon_0 f(z) e^{zr} \right) dz. \end{aligned}$$

Let us choose an integration path in such a way that the quantity

$$\left(z^2 f(z) + \delta_0 z f(z) + \epsilon_0 f(z) \right) e^{zr} \tag{4.177}$$

takes the same value at the integration limits. In order to satisfy Eq. (4.173), the following differential equation must hold:

$$\delta_1 z f(z) + \epsilon_1 f(z) - \frac{d}{dz} \left(z^2 f(z) + \delta_0 z f(z) + \epsilon_0 f(z) \right) = 0. \tag{4.178}$$

Equation (4.178) can be cast in the form

$$\frac{f'(z)}{f(z)} = \frac{(\delta_1 - 2)z + \epsilon_1 - \delta_0}{z^2 + \delta_0 z + \epsilon_0} = \frac{\beta_1}{z - c_1} + \frac{\beta_2}{z - c_2}, \tag{4.179}$$

where c_1 and c_2 are the roots of the equation

$$z^2 + \delta_0 z + \epsilon_0 = 0. \tag{4.180}$$

From Eq. (4.179) it follows that

$$\beta_1 + \beta_2 = \delta_1 - 2, \tag{4.181}$$

$$\beta_1 c_2 + \beta_2 c_1 = \delta_0 - \epsilon_1, \tag{4.182}$$

from which we get

$$\begin{aligned}\beta_1 &= \frac{c_1\delta_1 - 2c_1 - \delta_0 + \epsilon_1}{c_1 - c_2},\\ \beta_2 &= \frac{c_2\delta_1 - 2c_2 - \delta_0 + \epsilon_1}{c_2 - c_1},\end{aligned} \tag{4.183}$$

or, noting that $\delta_0 = -(c_1 + c_2)$ from (4.180),

$$\beta_1 = \frac{\epsilon_1 + \delta_1 c_1}{c_1 - c_2} - 1, \quad \beta_2 = \frac{\epsilon_1 + \delta_1 c_2}{c_2 - c_1} - 1. \tag{4.184}$$

On setting, for simplicity

$$\beta_1 = \alpha_1 - 1, \quad \beta_2 = \alpha_2 - 1, \tag{4.185}$$

we find

$$\beta_1 = \frac{\epsilon_1 + \delta_1 c_1}{c_1 - c_2}, \quad \beta_2 = \frac{\epsilon_1 + \delta_1 c_2}{c_2 - c_1}. \tag{4.186}$$

On integrating Eq. (4.179), we obtain

$$f(z) = (z - c_1)^{\alpha_1 - 1} (z - c_2)^{\alpha_2 - 1}. \tag{4.187}$$

Then the integral representation (4.173) takes the form

$$u = \int_L \mathrm{e}^{zr} (z - c_1)^{\alpha_1 - 1} (z - c_2)^{\alpha_2 - 1} \, \mathrm{d}z, \tag{4.188}$$

and the condition that must be satisfied by the integration path L so that the quantity in Eq. (4.177) assumes the same value at both ends of the integration interval, can be cast in the simple form

$$\int_L \frac{\mathrm{d}}{\mathrm{d}z} \left(\mathrm{e}^{zr} (z - c_1)^{\alpha_1} (z - c_2)^{\alpha_2} \right) \mathrm{d}z = 0. \tag{4.189}$$

12. POLARIZATION FORCES BETWEEN HYDROGEN ATOMS

In the following we adopt the usual electronic system of units, with $\hbar = e = m = 1$ and the energy unit $e^2/a_0 = 2Ry$. Let us consider two hydrogen atoms placed at a large distance R apart (R is normalized to the Bohr radius). Since the eigenfunctions of the two atoms fall off exponentially with distance, we can assume that the two atoms are perfectly separated and that their size is small compared with R. These assumptions are quite reasonable, given that we are interested in studying the interaction between these atoms and this declines as a finite negative power of R. In particular, there is no need for a distinction between the symmetric (with respect to the two electrons) and the antisymmetric solutions (given their mass, we consider fixed protons). In fact, the separation between ortho-states and para-states decreases exponentially with R. Since the atoms are neutral, in first approximation the interaction is zero; we then calculate the second approximation using the Ritz method. The unperturbed wavefunction of the two electrons is, apart from a normalization factor,

$$\psi_0 = \mathrm{e}^{-(r_1 + r_2)}, \tag{4.190}$$

where r_1 is the distance of the first electron from the first nucleus, and r_2 is that of the second electron from the second nucleus. Thus ψ_0 is the product of the eigenfunctions of the two electrons; this is because, in this limit, no resonance occurs. Note that the correct eigenfunctions are obtained from ψ_0 by interchanging r_1 and r_2 and taking the sum (symmetric eigenfunction) or the difference (antisymmetric eigenfunction) of the two expressions. In our units, the unperturbed Hamiltonian is given by

$$H = -\frac{1}{2}\nabla_1^2 - \frac{1}{2}\nabla_2^2 - \frac{1}{r_1} - \frac{1}{r_2}. \tag{4.191}$$

The perturbation arising from the coexistence of two atoms comes from their dipole interactions and, for large R, is

$$\delta H = -\frac{2x_1x_2 - y_1y_2 - z_1z_2}{R^3}, \tag{4.192}$$

where the meaning of each quantity is obvious. We will determine the perturbed eigenfunction by setting

$$\psi = \psi_0 + c\,\delta H\,\psi_0, \tag{4.193}$$

where c is a constant to be determined. To this end, let us consider the mean energy

$$W = \int \psi\,(H + \delta H)\,\psi\,\mathrm{d}\tau \Big/ \int \psi^2\,\mathrm{d}\tau. \tag{4.194}$$

On noting that $H\psi_0 = -\psi_0$ and using

$$\int \psi_0^2\,\delta H\,\mathrm{d}\tau = 0, \tag{4.195}$$

$$\int (\delta H)^2 \, \psi_0^2 \, \mathrm{d}\tau \;=\; \frac{6}{R^6} \int \psi_0^2 \, \mathrm{d}\tau, \tag{4.196}$$

$$\int (\delta H)^3 \, \psi_0^2 \, \mathrm{d}\tau \;=\; 0, \tag{4.197}$$

$$\int \psi_0 \, (\delta H) \; H \; (\delta H) \, \psi_0 \, \mathrm{d}\tau \;=\; - \int (\delta H)^2 \, \psi_0^2 \, \mathrm{d}\tau$$

$$+ \int \psi_0 \, (\delta H) \, (H \, (\delta H) \; - \; (\delta H) \, H) \, \psi_0 \, \mathrm{d}\tau \;=\; 0, \tag{4.198}$$

by simple calculations, it follows that

$$W \;=\; - \frac{1 \, - \, 12c/R^6}{1 \, + 6c^2/R^6}. \tag{4.199}$$

In our limit $R \to \infty$ we thus have

$$W \;=\; -1 + \frac{12}{R^6} \, c + \frac{6}{R^6} \, c^2. \tag{4.200}$$

From the condition $\mathrm{d}W/\mathrm{d}c \;=\; 0$, we deduce that $c = -1$, so that

$$W \;=\; -1 \, - \, 6/R^6. \tag{4.201}$$

In regular units our method then yields the expression $-6e^2/a_0R^6$ for the potential energy of the polarization forces; this is quite a good result compared to the exact expression $(-6.47e^2/a_0R^6)$ deduced by Landau with a different method. Note that, according to a known theorem on the minimization of the energy of a system in its ground state, our method gives a value exceeding the exact value. On setting $c = -1$ in Eq. (4.193), the perturbed eigenfunction is approximately given by

$$\psi \;=\; \mathrm{e}^{-(r_1 \, + \, r_2)} \, + \, (1/R^3)(2x_1x_2 \, - \, y_1y_2 \, - z_1z_2) \, \mathrm{e}^{-(r_1 \, + \, r_2)}. \tag{4.202}$$

Obviously, this approximation is worse than that for the eigenvalue.

13. INTEGRAL REPRESENTATION OF THE BESSEL FUNCTIONS

The differential equation obeyed by the Bessel functions,

$$y'' \, + \, \frac{1}{x} \, y' \, + \, \left(1 \, - \, \frac{\lambda^2}{x^2}\right) \, y \;=\; 0, \tag{4.203}$$

can be simplified by setting

$$y = x^{\lambda} u. \tag{4.204}$$

Indeed, substituting in Eq. (4.203) and dividing by x^{λ}, we obtain

$$u'' + \frac{2\lambda + 1}{x} u' + u = 0. \tag{4.205}$$

This is a particular case of the Laplace equation (4.172) with $\delta_0 = 0$, $\delta_1 = 2\lambda + 1$, $\epsilon_0 = 1$, and $\epsilon_1 = 0$. We can thus use the expansion (4.188), with constants given by [see Eqs. (4.180) and (4.186)]:

$$c_1 = i, \quad c_2 = -i, \quad \alpha_1 = \alpha_2 = \frac{2\lambda + 1}{2}. \tag{4.206}$$

Introducing an arbitrary multiplicative constant, we write

$$u = k \int_L e^{zx} \left(z^2 + 1\right)^{\lambda - 1/2} dz, \tag{4.207}$$

subject to the condition

$$\left[e^{zx} \left(z^2 + 1\right)^{\lambda + 1/2} \right]_A^B = 0, \tag{4.208}$$

where A and B are the integration limits. The points $+i$ and $-i$ are bifurcation points of the integrand function; on setting $z = it$, they assume the values ± 1 and (4.207) and (4.208) become

$$u = k \int_C e^{itx} \left(t^2 - 1\right)^{\lambda - 1/2} dt, \tag{4.209}$$

$$\left[e^{itx} \left(t^2 - 1\right)^{\lambda + 1/2} \right]_C = 0. \tag{4.210}$$

In order to define the quantity $\left(t^2 - 1\right)^{\lambda+1/2}$ in the complex plane, we must give a univocal definition of $\log(t^2 - 1)$.

To this end, let us divide the complex plane with two half-lines starting from the bifurcation points ± 1, and extending parallel to the positive imaginary axis, and define $\log(t^2 - 1)$ as positive (real) for $t > 1$, while in the other cases it takes the values obtained by requiring continuity without crossing the bifurcation lines. Then, let us define the Hankel functions H_{λ}^1:

$$H_{\lambda}^1 = \frac{\Gamma(1/2 - \lambda)\,(1/2x)^{\lambda}}{\pi\, i\, \Gamma(1/2)} \int e^{itx} \left(t^2 - 1\right)^{\lambda - 1/2} dt. \tag{4.211}$$

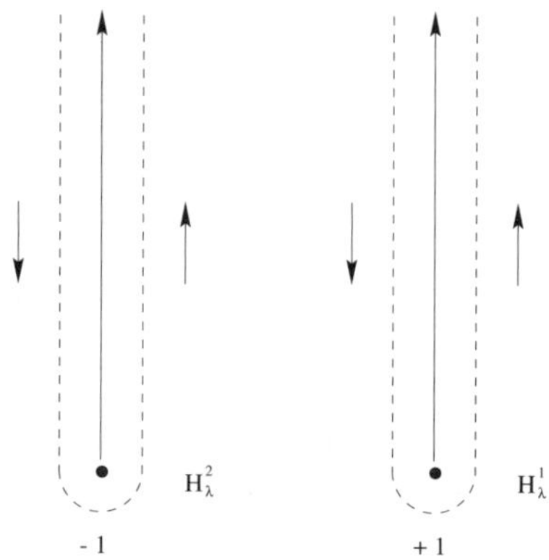

Fig. 4.1. The paths in the complex plane used to define the Hankel functions (see text).

Since this function satisfies the condition in (4.210), H^1_λ will be a solution of Eq. (4.203). Similarly, we define the function H^2_λ on the path starting from -1. For real x we have $H^1_\lambda = H^{1*}_\lambda$ and, in general, as can be deduced from the behavior for $x \to 0$:

$$I_\lambda = \frac{1}{2}(H^1_\lambda + H^2_\lambda), \quad N_\lambda = \frac{1}{2i}(H^1_\lambda - H^2_\lambda), \tag{4.212}$$

where I_λ and N_λ are the Bessel and Neumann functions, respectively. For real x, it follows that

$$H^1_\lambda(x) = I_\lambda(x) + i\,N_\lambda(x); \tag{4.213}$$

thus I_λ and N_λ are two real solutions of Eq. (4.203), the first one being regular for $x = 0$, while the sum of the squares of the two is regular for $x \to \infty$.

We now calculate the asymptotic behavior of $H^1_\lambda(x)$ for $x \to \infty$ (real x). Let us set

$$t = 1 + i\,\frac{s}{x}, \tag{4.214}$$

$$t^2 - 1 = 2i\,\frac{s}{x} - \frac{s^2}{x^2}, \tag{4.215}$$

where s goes from ∞ to 0 and then from 0 to ∞. Given the above conventions, in the first part of the path the quantity $\log(t^2-1)$ will take its principal value decreased by 2π, while in the second part it will assume its principal value (that is, the absolute value of imaginary part lower than π). After obvious manipulations, one finds

$$H^1_\lambda(x) = \sqrt{\frac{2}{\pi x}}\,\frac{\exp\left\{i\,(x - \lambda\pi/2 - \pi/4)\right\}}{\Gamma\,(\lambda + 1/2)}$$

$$\times \int_0^\infty e^{-s} s^{\lambda - 1/2} \left(1 + \frac{is}{2x}\right)^{\lambda - 1/2} ds. \quad (4.216)$$

For $x \to \infty$, we obtain the asymptotic behavior

$$H^1_\lambda(x) \sim \sqrt{\frac{2}{\pi x}} \exp\{i(x - \lambda\pi/2 - \pi/4)\}. \quad (4.217)$$

14. CUBIC SYMMETRY

The group of the 24 (proper) rotations that transform the x, y, z axes into themselves (except for the order and the direction) is holomorphic to the group of permutations of 4 objects. The holomorphic correspondence can be established in the following way:

I - Identity (1+)

Direction cosines of the rotation	*Rotation angle*	*Permutation*
	0	identity

II - Class 01 (6−)

Direction cosines of the rotation			*Rotation angle*	*Permutation*
0	$1/\sqrt{2}$	$1/\sqrt{2}$	180°	(14)
$1/\sqrt{2}$	0	$1/\sqrt{2}$	180°	(24)
$1/\sqrt{2}$	$1/\sqrt{2}$	0	180°	(34)
0	$1/\sqrt{2}$	$-1/\sqrt{2}$	180°	(23)
$-1/\sqrt{2}$	0	$1/\sqrt{2}$	180°	(31)
$1/\sqrt{2}$	$-1/\sqrt{2}$	0	180°	(12)

III - Class 02 (3+)

Direction cosines of the rotation			*Rotation angle*	*Permutation*
1	0	0	180°	(14) (23)
0	1	0	180°	(24) (31)
0	0	1	180°	(34) (12)

IV - Class 101 (8+)

Direction cosines of the rotation	*Rotation angle*	*Permutation*
$1/\sqrt{3}$ $1/\sqrt{3}$ $1/\sqrt{3}$	120°	(123)
$1/\sqrt{3}$ $1/\sqrt{3}$ $1/\sqrt{3}$	240°	(321)
		(234)
		(314)
		(124)
		(324)
		(134)
		(214)

V - Class 0001 (6−)

Direction cosines of the rotation	*Rotation angle*	*Permutation*
		(1234)
		(2314)
		(3124)
		(3214)
		(1324)
		(2134)

Since the considered group is equivalent to the group of permutations of 4 objects, it has 5 irreducible representations χ_s $(s = 1, 2, 3, 4, 5)$, whose characters are given in Sec. 4.7 $(f = 4)$. An irreducible representation $\mathcal{D}_j$ (with integer j) of the complete group of spatial rotations is also a (reducible) representation of the considered group. If n_s is the mean value of $\chi_j{\cdot}\chi_s^*$ over the elements of this group, the above-mentioned representation can be reduced to n_s representations χ_s. The characters $\mathcal{D}_j$ are given by $\sin(2j+1)\omega/\sin\omega$, where $\omega = \alpha/2$ is half of the rotation angle (see Sec. 3.20). Then, the values of χ_j for the 5 classes of the group under consideration are, respectively,

$$2j+1, \quad (-1)^j, \quad (-1)^j;$$

$$1 - \text{rest of } \frac{j}{3}, \quad 1 + \text{rest of } \frac{j}{2} - \text{rest of } \frac{j}{4}.$$

We can now evaluate the frequency n_s of each irreducible representation; keeping the same order as in the table of Sec. 4.7, for $f = 4$, we find

$$\begin{aligned} n_1 \;&=\; \frac{j}{12} + 1 - \frac{1}{2}\left(\text{rest of } \frac{j}{2}\right) - \frac{1}{3}\left(\text{rest of } \frac{j}{3}\right) \\ &\quad - \frac{1}{4}\left(\text{rest of } \frac{j}{4}\right), \end{aligned} \tag{4.218}$$

$$
\begin{aligned}
n_2 &= \frac{j}{4} - \frac{1}{2}\left(\text{rest of } \frac{j}{2}\right) + \frac{1}{4}\left(\text{rest of } \frac{j}{4}\right), && (4.219)\\
n_3 &= \frac{j}{6} - \frac{1}{2}\left(\text{rest of } \frac{j}{2}\right) + \frac{1}{3}\left(\text{rest of } \frac{j}{3}\right), && (4.220)\\
n_4 &= \frac{j}{4} + \left(\text{rest of } \frac{j}{2}\right) - \frac{1}{4}\left(\text{rest of } \frac{j}{4}\right), && (4.221)\\
n_5 &= \frac{j}{12} - \frac{1}{3}\left(\text{rest of } \frac{j}{3}\right) + \frac{1}{4}\left(\text{rest of } \frac{j}{4}\right). && (4.222)
\end{aligned}
$$

Noting that the degrees of the irreducible representations are 1,3,2,3,1, respectively, we obviously have

$$n_1 + 3\,n_2 + 2\,n_3 + 3\,n_4 + n_5 = 2\,j + 1. \tag{4.223}$$

We observe that, as in the normal representation, for large j the frequencies of appearance of the irreducible representations are proportional to their degrees. Moreover, if the values of n_s for a given value of j are known, we can obtain the values corresponding to $j + 12q$ from the following scheme:

$$
\begin{array}{lcl}
j' & = & j + 12\,q \\
\hline
n_1' & = & n_1 + 1\cdot q \\
n_2' & = & n_2 + 3\cdot q \\
n_3' & = & n_3 + 2\cdot q \\
n_4' & = & n_4 + 3\cdot q \\
n_5' & = & n_5 + 1\cdot q
\end{array}
$$

where the coefficients of q are exactly the degrees of the irreducible representations. It is then sufficient to evaluate only the values of n_s from $j = 0$ to $j = 11$. We summarize the results in the following table.

j	n_1	n_2	n_3	n_4	n_5
0	1	0	0	0	0
1	0	0	0	1	0
2	0	1	1	0	0
3	0	1	0	1	1
4	1	1	1	1	0
5	0	1	1	2	0
6	1	2	1	1	1
7	0	2	1	2	1
8	1	2	2	2	0
9	1	2	1	3	1
10	1	3	2	2	1
11	0	3	2	1	1
12	1+1	0+3	0+2	0+3	0+1
13	0+1	0+3	0+2	1+3	0+1
...	...	...	...	...	...

15. FORMULAE

(1) Volume and surface area of an n-dimensional sphere of radius R:

$$\begin{aligned} V_n &= \frac{\pi^{n/2}}{(n/2)!}\, R^n, \\ S_n &= \frac{n}{R}\, V_n = n\,\frac{\pi^{n/2}}{(n/2)!}\, R^{n-1} = \frac{2\pi}{R}\, V_{n-2}. \end{aligned} \tag{4.224}$$

n	V_n/R^n	V_n/V_{n-1}	S_n/R_{n-1}	S_n/S_{n-1}
1	2		2	
2	π	$\frac{1}{2}\pi$	2π	π
3	$\frac{4}{3}\pi$	$\frac{4}{3}$	4π	2
4	$\frac{1}{2}\pi^2$	$\frac{3}{8}\pi$	$2\pi^2$	$\frac{1}{2}\pi$
5	$\frac{8}{15}\pi^2$	$\frac{16}{15}$	$\frac{8}{3}\pi^2$	$\frac{4}{3}$
6	$\frac{1}{6}\pi^3$	$\frac{5}{16}\pi$	π^3	$\frac{3}{8}\pi$

$$V_n = \frac{R}{2\pi}\, S_{n+2}, \quad S_n = \frac{2\pi}{R}\, V_{n-2}; \tag{4.225}$$

$$V_n = \frac{R}{n}\, S_n, \quad S_n = \frac{n}{R}\, V_n; \tag{4.226}$$

$$V_n = \frac{2\pi}{n}\, V_{n-2}, \quad S_n = \frac{2\pi}{n-2}\, S_{n-2}. \tag{4.227}$$

(2) Let a and b be integer or half-integer positive semi-definite numbers and c one of the following numbers:

$$c = a+b, \quad a+b-1, \quad \ldots, \quad |a-b|. \tag{4.228}$$

The following identity holds:

$$\sum_{s=0}^{a+b-c} \frac{(c+a-b+s)!\,(2b-s)!}{s!\,(a+b-c-s)!} = \frac{(a+b+c+1)!\,(c+a-b)!\,(c+b-a)!}{(2c+1)!\,(a+b-c)!}. \qquad (4.229)$$

The r.h.s. of Eq. (4.229) - as well as the l.h.s - is symmetric under the change $a \to b$. This can be tested by changing $s \to a+b-c-s$. For simplicity let us denote the r.h.s. of Eq. (4.229) by $f(a,b,c)$ and assume that the identity is true for a given value a (and values lower than a); then we shall prove that it also holds for $a+1/2$. Indeed, we have

$$\begin{aligned} & f\,(a+1/2,\ b,\ c+1/2) \\ &= \sum_{s=0}^{a+b-c} (c+a-b+s+1)\,\frac{(c+a-b+s)!\,(2b-s)!}{s!\,(a+b-c-s)!} \\ &= (c+a-b+s)\ f(a,b,c) \\ &\quad + \sum_{s=0}^{a+b-c-1} \frac{(c+a-b+1+s)!\,(2b-1-s)!}{s!\,(a+b-1-1-s)!} \\ &= (c+a-b+s)\ f(a,b,c)\ +\ f\,(a,\ b-1/2,\ c+1/2)\,. \end{aligned} \qquad (4.230)$$

On substituting the l.h.s. of Eq. (4.229) into Eq. (4.230), we immediately see that the identity is true also for $f(a+1/2,\ b,\ c+1/2)$. If the condition (4.228) is not satisfied by one of the terms of its l.h.s., we can set $f(a,b,c)=0$, and thus (4.230) is again true. To complete the demonstration of the identity (4.229) it is sufficient to prove that it holds also for $a=0$ and, as a consequence, $b=c$. In such a case, the sum in Eq. (4.229) reduces to $(2c)!$, which is also the value of the l.h.s.

16. PLANE WAVES IN THE DIRAC THEORY

The first and second pair of components of the Dirac wavefunction ψ are relativistic invariants. By choosing the components in such a way that the equation

$$\left(\frac{W}{c} + \rho_3\,(\boldsymbol{\sigma}\cdot\mathbf{p}) + \rho_1\,m\,c\right)\psi = 0 \qquad (4.231)$$

is satisfied, for given values of p_x, p_y, p_z we have two positive waves (i.e., $W/c = \sqrt{p^2 + m^2c^2}$) and two negative waves (i.e., $W/c = -\sqrt{p^2 + m^2c^2}$). At a given point we can set

$$\begin{aligned} \psi' &= 1{\cdot}\psi_1 + 0{\cdot}\psi_2 - \frac{1}{mc}\left(\frac{W}{c} + p_z\right)\psi_3 - \frac{1}{mc}\left(p_x + ip_y\right)\psi_4, \\ \psi'' &= \frac{1}{mc}\left(p_x - ip_y\right)\psi_1 - \frac{1}{mc}\left(\frac{W}{c} + p_z\right)\psi_2 + 0{\cdot}\psi_3 + 1{\cdot}\psi_4, \end{aligned} \tag{4.232}$$

so that positive waves occur for $W > 0$ and negative waves for $W < 0$. The waves are orthogonal to each other and, furthermore, the transition current between the positive waves or between the negative ones vanishes.

Let us now choose the components of ψ in such a way that the original Dirac equation holds instead of Eq. (4.231) [10]:

$$\left(\frac{W}{c} + \rho_1\,(\boldsymbol{\sigma}{\cdot}\mathbf{p}) + \rho_3\, m\, c\right)\psi = 0 \tag{4.233}$$

(ψ_1 and ψ_2 are the small components for small velocities, while ψ_3 and ψ_4 are the large ones). At a given spatial point and instant of time, we can set

$$\begin{aligned} \psi' &= -\frac{p_z}{W/c + mc}\psi_1 - \frac{p_x + ip_y}{W/c + mc}\psi_2 + 1{\cdot}\psi_3 + 0{\cdot}\psi_4, \\ \psi'' &= -\frac{p_x - ip_y}{W/c + mc}\psi_1 + \frac{p_z}{W/c + mc}\psi_2 + 0{\cdot}\psi_3 + 1{\cdot}\psi_4; \end{aligned} \tag{4.234}$$

and for $W/c = \pm\sqrt{p^2 + m^2c^2}$ we have positive and negative waves, respectively.

On setting

$$\begin{aligned} \phi_1 &= (1\,,0\,,0\,,0), & \phi_2 &= (0\,,1\,,0\,,0), \\ \phi_3 &= (0\,,0\,,1\,,0), & \phi_4 &= (0\,,0\,,0\,,1) \end{aligned}$$

in the representation associated with Eq. (4.231), so that

$$\psi = (\psi_1,\, \psi_2,\, \psi_3,\, \psi_4) = \psi_1\,\phi_1 + \psi_2\,\phi_2 + \psi_3\,\phi_3 + \psi_4\,\phi_4,$$

and, in the same way,

$$\tilde{\psi} = \left(\tilde{\psi}_1,\, \tilde{\psi}_2,\, \tilde{\psi}_3,\, \tilde{\psi}_4\right) = \tilde{\psi}_1\,\tilde{\phi}_1 + \tilde{\psi}_2\,\tilde{\phi}_2 + \tilde{\psi}_3\,\tilde{\phi}_3 + \tilde{\psi}_4\,\tilde{\phi}_4,$$

[10] For $p = 0$, the first pair of components of ψ represents the negative states, the second pair the positive states

in the representation of Eq. (4.233), the relations between the eigenfunctions ϕ and $\tilde{\phi}$ will read (except for a phase factor) as follows:

$$\begin{aligned}
\tilde{\phi}_1 &= \frac{1}{\sqrt{2}}(\phi_1+\phi_3), & \phi_1 &= \frac{1}{\sqrt{2}}(\tilde{\phi}_1+\tilde{\phi}_3),\\
\tilde{\phi}_2 &= \frac{1}{\sqrt{2}}(\phi_2+\phi_4), & \phi_2 &= \frac{1}{\sqrt{2}}(\tilde{\phi}_2+\tilde{\phi}_4),\\
\tilde{\phi}_3 &= \frac{1}{\sqrt{2}}(\phi_1-\phi_3), & \phi_3 &= \frac{1}{\sqrt{2}}(\tilde{\phi}_1-\tilde{\phi}_3),\\
\tilde{\phi}_4 &= \frac{1}{\sqrt{2}}(\phi_2-\phi_4), & \phi_4 &= \frac{1}{\sqrt{2}}(\tilde{\phi}_2-\tilde{\phi}_4).
\end{aligned} \tag{4.235}$$

We then have $\tilde{\psi} = \epsilon\psi$ and $\psi = \epsilon^{-1}\tilde{\psi} = \epsilon\tilde{\psi}$, with

$$\epsilon = \begin{pmatrix} 1/\sqrt{2} & 0 & 1/\sqrt{2} & 0 \\ 0 & 1/\sqrt{2} & 0 & 1/\sqrt{2} \\ 1/\sqrt{2} & 0 & -1/\sqrt{2} & 0 \\ 0 & 1/\sqrt{2} & 0 & -1/\sqrt{2} \end{pmatrix} = \epsilon^{-1} = \frac{1}{\sqrt{2}}\left(\rho_1 + \rho_3\right). \tag{4.236}$$

Given the values $(p_x, p_y, p_z) = \mathbf{p}$, let us denote with y_p^1 and y_p^2 the positive plane waves in Eqs. (4.232) and with y_p^3 and y_p^4 the negative ones obtained by replacing W with $-W$; analogously, we denote by z_p^1 and z_p^2 the positive plane waves in Eqs. (4.234) and by z_p^3 and z_p^4 the corresponding negative ones. All these functions are normalized so that $\psi^*\psi = 1$. At a given time, the relations between y and z are the following:

$$z_p^* = \sum_i S_{ik}\, y_p^i, \tag{4.237}$$

$$y_p^* = \sum_i S_{ik}^{-1}\, z_p^i, \tag{4.238}$$

where

$$S = S^{-1} = \begin{pmatrix} \dfrac{A + \dfrac{p_z}{mc}}{\sqrt{2AB}} & \dfrac{\dfrac{p_x + ip_y}{mc}}{\sqrt{2AB}} & 0 & 0 \\ \dfrac{\dfrac{p_x - ip_y}{mc}}{\sqrt{2AB}} & -\dfrac{A + \dfrac{p_z}{mc}}{\sqrt{2AB}} & 0 & 0 \\ 0 & 0 & -\dfrac{A' - \dfrac{p_z}{mc}}{\sqrt{2A'B'}} & \dfrac{\dfrac{p_x + ip_y}{mc}}{\sqrt{2A'B'}} \\ 0 & 0 & \dfrac{\dfrac{p_x - ip_y}{mc}}{\sqrt{2A'B'}} & \dfrac{A' - \dfrac{p_z}{mc}}{\sqrt{2A'B'}} \end{pmatrix}, \tag{4.239}$$

with

$$\begin{aligned} A &= \frac{\sqrt{p^2 + m^2c^2} + mc}{mc}, \quad A' = \frac{\sqrt{p^2 + m^2c^2} - mc}{mc}, \\ B &= \frac{\sqrt{p^2 + m^2c^2} + p_z}{mc}, \quad B' = \frac{\sqrt{p^2 + m^2c^2} - p_z}{mc}. \end{aligned} \tag{4.240}$$

(It follows that $A + A' = B + B'$. For $p = 0$ we have $A = 2$, $A' = 0$, $B = 1$, $B' = 1$.)

In the following, R1 and R2 will denote the representations in which Eqs. (4.231) and (4.233) respectively hold. Note that the matrices σ_x, σ_y, σ_z obviously describe the same operators S_x, S_y, S_z both in R1 and in R2, due to the property

$$\sigma \, \frac{\rho_1 + \rho_3}{\sqrt{2}} = \frac{\rho_1 + \rho_3}{\sqrt{2}} \, \sigma. \tag{4.241}$$

By contrast, let us consider an operator γ, which is described by ρ_3 in R1 and by ρ_1 in R2, and another operator γ_1 which is instead described by ρ_1 in R1 and by ρ_3 in R2. Then the Dirac equation can be written as follows in both representations:

$$\left(\frac{W}{c} + \gamma \, (\boldsymbol{\xi} \cdot \mathbf{p}) + \gamma_1 \, m \, c \right) \psi = 0. \tag{4.242}$$

The operators ξ and γ transform any combination of 4 plane waves corresponding to a given value of p into another such combination. Obviously, the representation of the matrices describing these operators depends on whether we choose as orthogonal unitary vectors the normalized plane waves in Eq. (4.232) (i.e., the y_p^i) or those in Eq. (4.234) (i.e., the z_p^i). The matrices corresponding to the second case can be obtained from those corresponding to the first case by means of a transformation involving S [note that this is the matrix in (4.239) and not the spin $\mathbf{S} = (S_x, S_y, S_z)$].

In the first case (plane waves y_p^i) we have

$$\text{(i)}\quad S_z = \begin{pmatrix} s_z^a & s_z^b \\ s_z^{b\,\dagger} & s_z^c \end{pmatrix}, \quad S_x = \begin{pmatrix} s_x^a & s_x^b \\ s_x^{b\,\dagger} & s_x^c \end{pmatrix}, \quad S_y = \begin{pmatrix} s_y^a & s_y^b \\ s_y^{b\,\dagger} & s_y^c \end{pmatrix}, \tag{4.243}$$

where the sub-matrices are given by $[(a_{ij})^\dagger = (a_{ji})^*]$

$$s_z^a = \begin{pmatrix} \dfrac{2+B^2-BB'}{B(B+B')} & 2\dfrac{\frac{p_x - ip_y}{mc}}{B(B+B')} \\ 2\dfrac{\frac{p_x + ip_y}{mc}}{B(B+B')} & \dfrac{2+B^2-BB'}{B(B+B')} \end{pmatrix},$$

$$s_z^b = \begin{pmatrix} -2\dfrac{BB'-1}{(B+B')\sqrt{BB'}} & 2\dfrac{\frac{p_x - ip_y}{mc}}{(B+B')\sqrt{BB'}} \\ 2\dfrac{\frac{p_x + ip_y}{mc}}{(B+B')\sqrt{BB'}} & 2\dfrac{BB'-1}{(B+B')\sqrt{BB'}} \end{pmatrix},$$

$$s_z^c = \begin{pmatrix} \dfrac{2+B'^2-BB'}{B'(B+B')} & 2\dfrac{\frac{p_x - ip_y}{mc}}{B'(B+B')} \\ 2\dfrac{\frac{p_x + ip_y}{mc}}{B'(B+B')} & \dfrac{2+B'^2-BB'}{B'(B+B')} \end{pmatrix};$$

$$s_x^a = \begin{pmatrix} 2\dfrac{\frac{p_x}{mc}}{B+B'} & -\dfrac{2}{B+B'} \\ -\dfrac{2}{B+B'} & -2\dfrac{\frac{p_x}{mc}}{B+B'} \end{pmatrix},$$

$$s_x^b = \begin{pmatrix} \dfrac{2\dfrac{p_z p_x}{m^2c^2}}{(B+B')\sqrt{BB'}} + \dfrac{i\dfrac{p_y}{mc}}{\sqrt{BB'}} & -\dfrac{2\dfrac{p_z}{mc}}{(B+B')\sqrt{BB'}} \\ -\dfrac{2\dfrac{p_z}{mc}}{(B+B')\sqrt{BB'}} & \dfrac{2\dfrac{p_z p_x}{m^2c^2}}{(B+B')\sqrt{BB'}} + \dfrac{i\dfrac{p_y}{mc}}{\sqrt{BB'}} \end{pmatrix},$$

$$s_x^c = \begin{pmatrix} -2\dfrac{\dfrac{p_x}{mc}}{B+B'} & \dfrac{2}{B+B'} \\ \dfrac{2}{B+B'} & 2\dfrac{\dfrac{p_x}{mc}}{B+B'} \end{pmatrix};$$

$$s_y^a = \begin{pmatrix} 2\dfrac{\dfrac{p_y}{mc}}{B+B'} & i\dfrac{2}{B+B'} \\ -i\dfrac{2}{B+B'} & -2\dfrac{\dfrac{p_y}{mc}}{B+B'} \end{pmatrix},$$

$$s_y^b = \begin{pmatrix} \dfrac{2\dfrac{p_z p_y}{m^2c^2}}{(B+B')\sqrt{BB'}} - \dfrac{i\dfrac{p_x}{mc}}{\sqrt{BB'}} & i\dfrac{2\dfrac{p_z}{mc}}{(B+B')\sqrt{BB'}} \\ -i\dfrac{2\dfrac{p_z}{mc}}{(B+B')\sqrt{BB'}} & -\dfrac{2\dfrac{p_z p_y}{m^2c^2}}{(B+B')\sqrt{BB'}} - \dfrac{i\dfrac{p_x}{mc}}{\sqrt{BB'}} \end{pmatrix},$$

$$s_y^c = \begin{pmatrix} -2\dfrac{\dfrac{p_y}{mc}}{B+B'} & -i\dfrac{2}{B+B'} \\ i\dfrac{2}{B+B'} & 2\dfrac{\dfrac{p_y}{mc}}{B+B'} \end{pmatrix};$$

(ii) $$\gamma = \begin{pmatrix} \gamma^a & \gamma^b \\ \gamma^{b\,\dagger} & \gamma^c \end{pmatrix}, \quad \gamma_1 = \begin{pmatrix} \gamma_1^a & \gamma_1^b \\ \gamma_1^{b\,\dagger} & \gamma_1^c \end{pmatrix}, \qquad (4.244)$$

with

$$
\begin{aligned}
\gamma^a &= \begin{pmatrix} \dfrac{2}{B(B+B')} - 1 & 2\dfrac{\dfrac{p_x - ip_y}{mc}}{B(B+B')} \\ 2\dfrac{\dfrac{p_x + ip_y}{mc}}{B(B+B')} & 1 - \dfrac{2}{B(B+B')} \end{pmatrix}, \\
\gamma^b &= \begin{pmatrix} \dfrac{2}{(B+B')\sqrt{BB'}} & 2\dfrac{\dfrac{p_x - ip_y}{mc}}{(B+B')\sqrt{BB'}} \\ 2\dfrac{\dfrac{p_x + ip_y}{mc}}{(B+B')\sqrt{BB'}} & -\dfrac{2}{(B+B')\sqrt{BB'}} \end{pmatrix}, \\
\gamma^c &= \begin{pmatrix} \dfrac{2}{B'(B+B')} - 1 & 2\dfrac{\dfrac{p_x - ip_y}{mc}}{B'(B+B')} \\ 2\dfrac{\dfrac{p_x + ip_y}{mc}}{B(B+B')} & 1 - \dfrac{2}{B'(B+B')} \end{pmatrix}; \\
\gamma_1^a &= \begin{pmatrix} -\dfrac{2}{B+B'} & 0 \\ 0 & -\dfrac{2}{B+B'} \end{pmatrix}, \\
\gamma_1^b &= \begin{pmatrix} -\dfrac{2\dfrac{p_z}{mc}}{(B+B')\sqrt{BB'}} & -\dfrac{\dfrac{p_x - ip_y}{mc}}{\sqrt{BB'}}\,\dfrac{\dfrac{p_x + ip_y}{mc}}{\sqrt{BB'}} \\ -\dfrac{2\dfrac{p_z}{mc}}{(B+B')\sqrt{BB'}} & \end{pmatrix}, \\
\gamma_1^c &= \begin{pmatrix} \dfrac{2}{B+B'} & 0 \\ 0 & \dfrac{2}{B+B'} \end{pmatrix};
\end{aligned}
$$

(iii)
$$\gamma\, S_z = \begin{pmatrix} \gamma_z^a & \gamma_z^b \\ \gamma_z^{b\,\dagger} & \gamma_z^c \end{pmatrix}, \quad \gamma\, S_x = \begin{pmatrix} \gamma_x^a & \gamma_x^b \\ \gamma_x^{b\,\dagger} & \gamma_x^c \end{pmatrix},$$
$$\gamma\, S_y = \begin{pmatrix} \gamma_y^a & \gamma_y^b \\ \gamma_y^{b\,\dagger} & \gamma_y^c \end{pmatrix}, \tag{4.245}$$

with

$$\gamma_z^a = \begin{pmatrix} -\dfrac{\dfrac{p_z}{mc}}{B+B'} & 0 \\ 0 & -\dfrac{\dfrac{p_z}{mc}}{B+B'} \end{pmatrix},$$
$$\gamma_z^b = \begin{pmatrix} \dfrac{2\sqrt{BB'}}{B+B'} & 0 \\ 0 & \dfrac{2\sqrt{BB'}}{B+B'} \end{pmatrix},$$
$$\gamma_z^c = \begin{pmatrix} \dfrac{\dfrac{p_z}{mc}}{B+B'} & 0 \\ 0 & \dfrac{\dfrac{p_z}{mc}}{B+B'} \end{pmatrix};$$
$$\gamma_x^a = \begin{pmatrix} -\dfrac{\dfrac{p_x}{mc}}{B+B'} & 0 \\ 0 & -\dfrac{\dfrac{p_x}{mc}}{B+B'} \end{pmatrix},$$
$$\gamma_x^b = \begin{pmatrix} -\dfrac{2\dfrac{p_z p_x}{m^2c^2}}{(B+B')\sqrt{BB'}} - \dfrac{i\dfrac{p_y}{mc}}{\sqrt{BB'}} & \dfrac{1}{\sqrt{BB'}} \\ -\dfrac{1}{\sqrt{BB'}} & -\dfrac{2\dfrac{p_z p_x}{m^2c^2}}{(B+B')\sqrt{BB'}} + \dfrac{i\dfrac{p_y}{mc}}{\sqrt{BB'}} \end{pmatrix},$$

$$\gamma_x^c = \begin{pmatrix} \dfrac{\dfrac{p_x}{mc}}{B+B'} & 0 \\ 0 & \dfrac{\dfrac{p_x}{mc}}{B+B'} \end{pmatrix};$$

$$\gamma_y^a = \begin{pmatrix} -\dfrac{\dfrac{p_y}{mc}}{B+B'} & 0 \\ 0 & -\dfrac{\dfrac{p_y}{mc}}{B+B'} \end{pmatrix},$$

$$\gamma_y^b = \begin{pmatrix} -\dfrac{2\dfrac{p_z p_y}{m^2c^2}}{(B+B')\sqrt{BB'}} + \dfrac{i\dfrac{p_x}{mc}}{\sqrt{BB'}} & -\dfrac{i}{\sqrt{BB'}} \\ -\dfrac{i}{\sqrt{BB'}} & -\dfrac{2\dfrac{p_z p_y}{m^2c^2}}{(B+B')\sqrt{BB'}} - \dfrac{i\dfrac{p_x}{mc}}{\sqrt{BB'}} \end{pmatrix},$$

$$\gamma_y^c = \begin{pmatrix} \dfrac{\dfrac{p_y}{mc}}{B+B'} & 0 \\ 0 & \dfrac{\dfrac{p_y}{mc}}{B+B'} \end{pmatrix}.$$

On the other hand, adopting the plane waves in Eq. (4.233) (normalized to $\psi^*\psi = 1$) as the unitary vectors, the matrices representing the operators S_z, S_x, S_y, γ, γ_1, γS_z, γS_x, γS_y can be obtained from those listed in the previous pages by transforming them with S given in Eq. (4.239). However, it is simpler to calculate them directly; if we use the following notations [the velocity of the positive electron is (v_x, v_y, v_z) that of the negative electron $(-v_x, -v_y, -v_z)$; the speed is v in both cases]:

$$\begin{aligned} \beta &= \frac{v}{c} = \frac{p}{\sqrt{p^2+m^2c^2}}, \\ \beta_x &= \frac{v_x}{c} = \frac{p_x}{\sqrt{p^2+m^2c^2}}, \\ \beta_y &= \frac{v_y}{c} = \frac{p_y}{\sqrt{p^2+m^2c^2}}, \\ \beta_z &= \frac{v_z}{c} = \frac{p_z}{\sqrt{p^2+m^2c^2}}, \end{aligned}$$

then they are

$$S_z = \begin{pmatrix} s_z^a & s_z^b \\ s_z^{b\,\dagger} & s_z^c \end{pmatrix}, \quad S_x = \begin{pmatrix} s_x^a & s_x^b \\ s_x^{b\,\dagger} & s_x^c \end{pmatrix}, \quad S_y = \begin{pmatrix} s_y^a & s_y^b \\ s_y^{b\,\dagger} & s_y^c \end{pmatrix}; \tag{4.246}$$

with

$$s_z^a = \begin{pmatrix} 1 - \dfrac{\beta_x^2+\beta_y^2}{1+\sqrt{1-\beta^2}} & \dfrac{\beta_z(\beta_x - i\beta_y)}{1+\sqrt{1-\beta^2}}, \\ \dfrac{\beta_z(\beta_x + i\beta_y)}{1+\sqrt{1-\beta^2}} & -1 + \dfrac{\beta_x^2+\beta_y^2}{1+\sqrt{1-\beta^2}} \end{pmatrix},$$

$$s_z^b = \begin{pmatrix} \dfrac{\beta_x^2+\beta_y^2}{\beta} & -\dfrac{\beta_z(\beta_x - i\beta_y)}{\beta} \\ -\dfrac{\beta_z(\beta_x + i\beta_y)}{\beta} & -\dfrac{\beta_x^2+\beta_y^2}{\beta} \end{pmatrix},$$

$$s_z^c = \begin{pmatrix} 1 - \dfrac{\beta_x^2+\beta_y^2}{1-\sqrt{1-\beta^2}} & \dfrac{\beta_z(\beta_x - i\beta_y)}{1-\sqrt{1-\beta^2}} \\ \dfrac{\beta_z(\beta_x + i\beta_y)}{1-\sqrt{1-\beta^2}} & -1 + \dfrac{\beta_x^2+\beta_y^2}{1-\sqrt{1-\beta^2}} \end{pmatrix},$$

$$s_x^a = \begin{pmatrix} \dfrac{\beta_z\beta_x}{1+\sqrt{1-\beta^2}} & 1 - \dfrac{\beta^2 - \beta_x(\beta_x - i\beta_y)}{1+\sqrt{1-\beta^2}} \\ 1 - \dfrac{\beta^2 - \beta_x(\beta_x + i\beta_y)}{1+\sqrt{1-\beta^2}} & -1\dfrac{\beta_z\beta_x}{1+\sqrt{1-\beta^2}} \end{pmatrix},$$

$$s_x^b = \begin{pmatrix} -\dfrac{\beta_z\beta_x}{\beta} & \dfrac{\beta^2 - \beta_x(\beta_x - i\beta_y)}{\beta} \\ \dfrac{\beta^2 - \beta_x(\beta_x + i\beta_y)}{\beta} & -\dfrac{\beta_z\beta_x}{\beta} \end{pmatrix},$$

$$s_x^c = \begin{pmatrix} \dfrac{\beta_z\beta_x}{1-\sqrt{1-\beta^2}} & 1 - \dfrac{\beta^2 - \beta_x(\beta_x - i\beta_y)}{1-\sqrt{1-\beta^2}} \\ 1 - \dfrac{\beta^2 - \beta_x(\beta_x + i\beta_y)}{1-\sqrt{1-\beta^2}} & -\dfrac{\beta_z\beta_x}{1-\sqrt{1-\beta^2}} \end{pmatrix};$$

and so on.

Let us now write the Dirac equation with no (interacting) field:

$$\left(\frac{W}{c} + (\boldsymbol{\alpha}\cdot\mathbf{p}) + \beta\, m\, c\right)\psi = 0. \tag{4.247}$$

The spin functions associated with a plane wave of momentum (p_x, p_y, p_z) can be obtained from those associated with a wave of vanishing momentum by means of a relativistic transformation (a rotation in the plane $t-p$). From the known transformation laws for spinors, we find

$$u_p = \left[\sqrt{\frac{1+\sqrt{1+(p/mc)^2}}{2}} \mp \frac{\boldsymbol{\alpha}\cdot\mathbf{p}/mc}{\sqrt{2\left(1+\sqrt{1+(p/mc)^2}\right)}}\right] u_0, \tag{4.248}$$

where the upper sign applies to positive waves and the lower sign to negative ones. The spin functions thus obtained are normalized in the "invariant sense":

$$\left(u_p^\dagger u_p\right)^2 - \left(u_p^\dagger \alpha_x u_p\right)^2 - \left(u_p^\dagger \alpha_y u_p\right)^2 - \left(u_p^\dagger \alpha_z u_p\right)^2 = 1. \tag{4.249}$$

The spin functions normalized in the ordinary sense $(u_p'^\dagger u_p' = 1)$ are instead given by

$$u_p' = \frac{u_p}{\sqrt[4]{1+(p/mc)^2}} \tag{4.250}$$

$$= \left(\sqrt{\frac{1+\sqrt{1+(p/mc)^2}}{2\sqrt{1+(p/mc)^2}}} \mp \sqrt{\frac{-1+\sqrt{1+(p/mc)^2}}{2\sqrt{1+(p/mc)^2}}}\,\frac{\boldsymbol{\alpha}\cdot\mathbf{p}}{p}\right) u_0. \tag{4.251}$$

17. IMPROPER OPERATORS

Let $u(x, y, z)$ be an arbitrary function of x, y, z that can be expanded in harmonic components:

$$u(x, y, z) = \int \alpha(x, y, z)\, \mathrm{e}^{2\pi i\,(\gamma_1 x + \gamma_2 y + \gamma_3 z)}\, \mathrm{d}\gamma_1\, \mathrm{d}\gamma_2\, \mathrm{d}\gamma_3, \tag{4.252}$$

where

$$\alpha(x,y,z) = \int u(x,y,z)\,\mathrm{e}^{-2\pi i\,(\gamma_1 x + \gamma_2 y + \gamma_3 z)}\,\mathrm{d}x\,\mathrm{d}y\,\mathrm{d}z. \qquad (4.253)$$

Let F^r be the operator that transforms u into the function

$$v(x,y,z) = F^r\,u(x,y,z) \qquad (4.254)$$

defined by the Fourier integral expansion

$$v(x,y,z) = \int \lambda^r\,\alpha(x,y,z)\,\mathrm{e}^{2\pi i\,(\gamma_1 x + \gamma_2 y + \gamma_3 z)}\,\mathrm{d}\gamma_1\,\mathrm{d}\gamma_2\,\mathrm{d}\gamma_3, \qquad (4.255)$$

where

$$\lambda = \frac{1}{\gamma} = \frac{1}{\sqrt{\gamma_1^2 + \gamma_2^2 + \gamma_3^2}} \qquad (4.256)$$

is the wavelength of the $(\gamma_1,\ \gamma_2,\ \gamma_3)$ harmonic component. The following properties evidently hold:

$$F^r\,F^s = F^s\,F^r = F^{r+s}, \quad F^0 = 1. \qquad (4.257)$$

Disregarding possible convergence problems, we can set

$$v(x,y,z) = \int K_r(x,y,z;x',y',z')\,u(x',y',z')\,\mathrm{d}x'\,\mathrm{d}y'\,\mathrm{d}z'. \qquad (4.258)$$

Substituting Eq. (4.253) into Eq. (4.255), we find

$$v(x,y,z) = \int\!\!\int \lambda^r\,\mathrm{e}^{2\pi i\,[\gamma_1(x-x') + \gamma_2(y-y') + \gamma_3(z-z')]} \times u(x',y',z')\,\mathrm{d}\gamma_1\,\mathrm{d}\gamma_2\,\mathrm{d}\gamma_3\,\mathrm{d}x'\,\mathrm{d}y'\,\mathrm{d}z', \qquad (4.259)$$

from which we deduce

$$K_r(x,y,z,x',y',z') = \int \lambda^r\,\mathrm{e}^{2\pi i\,(\gamma_1\xi + \gamma_2\eta + \gamma_3\zeta)}\,\mathrm{d}\gamma_1\,\mathrm{d}\gamma_2\,\mathrm{d}\gamma_3, \qquad (4.260)$$

where

$$\xi = x - x', \quad \eta = y - y', \quad \zeta = z - z'. \qquad (4.261)$$

Performing the integration in (4.260) on a sphere of radius $D = 1/\lambda = \sqrt{\gamma_1^2 + \gamma_2^2 + \gamma_3^2}$ and setting

$$R = \sqrt{\xi^2 + \eta^2 + \zeta^2} = \sqrt{(x-x')^2 + (y-y')^2 + (z-z')^2}, \qquad (4.262)$$

we get

$$\begin{aligned} K_r(x,y,z,x',y',z') &= K_r(R) \\ = \int_0^\infty \frac{2\sin 2\pi sR}{R\, s^{r-1}}\,\mathrm{d}s &= \frac{(2\pi R)^{r-1}}{\pi R^2}\int_0^\infty \frac{\sin t}{t^{r-1}}\,\mathrm{d}t. \quad (4.263) \end{aligned}$$

This formula can be used for $1 \le r < 3$; the expression corresponding to the case $r = 1$ can be obtained from that corresponding to $r = 1+\epsilon$ by taking the limit $\epsilon \to 0$ or, in an equivalent way, by taking the mean value of the integral with arbitrary upper limit. We, thus, find

$$K_1 = 1/\pi R^2, \tag{4.264}$$

$$K_2 = \pi/R, \tag{4.265}$$

and

$$F^1\, u(x,y,z) = \int (1/\pi R^2)\, u(x',y',z')\,\mathrm{d}x'\,\mathrm{d}y'\,\mathrm{d}z', \tag{4.266}$$

$$F^2\, u(x,y,z) = \int (\pi/R)\, u(x',y',z')\,\mathrm{d}x'\,\mathrm{d}y'\,\mathrm{d}z'. \tag{4.267}$$

On applying the Laplace operator on the two sides of Eq. (4.267), we get

$$\boldsymbol{\nabla}^2\, F^2 = -\,4\pi^2 \tag{4.268}$$

from which, being F^2 invertible,

$$\boldsymbol{\nabla}^2 = -\,4\pi^2\, F^{-2}; \tag{4.269}$$

this relation immediately follows from Eq. (4.255). We can define the operator $\sqrt{\boldsymbol{\nabla}^2}$ by setting

$$\sqrt{\boldsymbol{\nabla}^2} = 2\pi i\, F^{-1}, \tag{4.270}$$

which can be written, using Eqs. (4.269) and (4.259), as

$$\sqrt{\boldsymbol{\nabla}^2} = 2\pi i\, F^1\, F^{-2} = \frac{1}{2\pi i}\, F^1\, \boldsymbol{\nabla}^2\,. \tag{4.271}$$

Then, from Eq. (4.266), it follows that

$$\sqrt{\boldsymbol{\nabla}^2}\, u(x,y,z) = \int \frac{1}{2\pi^2 R^2 i}\, \boldsymbol{\nabla}^2\, u(x',y',z')\,\mathrm{d}x'\,\mathrm{d}y'\,\mathrm{d}z'. \tag{4.272}$$

Moreover, from Eq. (4.270), we deduce

$$\frac{1}{\sqrt{\boldsymbol{\nabla}^2}} = \frac{1}{2\pi i}\, F^1, \tag{4.273}$$

from which we infer

$$\frac{1}{\sqrt{\nabla^2}}\, u(x,y,z) \;=\; \int \frac{1}{2\pi^2 R^2 i}\, u(x',y',z')\, \mathrm{d}x'\, \mathrm{d}y'\, \mathrm{d}z'. \qquad (4.274)$$

18. INTEGRAL REPRESENTATION OF HYDROGEN EIGENFUNCTIONS

In the following, we will use electronic units, in which $e = m = \hbar = 1$ and the energy unit is $2Ry$. We denote by $\chi(r)$ the radial part of the hydrogen eigenfunctions multiplied by r and by ℓ the azimuthal quantum number. The basic equation is

$$\chi'' + \left(2E + \frac{2}{r} - \frac{\ell(\ell+1)}{r^2}\right)\chi \;=\; 0. \qquad (4.275)$$

Let us set

$$\chi \;=\; r^{l+1}\, u; \qquad (4.276)$$

then, the differential equation for u becomes

$$u'' + 2\,\frac{\ell+1}{r}\, u' + \left(2E + \frac{2}{r}\right) u \;=\; 0. \qquad (4.277)$$

This equation has the form of the Laplace equation (see Sec. 4.11) with the following values for the constants:

$$\delta_0 \;=\; 0, \quad \delta_1 \;=\; 2\,(\ell+1), \quad \epsilon_0 \;=\; 2E, \quad \epsilon_1 \;=\; 2. \qquad (4.278)$$

The values of the constants appearing in Eq. (4.188) needed for the integral representation of u then are, in the considered case (assume $E > 0$),

$$c_1 \;=\; i\sqrt{2E}, \qquad c_2 \;=\; -i\sqrt{2E}, \qquad (4.279)$$

$$\alpha_1 \;=\; \ell + 1 - i/\sqrt{2}E, \qquad \alpha_2 \;=\; \ell + 1 + i/\sqrt{2}E. \qquad (4.280)$$

Substituting into Eq. (4.188), we get, except for a constant factor,

$$u \sim \int_C \mathrm{e}^{t\,r} \left(t - i\sqrt{2E}\right)^{\ell - i/\sqrt{2E}} \left(t + i\sqrt{2E}\right)^{\ell + i/\sqrt{2E}} \mathrm{d}t, \qquad (4.281)$$

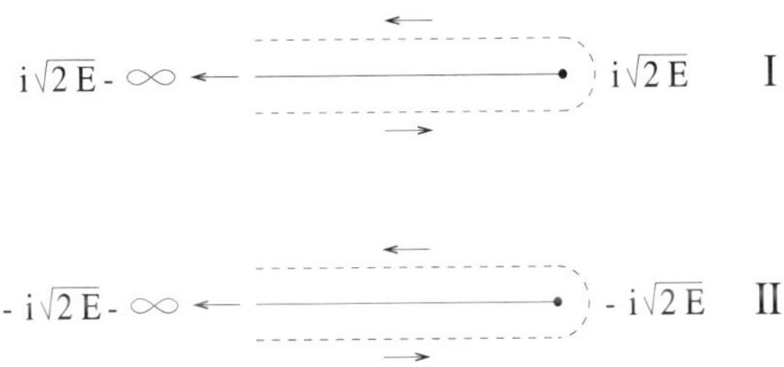

Fig. 4.2. The paths in the complex plane to define $\log(t-i\sqrt{2E})$ and $\log(t+i\sqrt{2E})$ (see text).

where the condition

$$\int_C \frac{\mathrm{d}}{\mathrm{d}t}\left[\mathrm{e}^{t\,r}\left(t-i\sqrt{2E}\right)^{\ell+1-i/\sqrt{2E}}\left(t+i\sqrt{2E}\right)^{\ell+1+i/\sqrt{2E}}\right]\mathrm{d}t = 0 \tag{4.282}$$

must be satisfied. For real and positive r, this happens if the limits of the integration domain lie at infinity along the negative real axis. This fixes the domain C.

Now, we still have to give a univocal definition of $\log(t-i\sqrt{2e})$ and $\log(t+i\sqrt{2e})$ for the integrand function to be determined. Let us assume that the imaginary part of such logarithms is less than or equal to π. In this case, we have that the discontinuity lines are the half-lines starting from $\pm i\sqrt{2E}$ and extending parallel to the negative real semi-axis. Let us denote by u_1 the integral in Eq. (4.281) performed along path I (see Fig. 4.2) and by χ_1 the corresponding solution of Eq. (4.275); similarly, u_2 and χ_2 correspond to path II. It is more convenient to introduce another integration variable by setting $t = i\sqrt{2e}t_1$. Then, changing t_1 in t, we have

$$u = k\int \mathrm{e}^{i\sqrt{2E}tr}\,(t-1)^{l-i/\sqrt{2E}}\,(t+1)^{l+i/\sqrt{2E}}\,\mathrm{d}t. \tag{4.283}$$

The integration paths I and II are shown in Fig. 4.3 . The logarithm of $t-1$ and $t+1$ is understood to be real for $t-1>0$ and $t+1>0$, respectively, the discontinuity lines being the half-lines $1+ai$ and $1-ai$ $(a>0)$, respectively.

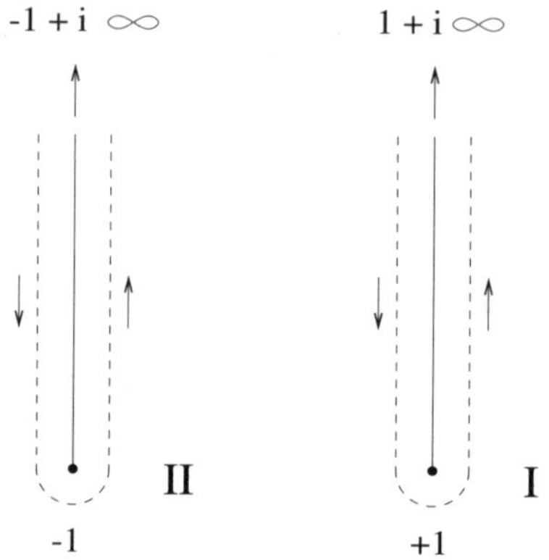

Fig. 4.3. The paths in the complex plane to define the function u in (4.283) (see text).

19. DEFLECTION OF AN ALPHA RAY INDUCED BY A HEAVY NUCLEUS (CLASSICAL MECHANICS)[11]

Substituting Eq. (4.139) into Eq. (4.134), we find

$$\frac{1}{\rho} = -\frac{k}{v^2\delta^2} - \frac{k}{v^2\delta^2}\cos\theta + \frac{1}{\delta}\sin\theta. \tag{4.284}$$

The envelope of the hyperbolas defined in (4.284) satisfies Eq. (4.284) as well as the equation obtained from this by differentiating with respect to δ. Introducing the minimum distance l from the nucleus [see Eq. (4.145)], and noting that here $W = Mv^2/2$, from Eq. (4.125), we have the following expression for k:

$$\frac{k}{v^2} = \frac{l}{2},$$

so that Eq. (4.284) becomes

$$\frac{1}{\rho} = -\frac{l}{2\delta^2} - \frac{l}{2\delta^2}\cos\theta + \frac{1}{\delta}\sin\theta. \tag{4.285}$$

Differentiating with respect to δ and equating to zero the resulting expression, we find

$$\frac{l}{\delta} + \frac{l}{\delta}\cos\theta - \sin\theta = 0, \tag{4.286}$$

[11]See Sec. 4.9.

from which we get

$$\frac{\delta}{l} = \frac{1 + \cos\theta}{\sin\theta}. \tag{4.287}$$

20. SCATTERING FROM A POTENTIAL OF THE FORM $a/r - b/r^2$

Let us consider a particle of mass 1 and speed k, travelling through a region in which a field is present. Its potential is

$$\frac{1}{r}\left(1 - \frac{r_0}{r}\right), \tag{4.288}$$

which is repulsive for $r > 2r_0$ and attractive for $r < 2r_0$. The problem is to find the cross section for the scattering of the particle at an angle θ. In classical mechanics, the equations of motion in polar coordinates are

$$r^2\,\dot{\theta} = c \tag{4.289}$$

$$\ddot{r} - r\,\dot{\theta}^2 = \frac{1}{r^2} - \frac{2r_0}{r^3} = \frac{1}{r^2}\left(1 - \frac{2r_0}{r}\right). \tag{4.290}$$

We have

$$\ddot{r} = -\frac{c^2}{r^2}\frac{\mathrm{d}^2}{\mathrm{d}\theta^2}\frac{1}{r}, \tag{4.291}$$

$$r\,\dot{\theta}^2 = c^2/r^3, \tag{4.292}$$

which leads to

$$\frac{\mathrm{d}^2}{\mathrm{d}\theta^2}\frac{1}{r} + \frac{1}{r} + \frac{1}{c^2}\left(1 - \frac{2r_0}{r}\right) = 0, \tag{4.293}$$

or

$$\frac{\mathrm{d}^2}{\mathrm{d}\theta^2}\frac{1}{r} + \left(1 - \frac{2r_0}{c^2}\right)\frac{1}{r} + \frac{1}{c^2} = 0. \tag{4.294}$$

It follows that, if $|c| > \sqrt{2r_0}$,

$$\frac{1}{r} = -\frac{1}{c^2\gamma^2} + A\cos\gamma\theta + B\sin\gamma\theta, \tag{4.295}$$

with

$$\gamma = \sqrt{1 - \frac{2r_0}{c^2}}. \tag{4.296}$$

Instead, if $|c| < \sqrt{2r_0}$, we have

$$\frac{1}{r} = \frac{1}{c^2\epsilon^2} + C\,\mathrm{e}^{\epsilon\theta} + D\,\mathrm{e}^{-\epsilon\theta}, \tag{4.297}$$

with

$$\epsilon = \sqrt{\frac{2r_0}{c^2} - 1}. \tag{4.298}$$

Finally, for $|c| = \sqrt{2r_0}$, we have

$$\frac{1}{r} = -\frac{1}{4r_0}\,\theta^2 + F\,\theta + G. \tag{4.299}$$

Let us set $z = r\cos\theta$, $\xi = r\sin\theta$ and assume that the particles of speed k are incident from $-\infty$ along the z axis, at a distance δ from this axis. Assume also that (for positive δ) the line $\xi = \delta$ is an asymptote for the trajectory; then, evidently, $c = -k\delta$. Moreover, for $\theta = \pi$, we have $r = \infty$ and

$$\dot{r} = \frac{\mathrm{d}r}{\mathrm{d}\theta}\,\dot{\theta} = -\,r^2\,\dot{\theta}\,\frac{\mathrm{d}}{\mathrm{d}\theta}\,\frac{1}{r} = -\,c\,\frac{\mathrm{d}}{\mathrm{d}\theta}\,\frac{1}{r} = -\,k, \tag{4.300}$$

that is,

$$\frac{\mathrm{d}}{\mathrm{d}\theta}\,\frac{1}{r} = -\,\frac{1}{\delta}, \qquad \theta = \pi. \tag{4.301}$$

Depending on the value of δ, it then follows that

(1) $\delta > \dfrac{\sqrt{2r_0}}{k}$:

$$\frac{1}{r} = \frac{-1}{k^2\delta^2 - 2r_0} + \frac{\cos\gamma(\pi-\theta)}{k^2\delta^2 - 2r_0} + \frac{\sin\gamma(\pi-\theta)}{\delta\,\gamma}, \tag{4.302}$$

$$\gamma = \sqrt{1 - \frac{2r_0}{k^2\delta^2}}. \tag{4.303}$$

(2) $\delta = \dfrac{\sqrt{2r_0}}{k}$:

$$\frac{1}{r} = -\,\frac{1}{4r_0}\,(\pi - \theta)^2 + \frac{1}{\delta}\,(\pi - \theta)\,. \tag{4.304}$$

(3) $\delta < \dfrac{\sqrt{2r_0}}{k}$:

$$\frac{1}{r} = \frac{-1}{2r_0 - k^2\delta^2} - \frac{\cosh\epsilon(\pi-\theta)}{2r_0 - k^2\delta^2} + \frac{\sinh\epsilon(\pi-\theta)}{\delta\,\epsilon}, \tag{4.305}$$

$$\epsilon = \sqrt{\frac{2r_0}{k^2\delta^2} - 1}. \tag{4.306}$$

The particle will be scattered in the direction θ of the second asymptote:

$$(1)\ \theta = \pi - \frac{2}{\gamma}\arctan\gamma k^2\delta = \frac{2}{\gamma}\arctan\frac{1}{\gamma k^2\delta} - \pi\left(\frac{1}{\gamma} - 1\right).$$

$$(2)\ \theta = \pi - 2k^2\delta = \pi - \frac{4r_0}{\delta}.$$

21. THE SET OF ORTHOGONAL FUNCTIONS DEFINED BY THE EQUATION $y_a'' = (x - a)y_a$

On setting $\xi = x - a$, so that $y''(\xi) = \xi y$, the secular solutions of the equation

$$y_a'' = (x - a)\, y_a \tag{4.307}$$

can be cast in the form

$$y_a(x) = y(x - a) = y(\xi) \tag{4.308}$$

or

$$y_a(\xi + a) = y(\xi), \tag{4.309}$$

and the determination of the complete set of regular solutions of Eq. (4.307), corresponding to the complete set of eigenvalues of a, reduces to the determination of the unique regular solution of

$$y''(\xi) = \xi\, y(\xi). \tag{4.310}$$

Requiring that y_a be normalized with respect to $\mathrm{d}a$, we have

$$\int_{-\infty}^{\infty} y_a^*(x)\,\mathrm{d}x \int_{a-\Delta}^{a+\Delta} y_a(x)\,\mathrm{d}a = 1, \tag{4.311}$$

and, using Eq. (4.308),

$$\int_{-\infty}^{+\infty} y^*(\xi)\,\mathrm{d}\xi \int_{-\Delta}^{\Delta} y(\xi + \epsilon)\,\mathrm{d}\epsilon = 1. \tag{4.312}$$

Since y vanishes exponentially for $\xi \to \infty$, the main contribution to the integral for $\Delta \to 0$ comes from large negative values of ξ. The asymptotic expression of y for $\xi \to \infty$ will read

$$\xi \to -\infty: \qquad y \sim \frac{A}{\sqrt[4]{-\xi}}\sin\left(\frac{2}{3}(-\xi)^{3/2} + \alpha\right). \tag{4.313}$$

For $\xi \to -\infty$ (and small values of ϵ), we have

$$(-\xi - \epsilon)^{3/2} \sim (-\xi)^{3/2} - \frac{3}{2}\epsilon\,(-\epsilon)^{1/2} + \ldots, \tag{4.314}$$

so that

$$\xi \to -\infty: \qquad y \sim \frac{A}{\sqrt[4]{-\xi}} \sin\left(\frac{2}{3}(-\xi)^{3/2} - \epsilon\,(-\epsilon)^{1/2} + \alpha\right) \tag{4.315}$$

and thus

$$\int_{-\Delta}^{\Delta} y(\xi+\epsilon)\,\mathrm{d}\epsilon \sim (-\xi)^{-3/4}\left[\cos\left(\frac{2}{3}(-\xi)^{3/2} - \Delta\,(-\xi)^{1/2} + \alpha\right)\right.$$
$$\left. - \cos\left(\frac{2}{3}(-\xi)^{3/2} + \Delta\,(-\xi)^{1/2} + \alpha\right)\right]. \tag{4.316}$$

Let us set

$$-\xi = \zeta^2, \quad \mathrm{d}\xi = -2\zeta\,\mathrm{d}\zeta. \tag{4.317}$$

For $\xi \to -\infty$, we have

$$y \sim \frac{A}{\sqrt{\zeta}} \sin\left(\frac{2}{3}\zeta^3 + \alpha\right), \tag{4.318}$$

$$\int_{-\Delta}^{\Delta} y(\xi+\epsilon)\,\mathrm{d}\epsilon \sim \frac{1}{\zeta^{3/2}}\left[\cos\left(\frac{2}{3}\zeta^3 - \Delta\zeta + \alpha\right)\right.$$
$$\left. - \cos\left(\frac{2}{3}\zeta^3 + \Delta\zeta + \alpha\right)\right]. \tag{4.319}$$

For $\Delta \to 0$, it easily follows that

$$\int_{-\infty}^{\infty} y^*(\xi)\,\mathrm{d}\xi \int_{-\Delta}^{\Delta} y(\xi+\epsilon)\,\mathrm{d}\epsilon = \pi\,A^*\,A. \tag{4.320}$$

In order to obtain the normalized solution according to Eq. (4.311), we can then assume that

$$A = 1/\sqrt{\pi}. \tag{4.321}$$

Taking the Laplace transform of Eq. (4.310), we can easily get the integral representation

$$y = \frac{i}{2\pi} \int_{\infty\,\mathrm{e}^{i\phi_1}}^{\infty\,\mathrm{e}^{i\phi_2}} \mathrm{e}^{-t^3/3}\,\mathrm{e}^{t\xi}\,dt, \tag{4.322}$$

with

$$\frac{\pi}{2} < \phi_1 < \frac{5}{6}\pi, \quad \frac{7}{6}\pi < \phi_2 < \frac{3}{2}\pi. \tag{4.323}$$

By choosing appropriately the integration path in Eq. (4.322), we find different representations for y and y' that are suitable for evaluating these functions in $\xi = 0$ (I) or for the asymptotic expansion when $\xi \to \infty$ (II) or for the asymptotic expansion for $\xi \to -\infty$ (III) [12]:

$$\text{(I)} \quad y = \frac{1}{\pi} \int_0^{\infty} \mathrm{e}^{-p^3/3 - p\xi/2} \left(\frac{\sqrt{3}}{2} \cos \frac{\sqrt{3}}{2} p\xi - \frac{1}{2} \sin \frac{\sqrt{3}}{2} p\xi \right) \mathrm{d}p.$$

$$\text{(II)} \quad y = \frac{\sqrt{\xi}}{2\pi} \mathrm{e}^{-2\xi^{3/2}/3} \int_{-\infty}^{\infty} \mathrm{e}^{-p^2 \xi^{3/2}} \cos \left(\frac{1}{3} p^3 \, \xi^{3/2} \right) \mathrm{d}p.$$

$$\text{(III)} \quad y = \frac{-2\xi}{\pi} \int_{-1}^{\infty} \mathrm{e}^{-\left(2p^2 + 2p^3/3\right)(-\xi)^{3/2}} \times \sin \left[\left(\frac{2}{3} + \frac{2}{3} p^3 \right) (-\,\xi)^{3/2} + \frac{\pi}{4} \right] \mathrm{d}p.$$

22. FOURIER INTEGRAL EXPANSIONS

(1) We have

$$\frac{1}{r} = \int \frac{1}{\pi \gamma^2} \mathrm{e}^{2\pi i \, \boldsymbol{\gamma} \cdot \mathbf{r}} \, \mathrm{d}\gamma, \tag{4.324}$$

where $\mathrm{d}\gamma = \mathrm{d}\gamma_1 \mathrm{d}\gamma_2 \mathrm{d}\gamma_3$.

(2) We have

$$\frac{\mathrm{e}^{ikr}}{r} = \int \left[\pi \left(\gamma^2 - \left(\frac{k + \epsilon i}{2\pi} \right)^2 \right) \right]^{-1} \mathrm{e}^{2\pi i \, \boldsymbol{\gamma} \cdot \mathbf{r}} \, \mathrm{d}\gamma, \tag{4.325}$$

$$\frac{\mathrm{e}^{-ikr}}{r} = \int \left[\pi \left(\gamma^2 - \left(\frac{k - \epsilon i}{2\pi} \right)^2 \right) \right]^{-1} \mathrm{e}^{2\pi i \, \boldsymbol{\gamma} \cdot \mathbf{r}} \, \mathrm{d}\gamma \tag{4.326}$$

(for $\epsilon > 0$, $\epsilon \to 0$). It follows that

$$\frac{\sin kr}{r} = \int \frac{8\pi \, k \, \epsilon}{(4\pi^2 \gamma^2 - k^2 + \epsilon^2)^2 + 4k^2 \epsilon^2} \mathrm{e}^{2\pi i \, \boldsymbol{\gamma} \cdot \mathbf{r}} \, \mathrm{d}\gamma \tag{4.327}$$

(for $\epsilon > 0$, $\epsilon \to 0$), or

$$\frac{\sin kr}{r} = \int \frac{1}{2\gamma} \delta \left(|\gamma| - \frac{k}{2\pi} \right) \mathrm{e}^{2\pi i \, \boldsymbol{\gamma} \cdot \mathbf{r}} \, \mathrm{d}\gamma$$

---

[12]@ In the original manuscript, this section is incomplete. It ends with the following sentence: "For ξ close to 0, we can expand the integrand function in I in power series of ξ; we have ..."

$$= \int \frac{k}{4\pi\gamma}\,\delta\left(|\gamma| - \frac{k}{2\pi}\right) \mathrm{e}^{2\pi i\,\boldsymbol{\gamma}\cdot\mathbf{r}}\,\mathrm{d}\gamma. \qquad (4.328)$$

(3) We have

$$\frac{1}{r^2} = \int \frac{\pi}{\gamma}\,\mathrm{e}^{2\pi i\,\boldsymbol{\gamma}\cdot\mathbf{r}}\,\mathrm{d}\gamma \qquad (4.329)$$

(which follows from (4.324) by taking the inverse of the Fourier integral).

(4) The mean value of the function

$$F = \begin{cases} 0, & \text{for } r < R, \\ 1/r, & \text{for } r > R, \end{cases} \qquad (4.330)$$

is

$$< F > = \int \frac{\cos 2\pi\gamma R}{\pi\gamma^2}\,\mathrm{e}^{2\pi i\,\boldsymbol{\gamma}\cdot\mathbf{r}}\,\mathrm{d}\gamma. \qquad (4.331)$$

(5) The mean value of the function

$$F = \begin{cases} 1, & \text{for } r < R, \\ 0, & \text{for } r > R, \end{cases} \qquad (4.332)$$

is

$$< F > = \int (1/2\pi^2\gamma^3)(\sin 2\pi\gamma R - 2\pi\gamma R\cos 2\pi\gamma R)\,\mathrm{e}^{2\pi i\,\boldsymbol{\gamma}\cdot\mathbf{r}}\,\mathrm{d}\gamma. \qquad (4.333)$$

(6) We have

$$\mathrm{e}^{-\alpha r^2} = \int \left(\frac{\pi}{\alpha}\right)^{3/2} \mathrm{e}^{-\pi^2\gamma^2/\alpha}\,\mathrm{e}^{2\pi i\,\boldsymbol{\gamma}\cdot\mathbf{r}}\,\mathrm{d}\gamma. \qquad (4.334)$$

(7) We have

$$\mathrm{e}^{-kr} = \int \frac{8\pi k}{(k^2 + 4\pi^2\gamma^2)}\,\mathrm{e}^{2\pi i\,\boldsymbol{\gamma}\cdot\mathbf{r}}\,\mathrm{d}\gamma. \qquad (4.335)$$

(8) Given

$$f(q) = \int \phi(\gamma)\,\mathrm{e}^{2\pi i\,\boldsymbol{\gamma}\cdot\mathbf{q}}\,\mathrm{d}\gamma, \qquad (4.336)$$

$$f'(q) = \int \mathcal{U}(|\gamma|)\,\phi(\gamma)\,\mathrm{e}^{2\pi i\,\boldsymbol{\gamma}\cdot\mathbf{q}}\,\mathrm{d}\gamma, \qquad (4.337)$$

with $\mathbf{q} = (q_1, q_2.q_3)$, $\boldsymbol{\gamma} = (\gamma_1, \gamma_2.\gamma_3)$, $Q = \sqrt{q_1^2 + q_2^2 + q_3^2}$, $\Gamma = \sqrt{\gamma_1^2 + \gamma_2^2 + \gamma_3^2}$, then we have

$$\begin{aligned} f'(q) &= \int \mathcal{U}(\Gamma)\,\phi(\gamma)\,\mathrm{e}^{2\pi i\,\boldsymbol{\gamma}\cdot\mathbf{r}}\,\mathrm{d}\gamma \\ &= \int\!\!\int \mathcal{U}(\Gamma)\,\mathrm{e}^{2\pi i\,\boldsymbol{\gamma}\cdot\mathbf{q}}\,f(q')\,\mathrm{e}^{-2\pi i\,\boldsymbol{\gamma}\cdot\mathbf{q}'}\,\mathrm{d}\gamma\,\mathrm{d}q' \\ &= \int f(q')\,\mathrm{d}q' \int \mathcal{U}(\Gamma)\,\mathrm{e}^{-2\pi i\,\boldsymbol{\gamma}\cdot(\mathbf{q}'-\mathbf{q})}\,\mathrm{d}\gamma. \end{aligned} \tag{4.338}$$

Using

$$\mathcal{U}(\Gamma) = \int Y(q)\,\mathrm{e}^{2\pi i\,\mathbf{q}\cdot\boldsymbol{\gamma}}\,\mathrm{d}q, \tag{4.339}$$

$$Y(q) = \int \mathcal{U}(\Gamma)\,\mathrm{e}^{-2\pi i\,\boldsymbol{\gamma}\cdot\mathbf{q}}\,\mathrm{d}\gamma, \tag{4.340}$$

we get

$$f'(q) = \int Y(q'-q)\,f(q')\,\mathrm{d}q'. \tag{4.341}$$

On setting, as we can do, $Y(q'-q) = y(|\mathbf{q}'-\mathbf{q}|)$, we finally have

$$f'(q) = \int y(|\mathbf{q}-\mathbf{q}'|)\,f(q')\,\mathrm{d}q'. \tag{4.342}$$

23. CIRCULAR INTEGRALS

$$\int_0^{2\pi} \frac{\mathrm{d}\phi}{a + b\cos\phi} = \frac{2\pi}{\sqrt{a^2-b^2}}, \quad [a > |b| > 0], \tag{4.343}$$

$$\int_0^{2\pi} \frac{\mathrm{d}\phi}{a^2 - b^2\cos^2\phi} = \frac{2\pi}{a\sqrt{a^2-b^2}}, \quad [a > b > 0], \tag{4.344}$$

$$\int_0^{2\pi} \frac{\mathrm{d}\phi}{a^2\cos^2\phi + b^2\sin^2\phi} = \frac{2\pi}{ab}, \quad [a\,,\, b > 0], \tag{4.345}$$

$$\begin{aligned} \int_0^{2\pi} \frac{\mathrm{d}\phi}{(a + b\cos\phi)^n} &= \frac{2\pi}{(a^2-b^2)^{n/2}} \\ &\times \sum_{r=0}^{n-1} \binom{n-1}{r} \binom{-n}{r} \left(\frac{-a+\sqrt{a^2-b^2}}{2\sqrt{a^2-b^2}}\right)^r, \end{aligned} \tag{4.346}$$

with $a > |b| > 0$.
Examples:

$$\begin{aligned}\int_0^{2\pi} \frac{d\phi}{a+b\cos\phi} &= \frac{2\pi}{\sqrt{a^2-b^2}},\\ \int_0^{2\pi} \frac{d\phi}{(a+b\cos\phi)^2} &= \frac{2\pi}{a^2-b^2}\left(1+\frac{-a+\sqrt{a^2-b^2}}{\sqrt{a^2-b^2}}\right)\\ &= \frac{2\pi a}{(a^2-b^2)^{3/2}}.\end{aligned} \tag{4.347}$$

24. OSCILLATION FREQUENCIES OF AMMONIA

The H atoms in the NH_3 molecule lie at the vertices of an equilateral triangle, while the N atom lies on the perpendicular axis outside the plane of the triangle. There are six linearly independent displacements giving rise to attractive elastic forces; these are obtained from the twelve shifts of each atom towards the other three atoms, with the constraint that the sum of the forces acting on each equilibrium position P_i is zero.

Let $q_1 = 1$, $q_2 = q_3 = \ldots = q_6 = 0$ be the shift such that the first H atom, H^1, moves in the direction NH^1 by $M_N/(M_N+M_H)$ and the N atom moves in the opposite direction by an amount $M_H/(M_N+M_H)$. In the same way, we define the shifts $q_i = \delta_{i2}$ and $q_i = \delta_{i3}$. Instead, let $q_i = \delta_{i4}$ be the shift such that the H^3 atom moves by $1/2$ in the direction H^2H^3 and the H^2 atom moves by $1/2$ in the opposite direction; with a circular permutation, we finally define the shifts $q_i = \delta_{i5}$ and $q_i = \delta_{i6}$.

Let us denote by α the angle (in the equilibrium position) $N\widehat{H^1}H^2$ and by β the angle $H^1\widehat{N}H^2$. If D and d are the equilibrium distances NH and H^1H^2, respectively, we then have

$$\cos\alpha = \frac{d}{2D}, \quad \cos\beta = 1 - \frac{d^2}{2D^2} \qquad \left(\text{implying } \sin\frac{1}{2}\beta = \frac{d}{2D}\right). \tag{4.348}$$

The kinetic energy is given by

$$\begin{aligned} T = \frac{1}{2}\left[\frac{M_H^2 M_N}{(M_N+M_H)^2}\left(\dot{q}_1^2 + \dot{q}_2^2 + \dot{q}_3^2 + 2\dot{q}_1\dot{q}_2\cos\beta + 2\dot{q}_2\dot{q}_3\cos\beta\right.\right.\\ \left. + 2\dot{q}_3\dot{q}_1\cos\beta\right) + \frac{M_N^2 M_H}{(M_N+M_H)^2}\left(\dot{q}_1^2 + \dot{q}_2^2 + \dot{q}_3^2\right) \end{aligned}$$

$$+ \frac{M_N M_H}{M_N + M_H} \cos\alpha \left(\dot{q}_1\dot{q}_5 + \dot{q}_1\dot{q}_6 + \dot{q}_2\dot{q}_6 + \dot{q}_2\dot{q}_4 + \dot{q}_3\dot{q}_4 + \dot{q}_3\dot{q}_5\right)$$
$$+ \frac{1}{2} M_H \left(\dot{q}_4 + \dot{q}_5 + \dot{q}_6 + \frac{1}{2}\dot{q}_4\dot{q}_5 + \frac{1}{2}\dot{q}_5\dot{q}_6 + \frac{1}{2}\dot{q}_6\dot{q}_4\right)\Bigg]. \quad (4.349)$$

For simplicity, we assume $M_H = 1$ and $M_N = 14$; then, setting

$$T = \frac{1}{2}\sum_{i,k} b_{ik}\,\dot{q}_1\dot{q}_k, \quad b_{ik} = b_{ki}, \quad (4.350)$$

we get

$$b_{11} = b_{22} = b_{33} = 14/15, \quad (4.351)$$
$$b_{44} = b_{55} = b_{66} = 1/2, \quad (4.352)$$
$$b_{12} = b_{23} = b_{31} = b_{21} = b_{32} = b_{23} = 14/225 \cos\beta, \quad (4.353)$$
$$b_{45} = b_{56} = b_{64} = b_{54} = b_{65} = b_{46} = 1/8, \quad (4.354)$$
$$b_{14} = b_{25} = b_{36} = b_{41} = b_{52} = b_{63} = 0, \quad (4.355)$$
$$b_{15} = b_{26} = b_{34} = b_{16} = b_{24} = b_{35} = b_{51}$$
$$= b_{62} = b_{43} = b_{61} = b_{42} = b_{53} = 7/15 \cos\alpha. \quad (4.356)$$

The fact that many elements are actually equal is due to obvious symmetry feature; thus we need to know only six typical elements:

$$b_{11} = B_1 = 14/15, \quad b_{44} = B_2 = 1/2, \quad b_{12} = B_3 = 14/225 \cos\beta,$$
$$b_{45} = B_4 = 1/8, \quad b_{14} = B_5 = 0, \quad b_{15} = B_6 = 7/15 \cos\alpha.$$

Analogously, the matrix defining the potential energy

$$V = \frac{1}{2}\sum_{ik} a_{ik}\, q_1 q_k \quad (4.357)$$

depends on the following six typical elements:

$$a_{11} = A_1, \; a_{44} = A_2, a_{12} = A_3,$$
$$a_{45} = A_4, \; a_{14} = A_5, \; a_{15} = A_6. \quad (4.358)$$

Let us perform the following transformation:

$$q_1 = \sqrt{\frac{1}{3}}Q_1 + \sqrt{\frac{2}{3}}Q_3, \quad (4.359)$$
$$q_2 = \sqrt{\frac{1}{3}}Q_1 - \sqrt{\frac{1}{6}}Q_3 + \sqrt{\frac{1}{2}}Q_3', \quad (4.360)$$
$$q_3 = \sqrt{\frac{1}{3}}Q_1 - \sqrt{\frac{1}{6}}Q_3 - \sqrt{\frac{1}{2}}Q_3', \quad (4.361)$$

$$q_4 = \sqrt{\frac{1}{3}}\,Q_2 + \sqrt{\frac{2}{3}}\,Q_4, \tag{4.362}$$

$$q_5 = \sqrt{\frac{1}{3}}\,Q_2 - \sqrt{\frac{1}{6}}\,Q_4 + \sqrt{\frac{1}{2}}\,Q_4', \tag{4.363}$$

$$q_6 = \sqrt{\frac{1}{3}}\,Q_2 - \sqrt{\frac{1}{6}}\,Q_4 - \sqrt{\frac{1}{2}}\,Q_4'. \tag{4.364}$$

Then, we find

$$q_1^2 + q_2^2 + q_3^2 = Q_1^2 + Q_3^2 + Q_3'^2, \tag{4.365}$$

$$q_4^2 + q_5^2 + q_6^2 = Q_2^2 + Q_4^2 + Q_4'^2, \tag{4.366}$$

$$q_1 q_2 + q_2 q_3 + q_3 q_1 = Q_1^2 - \frac{1}{2}Q_3^2 - \frac{1}{2}Q_3'^2, \tag{4.367}$$

$$q_4 q_5 + q_5 q_6 + q_6 q_4 = Q_2^2 - \frac{1}{2}Q_4^2 - \frac{1}{2}Q_4'^2, \tag{4.368}$$

$$q_1 q_4 + q_2 q_5 + q_3 q_6 = Q_1 Q_2 + Q_3 Q_4 + Q_3' Q_4', \tag{4.369}$$

$$\begin{aligned} & q_1 q_5 + q_2 q_6 + q_3 q_4 + q_1 q_6 + q_2 q_4 + q_3 q_5 \\ & = 2Q_1 Q_2 - Q_3 Q_4 - Q_3' Q_4'. \end{aligned} \tag{4.370}$$

Thus double the kinetic energy takes the following form in the new coordinates:

$$\begin{aligned} 2T &= B_1\left(\dot{Q}_1^2 + \dot{Q}_3^2 + \dot{Q}_3'^2\right) + B_2\left(\dot{Q}_1^2 + \dot{Q}_4^2 + \dot{Q}_4'^2\right) \\ &+ 2B_3\left(Q_1^2 - \frac{1}{2}Q_3^2 - \frac{1}{2}Q_3'^2\right) + 2B_4\left(Q_2^2 - \frac{1}{2}Q_4^2 - \frac{1}{2}Q_4'^2\right) \\ &+ 2B_5\left(Q_1Q_2 + Q_3Q_4 + Q_3'Q_4'\right) + 2B_6\left(2Q_1Q_2 - Q_3Q_4 - Q_3'Q_4'\right) \\ &= \left(B_1 + 2B_3\right)\dot{Q}_1^2 + 2\left(B_5 + 2B_6\right)\dot{Q}_1\dot{Q}_2 + \left(B_2 + 2B_4\right)\dot{Q}_2^2 \\ &+ \left(B_1 - B_3\right)\dot{Q}_3^2 + 2\left(B_5 - B_6\right)\dot{Q}_3\dot{Q}_4 + \left(B_2 - B_4\right)\dot{Q}_4^2 \\ &+ \left(B_1 - B_3\right)\dot{Q}_3'^2 + 2\left(B_5 - B_6\right)\dot{Q}_3'\dot{Q}_4' + \left(B_2 - B_4\right)\dot{Q}_4'^2. \end{aligned} \tag{4.371}$$

In the same way we calculate

$$\begin{aligned} 2V &= \left(A_1 + 2A_3\right)Q_1^2 + 2\left(A_5 + 2A_6\right)Q_1Q_2 + \left(A_2 + 2A_4\right)Q_2^2 \\ &+ \left(A_1 - A_3\right)Q_3^2 + 2\left(A_5 - A_6\right)Q_3Q_4 + \left(A_2 - A_4\right)Q_4^2 \\ &+ \left(A_1 - A_3\right)Q_3'^2 + 2\left(A_5 - A_6\right)Q_3'Q_4' + \left(A_2 - A_4\right)Q_4'^2. \end{aligned} \tag{4.372}$$

We then find two simple oscillations related to the coordinates Q_1,Q_2 and two double ones related to the coordinates Q_3,Q_4 or Q_3',Q_4'. The squared angular velocities

$$\lambda = 4\pi^2\,\nu^2 \tag{4.373}$$

of the simple oscillations are obtained from the following secular equation:

$$\det\begin{pmatrix} A_1+2A_3-\lambda(B_1+2B_3) & A_5+2A_6-\lambda(B_5+2B_6) \\ A_5+2A_6-\lambda(B_5+2B_6) & A_2+2A_4-\lambda(B_2+2B_4) \end{pmatrix} = 0, \tag{4.374}$$

while those corresponding to degenerate oscillations are obtained from

$$\det\begin{pmatrix} A_1-A_3-\lambda(B_1-B_3) & A_5-A_6-\lambda(B_5-B_6) \\ A_5-A_6-\lambda(B_5-B_6) & A_2-A_4-\lambda(B_2-B_4) \end{pmatrix} = 0. \tag{4.375}$$

25. SPHERICAL FUNCTIONS WITH SPIN ONE

They are three-component functions of θ and ϕ following the transformation rules of $\mathcal{D}_1$. They correspond to particular values of j, l, m, the total angular momentum j and the orbital momentum taking the values $0, 1, 2, \ldots$ and $j-1$, j, $j+1$, respectively; for the case $j=0$, we instead only have the value $l=1$ for the orbital momentum.

We can set

$$\begin{aligned}
\varphi_{j,j-1}^m &= \left(\sqrt{\frac{(j+m)(j+m-1)}{2j(2j-1)}}\,\varphi_{j-1}^{m-1}, \right.\\
&\qquad \sqrt{\frac{(j+m)(j-m)}{j(2j-1)}}\,\varphi_{j-1}^{m}, \\
&\qquad \left.\sqrt{\frac{(j-m)(j-m-1)}{2j(2j-1)}}\,\varphi_{j-1}^{m+1}\right), \\
\varphi_{j,j}^m &= \left(\sqrt{\frac{(j+m)(j-m+1)}{2j(2j+1)}}\,\varphi_{j}^{m-1}, \right.\\
&\qquad -\frac{m}{\sqrt{j(j+1)}}\,\varphi_{j}^{m}, \\
&\qquad \left.-\sqrt{\frac{(j+m+1)(j-m)}{2j(2j+1)}}\,\varphi_{j}^{m+1}\right), \\
\varphi_{j,j+1}^m &= \left(\sqrt{\frac{(j-m+1)(j-m+2)}{2(j+1)(2j+3)}}\,\varphi_{j+1}^{m-1}, \right.\\
&\qquad -\sqrt{\frac{(j+m+1)(j-m+1)}{(j+1)(2j+3)}}\,\varphi_{j+1}^{m},
\end{aligned} \tag{4.376}$$

$$\left. \sqrt{\frac{(j+m+1)(j+m+2)}{2(j+1)(2j+3)}}\, \varphi_{j+1}^{m+1} \right).$$

The functions obtained are normalized and give rise to the ordinary representations of the angular momentum. Here φ_l^m are the usual normalized harmonics

$$\varphi_l^m = \frac{1}{2^l\, l!} \sqrt{\frac{(2l+1)(l+m)!}{4\pi(l-m)!}}\, (\sin\theta)^{-m}\, \frac{\mathrm{d}^{l-m}\left(\cos^2\theta - 1\right)^l}{(\mathrm{d}\cos\theta)^{l-m}}\, \mathrm{e}^{im\phi}. \tag{4.377}$$

There are some frequently used relations between the spherical functions with spin $\varphi_{j,l}^m$ for given values of j and m with $l = j-1, j.j+1$. Let us consider, for example, the following operator:

$$\begin{aligned} s_r &= \frac{x}{r}\, s_x + \frac{y}{r}\, s_y + \frac{z}{r}\, s_z \\ &= \frac{1}{2}\, \frac{x-iy}{r}\, (s_x + i s_y) + \frac{1}{2}\, \frac{x+iy}{r}\, (s_x - i s_y) + \frac{z}{r}\, s_z, \end{aligned} \tag{4.378}$$

where, as usual,

$$s_x = \begin{pmatrix} 0 & 1/\sqrt{2} & 0 \\ 1/\sqrt{2} & 0 & 1/\sqrt{2} \\ 0 & 1/\sqrt{2} & 0 \end{pmatrix}, \quad s_y = \begin{pmatrix} 0 & -i/\sqrt{2} & 0 \\ i/\sqrt{2} & 0 & -i/\sqrt{2} \\ 0 & i/\sqrt{2} & 0 \end{pmatrix},$$

$$s_z = \begin{pmatrix} 1 & 0 & 0 \\ 0 & 0 & 0 \\ 0 & 0 & -1 \end{pmatrix}. \tag{4.379}$$

Evidently, this operator is a scalar one, so that its commutator with j and m is zero. The following relations hold:

$$\begin{aligned} s_r\, \varphi_{j,j-1}^m &= \sqrt{\frac{j+1}{2j+1}}\, \varphi_{j,j}^m, \\ s_r\, \varphi_{j,j}^m &= \sqrt{\frac{j+1}{2j+1}}\, \varphi_{j,j-1}^m + \sqrt{\frac{j}{2j+1}}\, \varphi_{j,j+1}^m, \\ s_r\, \varphi_{j,j+1}^m &= \sqrt{\frac{j}{2j+1}}\, \varphi_{j,j}^m. \end{aligned} \tag{4.380}$$

The eigenvalues of s_r, i.e, the eigenvalues of matrices of the form

$$\left(\begin{array}{ccc} 0 & \sqrt{\dfrac{j+1}{2j+1}} & 0 \\ \sqrt{\dfrac{j+1}{2j+1}} & 0 & \sqrt{\dfrac{j}{2j+1}} \\ 0 & \sqrt{\dfrac{j}{2j+1}} & 0 \end{array} \right), \tag{4.381}$$

obviously are $\pm 1, 0$, as for a component of the spin along a fixed direction. For $j = 0$, the only allowed rotational state corresponds to $s_r = 0$.

Let us now consider three-valued functions depending on θ, ϕ and r and introduce the operator

$$\begin{aligned} \frac{1}{i}\,\mathbf{s}\cdot\boldsymbol{\nabla} \;&=\; \frac{1}{2i}\left(s_x + i s_y\right)\left(\frac{\partial}{\partial x} - i\frac{\partial}{\partial y}\right) + \frac{1}{2i}\left(s_x - i s_y\right)\left(\frac{\partial}{\partial x} + i\frac{\partial}{\partial y}\right) \\ &\quad + \frac{1}{i}\,s_z\,\frac{\partial}{\partial z}. \end{aligned} \tag{4.382}$$

Setting, for brevity

$$p_x \;=\; \frac{1}{i}\,\frac{\partial}{\partial x}, \qquad p_y \;=\; \frac{1}{i}\,\frac{\partial}{\partial y}, \qquad p_z \;=\; \frac{1}{i}\,\frac{\partial}{\partial z},$$

we find $(1/i)\mathbf{s}\cdot\boldsymbol{\nabla} \;=\; \mathbf{s}\cdot\mathbf{p}$. Noting that

$$\left(\frac{x}{r}\,p_x + \frac{y}{r}\,p_y + \frac{z}{r}\,p_z\right)\varphi_{j,l}^m \;=\; 0, \tag{4.383}$$

we get

$$\begin{aligned} (\mathbf{s}\cdot\mathbf{p})\,\varphi_{j,l}^m \;&=\; \left[\frac{x^2+y^2+z^2}{r^2}\,(\mathbf{s}\cdot\mathbf{p}) - \frac{1}{r}\left(\frac{x}{r}\,p_x + \frac{y}{r}\,p_y + \frac{z}{r}\,p_z\right)s_r\right]\varphi_{j,l}^m \\ &=\; \frac{1}{r}\left[\left(\frac{x}{r}s_y - \frac{y}{r}s_x\right)(xp_y - yp_x) + \left(\frac{y}{r}s_z - \frac{x}{r}s_y\right)(yp_z - zp_y)\right. \\ &\qquad \left. + \left(\frac{z}{r}s_x - \frac{x}{r}s_z\right)(zp_x - xp_z)\right]\varphi_{j,l}^m, \end{aligned}$$

or

$$\begin{aligned} (\mathbf{s}\cdot\mathbf{p})\,\varphi_{j,l}^m \;&=\; \frac{1}{r}\left[\left(\frac{x}{r}s_y - \frac{y}{r}s_x\right)l_z + \left(\frac{y}{r}s_z - \frac{x}{r}s_y\right)l_x\right. \\ &\qquad \left. + \left(\frac{z}{r}s_x - \frac{x}{r}s_z\right)l_y\right]\varphi_{j,l}^m. \end{aligned} \tag{4.384}$$

This can also be cast in the more convenient form

$$\begin{aligned}(\mathbf{s}\cdot\mathbf{p})\,\varphi_{j,l}^m &= \frac{1}{ir}\left\{s_z\left[\frac{1}{2}\frac{x+iy}{r}\,(l_x-il_y)-\frac{1}{2}\frac{x-iy}{r}\,(l_x+il_y)\right]\right.\\ &+\frac{1}{2}\,(s_x+is_y)\left[\frac{x-iy}{r}\,l_z-\frac{z}{r}\,(l_x-il_y)\right] \qquad (4.385)\\ &\left.+\,\frac{1}{2}\,(s_x-is_y)\left[\frac{z}{r}\,(l_x+il_y)-\frac{x+iy}{r}\,l_z\right]\right\}\varphi_{j,l}^m.\end{aligned}$$

Note that l_x, l_y, l_z are the components of the orbital angular momentum measured in units of $h/2\pi$. We thus find:

$$\begin{aligned}\frac{1}{i}\,(\mathbf{s}\cdot\boldsymbol{\nabla})\,\varphi_{j,j-1}^m &= \frac{i}{r}\,(j-1)\sqrt{\frac{j+i}{2j+1}}\,\varphi_{j,j}^m,\\ \frac{1}{i}\,(\mathbf{s}\cdot\boldsymbol{\nabla})\,\varphi_{j,j}^m &= -\frac{i}{r}\,(j+1)\sqrt{\frac{j+i}{2j+1}}\,\varphi_{j,j-1}^m\\ &\quad+\frac{i}{r}\,j\sqrt{\frac{j}{2j+1}}\,\varphi_{j,j+1}^m, \qquad (4.386)\\ \frac{1}{i}\,(\mathbf{s}\cdot\boldsymbol{\nabla})\,\varphi_{j,j+1}^m &= -\frac{i}{r}\,(j+2)\sqrt{\frac{j}{2j+1}}\,\varphi_{j,j}^m.\end{aligned}$$

Equations (4.380) and (4.386) can be generalized by applying the operators s_r and $(1/i)\mathbf{s}\cdot\boldsymbol{\nabla}$ to functions of the type $f(r)\varphi_{j,l}^m$, since we evidently have

$$\begin{aligned}s_r\,f(r)\,\varphi_{j,l}^m &= f(r)\,s_r\,\varphi_{j,l}^m,\\ \frac{1}{i}\,(\mathbf{s}\cdot\boldsymbol{\nabla})\,f(r)\,\varphi_{j,l}^m &= f(r)\,\frac{1}{i}\,(\mathbf{s}\cdot\boldsymbol{\nabla})\,\varphi_{j,l}^m+\frac{1}{i}\,f'(r)\,s_r\,\varphi_{j,l}^m.\end{aligned} \qquad (4.387)$$

Let us now turn to an application of the spherical functions with spin. The problem is to find the eigenfunctions defined by the differential equation

$$\frac{1}{i}\,(\mathbf{s}\cdot\boldsymbol{\nabla})\,\psi+k\,\psi=0. \qquad (4.388)$$

On setting

$$\begin{aligned}\psi_1 &= \frac{-\psi_x+i\psi_y}{\sqrt{2}}, &\qquad \psi_x &= \frac{\psi_3-\psi_1}{\sqrt{2}},\\ \psi_2 &= \psi_z, & \psi_y &= \frac{\psi_1+\psi_3}{\sqrt{2}}, \qquad (4.389)\\ \psi_3 &= \frac{\psi_x+i\psi_y}{\sqrt{2}}, & \psi_z &= \psi_2,\end{aligned}$$

and regarding ψ_x, ψ_y, ψ_z as components of ψ, we get

$$\frac{1}{i}\,(\mathbf{s}\cdot\boldsymbol{\nabla}\,) \;\equiv\; \boldsymbol{\nabla}\times, \tag{4.390}$$

and Eq. (4.388) simply reads

$$\boldsymbol{\nabla}\times\psi \,+\, k\,\psi \;=\; 0. \tag{4.391}$$

There are two kinds of solutions of Eq.(4.391): For $k \neq 0$, we have $\boldsymbol{\nabla}\cdot\psi = 0$, while for $k = 0$ we have $\boldsymbol{\nabla}\times\psi = 0$ and then $\psi = \boldsymbol{\nabla}\,\Phi$, where Φ is completely arbitrary. In the first case, noting that

$$\boldsymbol{\nabla}\times\boldsymbol{\nabla}\times \;=\; \boldsymbol{\nabla}\,(\boldsymbol{\nabla}\,\cdot) \,-\, \boldsymbol{\nabla}^2, \tag{4.392}$$

we get

$$\boldsymbol{\nabla}^2\,\psi \,+\, k^2\,\psi \;=\; 0, \tag{4.393}$$

with the further condition $\boldsymbol{\nabla}\cdot\psi = 0$. The solutions of Eq. (4.393), which are orthogonal to the ones with $\boldsymbol{\nabla}\cdot\psi = 0$, can be cast in the form $\psi_k = \boldsymbol{\nabla}\,\Phi_k$, with

$$\boldsymbol{\nabla}^2\,\Phi_k \,+\, k^2\,\Phi_k \;=\; 0, \tag{4.394}$$

and all these solutions satisfy Eq. (4.391) with only one eigenvalue $k = 0$. (Indeed, considering the solutions of Eq. (4.393) for a given eigenvalue $k \neq 0$, we have $(\boldsymbol{\nabla}\,(\boldsymbol{\nabla}\,\cdot))^2 = k^2\boldsymbol{\nabla}\,(\boldsymbol{\nabla}\,\cdot)$, so that the eigenvalues of $\boldsymbol{\nabla}\,(\boldsymbol{\nabla}\,\cdot)$ are k^2 or 0. In the second case, $(\boldsymbol{\nabla}\,\times)^2 = k^2$ and then $\boldsymbol{\nabla}\times\, = \pm k$, so that we have solutions of Eq. (4.391) with $k \neq 0$ and thus $\boldsymbol{\nabla}\cdot\psi = 0$. In the first case we have instead $(\boldsymbol{\nabla}\,\times)^2 = 0$ and then $\boldsymbol{\nabla}\times\psi_k = 0$, so that $\psi_k = \boldsymbol{\nabla}\,\Phi_k$.)

Let us now return to the first considered representation of the components of ψ and consider Eq. (4.388), assuming $k \neq 0$. From the above digression, it follows that we must have $\boldsymbol{\nabla}\cdot\psi = 0$ or, in the considered representation,

$$-\,\frac{1}{\sqrt{2}}\left(\frac{\partial}{\partial x} \,+\, i\,\frac{\partial}{\partial y}\right)\psi_1 \,+\, \frac{\partial}{\partial z}\,\psi_2 \,+\, \frac{1}{\sqrt{2}}\left(\frac{\partial}{\partial x} \,-\, i\,\frac{\partial}{\partial y}\right)\psi_3 \;=\; 0. \tag{4.395}$$

A solution of Eq. (4.388) corresponding to given values of j and m can be cast in the form

$$\psi \;=\; \frac{u}{r}\,\varphi^m_{j,j-1} \,+\, i\,\frac{v}{r}\,\varphi^m_{j,j} \,+\, \frac{w}{r}\,\varphi^m_{j,j+1}. \tag{4.396}$$

Due to Eq. (4.393), we can expect that, apart from the common factor $\sqrt{r}$ and constant factors, u, v, w are Bessel or Hankel functions of order $j - 1/2$, $j + 1/2$, $j + 3/2$, respectively. In fact, substituting Eq. (4.396) in Eq. (4.388) and using Eqs. (4.387), (4.386), and (4.380), we find [13]

$$
\begin{aligned}
k\,u + \sqrt{\frac{j+1}{2j+1}}\left(v' + \frac{j}{r}\,v\right) &= 0, \\
k\,v - \sqrt{\frac{j+1}{2j+1}}\left(u' - \frac{j}{r}\,u\right) - \sqrt{\frac{j}{2j+1}}\left(w' + \frac{j+1}{r}\,w\right) &= 0, \quad (4.397) \\
k\,w + \sqrt{\frac{j}{2j+1}}\left(v' - \frac{j+1}{r}\,v\right) &= 0.
\end{aligned}
$$

For $k \neq 0$, combining the first and the third equations (4.397) and their derivatives, it follows that

$$
\sqrt{j}\left(u' - \frac{j}{r}\,u\right) - \sqrt{j+1}\left(w' + \frac{j+1}{r}\,w\right) = 0, \qquad (4.398)
$$

which equation is analogous to Eq. (4.395). Given the initial values of u and v, for example, from Eqs. (4.397), and (4.398) w, u', v', w' are algebraically determined, so that the set of equations (4.397) has only two independent solutions. We can eliminate the term $w' + (w/r)(\mathrm{j}{+}1)$ using Eq. (4.398), thus obtaining

$$
\begin{aligned}
k\,u + \sqrt{\frac{j+1}{2j+1}}\left(v' + \frac{j}{r}\,v\right) &= 0, \\
k\,v - \sqrt{\frac{2j+1}{j+1}}\left(u' + \frac{j}{r}\,u\right) &= 0.
\end{aligned} \qquad (4.399)
$$

Finally, eliminating u, we get

$$
v'' + \left(k^2 - \frac{j(j+1)}{r^2}\right) v = 0. \qquad (4.400)
$$

The only regular solution of Eq. (4.400) is $\sqrt{r}\, J_{j+1/2}(|k|r)$; substituting in the first and in the last one of Eqs. (4.397), we immediately obtain u and w. It is useful to remember the following relations:

$$
\begin{aligned}
I_n'(x) + \frac{n}{x}\,I_n(x) &= I_{n-1}(x), \\
I_n'(x) - \frac{n}{x}\,I_n(x) &= -\,I_{n+1}(x),
\end{aligned} \qquad (4.401)
$$

[13] For $j = 0$, u and v do not exist, and we simply have $kw = 0$ or, for $k \neq 0$, $w = 0$.

or, setting $F = \sqrt{x}\, I$,

$$
\begin{aligned}
F_n'(x) + \left(n - \frac{1}{2}\right) \frac{F_n(x)}{x} &= F_{n-1}(x), \\
F_n'(x) - \left(n + \frac{1}{2}\right) \frac{F_n(x)}{x} &= -F_{n+1}(x).
\end{aligned}
\tag{4.402}
$$

It follows that the only solution of Eqs. (4.397) that is regular for $r = 0$ is given, except for a constant factor, by

$$
\begin{aligned}
u &= -\sqrt{\frac{j+1}{2j+1}}\, \sqrt{r}\, I_{j-1/2}(|k|r) \cdot \frac{k}{|k|}, \\
v &= \sqrt{r}\, I_{j+1/2}(|k|r), \\
w &= \sqrt{\frac{j}{2j+1}}\, \sqrt{r}\, I_{j+3/2}(|k|r) \cdot \frac{k}{|k|}.
\end{aligned}
\tag{4.403}
$$

Two independent singular solutions of Eqs. (4.397) are obviously obtained on replacing the Bessel functions with the Hankel functions of the first or second kind:

$$
\begin{aligned}
u^{1,2} &= -\sqrt{\frac{j+1}{2j+1}}\, \sqrt{r}\, H^{1,2}_{j-1/2}(|k|r) \cdot \frac{k}{|k|}, \\
v^{1,2} &= \sqrt{r}\, H^{1,2}_{j+1/2}(|k|r), \\
w^{1,2} &= \sqrt{\frac{j}{2j+1}}\, \sqrt{r}\, H^{1,2}_{j+3/2}(|k|r) \cdot \frac{k}{|k|}.
\end{aligned}
\tag{4.404}
$$

Let us consider the simplest case in which $j = 1$ (for $j = 0$ there are no solutions of Eq. (4.388) with $k \neq 0$). The functions $\varphi^m_{1,0}, \varphi^m_{1,1}, \varphi^m_{1,2}$, the Bessel and Hankel functions of order 1/2, 3/2, 5/2, enter into the expression (4.396) for ψ. Here we explicitly list these functions:

$$
\varphi^1_{1,0} = \sqrt{\frac{1}{4\pi}}\, (1\,,\, 0\,,\, 0)\,, \tag{4.405}
$$

$$
\varphi^0_{1,0} = \sqrt{\frac{1}{4\pi}}\, (0\,,\, 1\,,\, 0)\,, \tag{4.406}
$$

$$
\varphi^{-1}_{1,0} = \sqrt{\frac{1}{4\pi}}\, (0\,,\, 0\,,\, 1)\,; \tag{4.407}
$$

$$
\varphi^1_{1,1} = \sqrt{\frac{1}{4\pi}} \left(\sqrt{\frac{3}{2}} \cos\theta\,,\, \sqrt{\frac{3}{4}} \sin\theta\, \mathrm{e}^{i\phi}\,,\, 0 \right), \tag{4.408}
$$

$$\begin{aligned}
\varphi_{1,1}^{0} &= \sqrt{\frac{1}{4\pi}}\left(\sqrt{\frac{3}{4}}\sin\theta\,\mathrm{e}^{-i\phi}\,,\;0\,,\;\sqrt{\frac{3}{4}}\sin\theta\,\mathrm{e}^{i\phi}\right), && (4.409)\\
\varphi_{1,1}^{-1} &= \sqrt{\frac{1}{4\pi}}\left(0\,,\;\sqrt{\frac{3}{4}}\sin\theta\,\mathrm{e}^{-i\phi}\,,\;-\sqrt{\frac{3}{2}}\cos\theta\right); && (4.410)\\
\varphi_{1,2}^{1} &= \sqrt{\frac{1}{4\pi}}\left(\sqrt{\frac{9}{8}}\cos^2\theta-\sqrt{\frac{1}{8}}\,,\;\frac{3}{2}\sin\theta\,\cos\theta\,\mathrm{e}^{i\phi}\,,\right.\\
&\qquad\left.\sqrt{\frac{9}{8}}\sin^2\theta\,\mathrm{e}^{2i\phi}\right), && (4.411)\\
\varphi_{1,2}^{0} &= \sqrt{\frac{1}{4\pi}}\left(\frac{3}{2}\sin\theta\,\cos\theta\,\mathrm{e}^{-i\phi}\,,\;-\sqrt{\frac{9}{2}}\cos^2\theta+\sqrt{\frac{1}{2}}\,,\right.\\
&\qquad\left.-\frac{3}{2}\sin\theta\,\cos\theta\,\mathrm{e}^{i\phi}\right), && (4.412)\\
\varphi_{1,2}^{-1} &= \sqrt{\frac{1}{4\pi}}\left(\sqrt{\frac{9}{8}}\sin^2\theta\,\mathrm{e}^{-2i\phi}\,,\;-\frac{3}{2}\sin\theta\,\cos\theta\,\mathrm{e}^{-i\phi}\,,\right.\\
&\qquad\left.\sqrt{\frac{9}{8}}\cos^2\theta-\sqrt{\frac{1}{8}}\right); && (4.413)
\end{aligned}$$

$$\begin{aligned}
I_{1/2}(x) &= \sqrt{\frac{2}{\pi x}}\sin x, && (4.414)\\
I_{3/2}(x) &= \sqrt{\frac{2}{\pi x}}\left(-\cos x+\frac{\sin x}{x}\right), && (4.415)\\
I_{5/2}(x) &= \sqrt{\frac{2}{\pi x}}\left(-\sin x-3\frac{\cos x}{x}+3\frac{\sin x}{x^2}\right); && (4.416)
\end{aligned}$$

$$\begin{aligned}
H_{1/2}^{1}(x) &= -i\sqrt{\frac{2}{\pi x}}\,\mathrm{e}^{ix}, && (4.417)\\
H_{3/2}^{1}(x) &= \sqrt{\frac{2}{\pi x}}\,\mathrm{e}^{ix}\left(-1-\frac{i}{x}\right), && (4.418)\\
H_{5/2}^{1}(x) &= \sqrt{\frac{2}{\pi x}}\,\mathrm{e}^{ix}\left(i-\frac{3}{x}-\frac{3i}{x^2}\right); && (4.419)\\
H_{1/2}^{2}(x) &= i\sqrt{\frac{2}{\pi x}}\,\mathrm{e}^{-ix}, && (4.420)\\
H_{3/2}^{2}(x) &= \sqrt{\frac{2}{\pi x}}\,\mathrm{e}^{-ix}\left(-1+\frac{i}{x}\right), && (4.421)\\
H_{5/2}^{2}(x) &= \sqrt{\frac{2}{\pi x}}\,\mathrm{e}^{-ix}\left(-i-\frac{3}{x}+\frac{3i}{x^2}\right). && (4.422)
\end{aligned}$$

Substituting these expressions into Eq. (4.396) and neglecting a constant factor, for the solution that is regular at 0 we have ($\xi = |kr|$; the upper sign refers to $k > 0$ and the lower one to $k < 0$):

(a) $m = 1$:

$$\begin{aligned}
\psi_1 &= \frac{\sin\xi}{r}\left[\mp\sqrt{\frac{3}{8}}\left(1+\cos^2\theta\right) + \frac{i}{\xi}\sqrt{\frac{3}{2}}\cos\theta\right.\\
&\quad\left.\pm\frac{1}{\xi^2}\sqrt{\frac{3}{2}}\left(\frac{3}{2}\cos^2\theta-\frac{1}{2}\right)\right] + \frac{\cos\xi}{r}\left[-i\sqrt{\frac{3}{2}}\cos\theta\right.\\
&\quad\left.\mp\frac{1}{\xi}\sqrt{\frac{3}{2}}\left(\frac{3}{2}\cos^2\theta-\frac{1}{2}\right)\right], \qquad (4.423)\\
\psi_2 &= \frac{\sin\xi}{r}\left[\mp\sqrt{\frac{3}{4}}\sin\theta\cos\theta\,\mathrm{e}^{i\phi} + \frac{i}{\xi}\sqrt{\frac{3}{4}}\sin\theta\,\mathrm{e}^{i\phi}\right.\\
&\quad\left.\pm\frac{1}{\xi^2}\sqrt{\frac{27}{4}}\sin\theta\cos\theta\,\mathrm{e}^{i\phi}\right] + \frac{\cos\xi}{r}\left[-i\sqrt{\frac{3}{4}}\sin\theta\,\mathrm{e}^{i\phi}\right.\\
&\quad\left.\mp\frac{1}{\xi}\sqrt{\frac{27}{4}}\sin\theta\cos\theta\,\mathrm{e}^{i\phi}\right], \qquad (4.424)\\
\psi_3 &= \frac{\sin\xi}{r}\left[\mp\sqrt{\frac{3}{8}}\sin^2\theta\,\mathrm{e}^{2i\phi}\pm\frac{1}{\xi^2}\sqrt{\frac{27}{8}}\sin^2\theta\,\mathrm{e}^{2i\phi}\right]\\
&\quad+\frac{\cos\xi}{r}\left[\mp\frac{1}{\xi}\sqrt{\frac{27}{8}}\sin^2\theta\,\mathrm{e}^{2i\phi}\right]. \qquad (4.425)
\end{aligned}$$

(b) $m = 0$:

$$\begin{aligned}
\psi_1 &= \frac{\sin\xi}{r}\left[\mp\sqrt{\frac{3}{4}}\sin\theta\cos\theta\,\mathrm{e}^{-i\phi} + \frac{i}{\xi}\sqrt{\frac{3}{4}}\sin\theta\,\mathrm{e}^{-i\phi}\right.\\
&\quad\left.\pm\frac{1}{\xi^2}\sqrt{\frac{27}{4}}\sin\theta\cos\theta\mathrm{e}^{-i\phi}\right] + \frac{\cos\xi}{r}\left[-i\sqrt{\frac{3}{4}}\sin\theta\,\mathrm{e}^{-i\phi}\right.\\
&\quad\left.\mp\frac{1}{\xi}\sqrt{\frac{27}{4}}\sin\theta\cos\theta\,\mathrm{e}^{-i\phi}\right], \qquad (4.426)\\
\psi_2 &= \frac{\sin\xi}{r}\left[\mp\sqrt{\frac{3}{2}}\left(1-\cos^2\theta\right)\mp\frac{1}{\xi^2}\sqrt{6}\left(\sqrt{\frac{3}{2}}\cos^2\theta-\frac{1}{2}\right)\right]\\
&\quad+\frac{\cos\xi}{r}\left[\frac{1}{\xi}\sqrt{6}\left(\frac{3}{2}\cos^2\theta-\frac{1}{2}\right)\right], \qquad (4.427)\\
\psi_3 &= \frac{\sin\xi}{r}\left[\pm\sqrt{\frac{3}{4}}\sin\theta\cos\theta\,\mathrm{e}^{i\phi} + \frac{i}{\xi}\sqrt{\frac{3}{4}}\sin\theta\,\mathrm{e}^{i\phi}\right.
\end{aligned}$$

$$
\mp \frac{1}{\xi^2}\sqrt{\frac{27}{4}}\, \sin\theta\cos\theta e^{i\phi}\bigg] + \frac{\cos\xi}{r}\left[-i\sqrt{\frac{3}{4}}\,\sin\theta\, e^{i\phi} \pm \frac{1}{\xi}\sqrt{\frac{27}{4}}\,\sin\theta\cos\theta\, e^{i\phi}\right]. \tag{4.428}
$$

(c) $m = -1$:

$$
\begin{aligned}
\psi_1 &= \frac{\sin\xi}{r}\left[\mp\sqrt{\frac{3}{8}}\,\sin^2\theta\, e^{-2i\phi} \pm \frac{1}{\xi^2}\sqrt{\frac{27}{8}}\,\sin^2\theta\, e^{-2i\phi}\right] \\
&\quad + \frac{\cos\xi}{r}\left[\mp\frac{1}{\xi}\sqrt{\frac{27}{8}}\,\sin^2\theta\, e^{-2i\phi}\right],
\end{aligned} \tag{4.429}
$$

$$
\begin{aligned}
\psi_2 &= \frac{\sin\xi}{r}\left[\pm\sqrt{\frac{3}{4}}\,\sin\theta\cos\theta\, e^{-i\phi} + \frac{i}{\xi}\sqrt{\frac{3}{4}}\,\sin\theta\, e^{-i\phi} \right.\\
&\quad \left. \mp \frac{1}{\xi^2}\sqrt{\frac{27}{4}}\,\sin\theta\cos\theta e^{-i\phi}\right] + \frac{\cos\xi}{r}\left[-i\sqrt{\frac{3}{4}}\,\sin\theta\, e^{i\phi} \right.\\
&\quad \left. \pm \frac{1}{\xi}\sqrt{\frac{27}{4}}\,\sin\theta\cos\theta\, e^{-i\phi}\right],
\end{aligned} \tag{4.430}
$$

$$
\begin{aligned}
\psi_3 &= \frac{\sin\xi}{r}\left[\mp\sqrt{\frac{3}{8}}\left(1+\cos^2\theta\right) - \frac{i}{\xi}\sqrt{\frac{3}{2}}\,\cos\theta \right.\\
&\quad \left. \pm \frac{1}{\xi^2}\sqrt{\frac{3}{2}}\left(\frac{3}{2}\cos^2\theta - \frac{1}{2}\right)\right] + \frac{\cos\xi}{r}\left[i\sqrt{\frac{3}{2}}\,\cos\theta \right.\\
&\quad \left. \mp \frac{1}{\xi}\sqrt{\frac{3}{2}}\left(\frac{3}{2}\cos^2\theta - \frac{1}{2}\right)\right].
\end{aligned} \tag{4.431}
$$

The wavefunction $\boldsymbol{\psi}$ defines two real vector fields in ordinary space. Indeed, using Eqs. (4.389), we can consider the components of ψ along the Cartesian axes x, y, z by setting

$$
\boldsymbol{\psi} = \mathbf{A} + i\,\mathbf{B}, \tag{4.432}
$$

where $\mathbf{A}$ and $\mathbf{B}$ are real vectors. In other words,

$$
\begin{aligned}
\psi_x &= A_x + i\,B_x, && (4.433)\\
\psi_y &= A_y + i\,B_y, && (4.434)\\
\psi_z &= A_z + i\,B_z. && (4.435)
\end{aligned}
$$

Substituting these expressions in Eq. (4.389), we get, apart from a constant ($\pm\sqrt{3/4}$) factor ($\xi = |kr|$; the upper sign refers to $k > 0$ the lower one to $k < 0$):

(a) $m = 1$:

$$
\begin{aligned}
A_x &= \frac{\sin\xi}{r}\left[1 - \sin^2\theta\cos^2\phi - \frac{1}{\xi^2}\left(1 - 3\sin^2\theta\cos^2\phi\right)\right] \\
&+ \frac{\cos\xi}{\xi r}\left(1 - 3\sin^2\theta\cos^2\phi\right), \\
A_y &= \frac{\sin\xi}{r}\left(-\sin^2\theta\sin\phi\cos\phi \pm \frac{1}{\xi}\cos\theta\right. \\
&\left. + \frac{3}{\xi^2}\cdot\sin^2\theta\sin\phi\cos\phi\right) + \frac{\cos\xi}{r}\left(\mp\cos\theta - \frac{3}{\xi}\cdot\sin^2\theta\sin\phi\cos\phi\right), \\
A_z &= \frac{\sin\xi}{r}\left(-\sin\theta\cos\theta\cos\phi \mp \frac{1}{\xi}\sin\theta\sin\phi\right. \\
&\left. + \frac{3}{\xi^2}\cdot\sin\theta\cos\theta\cos\phi\right) + \frac{\cos\xi}{r}\left(\pm\sin\theta\sin\phi - \frac{3}{\xi}\cdot\sin\theta\cos\theta\cos\phi\right); \\
B_x &= \frac{\sin\xi}{r}\left(-\sin^2\theta\sin\phi\cos\phi \mp \frac{1}{\xi}\cos\theta\right. \\
&\left. + \frac{3}{\xi^2}\cdot\sin^2\theta\sin\phi\cos\phi\right) + \frac{\cos\xi}{r}\left(\pm\cos\theta - \frac{3}{\xi}\cdot\sin^2\theta\sin\phi\cos\phi\right), \\
B_y &= \frac{\sin\xi}{r}\left[1 - \sin^2\theta\sin^2\phi - \frac{1}{\xi^2}\left(1 - 3\sin^2\theta\sin^2\phi\right)\right] \\
&+ \frac{\cos\xi}{\xi r}\left(1 - 3\sin^2\theta\sin^2\phi\right), \\
B_z &= \frac{\sin\xi}{r}\left(-\sin\theta\cos\theta\sin\phi \pm \frac{1}{\xi}\sin\theta\cos\phi\right. \\
&\left. + \frac{3}{\xi^2}\cdot\sin\theta\cos\theta\sin\phi\right) + \frac{\cos\xi}{r}\left(\mp\sin\theta\cos\phi - \frac{3}{\xi}\cdot\sin\theta\cos\theta\sin\phi\right).
\end{aligned}
$$

Similar relations hold for the other two cases $m = 0$, $m = -1$.

The following are useful formulae involving the ordinary spherical functions:

$$
\begin{aligned}
&\left(\frac{\partial}{\partial x} - i\frac{\partial}{\partial y}\right) f(r)\,\varphi_l^m \\
&= -\left(f'(r) + \frac{l+1}{r}f(r)\right)\sqrt{\frac{(l+m)(l+m-1)}{(2l+1)(2l-1)}}\,\varphi_{l-1}^{m-1} \\
&\quad + \left(f'(r) - \frac{l}{r}f(r)\right)\sqrt{\frac{(l-m+1)(l-m+2)}{(2l+1)(2l+3)}}\,\varphi_{l+1}^{m-1},
\end{aligned}
$$

$$\begin{aligned}
&\left(\frac{\partial}{\partial x}+i\,\frac{\partial}{\partial y}\right) f(r)\,\varphi_l^m \\
&= \left(f'(r)+\frac{l+1}{r}f(r)\right)\sqrt{\frac{(l-m)(l-m-1)}{(2l+1)(2l-1)}}\,\varphi_{l-1}^{m+1}, \qquad (4.436)\\
&\quad - \left(f'(r)-\frac{l}{r}f(r)\right)\sqrt{\frac{(l+m+1)(l+m+2)}{(2l+1)(2l+3)}}\,\varphi_{l+1}^{m+1},\\
&\frac{\partial}{\partial z}\,f(r)\,\varphi_l^m \\
&= \left(f'(r)+\frac{l+1}{r}f(r)\right)\sqrt{\frac{(l+m)(l-m)}{(2l+1)(2l-1)}}\,\varphi_{l-1}^{m}\\
&\quad + \left(f'(r)-\frac{l}{r}f(r)\right)\sqrt{\frac{(l+m+1)(l-m+1)}{(2l+1)(2l+3)}}\,\varphi_{l+1}^{m}.
\end{aligned}$$

Given a single-valued function u, we can set $\psi = (\psi_1, \psi_2, \psi_3) = \boldsymbol{\nabla}\, u$, with

$$\begin{aligned}
\psi_1 &= -\frac{1}{\sqrt{2}}\left(\frac{\partial u}{\partial x}-i\,\frac{\partial u}{\partial y}\right),\\
\psi_2 &= \frac{\partial u}{\partial z},\\
\psi_3 &= \frac{1}{\sqrt{2}}\left(\frac{\partial u}{\partial x}+i\,\frac{\partial u}{\partial y}\right).
\end{aligned} \qquad (4.437)$$

Then, from the previous formulae and from Eqs. (4.376), it follows that

$$\begin{aligned}
\boldsymbol{\nabla}\, f(r)\,\varphi_l^m &= \sqrt{\frac{l}{2l+1}}\left(f'(r)+\frac{l+1}{r}f(r)\right)\varphi_{l,l-1}^m\\
&\quad -\sqrt{\frac{l+1}{2l+1}}\left(f'(r)\frac{l}{r}f(r)\right)\varphi_{l,l+1}^m
\end{aligned} \qquad (4.438)$$

(this discussion continues in Sec. 4.29).

26. SCATTERING OF FAST ELECTIONS: RELATIVISTIC BORN METHOD

Let us consider the field-free Dirac equation

$$\left(\frac{W}{c}+\rho_1\,\boldsymbol{\sigma}\cdot\mathbf{p}+\rho_3\,mc\right)\psi = 0. \qquad (4.439)$$

First of all, given a 4-valued function $P(q)$, we want to solve the problem of finding the solutions of the differential equation

$$\left(\frac{W}{c} + \rho_1\, \boldsymbol{\sigma}\cdot\mathbf{p} + \rho_3\, mc\right) \psi \;=\; P \tag{4.440}$$

for constant W with the limiting condition that ψ is a diverging wave at infinity. Let us apply the operator $\frac{W}{c} - \rho_1\, \boldsymbol{\sigma}\cdot\mathbf{p} - \rho_3\, mc$ on both sides of Eq. (4.440); we find

$$\left(\frac{W^2}{c^2} - m^2c^2 - p^2\right) \psi \;=\; \left(\frac{W}{c} - \rho_1\, \boldsymbol{\sigma}\cdot\mathbf{p} - \rho_3\, mc\right) P. \tag{4.441}$$

Writing down the operator $\mathbf{p}$ and introducing the constant

$$k \;=\; \frac{1}{\hbar}\sqrt{W^2/c^2 - m^2c^2} \;=\; \frac{1}{\hbar}\,|\mathbf{p}|, \tag{4.442}$$

we get

$$\boldsymbol{\nabla}^2\psi + k^2\psi \;=\; \left[\frac{1}{\hbar^2}\left(\frac{W}{c} - \rho_3\, mc\right) + \frac{i}{\hbar}\,\rho_1\, \boldsymbol{\sigma}\cdot\boldsymbol{\nabla}\right] P. \tag{4.443}$$

As is known, it follows from this that the solution obeying the constraints defined above has the form

$$\psi(q) \;=\; -\frac{1}{4\pi}\int \frac{\mathrm{e}^{ik|\mathbf{q}-\mathbf{q}'|}}{|\mathbf{q}-\mathbf{q}'|}\left[\frac{1}{\hbar^2}\left(\frac{W}{c} - \rho_3\, mc\right) + \frac{i}{\hbar}\,\rho_1\, \boldsymbol{\sigma}\cdot\boldsymbol{\nabla}\right] P(q')\,\mathrm{d}q', \tag{4.444}$$

which can be simplified by integrating by parts. Using the relation ($\boldsymbol{\nabla}$ acts on the independent variable q')

$$\boldsymbol{\nabla}\,\frac{\mathrm{e}^{ik|\mathbf{q}-\mathbf{q}'|}}{|\mathbf{q}-\mathbf{q}'|} \;=\; -\,\frac{\mathbf{q}-\mathbf{q}'}{|\mathbf{q}-\mathbf{q}'|}\left(ik - \frac{1}{|\mathbf{q}-\mathbf{q}'|}\right)\frac{\mathrm{e}^{ik|\mathbf{q}-\mathbf{q}'|}}{|\mathbf{q}-\mathbf{q}'|}, \tag{4.445}$$

we thus find the desired solution

$$\begin{aligned}\psi(q) \;=\; & -\frac{1}{4\pi}\int \frac{\mathrm{e}^{ikr}}{r}\left[\frac{1}{\hbar^2}\left(\frac{W}{c} - \rho_3\, mc\right)\right.\\ & \left. -\,\frac{1}{\hbar}\,\frac{k+i/r}{r}\,\rho_1\, \boldsymbol{\sigma}\cdot(\mathbf{q}-\mathbf{q}')\right] P(q')\,\mathrm{d}q'.\end{aligned} \tag{4.446}$$

Note that, by changing the sign of k, we get the solution of Eq. (4.440) representing a converging wave at infinity.

Let us now assume that an electron plane wave interacts with a field whose potential is V (if this field is described by a scalar potential we have $V = -e\phi$). The Dirac equation can be written as

$$\left(\frac{W}{c} + \rho_1\, \boldsymbol{\sigma}\cdot\mathbf{p} + \rho_3\, mc\right) \psi \;=\; \frac{V}{c}\,\psi. \tag{4.447}$$

This equation can be solved, by means of re-iterated approximations using the Born method, by setting

$$\psi = \psi_0 + \psi_1 + \psi_2 + \dots, \tag{4.448}$$

where ψ_0 is the unperturbed wave and $\psi_1, \psi_2, \dots$ are calculated iteratively by solving the differential equation

$$\left(\frac{W}{c} + \rho_1 \boldsymbol{\sigma}\cdot\mathbf{p} + \rho_3 mc\right) \psi_n = \frac{V}{c} \psi_{n-1} \tag{4.449}$$

in the mentioned way.

Let us study the first approximation taking ψ_0 to be a plane wave along the direction of the z axis:

$$\psi_0 = u\, \mathrm{e}^{ikz}, \quad k = \frac{p}{\hbar}, \tag{4.450}$$

where u is a spin function assumed to be normalized. If the scattering center is 0, we want to determine ψ_1 at a large distance R from the center in the direction specified by θ, ϕ. Let us denote with $\mathbf{t}$ and $\mathbf{t}_1$ unit vectors in the directions of z and θ, ϕ, respectively. We have $\psi_0(q') = u \exp\{ik\mathbf{q}'\cdot\mathbf{t}\}$ and, for large R,

$$|\mathbf{q} - \mathbf{q}'| = R - \mathbf{q}'\cdot\mathbf{t}_1, \quad R \to \infty. \tag{4.451}$$

Substituting into Eq. (4.446), with ψ replaced by ψ_1 and P by $(V/c)\, \psi_0$, for $R \to \infty$ we find

$$\begin{aligned} \psi_1(R;\theta,\phi) = & -\frac{\mathrm{e}^{ikR}}{4\pi R} \int \mathrm{e}^{-i\mathbf{q}'\cdot(k\mathbf{t}_1 - k\mathbf{t})} \left[\frac{1}{\hbar^2}\left(\frac{W}{c} - \rho_3 mc\right)\right. \\ & \left. - \frac{1}{\hbar} k \rho_1 \boldsymbol{\sigma}\cdot\mathbf{t}_1\right] \frac{V(q')}{c} u\, \mathrm{d}q'. \end{aligned} \tag{4.452}$$

Let us assume, for simplicity, that in the scattering field only the scalar potential is different from zero. In this case, we have $V = -e\phi$, and the potential does not depend on the spin variables. Thus, we can replace $V(q')u$ with $uV(q')$ and take the constant part outside the integral. We then deduce

$$\begin{aligned} \psi_1(R;\theta,\phi) = & -\frac{\mathrm{e}^{ikR}}{4\pi R} \int \mathrm{e}^{-i\mathbf{q}'\cdot(k\mathbf{t}_1 - k\mathbf{t})} V(q')\, \mathrm{d}q' \\ & \times \left[\frac{1}{\hbar^2}\left(\frac{W}{c^2} - \rho_3 m\right) - \frac{1}{\hbar}\frac{k}{c} \rho_1 \boldsymbol{\sigma}\cdot\mathbf{t}_1\right] u. \end{aligned} \tag{4.453}$$

We now have to remember that u is a spin function of a plane wave with momentum $p_x = p_y = 0$, $p_z = \hbar k$, so that

$$\left(\frac{W}{c} + \hbar k \rho_1 \sigma_z + \rho_3 mc\right) u = 0. \tag{4.454}$$

Let us set $u = (a, b)$, a and b being the first and the second pair of values of u, respectively. In this case, Eq. (4.454) becomes

$$\begin{aligned} \left(\frac{W}{c} + mc\right) a + \hbar k\, \sigma_z\, b &= 0, \\ \left(\frac{W}{c} - mc\right) b + \hbar k\, \sigma_z\, a. &= 0. \end{aligned} \tag{4.455}$$

We then infer that

$$a = -\,\hbar\, \frac{k\, \sigma_z}{W/c + mc}\, b, \quad b = -\,\hbar\, \frac{k\, \sigma_z}{W/c - mc}\, a. \tag{4.456}$$

Furthermore,

$$\begin{aligned} &\left[\frac{1}{\hbar^2}\left(\frac{W}{c^2} - \rho_3\, m\right) - \frac{1}{\hbar}\,\frac{k}{c}\,\rho_1\, \boldsymbol{\sigma}\cdot\mathbf{t}_1\right] u \\ &= \left(\frac{1}{\hbar^2}\,\frac{W - mc^2}{c^2}\, a - \frac{1}{\hbar}\,\frac{k\, \boldsymbol{\sigma}\cdot\mathbf{t}_1}{c}\, b,\ \frac{1}{\hbar^2}\,\frac{W + mc^2}{c^2}\, b - \frac{1}{\hbar}\,\frac{k\, \boldsymbol{\sigma}\cdot\mathbf{t}_1}{c}\, a\right) \\ &= \frac{1}{\hbar^2}\left(\frac{W - mc^2 + (W + mc^2)\boldsymbol{\sigma}\cdot\mathbf{t}_1\, \sigma_z}{c^2}\, a,\right. \\ &\qquad \left.\frac{W + mc^2 + (W - mc^2)\boldsymbol{\sigma}\cdot\mathbf{t}_1\, \sigma_z}{c^2}\, b\right); \end{aligned} \tag{4.457}$$

and, on setting, for simplicity, $\boldsymbol{\sigma}\cdot\mathbf{t}_1 = \sigma_R$ and

$$\begin{aligned} a' &= \frac{W - mc^2 + (W + mc^2)\sigma_R\, \sigma_z}{\hbar^2\, c^2}\, a, \\ b' &= \frac{W + mc^2 + (W - mc^2)\sigma_R\, \sigma_z}{\hbar^2\, c^2}\, b, \end{aligned} \tag{4.458}$$

we get

$$\left[\frac{1}{\hbar^2}\left(\frac{W}{c^2} - \rho_3\, m\right) - \frac{1}{\hbar}\,\frac{k}{c}\,\rho_1\, \sigma_R\right] u = (a',\ b'). \tag{4.459}$$

Noting that $\sigma_R\sigma_z + \sigma_z\sigma_R = 2\cos\theta$, we also have

$$\begin{aligned} a'^{\dagger} a' &= \frac{1}{\hbar^4}\left[\left(\frac{W \quad mc^2}{c^2}\right)^2 + \left(\frac{W + mc^2}{c^2}\right)^2\right. \\ &\qquad \left. + 2\,\frac{W^2 - m^2c^4}{c^4}\cos\theta\right] a^{\dagger} a \\ &= \frac{2m^2}{\hbar^4}\left[\frac{W^2}{m^2c^4}\,(1 + \cos\theta) + (1 - \cos\theta)\right] a^{\dagger} a. \end{aligned} \tag{4.460}$$

Analogously,

$$b'^{\dagger} b' = \frac{2m^2}{\hbar^4} \left[\frac{W^2}{m^2 c^4} (1 + \cos\theta) + (1 - \cos\theta) \right] b^{\dagger} b. \qquad (4.461)$$

The cross section for scattering along the direction θ, ϕ per unit solid angle is given by

$$S(\theta, \phi) = R^2 \frac{|\psi_1|^2}{|\psi_0|^2}, \quad R \to \infty. \qquad (4.462)$$

On substituting from Eq. (4.453) we find

$$\begin{aligned} S(\theta, \phi) &= \frac{m^2}{8\pi^2 \hbar^4} \left[\frac{W^2}{m^2 c^4} (1 + \cos\theta) + (1 - \cos\theta) \right] \\ &\quad \cdot \left| \int \mathrm{e}^{-i\mathbf{q}\cdot(k\mathbf{t}_1 - k\mathbf{t}} \, V(q) \, \mathrm{d}q \right|^2 . \end{aligned} \qquad (4.463)$$

In the non-relativistic case, this formula takes the well-known simple form

$$S(\theta, \phi) = \frac{m^2}{4\pi^2 \hbar^4} \left| \int \mathrm{e}^{-i\mathbf{q}\cdot(k\mathbf{t}_1 - k\mathbf{t}} \, V(q) \, \mathrm{d}q \right|^2, \quad k = \frac{p}{\hbar}. \qquad (4.464)$$

Let us now turn to Eq. (4.463) and consider the Coulomb field

$$V = -\frac{Ze^2}{r}. \qquad (4.465)$$

It is known that:

$$\begin{aligned} \int \mathrm{e}^{-ik\mathbf{q}\cdot(\mathbf{t}_1 - \mathbf{t}} \, V(q) \, \mathrm{d}q &= -\frac{4\pi \, Ze^2}{k^2 |\mathbf{t}_1 - \mathbf{t}|^2} = -\frac{4\pi \, Ze^2}{k^2 \cdot 4 \sin^2 \theta/2} \\ &= \frac{-\pi \, Ze^2}{k^2 \sin^2 (\theta/2)}. \end{aligned} \qquad (4.466)$$

Introducing the free-electron momentum $p = \hbar k$ and the speed given by

$$v = \frac{c^2}{W} p, \quad \text{with} \quad W^2 = \frac{m^2 c^4}{1 - v^2/c^2}, \qquad (4.467)$$

we finally conclude that

$$S(\theta) = \frac{Z^2 e^4}{4p^2 v^2 \sin^4 (\theta/2)} \left(\frac{2c^2 - v^2}{2c^2} + \frac{v^2}{2c^2} \cos\theta \right). \qquad (4.468)$$

This expression has to be compared with the corresponding classical analog

$$S_{\rm cl}(\theta) = \frac{Z^2 e^4}{4p^2 v^2 \sin^4 \theta/2}. \tag{4.469}$$

We observe that, for small angles, the relativistic scattering coincides with the scattering for a classical electron with the same value of pv. On the contrary, for large angles the relativistic scattering is significantly lower than the classical scattering. We can more easily compare the classical and relativistic scattering (the last one studied using only the first approximation of the Born method) for an electron with given energy. Denoting with E the total energy minus the rest energy,

$$E = W - mc^2 = \frac{mc^2}{\sqrt{1 - v^2/c^2}} - mc^2, \tag{4.470}$$

and setting, for brevity,

$$s = \sqrt{1 - v^2/c^2} = \frac{mc^2}{mc^2 + E}, \tag{4.471}$$

we find

$$pv = E\,(1 + s)\,; \tag{4.472}$$

and, substituting in Eq. (4.468), we get

$$S(\theta) = \frac{Z^2 e^4}{16 E^2 \sin^4 \theta/2} \left(\frac{2 + 2s^2}{(1+s)^2} + \frac{2 - 2s^2}{(1+s)^2} \cos\theta \right). \tag{4.473}$$

The classical expression ($s = 1$) instead is

$$S_{\rm cl}(\theta) = \frac{Z^2 e^4}{16 E^2 \sin^4 (\theta/2)}. \tag{4.474}$$

For given energy, the relativistic scattering is thus greater than the classical value for small scattering angles but smaller than it for large angles. Since

$$\frac{S}{S_{\rm cl}} = \frac{2 + 2s^2}{(1+s)^2} + \frac{2 - 2s^2}{(1+s)^2} \cos\theta, \tag{4.475}$$

the scattering angle θ for which the classical and relativistic formula agree is given by

$$\cos\theta_0 = -\frac{1 - s}{2(1+s)}, \tag{4.476}$$

so that $\theta_0 = 90^o$ for $s \to 1$ and $\theta_0 = 120^o$ for $s \to 0$.

Table 4.2. The ratio between the relativistic and the classical electron scattering cross sections for several values of the scattering angle θ and of the relativistic factor s (see text).

	$s=1$	$s=1/2$	$s=1/3$	$s=0$
$\theta=0^o$	1	1.78	2.25	4.00
$\theta=30^o$	1	1.69	2.12	3.73
$\theta=60^o$	1	1.44	1.75	3.00
$\theta=90^o$	1	1.11	1.25	2.00
$\theta=120^o$	1	0.78	0.75	1.00
$\theta=150^o$	1	0.53	0.38	0.27
$\theta=180^o$	1	0.44	0.25	0.00

In Table 4.2 we list the ratio between the relativistic and the classical scattering cross sections $S(\theta)/S_{\rm cl}(\theta)$ for several values of s and for particular values of θ.

27. FREQUENTLY USED ATOMIC QUANTITIES

(1) *Harmonic oscillator.* Denoting with ν the oscillation frequency in cm^{-1} and with A/N the mass of the oscillating particle (N being the Avogadro number), the largest classical elongation a in an orbit characterized by the quantum number n is given by

$$a = \sqrt{\frac{n}{A}}\sqrt{\frac{N\,\hbar}{\pi\,c\,\nu}} = \sqrt{\frac{n}{A}\,\frac{6.7}{\nu}}\cdot 10^{-8}\ \mathrm{cm}. \tag{4.477}$$

For example, for the hydrogen molecule, with reduced mass $A = 1/2$ and $\nu \sim 4400$, we have $a \sim 0.175\,\sqrt{n}\cdot 10^{-8}$; this result is valid only for very small values of n.

(2) *Energy-wavelength relationship.*[14] Energy of an α particle with wavelength $\lambda_0 = 10^{-12}$cm:

$$E_0 = \frac{300\cdot N\pi^2\hbar^2}{2\lambda_0^2\cdot e}\,\mathrm{V} = 2.09\cdot 10^6\,\mathrm{V}. \tag{4.478}$$

[14]@ The numerical values of the energies reported in the following differ slightly from those found in the original manuscript ($2.05\cdot 10^6$, 150, and $2.08\cdot 10^6$, respectively).

Energy of an electron with wavelength $\lambda_0 = 10^{-8}$cm:

$$E_0 = \frac{2\pi^2\hbar^2\cdot 300}{m\lambda_0^2\cdot e}\,\mathrm{V} = 153\,\mathrm{V}. \tag{4.479}$$

(3) *Energy-velocity relationship.* Energy of an α particle with speed $v = 10^9$cm/s:

$$E_0 = \frac{3.3\cdot 10^{-6}}{1.59\cdot 10^{-12}} = 2.108\cdot 10^6\,\mathrm{V}. \tag{4.480}$$

28. QUASI-STATIONARY STATES

Let us consider an unperturbed system for which a discrete state ψ_0 with energy E_0 exists together with a continuum spectrum ψ_W of energy E_0+W. Now introduce a perturbation linking the discrete state ψ_0 with the continuum states defined by

$$I_W = \int \overline{\psi}_0\, H_p\, \psi_W\, \mathrm{d}\tau. \tag{4.481}$$

Due to this perturbation, the discrete state ψ_0 will be absorbed by the continuum spectrum. The problem is to find the perturbed eigenfunction ψ'_W. Denoting by H the total Hamiltonian, we have

$$\begin{aligned} H\,\psi_0 &= E_0\,\psi_0 + \int \overline{I}_W\,\psi_W\,\mathrm{d}W, \\ H\,\psi_W &= (E_0 + W)\,\psi_W + I_W\,\psi_0, \end{aligned} \tag{4.482}$$

$$H\,\psi'_W = (E_0 + W)\,\psi'_W. \tag{4.483}$$

The problem can be solved exactly; the perturbed eigenfunctions ψ'_W, which are normalized with respect to $\mathrm{d}W$ as assumed for ψ_W, are as follows:

$$\begin{aligned} \psi'_W = {} & \frac{1}{\sqrt{|a|^2+|b|^2}}\,\psi_0 + \frac{a}{\sqrt{|a|^2+|b|^2}}\,\psi_W \\ & - \frac{1}{\sqrt{|a|^2+|b|^2}} \int \overline{I}_{W'}\,\frac{\psi_{W'}}{W'-W}\,\mathrm{d}W', \end{aligned} \tag{4.484}$$

where the integral takes its principal value and

$$a = I_W^{-1}\left(W + \int |I_{W'}|^2\,\frac{\mathrm{d}W'}{W'-W}\right); \quad b = \pi\, I_W \tag{4.485}$$

(the integral again taking its principal value). On setting

$$N_w = \sqrt{|a|^2 + |b|^2}, \tag{4.486}$$

Eq. (4.484) becomes

$$\psi'_W = \frac{1}{N_W}\left(\psi_0 - \int \overline{I}_{W'}\,\frac{\psi_{W'}}{W'-W}\,\mathrm{d}W' + a\,\psi_W\right), \tag{4.487}$$

We can expand the discrete state ψ_0 in the states ψ'_W; getting

$$\psi_0 = \int \frac{1}{N_W}\,\psi'_W\,\mathrm{d}W. \tag{4.488}$$

Let us now consider some approximations by neglecting third-order terms in I_W. Since the interesting values of W, that is, the values entering Eq. (4.488) in a relevant way, are of the same order as I_W^2, we can treat the quantities in the above formulae

$$I_W = I, \quad \int |I_{W'}|^2\,\frac{\mathrm{d}W'}{W'-W} = k, \tag{4.489}$$

as constants. Also setting

$$W = \epsilon - k, \quad \epsilon = W + k, \tag{4.490}$$

we have, in our approximation

$$\begin{aligned} a &= \frac{\epsilon}{I}, \quad b = \pi I, \quad N = \sqrt{\epsilon^2/|I|^2 + \pi^2|I|^2}, \\ \psi'_W &= \frac{1}{\sqrt{\epsilon^2/|I|^2 + \pi^2|I|^2}}\left(\psi_0 - \overline{I}\int \frac{\psi_{W'}}{W'-W}\,\mathrm{d}W' + \frac{\epsilon}{I}\,\psi_W\right), \\ \psi_0 &= \int \frac{\psi'_W}{\sqrt{\epsilon^2/|I|^2 + \pi^2|I|^2}}\,\mathrm{d}W. \end{aligned} \tag{4.491}$$

Obviously, we obtain an identity on substituting the second equation of (4.491) into the third one.

Let us next consider the time dependence of the eigenfunctions, noting that the above relations hold for $t = 0$. We then take the following expression as the factor defining the time dependence of ψ'_W:

$$\mathrm{e}^{-iEt/\hbar} = \mathrm{e}^{-i(E_0-k)t/\hbar}\;\mathrm{e}^{-i\epsilon t/\hbar} \tag{4.492}$$

and assume that at $t = 0$ the system is in the state ψ_0. Using the last equation in (4.491) and (4.492) and factoring out the time-dependent factor in ψ'_W,we have at an arbitrary time,

$$\psi = e^{-i(E_0-k)t/\hbar} \int \frac{e^{-i\epsilon t/\hbar}}{\sqrt{\epsilon^2/|I|^2 + \pi^2|I|^2}} \psi'_W \, d\epsilon, \tag{4.493}$$

so that, on substitution of the second equation into (4.491),

$$\begin{aligned} \psi &= e^{-i(E_0-k)t/\hbar} \Big[e^{-t/2T} \psi_0 \\ &+ \int \frac{\overline{I}}{\epsilon + i\pi|I|^2} \left(e^{-i\epsilon t/\hbar} - e^{-t/2T} \right) \psi_W \, d\epsilon \Big], \end{aligned} \tag{4.494}$$

where $1/T = (2\pi/\hbar)|I|^2$. It is natural to wonder whether Eq. (4.494) can be directly deduced from Eq. (4.482) without using the stationary states ψ'_W and by setting $I_W = I =$ constant right from the start. In such an approach, the quantity k is undetermined (see Eq. (4.489)), so that one could obtain Eq. (4.494) except for an indeterminacy [15]. Equation (4.494) can be written as ($\epsilon = W + k$) [16]

$$\begin{aligned} \psi &= e^{-iE_0t/\hbar} e^{ikt/\hbar} e^{-t/2T} \psi_0 \\ &+ \int \frac{\overline{I}\psi_W}{(W+k) + i\pi|I|^2} e^{-iEt/\hbar} \left(1 - e^{i(W+k)t/\hbar} e^{-t/2T} \right) dW; \end{aligned} \tag{4.495}$$

and, on setting

$$\psi = c\,\psi_0 \, e^{-iE_0t/\hbar} \int c_W \, \psi_W e^{-iEt/\hbar} \, dW \tag{4.496}$$

and replacing in Eq. (4.482) I_W with I, we get

$$\dot{c} = -\frac{i}{\hbar} I \int e^{-iWt/\hbar} c_W \, dW, \; \dot{c}_W = -\frac{i}{\hbar} \overline{I} \, e^{-iWt/\hbar} c. \tag{4.497}$$

We can find solutions of these equations in the form

$$\begin{aligned} c &= e^{ixt/\hbar} e^{-t/2T}, \\ c_W &= \frac{\overline{I}}{W + x + i\pi|I|^2} \left(1 - e^{i(W+x)t/\hbar} e^{-t/2T} \right), \end{aligned} \tag{4.498}$$

with an arbitrary x, although the initial conditions are determined ($c = 1$, $c_W = 0$). This indeterminacy depends on the fact that the integral

---

[15]@ In other words, using the "direct" method, the (perturbed) energy eigenvalues remain undetermined.

[16]@ In the original manuscript, the integration differential $d\epsilon$ appears in the following equation; however, it is evident that it must rather be dW.

in the first equation (4.497) does not converge. Equations (4.498) give an expression for the time-dependent ψ that coincides with the one in Eq. (4.495), except for the substitution of the arbitrary quantity x by the determined quantity k.

Let us now assume that for the unperturbed system there exist a discrete state ψ_0 of energy E_0 and *two* continuum spectra ψ_W and ϕ_W with energy E_0+W. Consider then a perturbation connecting the state ψ_0 with both set of states ψ_W and ϕ_W:

$$I_W = \int \overline{\psi}_0\, H_p\, \psi_W\, \mathrm{d}\tau, \;\; L_W = \int \overline{\psi}_0\, H_p\, \phi_W\, \mathrm{d}\tau. \tag{4.499}$$

Instead of Eqs. (4.482), we now have

$$\begin{aligned} H\,\psi_0 &= E_0\,\psi_0 + \int \overline{I}_W\,\psi_W\,\mathrm{d}W + \int \overline{L}_W\,\phi_W\,\mathrm{d}W, \\ H\,\psi_W &= (E_0 + W)\,\psi_W + I_W\,\psi_0, \\ H\,\phi_W &= (E_0 + W)\,\phi_W + L_W\,\psi_0. \end{aligned} \tag{4.500}$$

Due to the induced perturbation, the discrete state ψ_0 will be absorbed in the continuum spectrum, but now, for each value of W, we have two stationary states Z_W^1 and Z_W^2:

$$H\,Z_W^1 = (E_0 + W)\,Z_W^1, \;\; H\,Z_W^2 = (E_0 + W)\,Z_W^2. \tag{4.501}$$

We can choose Z_W^1 and Z_W^2 to be orthogonal and normalized as follows:

$$\begin{aligned} Z_W^1 &= \frac{1}{N_W'}\left(\psi_0 + a\,\psi_W + A\,\phi_W - \int \frac{\overline{I}_{W'}\psi_{W'}}{W'-W}\,\mathrm{d}W' \right. \\ &\quad \left. - \int \frac{\overline{L}_{W'}\phi_{W'}}{W'-W}\,\mathrm{d}W'\right), \\ Z_W^2 &= \frac{L_W\psi_W}{\sqrt{|I_W|^2+|L_W|^2}} - \frac{I_W\phi_W}{\sqrt{|I_W|^2+|L_W|^2}}, \end{aligned} \tag{4.502}$$

where, as usual, the integrals take their principal values and

$$\begin{aligned} a &= \frac{\overline{I}_W}{|I_W|^2+|L_W|^2}\left(W + \int |I_{W'}|^2\,\frac{\mathrm{d}W'}{W'-W} \right. \\ &\quad \left. + \int |L_{W'}|^2\,\frac{\mathrm{d}W'}{W'-W}\right), \end{aligned} \tag{4.503}$$

$$
\begin{aligned}
A &= \frac{\overline{L}_W}{|I_W|^2+|L_W|^2}\left(W + \int |I_{W'}|^2\,\frac{\mathrm{d}W'}{W'-W}\right.\\
&\qquad \left. + \int |L_{W'}|^2\,\frac{\mathrm{d}W'}{W'-W}\right), \qquad (4.504)
\end{aligned}
$$

$$
N'_W = \sqrt{|a|^2 + |A|^2 + \pi^2|I_W|^2 + \pi^2|L_W|^2}. \qquad (4.505)
$$

The states Z^2_W are orthogonal to ψ_0, so that ψ_0 can be expanded in the same way we did in Eq. (4.488), using only the states Z^1_W:

$$
\psi_0 = \int Z^1_W/N'_W\,\mathrm{d}W. \qquad (4.506)
$$

Let us now use some approximations analogous to those in Eqs. (4.489), (4.490), and (4.491) by setting

$$
\begin{aligned}
&I_W = I, \quad L_W = L,\\
&\int |I_{W'}|^2\,\frac{\mathrm{d}W'}{W'-W} + \int |L_{W'}|^2\,\frac{\mathrm{d}W'}{W'-W} = k,
\end{aligned} \qquad (4.507)
$$

$$
W = \epsilon - k, \quad \epsilon = W + k, \qquad (4.508)
$$

from which

$$
\begin{aligned}
&a = \frac{\overline{I}\epsilon}{|I|^2+|L|^2}, \quad A = \frac{\overline{L}\epsilon}{|I|^2+|L|^2},\\
&N'_W = \sqrt{\frac{\epsilon^2}{|I|^2+|L|^2} + \pi^2(|I|^2+|L|^2)},
\end{aligned} \qquad (4.509)
$$

$$
\begin{aligned}
Z^1_W &= \frac{1}{N'_W}\left(\psi_0 + \frac{\epsilon\,\overline{I}}{|I|^2+|L|^2}\,\psi_W - \overline{I}\int \frac{\psi_{W'}}{W'-W}\,\mathrm{d}W'\right.\\
&\qquad \left. + \frac{\epsilon\,\overline{L}}{|I|^2+|L|^2}\,\phi_W - \overline{L}\int \frac{\phi_{W'}}{W'-W}\,\mathrm{d}W'\right), \qquad (4.510)
\end{aligned}
$$

$$
\begin{aligned}
Z^2_W &= \frac{L\,\psi_W - I\,\phi_W}{\sqrt{|I|^2+|L|^2}},\\
\psi_0 &= \int \frac{1}{\sqrt{\epsilon^2/(|I|^2+|L|^2) + \pi^2(|I|^2+|L|^2)}}\,Z^1_W\,\mathrm{d}\epsilon. \qquad (4.511)
\end{aligned}
$$

Equations (4.510) and (4.511) are in strict analogy with Eqs. (4.491); we can immediately deduce that, if the system is initially in the state

ψ_0, its eigenfunction at time t will be expressed, in a way analogous to that in Eq.(4.495), by

$$\begin{aligned}\psi &= \mathrm{e}^{-i(E_0-k)t/\hbar}\,\mathrm{e}^{-t/2T}\,\psi_0 \\ &+ \overline{I}\int \frac{\psi_W\,\mathrm{e}^{-iEt/\hbar}}{W+k+i\pi(|I|^2+|L|^2)}\left(1-\mathrm{e}^{i(W+k)t/\hbar}\,\mathrm{e}^{-t/2T}\right)\mathrm{d}W \qquad (4.512) \\ &+ \overline{L}\int \frac{\phi_W\,\mathrm{e}^{-iEt/\hbar}}{W+k+i\pi(|I|^2+|L|^2)}\left(1-\mathrm{e}^{i(W+k)t/\hbar}\,\mathrm{e}^{-t/2T}\right)\mathrm{d}W,\end{aligned}$$

where now

$$\frac{1}{T} = \frac{2\pi}{\hbar}\left(|I|^2+|L|^2\right). \qquad (4.513)$$

The probability per unit time for the transition between the state ψ_0 and the states ψ_W is then $2\pi|I|^2/\hbar$, while that for the transition between the state ψ_0 and the states ϕ_W is $2\pi|I|^2/\hbar$, as expected.

We now consider another problem. Let us assume that the system is initially in the continuum state ψ_W and let us calculate the relative probability for the system to be at time t in the state ψ_0 or in the states $\phi_{W'}$ or $\psi_{W'}$, with W' different from W, We may use the usual interpretation, that is, the probability for the system to be in the arbitrary state Y is unity if

$$\left|\int \overline{Y}\,\psi\,\mathrm{d}\tau\right|^2 = 1,$$

so that $|\psi|^2$ is the probability density in the configuration space τ. However, we prefer to use the concept of "number of systems" in a considered state rather than that of the relative probability for the system to be in that state. Now, although the continuum state ψ_W is not strictly stationary and represents an infinite number of systems, only a finite number of these has an energy that differs from E_0+W by a finite quantity. Thus we can expect that only transitions to states next to ψ_W and ϕ_W will increase in number with time (and we can presume this increase to be linear). We face the problem by using the stationary states Z_W^1, Z_W^2 and the approximations in Eqs. (4.507) and (4.509). Expanding ψ_W in the states Z_W^1 and Z_W^2, we have

$$\begin{aligned}\psi_W &= \frac{1}{N'_W}\,\frac{\epsilon\,I}{|I|^2+|L|^2}\,Z_W^1 + I\int\frac{Z_{W'}^1}{N'_{W'}(W'-W)}\,\mathrm{d}W' \\ &+ \frac{\overline{L}}{|I|^2+|L|^2}\,Z_W^2. \qquad (4.514)\end{aligned}$$

As usual, the integral takes its principal value. If at time $t=0$ we have $\psi=\psi_W$, we can immediately calculate ψ at time t using the expansion in Eq.(4.514):

$$\psi = \frac{1}{N'_W} \frac{\epsilon I}{|I|^2 + |L|^2} \mathrm{e}^{-iEt/\hbar} Z^1_W + I \int \mathrm{e}^{-iE't/\hbar} \frac{Z^1_{W'}}{N'_{W'}(W' - W)} \mathrm{d}W' + \frac{\overline{L}}{|I|^2 + |L|^2} \mathrm{e}^{-iEt/\hbar} Z^2_W, \tag{4.515}$$

where $E = E_0 + W$, $E' = E_0 + W'$. Substituting (4.510) into (4.515), we can obtain the expression for ψ in terms of the unperturbed states ψ_0, ψ_W, ϕ_W. Note that, to avoid some difficulties stemming from the singularities in the above integrals, it is useful to replace there the expression $(1/W' - W)$ by

$$\frac{W' - W}{(W' - W)^2 + \alpha^2}$$

and then to take the limit $\alpha \to 0$. For $t > 0$ it is also useful to express ψ as the sum of two particular solutions, $\psi = \psi_1 + \psi_2$, such that $\psi_1 + \psi_2 = \psi_W$ for $t = 0$ and that the state ψ_1 substantially describes the phenomenon for sufficiently large values of time, while ψ_2 is one of the discrete states of the form given in (4.512). In this way we have

$$t > 0, \quad \psi = \psi_1 + \psi_2,$$

$$\begin{aligned} \psi_1 &= \mathrm{e}^{-iEt/\hbar} \psi_W + \frac{I}{\epsilon + i\pi Q^2} \mathrm{e}^{-iEt/\hbar} \psi_0 \\ &\quad - \frac{I}{\epsilon + i\pi Q^2} \int \frac{\overline{I}\psi_{W'} + \overline{L}\phi_{W'}}{\epsilon' - \epsilon} \mathrm{e}^{-iEt/\hbar} \left(1 - \mathrm{e}^{i(E-E')t/\hbar}\right) \mathrm{d}E', \\ \psi_2 &= \frac{I}{\epsilon + i\pi Q^2} \left[\mathrm{e}^{i(E_0-k)t/\hbar} \mathrm{e}^{-t/2T} \psi_0 \right. \\ &\quad \left. + \int \frac{\overline{I}\psi_{W'} + \overline{L}\phi_{W'}}{\epsilon' + i\pi Q^2} \mathrm{e}^{-iE't/\hbar} \left(1 - \mathrm{e}^{i(E'-E_0+k)t/\hbar} \mathrm{e}^{-t/2T}\right) \mathrm{d}E' \right], \end{aligned} \tag{4.516}$$

with

$$Q = \sqrt{|I|^2 + |L|^2}, \quad \frac{1}{T} = \frac{2\pi}{\hbar} Q^2,$$

$$\epsilon = E - E_0 + k, \quad \epsilon' = E' - E_0 + k.$$

The number of transitions per unit time from the state ψ_W to the states $\psi_{W'}$ and $\phi_{W'}$ with energy close to E depends on the resonance denominator $1/(\epsilon' - \epsilon)$ in the expression for ψ_1 for sufficiently large values of the time. Denoting with A the number of transitions per unit time to states $\psi_{W'}$ $(W' \neq W)$ and with B the same number for the states $\phi_{W'}$, we find

$$A = \frac{2\pi}{\hbar}\,|I|^2\,\frac{|I|^2}{\epsilon^2+\pi^2 Q^4}, \qquad B = \frac{2\pi}{\hbar}\,|L|^2\,\frac{|I|^2}{\epsilon^2+\pi^2 Q^4}. \tag{4.517}$$

Let us now turn to the exact equations (4.502) and introduce some simplifying notations. We set

$$\begin{aligned} \epsilon_W &= W + \int |I_{W'}|^2\,\frac{\mathrm{d}W'}{W'-W} + \int |L_{W'}|^2\,\frac{\mathrm{d}W'}{W'-W} \\ &= W + k_W, \end{aligned} \tag{4.518}$$

$$Q_W = \sqrt{|I_W|^2 + |L_W|^2}, \tag{4.519}$$

so that Eqs. (4.503), (4.504), and (4.505) become

$$a = \epsilon_W\,\frac{\overline{I}_W}{Q_W^2}, \qquad A = \epsilon_W\,\frac{\overline{L}_W}{Q_W^2}, \qquad N'_W = \sqrt{\frac{\epsilon_W^2}{Q_W^2} + \pi^2 Q_W^2}, \tag{4.520}$$

while Eqs. (4.502) become

$$\begin{aligned} Z_W^1 &= \frac{1}{N'_W}\left(\psi_0 + \epsilon_W\,\frac{\overline{I}_W}{Q_W^2}\,\psi_W + \epsilon_W\,\frac{\overline{L}_W}{Q_W^2}\,\phi_W \right. \\ &\qquad \left. - \int \overline{I}_{W'}\,\frac{\psi_{W'}}{W'-W}\,\mathrm{d}W' - \int \overline{L}_{W'}\,\frac{\phi_{W'}}{W'-W}\,\mathrm{d}W'\right), \\ Z_W^2 &= \frac{L_W}{Q_W}\,\psi_W - \frac{I_W}{Q_W}\,\phi_W. \end{aligned} \tag{4.521}$$

It is useful to introduce some particular combinations of the states ψ_W and ϕ_W that in several applications have a precise physical meaning. We thus set

$$\psi_W = u_W^1 + u_W^2, \qquad \phi_W = v_W^1 + v_W^2, \tag{4.522}$$

where

$$\begin{aligned} u_W^1 &= \frac{1}{2}\,\psi_W - \frac{i}{2\pi}\int \frac{\overline{I}_{W'}}{\overline{I}_W}\,\frac{\psi_{W'}}{W'-W}\,\mathrm{d}W', \\ u_W^2 &= \frac{1}{2}\,\psi_W + \frac{i}{2\pi}\int \frac{\overline{I}_{W'}}{\overline{I}_W}\,\frac{\psi_{W'}}{W'-W}\,\mathrm{d}W'; \end{aligned} \tag{4.523}$$

$$
\begin{aligned}
v_W^1 &= \frac{1}{2}\,\phi_W - \frac{i}{2\pi}\int \frac{\overline{L}_{W'}}{\overline{L}_W}\,\frac{\phi_{W'}}{W'-W}\,\mathrm{d}W', \\
v_W^2 &= \frac{1}{2}\,\phi_W + \frac{i}{2\pi}\int \frac{L_{W'}}{\overline{L}_W}\,\frac{\phi_{W'}}{W'-W}\,\mathrm{d}W'.
\end{aligned}
\tag{4.524}
$$

Along with Eqs. (4.522), the following relations also hold:

$$
\begin{aligned}
\int \overline{I}_{W'}\,\frac{\psi_{W'}}{W'-W}\,\mathrm{d}W' &= i\pi\,\overline{I}_W\left(u_W^1 - u_W^2\right), \\
\int \overline{L}_{W'}\,\frac{\phi_{W'}}{W'-W}\,\mathrm{d}W' &= i\pi\,\overline{L}_W\left(v_W^1 - v_W^2\right).
\end{aligned}
\tag{4.525}
$$

On substituting these relations and Eqs. (4.522), Eqs. (4.521) become

$$
\begin{aligned}
Z_W^1 &= \frac{1}{N'_W}\,\psi_0 + \frac{\overline{I}_W}{N'_W}\left(\frac{\epsilon_W}{Q_W^2} - i\pi\right)u_W^1 + \frac{\overline{I}_W}{N'_W}\left(\frac{\epsilon_W}{Q_W^2} + i\pi\right)u_W^2 \\
&+ \frac{\overline{L}_W}{N'_W}\left(\frac{\epsilon_W}{Q_W^2} - i\pi\right)v_W^1 + \frac{\overline{L}_W}{N'_W}\left(\frac{\epsilon_W}{Q_W^2} + i\pi\right)v_W^2, \qquad (4.526) \\
Z_W^2 &= \frac{L_W}{Q_W}\,u_W^1 + \frac{L_W}{Q_W}\,u_W^2 - \frac{I_W}{Q_W}\,v_W^1 - \frac{I_W}{Q_W}\,v_W^2.
\end{aligned}
$$

The most general stationary state corresponding to the energy $E_0 + W$ is a combination of Z_W^1 and Z_W^2:

$$
Z_W = \lambda\,Z_W^1 + \mu\,Z_W^2. \tag{4.527}
$$

We can then set

$$
Z_W = c\,\psi_0 + c_1\,u_W^1 + c_2\,u_W^2 + C_1\,v_W^1 + C_2\,v_W^2, \tag{4.528}
$$

where

$$
\begin{aligned}
c &= \frac{\lambda}{N'_W}, \\
c_1 &= \lambda\,\frac{\overline{I}_W}{N'_W}\left(\frac{\epsilon_W}{Q_W^2} - i\pi\right) + \mu\,\frac{L_W}{Q_W}, \\
c_2 &= \lambda\,\frac{\overline{I}_W}{N'_W}\left(\frac{\epsilon_W}{Q_W^2} + i\pi\right) + \mu\,\frac{L_W}{Q_W}, \\
C_1 &= \lambda\,\frac{\overline{L}_W}{N'_W}\left(\frac{\epsilon_W}{Q_W^2} - i\pi\right) - \mu\,\frac{I_W}{Q_W}, \\
C_2 &= \lambda\,\frac{\overline{L}_W}{N'_W}\left(\frac{\epsilon_W}{Q_W^2} + i\pi\right) - \mu\,\frac{I_W}{Q_W}.
\end{aligned}
\tag{4.529}
$$

Note that the following identity holds:

$$|c_1|^2 + |C_1|^2 = |c_2|^2 + |C_2|^2 = |\lambda|^2 + |\mu|^2. \tag{4.530}$$

Let us now consider stationary states of the form (4.528) with $C_2 = 0$. It is sufficient to set

$$\lambda = \frac{I_W}{Q_W}, \quad \mu = \frac{\overline{L}_W}{N'_W}\left(\frac{\epsilon_W}{Q_W^2} + i\pi\right) \tag{4.531}$$

in Eq. (4.529). In order to apply Eqs. (4.530), we observe that, since

$$\left|\frac{\epsilon_W}{Q_W^2} + i\pi\right| = \frac{N'_W}{Q_W},$$

from Eqs. (4.531), we have

$$|\lambda| = \frac{|I_W|}{Q_W}, \quad |\mu| = \frac{|L_W|}{Q_W}, \quad |\lambda|^2 + |\mu|^2 = 1, \tag{4.532}$$

so that (4.530) becomes (given $C_2 = 0$)

$$|c_1|^2 + |C_1|^2 = 1, \quad |c_2|^2 = 1. \tag{4.533}$$

The expression for the considered state will be of the form

$$Z_W = c\,\psi_0 + c_1\,u_W^1 + c_2\,u_W^2 + C_1\,v_W^1, \tag{4.534}$$

in which the values of the constants are obtained by substituting Eqs. (4.531) into (4.529):

$$\begin{aligned}
c &= \frac{I_W}{N'_W Q_W}, \\
c_1 &= \frac{1}{N'_W Q_W}\left[\epsilon_W - i\pi\left(|I_W|^2 - |L_W|^2\right)\right], \\
C_1 &= -2i\pi\,\frac{I_W \overline{L}_W}{N'_W Q_W}, \\
c_2 &= \frac{1}{N'_W Q_W}\left(\epsilon_W + i\pi\,Q_W^2\right).
\end{aligned} \tag{4.535}$$

Table 4.3. The maximum value of the ratio $|C_1|^2/|c_2|^2$ as given by Eq. (4.539).

k	p_0
1	1
2 ; 1/2	0.889
3 ; 1/3	0.750
6 ; 1/6	0.490
10 ; 1/10	0.331
100 ; 1/100	0.039

It follows that

$$
\begin{aligned}
|c_1|^2 &= \frac{\epsilon_W^2 + \pi^2(|I_W|^2 - |L_W|^2)^2}{\epsilon_W^2 + \pi^2 Q_W^4}, \\
|C_1|^2 &= \frac{4\pi^2 |I_W|^2 |L_W|^2}{\epsilon_W^2 + \pi^2 Q_W^4}, \\
|c_2|^2 &= 1, \quad |c_1|^2 + |C_1|^2 = 1.
\end{aligned} \tag{4.536}
$$

In the approximation for which we can assume the terms $I_W = I$ and $L_W = L$ to be constant, the ratio $|C_1|^2/|c_2|^2$ takes its maximum value for $\epsilon_W = 0$. This value is given by

$$
\left(\frac{|C_1|^2}{|c_2|^2}\right)_0 = \frac{4|I|^2|L|^2}{Q^4} = \frac{4|I|^2|L|^2}{(|I|^2 + |L|^2)^2} = 1 - \left(\frac{|I|^2 - |L|^2}{|I|^2 + |L|^2}\right)^2 \tag{4.537}
$$

and always is less than 1 except if $|I|^2 = |L|^2$, when it equals 1. Let us set

$$
p_0 = \left(\frac{|C_1|^2}{|c_2|^2}\right)_0, \quad k = \frac{|I|^2}{|L|^2}. \tag{4.538}
$$

We then have

$$
p_0 = \frac{4k}{(k+1)^2} = \frac{4}{(1+k)(1+1/k)} = \frac{4}{k + 2 + 1/k}, \tag{4.539}
$$

so that $p_0(k) = p_0(1/k)$.

We can view $|C_1|^2/|c_2|^2$ as a function of ϵ and put

$$
p = p(\epsilon) = \frac{|C_1|^2}{|c_2|^2}. \tag{4.540}
$$

In the approximation for which $I_W = I$, $L_W = L$, we have

$$
p = \frac{4\pi^2 |I|^2 |L|^2}{\epsilon^2 + \pi^2 Q^4}. \tag{4.541}
$$

The integral $\int p(\epsilon) \mathrm{d}\epsilon$ plays a special role in some applications. We immediately find

$$\int p(\epsilon)\, \mathrm{d}\epsilon = \frac{4\pi^2 |I|^2 |L|^2}{Q^2} = 4\pi^2 \frac{|I|^2|L|^2}{Q^4} Q^2. \tag{4.542}$$

Introducing the annihilation probability

$$\frac{1}{T} = \frac{2\pi}{\hbar} Q^2$$

for the unstable state ψ_0 and the partial annihilation probabilities

$$\frac{1}{T_1} = \frac{2\pi}{\hbar} |I|^2, \quad \frac{1}{T_2} = \frac{2\pi}{\hbar} |L|^2$$

for transitions to the states ψ_W and ϕ_W, respectively, we get

$$\frac{1}{T} = \frac{1}{T_1} + \frac{1}{T_2}, \quad \frac{1}{T_1} = \frac{k}{k+1} \frac{1}{T}, \quad \frac{1}{T_2} = \frac{1}{k+1} \frac{1}{T}, \tag{4.543}$$

and Eq. (4.542) can be written as

$$\begin{aligned} \int p(\epsilon)\, \mathrm{d}\epsilon &= \frac{2\pi\hbar}{T} \frac{p_0}{4} = \frac{2\pi\hbar}{T} \frac{k}{(k+1)^2} \\ &= \frac{2\pi\hbar}{T_1} \frac{1}{k+1} = \frac{2\pi\hbar}{T_2} \frac{k}{k+1}. \end{aligned} \tag{4.544}$$

Let us now turn to an application of the previous formulae to the problem of disintegration of light nuclei resonances by α-particle capture with proton emission [17]. Let us then consider the simplest case of an unstable state ψ_0 of the system "nucleus + α-particle" that spontaneously emits an α-particle or a proton. We assume, for simplicity, that the particle coming from the disintegration of ψ_0 is emitted as an s-wave and that the daughter nucleus is always in its ground state. For the problem to be mathematically well-posed, besides the state ψ_0, we also have to consider some states ψ_W representing the parent nucleus and the α-particle in a hyperbolic s orbit, and some other states ϕ_W that describe the daughter nucleus and the free proton in a s orbit. The state ψ_0 is thus coupled to both ψ_W and ϕ_W states by a perturbation H_p defined by Eq. (4.499). Assuming ψ_W to be normalized with respect to dW (and disregarding the peculiarities of the nucleus), it is simple to see that it represents a converging or diverging flux of α-particles equal

[17]@ In modern language this means an (α, p) reaction: $N + \alpha \rightarrow p + N'$.

to $1/2\pi\hbar$ (number of particles per unit time [18]) and, in the same way, ϕ_W represents an ingoing or outgoing flux of protons equal to $1/2\pi\hbar$. By contrast, the non-stationary states u_W^1 and u_W^2 defined in Eq. (4.523) represent at large distances only outgoing or ingoing flux of α-particles, respectively, whose intensity is again $1/2\pi\hbar$. Analogously, v_W^1 and v_W^2 defined in Eq. (4.524) represent an outgoing or ingoing flux of protons. Let us now assume that a plane wave of α-particles with definite energy, representing a unitary flux per unit area, interacts with the parent nucleus; the problem is to study the scattering of the α-particles and to determine for what number of α-particles the nucleus disintegrates. To this end, we have to consider a stationary state representing the incident plane wave plus a diverging spherical wave of α-particles plus a diverging spherical wave of protons. Such a state can be obtained as a sum of particular solutions. The particular solutions corresponding to the parent nucleus plus α-particles with azimuthal quantum number different from zero represent the usual scattering processes, which have the well-known form given by the theory of scattering from a Coulomb field. However, the considered state must be composed also of particular solutions representing incident α-particles with $l = 0$ as well as of a diverging wave of α-particles with $l = 0$. Moreover, due to the coupling with ψ_0 and with the states ϕ_W, this state must also be composed of an excited ψ_0 state (at a certain degree of excitation) as well as of a diverging wave of protons. Such a particular solution will have the same form as Eq. (4.534), with the constants taking the values as in Eq. (4.535) except for a proportionality factor. Now, c_2 can be determined from the condition that the incoming flux of α-particles is due to the incident plane wave. This incoming flux equals $|c_2|^2/2\pi\hbar$. On the other hand, the number of α-particles with $l = 0$ impinging on the nucleus per unit time is equal to the flux through a circular cross section, normal to the propagation direction of the wave, with radius $\lambda/2\pi$ (λ being the wavelength of the α-particles). Since the incident wave represents a unit flux per unit area, we have

$$\frac{|c_2|^2}{2\pi\hbar} = \pi \left(\frac{\lambda}{2\pi}\right)^2 = \frac{\lambda^2}{4\pi} = \frac{\pi\hbar^2}{M^2 v^2}, \tag{4.545}$$

from which, apart for a phase constant,

$$c_2 = \sqrt{2\pi^2\hbar}\,\frac{\lambda}{2\pi}. \tag{4.546}$$

We can then obtain c, c_1 and C_1 from Eqs. (4.535) by multiplying the values given there by the value of c_2 in Eqs. (4.546) and dividing by the

[18]@ This point is quite obscure in the original manuscript.

value of c_2 corresponding to (4.535). However, we are interested in the moduli of c_1 and C_1, since we only study the frequency of disintegration processes, while we disregard scattering anomalies that also depend on the phase of c_1. From Eqs. (4.536) and (4.546), it follows that

$$
\begin{aligned}
|c_1|^2 &= \frac{\hbar\lambda^2}{2} \frac{\epsilon^2 + \pi^2(|I_W|^2 - |L_W|^2)^2}{\epsilon^2 + \pi^2 Q_W^4}, \\
|C_1|^2 &= \frac{\hbar\lambda^2}{2} \frac{4\pi^2 |I_W|^2 |L_W|^2}{\epsilon^2 + \pi^2 Q_W^4}, \qquad (4.547) \\
|c_2|^2 &= \frac{\hbar\lambda^2}{2}, \qquad \lambda = 2\pi\hbar/Mv. \qquad (4.548)
\end{aligned}
$$

The outgoing proton flux is given by $|C_1|^2/2\pi\hbar$, and it represents the cross section $S(\epsilon)$ for the disintegration process:

$$
S(\epsilon) = \frac{\lambda^2}{4\pi} \frac{4\pi^2 |I|^2 |L|^2}{\epsilon^2 + \pi^2 Q^4}, \qquad (4.549)
$$

or, introducing $p(\epsilon)$ from (4.541),

$$
S(\epsilon) = \frac{\lambda^2}{4\pi} p(\epsilon). \qquad (4.550)
$$

Since $\lambda^2/4\pi$ gives the cross section for α-particles with vanishing azimuthal quantum number, $p(\epsilon)$ is the probability that one of such particles will induce a disintegration. For $\epsilon = 0$, that is, the most favorable value for the energy, this probability reaches a maximum; the expression for $p(0)$ is given in Eq. (4.539). It is interesting to note that $p(0)$ can take the value 1 for $k = 1$. In other words, if the state ψ_0 has the same probability to emit a proton or an α-particle and the energy of the incident α-particles takes its most favorable value, then *all* the incident particles with vanishing azimuthal quantum number will induce disintegration.

It is often impossible to measure the cross section $S(\epsilon)$ for particles with definite energy $E_0 + k + \epsilon$. In these cases, only the quantity $\int S(\epsilon) d\epsilon$ is measurable. Using Eq. (4.544), we get

$$
\int S(\epsilon)\, d\epsilon = \frac{\hbar\lambda^2}{2T} \frac{p(0)}{4} = \frac{\hbar\lambda^2}{2T} \frac{k}{(k+1)^2} = \frac{\lambda^2}{4\pi} \frac{\pi\hbar}{2T} p(0). \qquad (4.551)
$$

29. SPHERICAL FUNCTIONS WITH SPIN (II)

Given a three-valued function obeying the transformation rule of $\mathcal{D}_1$, Eqs. (4.389) allow us to turn to Cartesian coordinates. Sometimes it is also useful to know the same function in spherical coordinates (r, θ, ϕ). Evidently, from Eqs. (4.389), we have

$$\begin{aligned}
\psi_r &= \frac{x}{r}\psi_x + \frac{y}{r}\psi_y + \frac{z}{r}\psi_z \\
&= -\frac{1}{\sqrt{2}}\frac{x+iy}{r}\psi_1 + \frac{z}{r}\psi_2 + \frac{1}{\sqrt{2}}\frac{x-iy}{r}\psi_3, \\
\psi_\theta &= \cos\theta\,\cos\phi\,\psi_x + \cos\theta\,\sin\phi\,\psi_y - \sin\theta\psi_z \\
&= -\frac{1}{\sqrt{2}}\frac{x}{r}\frac{x+iy}{\sqrt{x^2+y^2}}\psi_1 - \frac{\sqrt{x^2+y^2}}{r}\psi_2 + \frac{1}{\sqrt{2}}\frac{z}{r}\frac{x-iy}{\sqrt{x^2+y^2}}\psi_3, \\
\psi_\phi &= -\sin\phi\,\psi_x + \cos\phi\,\psi_y \\
&= -\frac{i}{\sqrt{2}}\frac{x+iy}{\sqrt{x^2+y^2}}\psi_1 - \frac{i}{\sqrt{2}}\frac{x-iy}{\sqrt{x^2+y^2}}\psi_3.
\end{aligned} \tag{4.552}$$

The (r, θ, ϕ) components of the spherical harmonics (4.376) can be obtained from Eqs. (5.163); it is then useful to cast Eqs. (4.552) in the form

$$\begin{aligned}
\psi_r &= -\frac{1}{\sqrt{2}}\frac{x+iy}{r}\psi_1 + \frac{z}{r}\psi_2 + \frac{1}{\sqrt{2}}\frac{x-iy}{r}\psi_3, \\
\psi_\theta &= \frac{1}{\sin\theta}\left(-\frac{1}{\sqrt{2}}\frac{x}{r}\frac{x+iy}{r}\psi_1 - \frac{x^2+y^2}{r^2}\psi_2 + \frac{1}{\sqrt{2}}\frac{z}{r}\frac{x-iy}{r}\psi_3\right), \\
\psi_\phi &= \frac{1}{\sin\theta}\left(-\frac{i}{\sqrt{2}}\frac{x+iy}{r}\psi_1 - \frac{i}{\sqrt{2}}\frac{x-iy}{r}\psi_3\right).
\end{aligned} \tag{4.553}$$

5

VOLUMETTO V

1. REPRESENTATIONS OF THE LORENTZ GROUP

The group of the real Lorentz transformations acting on the variables ct, x, y, z can be constructed from the infinitesimal transformations

$$S_x = \begin{pmatrix} 0 & 0 & 0 & 0 \\ 0 & 0 & 0 & 0 \\ 0 & 0 & 0 & -1 \\ 0 & 0 & 1 & 0 \end{pmatrix}, \quad S_y = \begin{pmatrix} 0 & 0 & 0 & 0 \\ 0 & 0 & 0 & 1 \\ 0 & 0 & 0 & 0 \\ 0 & -1 & 0 & 0 \end{pmatrix},$$

$$S_z = \begin{pmatrix} 0 & 0 & 0 & 0 \\ 0 & 0 & -1 & 0 \\ 0 & 1 & 0 & 0 \\ 0 & 0 & 0 & 0 \end{pmatrix},$$

$$T_x = \begin{pmatrix} 0 & 1 & 0 & 0 \\ 1 & 0 & 0 & 0 \\ 0 & 0 & 0 & 0 \\ 0 & 0 & 0 & 0 \end{pmatrix}, \quad T_y = \begin{pmatrix} 0 & 0 & 1 & 0 \\ 0 & 0 & 0 & 0 \\ 1 & 0 & 0 & 0 \\ 0 & 0 & 0 & 0 \end{pmatrix},$$

$$T_z = \begin{pmatrix} 0 & 0 & 0 & 1 \\ 0 & 0 & 0 & 0 \\ 0 & 0 & 0 & 0 \\ 1 & 0 & 0 & 0 \end{pmatrix},$$

which obey the following commutation relations:

$$\begin{aligned} S_x S_y - S_y S_x &= S_z, \\ T_x T_y - T_y T_x &= -S_z, \\ S_x T_x - T_x S_x &= 0, \\ S_x T_y - T_y S_x &= T_z, \\ S_x T_z - T_z S_x &= -T_y, \\ &\text{etc.} \end{aligned} \tag{5.1}$$

There exist two inequivalent irreducible representations of the Lorentz group that make use of rank-2 matrices with unit determinant. We shall refer to these as $\mathcal{D}_{1/2}$ and $\mathcal{D}'_{1/2}$, respectively. An irreducible representation $\mathcal{D}_j$ of the matrices belonging to $\mathcal{D}_{1/2}$ is still an irreducible representation of the Lorentz group. In the same way, the irreducible representations $\mathcal{D}'_{1/2}$ can also be constructed. Probably the most general irreducible representation of the Lorentz group is given by

$$\mathcal{D}_{jj'} = \mathcal{D}_j \times \mathcal{D}'_{j'}, \quad j, j' = 0, \frac{1}{2}, 1, \frac{3}{2}, \ldots \tag{5.2}$$

(Note that if $j + j'$ is an integer then we have univocal representations, otherwise we have double-valued representations.)

Let us now see how $\mathcal{D}_{1/2}$ and $\mathcal{D}'_{1/2}$ can be constructed explicitly. Consider a vector $p = (p_0, p_x, p_y, p_z)$ whose components transform in the same way as ct, x, y, z, and let us associate with it a rank-2 matrix, which, we shall again denote by p:

$$p = p_0 + p_x\sigma_x + p_y\sigma_y + p_z\sigma_z = \begin{pmatrix} p_0 + p_z & p_x - i\,p_y \\ p_x + i\,p_y & p_0 - p_z \end{pmatrix}. \tag{5.3}$$

It is also clear that, conversely, there is a well-determined 4-vector p that we can associate with every rank-2 matrix. We shall have

$$\det p = p_0^2 - p_x^2 - p_y^2 - p_z^2. \tag{5.4}$$

Now, let S and T be two rank-2 matrices with unit determinant[1]:

$$\begin{array}{rclr} S &=& S_0 + S_x\sigma_x + S_y\sigma_y + S_z\sigma_z, & S_0^2 - S_x^2 - S_y^2 - S_z^2 = 1, \\ T &=& T_0 + T_x\sigma_x + T_y\sigma_y + T_z\sigma_z, & T_0^2 - T_x^2 - T_y^2 - T_z^2 = 1. \end{array} \tag{5.5}$$

The $p \to p'$ transformation is a Lorentz transformation if the following relation holds for the corresponding matrices:

$$p' = S\,p\,T. \tag{5.6}$$

In fact, we have

$$\det p' = \det S\, \det p\, \det T = \det p, \tag{5.7}$$

i.e.,

$$p_0'^2 - p_x'^2 - p_y'^2 - p_z'^2 = p_0^2 - p_x^2 - p_y^2 - p_z^2. \tag{5.8}$$

[1]@ In the original manuscript, there is a note here: "Instead of S and T, it is more convenient to use two different symbols, for example, P and Q." However, although such a notation may be quite confusing, we prefer to use the original notation since the matrices considered here are 2x2, while those considered at the beginning are 4x4.

One can show that the most general Lorentz transformation can be derived from Eq. (5.6) and that each transformation is obtained twice, by simultaneous sign change of S and T. Transformations (5.6) are all the proper transformations that satisfy Eq. (5.8), so that they are all the real or imaginary Lorentz transformations. If we want to restrict ourselves to study only real transformations, we have to impose some relations between S and T. The most general rank-2 Hermitian matrix corresponds to the most general real 4-vector p; thus, if we want Eq. (5.6) to define a real transformation, p' must be Hermitian if p is Hermitian as well. Then, for an arbitrary Hermitian matrix p, we have

$$S\,p\,T \;=\; (S\,p\,T)^\dagger \;=\; T^\dagger\,p^\dagger\,S^\dagger \;=\; T^\dagger\,p\,S^\dagger, \tag{5.9}$$

i.e.,

$$\left(T^\dagger\right)^{-1}\,S\,p \;=\; p\,S^\dagger\,T^{-1}, \tag{5.10}$$

and, on setting $R = S^\dagger\,T^{-1}$, $R^\dagger = \left(T^\dagger\right)^{-1}\,S$:

$$R^\dagger\,p \;=\; p\,R. \tag{5.11}$$

On setting $p = 1$, we find $R = R^\dagger$, i.e, R is Hermitian as well. Then, it follows that $Rp = pR$ for an arbitrary Hermitian p matrix, so that R must be proportional to the identity matrix. Moreover, we also have that $\det R = \det S/\det T = 1$, so that we finally get $R = \pm 1$. We shall then have real Lorentz transformations in the two cases:

$$T \;=\; S^\dagger, \tag{5.12}$$

$$T \;=\; -\,S^\dagger. \tag{5.13}$$

However, the second case has no physical relevance, since the related transformations reverse the time axis. The most general physically interesting transformation is then

$$p' \;=\; S\,p\,S^\dagger, \tag{5.14}$$

with the only constraint on S being $\det S = 1$.

The sub-group of the real or imaginary spatial rotations is obtained for $T = S^{-1}$ since then we have identically $p_0' = p_0$. For real rotations, since $T = S^\dagger$, S must be the most general unitary matrix with unit determinant.

A real Lorentz transformation then determines (apart from its sign) a matrix S of the group $SU(2)$ [2]. The S matrices clearly form a (double-valued) irreducible representation of the Lorentz group, which we shall

---

[2]@ In the original manuscript, this group is denoted by u_2; however we use the modern notation of $SU(2)$.

denote by $\mathcal{D}'_{1/2}$. A second inequivalent irreducible representation of dimension two is given by the matrices $(S^\dagger)^{-1}$; we shall call this $\mathcal{D}_{1/2}$. Since the matrices S are unitary, so that $S = (S^\dagger)^{-1}$, the two representations coincide. It is easy to derive the expressions for the infinitesimal transformations in $\mathcal{D}_{1/2}$ and $\mathcal{D}'_{1/2}$. We find:

(a) Representations of $\mathcal{D}_{1/2}$:

$$
\begin{aligned}
S_x &= \frac{1}{2i}\begin{pmatrix} 0 & 1 \\ 1 & 0 \end{pmatrix}, \quad S_y = \frac{1}{2i}\begin{pmatrix} 0 & -i \\ i & 0 \end{pmatrix}, \\
S_z &= \frac{1}{2i}\begin{pmatrix} 1 & 0 \\ 0 & -1 \end{pmatrix}, \\
T_x &= -\frac{1}{2}\begin{pmatrix} 0 & 1 \\ 1 & 0 \end{pmatrix}, \quad T_y = -\frac{1}{2}\begin{pmatrix} 0 & -i \\ i & 0 \end{pmatrix}, \\
T_z &= -\frac{1}{2}\begin{pmatrix} 1 & 0 \\ 0 & -1 \end{pmatrix}.
\end{aligned}
\tag{5.15}
$$

(b) Representations of $\mathcal{D}'_{1/2}$:

$$
\begin{aligned}
S_x &= \frac{1}{2i}\begin{pmatrix} 0 & 1 \\ 1 & 0 \end{pmatrix}, \quad S_y = \frac{1}{2i}\begin{pmatrix} 0 & -i \\ i & 0 \end{pmatrix}, \\
S_z &= \frac{1}{2i}\begin{pmatrix} 1 & 0 \\ 0 & -1 \end{pmatrix}, \\
T_x &= +\frac{1}{2}\begin{pmatrix} 0 & 1 \\ 1 & 0 \end{pmatrix}, \quad T_y = +\frac{1}{2}\begin{pmatrix} 0 & -i \\ i & 0 \end{pmatrix}, \\
T_z &= +\frac{1}{2}\begin{pmatrix} 1 & 0 \\ 0 & -1 \end{pmatrix}.
\end{aligned}
\tag{5.16}
$$

The relations between the infinitesimal spatial rotations and the infinitesimal space-time rotations are as follows:

$$
\begin{aligned}
\mathcal{D}_{1/2}: &\qquad (T_x, T_y, T_z) = -i\,(S_x, S_y, S_z)\,, \\
\mathcal{D}'_{1/2}: &\qquad (T_x, T_y, T_z) = +i\,(S_x, S_y, S_z)\,.
\end{aligned}
\tag{5.17}
$$

Let $\psi = (\psi_1, \psi_2)$ be a vector that transforms according to $\mathcal{D}_{1/2}$, that is $\psi' = (S^\dagger)^{-1}\psi$. Let us set

$$\phi = \sigma_y \psi^*, \quad \psi^* = \sigma_y \phi. \tag{5.18}$$

Then

$$\phi' = \sigma_y (S^T)^{-1} \psi^* = \sigma_y (S^T)^{-1} \sigma_y \phi \tag{5.19}$$

obtains; and, since $\det S = 1$ and from

$$S = S_0 + S_x\sigma_x + S_y\sigma_y + S_z\sigma_z, \tag{5.20}$$

we get

$$S^{-1} = S_0 - S_x\sigma_x - S_y\sigma_y - S_z\sigma_z, \tag{5.21}$$

$$(S^T)^{-1} = S_0 - S_x\sigma_x + S_y\sigma_y - S_z\sigma_z, \tag{5.22}$$

so that

$$\sigma_y (S^T)^{-1} \sigma_y = S \tag{5.23}$$

and thus

$$\phi' = S\phi, \tag{5.24}$$

i.e., ϕ transforms according to $\mathcal{D}'_{1/2}$. Conversely, if ϕ transforms as $\mathcal{D}'_{1/2}$, then $\sigma_y\phi^*$ transforms as $\mathcal{D}_{1/2}$.

Let us set

$$p = \phi\phi^\dagger = \frac{1}{2}\left(\phi^\dagger\phi + \phi^\dagger\sigma_x\phi\,\sigma_x + \phi^\dagger\sigma_y\phi\,\sigma_y + \phi^\dagger\sigma_z\phi\,\sigma_z\right), \tag{5.25}$$

so that

$$p' = S\phi\phi^\dagger S^\dagger = SpS^\dagger. \tag{5.26}$$

From Eq. (5.14), it follows that the 4-vectors associated with p and p' are connected by Lorentz transformations. If $\phi^\dagger\phi = \psi^\dagger\psi$, $\phi^\dagger\sigma_x\phi = -\psi^\dagger\sigma_x\psi$, $\phi^\dagger\sigma_y\phi = -\psi^\dagger\sigma_y\psi$, $\phi^\dagger\sigma_z\phi = -\psi^\dagger\sigma_z\psi$, we shall have the table

$$\begin{array}{cccc} \psi^\dagger\psi, & -\psi^\dagger\sigma_x\psi, & -\psi^\dagger\sigma_y\psi, & -\psi^\dagger\sigma_z\psi, \\ \phi^\dagger\psi, & \phi^\dagger\sigma_x\phi, & \phi^\dagger\sigma_y\phi, & \phi^\dagger\sigma_z\phi, \\ ct, & x, & y, & z, \end{array} \tag{5.27}$$

where it is useful to recall that ψ is an arbitrary vector that transforms according to $\mathcal{D}_{1/2}$ ($\psi' = (S^\dagger)^{-1}\psi$) and ϕ is a vector that transforms according to $\mathcal{D}'_{1/2}$ ($\phi' = S\phi$).

Now, let ψ transform according to $\mathcal{D}_{1/2}$, and let us furthermore assume it to be a function of ct, x, y, z. We shall then have that

$$\phi = \left(\frac{1}{c}\frac{\partial}{\partial t} - \frac{\partial}{\partial x}\sigma_x - \frac{\partial}{\partial y}\sigma_y - \frac{\partial}{\partial z}\sigma_z\right)\psi \tag{5.28}$$

transforms according to $\mathcal{D}'_{1/2}$. In fact, let us consider a constant vector χ of the $\mathcal{D}_{1/2}$ kind. If we left-multiply both sides of Eq. (5.28) by $\chi^\dagger$, we find

$$\chi^\dagger \phi = \frac{1}{c}\frac{\partial}{\partial t}\left(\chi^\dagger \phi\right) + \frac{\partial}{\partial x}\left(-\chi^\dagger \sigma_x \psi\right) \tag{5.29}$$

$$+\frac{\partial}{\partial y}\left(-\chi^\dagger \sigma_y \psi\right) + \frac{\partial}{\partial z}\left(-\chi^\dagger \sigma_z \psi\right). \tag{5.30}$$

From the first row in (5.27) (which clearly hold for any vector transforming like $\psi^\dagger$ and $\phi^\dagger$), we can see that the r.h.s. of Eq. (5.30) is the divergence of a vector and thus an invariant. It then follows that also $\chi^\dagger \phi$ is an invariant, that is,

$$\chi^\dagger S^{-1} \phi' = \chi^\dagger \phi \tag{5.31}$$

for any χ, so that

$$\phi' = S\phi, \tag{5.32}$$

which is what we wanted to show.

Analogously, if ϕ transforms according to $\mathcal{D}'_{1/2}$, then

$$\psi = \left(\frac{1}{c}\frac{\partial}{\partial t} + \frac{\partial}{\partial x}\sigma_x + \frac{\partial}{\partial y}\sigma_y + \frac{\partial}{\partial z}\sigma_z\right)\phi \tag{5.33}$$

transforms according to $\mathcal{D}_{1/2}$.

In the Dirac equations

$$\begin{aligned} \left(\frac{W}{c} + \frac{e}{c}A_0\right)\psi + \boldsymbol{\sigma}\cdot\left(\mathbf{p} + \frac{e}{c}\mathbf{A}\right)\psi + mc\,\phi &= 0, \\ \left(\frac{W}{c} + \frac{e}{c}A_0\right)\phi - \boldsymbol{\sigma}\cdot\left(\mathbf{p} + \frac{e}{c}\mathbf{A}\right)\phi + mc\,\psi &= 0, \end{aligned} \tag{5.34}$$

the first couple of functions ψ transforms as $\mathcal{D}_{1/2}$, while the second couple ϕ transforms as $\mathcal{D}'_{1/2}$. Equations (5.34) can be written in short as

$$\left(\frac{W}{c} + \frac{e}{c}A_0\right)\psi + \rho_3\,\boldsymbol{\sigma}\cdot\left(\mathbf{p} + \frac{e}{c}\mathbf{A}\right)\psi + \rho_1\, mc\,\psi = 0. \tag{5.35}$$

(This discussion continues in Sec. 5.6.)

2. PROTON–NEUTRON SCATTERING

Let us consider the relative motion between a neutron and a proton and assume that it is possible to neglect the spin of the proton and, if it exists, the spin of the neutron. Let m be the reduced mass of the system ($m \sim 1/2M_N$), and assume that the interaction between these two particles can be described by a potential $V(r)$ that depends on the distance r between the particles. The radial Schrödinger equation for the azimuthal quantum number ℓ reads [3]

$$u'' + \frac{2}{r}u' + \left(\frac{2m}{\hbar^2}(E-V) - \frac{\ell(\ell+1)}{r^2}\right) u = 0. \tag{5.36}$$

Let us make the following oversimplified assumption for V:

$$\begin{aligned} V &= -A, \quad \text{for } r < R, \\ V &= 0, \quad \text{for } r > R. \end{aligned} \tag{5.37}$$

For $r < R$, a solution of Eq. (5.36) that is regular at the origin is then

$$r < R, \quad u = \frac{1}{\sqrt{r}} \mathcal{I}_{\ell+1/2}\left(\sqrt{\frac{2m}{\hbar^2}(E+A)}\, r\right), \tag{5.38}$$

while, for $r > R$, we have to find the solution among the linear combinations of

$$\begin{aligned} &\frac{1}{\sqrt{r}} \mathcal{I}_{\ell+1/2}\left(\sqrt{\frac{2m}{\hbar^2}E}\, r\right), \\ &\frac{1}{\sqrt{r}} \mathcal{N}_{\ell+1/2}\left(\sqrt{\frac{2m}{\hbar^2}E}\, r\right), \end{aligned} \tag{5.39}$$

with the constraint that the solution reduces to Eq. (5.38) at R. If we set, for brevity,

$$k^2 = \frac{2m}{\hbar^2}E, \quad k_0^2 = \frac{2m}{\hbar^2}(E+A) \tag{5.40}$$

---

[3]@ In the original manuscript, the old notation $h/2\pi$ is used for $\hbar$.

and introduce an arbitrary constant factor, we find the following solution of Eq. (5.36), which is regular at the origin:

$$\begin{aligned} u_\ell &= \frac{C_\ell}{\sqrt{r}}\, \mathcal{I}_{\ell+1/2}(k_0 r), \qquad r < R, \\ u_\ell &= \frac{C_\ell}{\sqrt{r}} \left(a\, \mathcal{I}_{\ell+1/2}(kr) + b\, \mathcal{N}_{\ell+1/2}(kr) \right), \qquad r > R, \end{aligned} \tag{5.41}$$

the constants a and b having the values

$$\begin{aligned} a &= \frac{\pi x}{2} \left(\mathcal{I}_{\ell+1/2}(k_0 r)\, \mathcal{N}'_{\ell+1/2}(kr) - \frac{k_0}{k}\, \mathcal{I}'_{\ell+1/2}(k_0 r)\, \mathcal{N}_{\ell+1/2}(kr) \right), \\ b &= \frac{\pi x}{2} \left(\frac{k}{k_0}\, \mathcal{I}_{\ell+1/2}(kr)\, \mathcal{I}'_{\ell+1/2}(k_0 r) - \mathcal{I}'_{\ell+1/2}(kr)\, \mathcal{I}_{\ell+1/2}(k_0 r) \right). \end{aligned} \tag{5.42}$$

We shall determine the constants C_ℓ in such a way that, far from the origin, the quantity

$$u = \sum_{\ell=0}^{\infty} u_\ell\, P_\ell(\cos\theta) \tag{5.43}$$

describes a plane wave (I) $\mathrm{e}^{ikz} = \mathrm{e}^{ikr\cos\theta}$ plus a diverging wave (S). As is known,

$$I = \sum_{\ell=0}^{\infty} i^\ell\, (2\ell+1) \sqrt{\frac{\pi}{2kr}}\, \mathcal{I}_{\ell+1/2}(kr)\, P_\ell(\cos\theta), \tag{5.44}$$

and $S = u - I$, for $r > R$, must have the form

$$S = \sum_{\ell=0}^{\infty} \frac{\epsilon_\ell}{\sqrt{r}}\, H^1_{\ell+1/2}(kr)\, P_\ell(\cos\theta), \tag{5.45}$$

with $H^1_{\ell+1/2} = \mathcal{I}_{\ell+1/2} + i\mathcal{N}_{\ell+1/2}$. We thus infer

$$\begin{aligned} C_\ell &= \frac{i^\ell}{a+ib}\, (2\ell+1) \sqrt{\frac{\pi}{2k}}, \\ \epsilon_\ell &= -\frac{2ibi^\ell}{a+ib}\, \frac{2\ell+1}{2} \sqrt{\frac{\pi}{2k}}. \end{aligned} \tag{5.46}$$

The effect of the scattering center on the ℓ-th order spherical wave is completely determined by the angle θ_ℓ describing the relative phase between u_ℓ and $\mathcal{I}_{\ell+1/2}$ at large distances:

$$\tan\theta_\ell = -\, b_\ell / a_\ell, \tag{5.47}$$

since the last of Eqs. (5.46) can be written as

$$\epsilon_\ell = \left(e^{2i\theta_\ell} - 1\right) i^\ell \frac{2\ell+1}{2} \sqrt{\frac{\pi}{2k}}. \tag{5.48}$$

For convenience, we here list the first half-order Bessel and Neumann functions :

$$\begin{aligned}
\mathcal{I}_{1/2}(x) &= \sqrt{\frac{2}{\pi x}} \sin x, \\
\mathcal{I}_{3/2}(x) &= \sqrt{\frac{2}{\pi x}} \left(- \cos x + \frac{\sin x}{x}\right), \\
\mathcal{I}_{5/2}(x) &= \sqrt{\frac{2}{\pi x}} \left(- \sin x - 3\frac{\cos x}{x} + 3\frac{\sin x}{x^2}\right);
\end{aligned} \tag{5.49}$$

$$\begin{aligned}
\mathcal{N}_{1/2}(x) &= -\sqrt{\frac{2}{\pi x}} \cos x, \\
\mathcal{N}_{3/2}(x) &= \sqrt{\frac{2}{\pi x}} \left(- \sin x - \frac{\cos x}{x}\right), \\
\mathcal{N}_{5/2}(x) &= \sqrt{\frac{2}{\pi x}} \left(\cos x - 3\frac{\sin x}{x} - 3\frac{\cos x}{x^2}\right);
\end{aligned} \tag{5.50}$$

$$\begin{aligned}
H^{1,2}_{1/2} &= \mp i \sqrt{\frac{2}{\pi x}}\, e^{\pm ix}, \\
H^{1,2}_{3/2} &= \sqrt{\frac{2}{\pi x}}\, e^{\pm ix} \left(-1 \mp \frac{i}{x}\right), \\
H^{1,2}_{5/2} &= \sqrt{\frac{2}{\pi x}}\, e^{\pm ix} \left(\pm i - \frac{3}{x} \mp \frac{3i}{x^2}\right),
\end{aligned} \tag{5.51}$$

where the upper sign applies to the Hankel functions of the first kind, and the lower sign to those of the second kind.

3. ZEROS OF HALF-ORDER BESSEL FUNCTIONS

Here are the numerical values of x_i that are solutions of $\mathcal{I}_{\ell+1/2}(\pi x_i) = 0$, apart from $x_i = 0$:

$$
\begin{array}{lllll}
\mathcal{I}_{1/2}: & 1.000, & 2.000, & 3.000, & 4.000; \\[2ex]
\mathcal{I}_{3/2}: & \dfrac{4.494}{\pi}, & \dfrac{7.726}{\pi}, & \dfrac{10.904}{\pi}, & \dfrac{14.066}{\pi}; \\[2ex]
\mathcal{I}_{5/2}: & \dfrac{5.763}{\pi}, & \dfrac{9.095}{\pi}, & \dfrac{12.324}{\pi}; & \\[2ex]
\mathcal{I}_{7/2}: & \dfrac{6.985}{\pi}, & \dfrac{10.416}{\pi}. & &
\end{array}
$$

4. STATISTICS AND THERMODYNAMICS

4.1 Entropy of a System in Equilibrium

Let $E_0, E_1, E_2, \cdots$ be the energies of the stationary states, and denote by E the mean energy. We will then have

$$E = \Sigma'/\Sigma, \tag{5.52}$$

with

$$\Sigma = \sum_i \mathrm{e}^{-E_i/kT}, \tag{5.53}$$

$$\Sigma' = \sum_i E_i \, \mathrm{e}^{-E_i/kT}, \tag{5.54}$$

where k is Boltzmann's constant. The probability to find the system in the state i will be

$$P_i = A\,\mathrm{e}^{-E_i/kT} = P(E_i), \tag{5.55}$$

where, clearly,

$$A = 1/\Sigma. \tag{5.56}$$

Let us now define the entropy S as

$$S = \int_0^T \frac{1}{T}\frac{\mathrm{d}E}{\mathrm{d}T}\,\mathrm{d}T. \tag{5.57}$$

Knowing that

$$\Sigma' = kT^2 \frac{\mathrm{d}\Sigma}{\mathrm{d}T}, \tag{5.58}$$

the integration can readily be performed; we find

$$\begin{aligned} S = \int \frac{1}{T}\frac{\mathrm{d}E}{\mathrm{d}T}\,\mathrm{d}T &= \frac{E}{T} + \int \frac{1}{T^2}\,E\,\mathrm{d}T = \frac{E}{T} + \int \frac{1}{T^2}\,\frac{\Sigma'}{\Sigma}\,\mathrm{d}T \\ &= \frac{E}{T} + k\int \frac{\mathrm{d}\Sigma}{\Sigma} = k\,\log\Sigma + \frac{E}{T}. \end{aligned} \tag{5.59}$$

Since this expression is zero for $T = 0$, as one can easily check, we simply obtain

$$\begin{aligned} S &= k\,\log\Sigma + \frac{E}{T} = k\,\log\,\frac{1}{A}\,\mathrm{e}^{E/kT} \\ &= k\,\log\,\frac{1}{P(E)}. \end{aligned} \tag{5.60}$$

We have thus found that S/k corresponds to the number of different quantum states that alternate during the life of the system in thermal equilibrium.

4.2 Perfect Gases

The number of particles in a quantum state of energy E_s obeys the Fermi or the Bose statistics, respectively:

$$\frac{n_s}{1-n_s} = A\,\mathrm{e}^{-E_s/kT}, \quad n_s = \frac{1}{\frac{1}{A}\mathrm{e}^{-E_s/kT}+1} \quad \text{(Fermi)}, \tag{5.61}$$

$$\frac{n_s}{1+n_s} = A\,\mathrm{e}^{-E_s/kT}, \quad n_s = \frac{1}{\frac{1}{A}\mathrm{e}^{-E_s/kT}-1} \quad \text{(Bose)}. \tag{5.62}$$

The entropy of the gas then becomes

$$S = k\sum_s\left(\log\frac{1}{1-n_s} - n_s\,\log\frac{n_s}{1-n_s}\right) \quad \text{(Fermi)}, \tag{5.63}$$

$$S = k\sum_s\left(\log\frac{1+n_s}{1} + n_s\,\log\frac{1+n_s}{n_s}\right) \quad \text{(Bose)}. \tag{5.64}$$

At high temperatures and low densities ($n_s \to 0$), on considering one mole (N particles, $R = Nk$, $U = \sum_s n_s E_s$ is the energy of the gas), both statistics yield

$$S = R\,(1 - \log A) + \frac{U}{T}. \tag{5.65}$$

In this limit, the particles can be considered as independent, and thus the entropy of the gas simply is the sum of the single particle entropies minus the quantity $k \log N!$, which takes into account the decrease in the number of quantum states due to the identity of the particles. From Eq. (5.60), the single particle entropy is

$$S' = -k \log \frac{A}{N} \mathrm{e}^{-U/NkT} = k (\log N - \log A) + \frac{U}{NT}, \tag{5.66}$$

so that the entropy of the gas becomes, retaining only quantities of the order of N, which are relevant for the entropy itself:

$$\begin{aligned} S &= N S' - k \log N! = R (\log N - \log A) + \frac{U}{T} - R \log N + R \\ &= R (1 - \log A) + \frac{U}{T}, \end{aligned} \tag{5.67}$$

which is the same as the result (5.65).

4.3 Monoatomic Gas

Let us assume that the ground state is far away from the other energy levels, and let g be its degeneracy [$g = (2j+1)$ or $g = (2j+1)(2i+1)$ if we have a weakly coupled nuclear spin]. It is well known that, at high temperatures and sufficiently low densities,

$$A = \frac{N h^3}{g v (2\pi m kT)^{3/2}} \tag{5.68}$$

and

$$U = \frac{3}{2} RT. \tag{5.69}$$

From Eq. (5.67), it follows that

$$\begin{aligned} S = R \Big(&\frac{3}{2} \log T + \log v + \log g + \frac{5}{2} \\ &+ \frac{3}{2} \log (2\pi m k) - \log N - 3 \log h \Big). \end{aligned} \tag{5.70}$$

4.4 Diatomic Gas

Let us assume that the electric moment vanishes, while there could be a non-vanishing nuclear spin contribution. In this case, at sufficiently high temperatures or sufficiently low densities ($n_s \to 0$), we have

$$A = \frac{N\,h^3}{v\,(2\pi\,m\,kT)^{3/2}\,(g_0\Sigma_0 + g_1\Sigma_1)} \tag{5.71}$$

$$U = \frac{g_0\Sigma_0 U_0^R + g_1\Sigma_1 U_1^R}{g_0\Sigma_0 + g_1\Sigma_1} + \frac{3}{2}RT = U^R + \frac{3}{2}RT. \tag{5.72}$$

In the previous relations, Σ_0 and Σ_1 are the state sums relative to the even and the odd rotational states, respectively; U_0^R and U_1^R are the rotational energies that would result if there existed only even or odd energy levels, respectively; and, finally, g_0 and g_1 are the statistical weights of the even or odd energy levels (depending on the nuclear spin), respectively. For nuclei of different kinds, we then have

$$g_0 = g_1 = (2i+1)(2i'+1), \tag{5.73}$$

while, for identical nuclei, one of the following two cases applies:

$$\left\{ \begin{array}{rcl} g_0 & = & i(2i+1), \\ g_1 & = & (i+1)(2i+1), \end{array} \right. \quad \text{or} \quad \left\{ \begin{array}{rcl} g_0 & = & (i+1)(2i+1), \\ g_1 & = & i(2i+1), \end{array} \right. \tag{5.74}$$

depending on the statistics followed by the two nuclei and on the parity of the electronic term. Both Σ_0 and Σ_1, as well as U_0^R/RT and U_1^R/RT, are functions of

$$\epsilon = T_0/T, \tag{5.75}$$

with T_0 being defined by

$$k\,T_0 = h^2/8\pi^2\mathcal{I}, \tag{5.76}$$

i.e., it is the temperature corresponding to the second half-difference between the rotational energy levels. In Table 5.1 we give approximate values for the quantities mentioned above, from which we can get an idea of their behavior at high ϵ (low temperatures).

For high temperatures ($\epsilon \to 0$), it is easy to derive the asymptotic expressions of such quantities. Considering only the first terms, we find

$$\begin{aligned} \Sigma_0 &= \frac{1}{2\epsilon} + \frac{1}{6} + \ldots, \\ \Sigma_1 &= \frac{1}{2\epsilon} + \frac{1}{6} + \ldots, \\ \Sigma &= \Sigma_0 + \Sigma_1 = \frac{1}{\epsilon} + \frac{1}{3} + \ldots, \end{aligned} \tag{5.77}$$

Table 5.1. Some thermodynamic quantities for a diatomic gas (see text).

$\epsilon = \dfrac{T_0}{T}$	Σ_0	Σ_1	$\Sigma_0+\Sigma_1$	$\dfrac{U_0^R}{RT}$	$\dfrac{U_1^R}{RT}$	$\dfrac{\Sigma_0 U_0^R + \Sigma_1 U_1^R}{(\Sigma_0+\Sigma_1)RT}$
∞	1	0	1	0	∞	0
1	1.01	0.41	1.42	0.08	2.00	0.63
0.8	1.04	0.61	1.65	0.19	1.60	0.71
0.6	1.14	0.91	2.05	0.44	1.23	0.79
0.4	1.46	1.41	3.87	0.77	0.96	0.86
0.2	2.68	2.67	5.35	0.93	0.94	0.93

$$\begin{aligned} U_0^R &= RT\left(1 - \frac{\epsilon}{3} - \ldots\right), \\ U_1^R &= RT\left(1 - \frac{\epsilon}{3} + \ldots\right). \end{aligned} \tag{5.78}$$

Thus, neglecting infinitesimal quantities in the limit $T \to \infty$, we obtain:

$$U^R = RT - \frac{1}{3} RT_0, \tag{5.79}$$

while the total energy is

$$U = RT\left(\frac{5}{2} - \frac{\epsilon}{3}\right) + \ldots. \tag{5.80}$$

The entropy may be computed from Eq. (5.65), neglecting terms that vanish faster than T_0/T:

$$\begin{aligned} S = R\Big(&\frac{3}{2}\log T + \log\frac{T}{T_0}\log v + \log g \\ &+\frac{7}{2} + \frac{3}{2}\log(2\pi\, m\, K) - \log N - 3\log h\Big), \end{aligned} \tag{5.81}$$

where

$$\begin{aligned} g &= (2i+1)(2i'+1), \quad \text{for different nuclei,} \\ g &= \frac{1}{2}(2i+1)^2, \quad \text{for identical nuclei.} \end{aligned} \tag{5.82}$$

4.5 Numerical Expressions for the Entropy of a Gas

The entropy (5.70) of a mole of a monoatomic gas can be cast in the form

$$S = R\left(\frac{3}{2}\log T + \log v + B\right). \tag{5.83}$$

The constant R is 1.97 cal mol^{-1} K^{-1}, [4] while the numerical constant B depends on g and on the atomic weight $P = Nm$. On substituting the numerical values, we get [5]

$$B = -5.575 + \log g + \frac{3}{2}\log P. \tag{5.84}$$

For atomic hydrogen, for example, we have

$$g = 4, \quad P = 1, \quad B = -4.189. \tag{5.85}$$

For helium:

$$g = 1, \quad P = 4, \quad B = -3.496. \tag{5.86}$$

For atomic sodium, contributions arising from nuclear spin are

$$g = 2, \quad P = 23, \quad B = -0.179. \tag{5.87}$$

The entropy of a diatomic gas at high temperatures (5.81) can also be written as

$$S = R\left(\frac{5}{2}\log T + \log v + B\right). \tag{5.88}$$

Now, the constant B depends on g (see Eq. (5.82)), on the molecular weight $P = NM = M/M_H$, and on the temperature T_0 at which the rotational degrees of freedom unfreeze:

$$B = -4.575 - \log T_0 + \log g + \frac{3}{2}\log P. \tag{5.89}$$

For the hydrogen molecule, for example, we have

$$g = 2, \quad P = 2, \quad T_0 \simeq 85\,K, \quad B = -7.28. \tag{5.90}$$

[4]@ In the original manuscript, the units of R are generically stated as cal/degree.

[5]@ Note that the numerical value -5.575 in Eq. (5.84) is obtained by using $R = 8.31{\cdot}10^7$ erg mol^{-1} K^{-1}, $N = 6.022{\cdot}10^{23}$, and $h = 6.626{\cdot}10^{-27}$ erg s.

The constant A appearing in the distribution function $u_s = A\mathrm{e}^{-E_s/KT}$ gives, in our normalization scheme for the energy, the occupation number for the (individual) states of minimum energy. Since the previous formulae are based on classical statistics, they only hold for $A << 1$.

On substituting numerical values into Eq. (5.68), for the monoatomic gas we find

$$A = \frac{3212}{g\, P^{3/2}\, v\, T^{3/2}}, \tag{5.91}$$

where P is the atomic weight. For diatomic gases the same expression applies with g_0 replacing g at the very low temperatures at which A can become of order unity since the rotational degrees of freedom are frozen. More precisely, note that for diatomic molecules with identical nuclei for $g_0 = 0$, the occupation number of the deepest levels is not given by A but rather by

$$A \exp\left(-2h^2/8\pi^2 \mathcal{I} kT\right), \tag{5.92}$$

and still is of the form of Eq. (5.91) with g_1 replacing g. Also note that, at very low temperatures, the nuclear momentum can be coupled to the electric momentum.

If p' is the pressure expressed in atmospheres ($1\,\mathrm{atm} = 1.013\,\mathrm{dyne/cm^2}$), on eliminating v in Eq. (5.91) from the law for perfect gases, we find

$$A = \frac{39.5}{g\, P^{3/2}} \frac{p'}{T^{5/2}}. \tag{5.93}$$

4.6 Free Energy of Diatomic Gases

It is given by

$$\begin{aligned} u - T s &= \frac{\partial}{\partial N} (U - T S) = \frac{1}{N} (U - T S + P V) \\ &= - kT \left(\frac{5}{2} \log \frac{T}{T_0} + \epsilon \log \frac{T}{T_1} - \log P \right). \end{aligned} \tag{5.94}$$

Here P is the pressure expressed in atmospheres, $T_0 = 4.31 \cdot M^{-3/5}\mathrm{K}$ (M being the molecular weight); $\epsilon = 0, 1$, or $3/2$ for monoatomic or diatomic and poli-atomic molecules, respectively. For diatomic molecules, we have

$$kT_1 = \frac{h^2}{8\pi^2 \mathcal{I}}. \tag{5.95}$$

When we consider many molecules, even at ordinary temperatures Eq. (5.91) should be corrected by including terms depending on the oscillations.

5. FREQUENTLY USED POLYNOMIALS

5.1 Legendre Polynomials

$$P_n(x) = \frac{1}{2^n n!} \frac{\mathrm{d}^n (x^2 - 1)^n}{\mathrm{d}x^n}, \tag{5.96}$$

$$P_0(x) = 1, \tag{5.97}$$

$$P_1(x) = x, \tag{5.98}$$

$$P_2(x) = \frac{3}{2}x^2 - \frac{1}{2}, \tag{5.99}$$

$$P_3(x) = \frac{5}{2}x^3 - \frac{3}{2}x, \tag{5.100}$$

$$P_4(x) = \frac{35}{8}x^4 - \frac{15}{4}x^2 + \frac{3}{8}, \tag{5.101}$$

$$P_5(x) = \frac{63}{8}x^5 - \frac{35}{4}x^3 + \frac{15}{8}x, \tag{5.102}$$

$$P_6(x) = \frac{231}{16}x^6 - \frac{315}{16}x^4 + \frac{105}{16}x^2 - \frac{5}{16}, \tag{5.103}$$

$$P_7(x) = \frac{429}{16}x^7 - \frac{693}{16}x^5 + \frac{315}{16}x^3 - \frac{35}{16}x, \tag{5.104}$$

$$P_8(x) = \frac{6425}{128}x^8 - \frac{3003}{32}x^6 + \frac{3465}{64}x^4 - \frac{315}{32}x^2 + \frac{35}{128}. \tag{5.105}$$

6. SPINOR TRANSFORMATIONS

Let us consider again the formulae of Sec. 5.1 to complete the discussion. Let the 4-vector

$$p = (p_0, p_x, p_y, p_z) \tag{5.106}$$

be associated with the rank-2 matrix

$$p = p_0 + p_x \sigma_x + p_y \sigma_y + p_z \sigma_z. \tag{5.107}$$

The most general real Lorentz transformation can be obtained by associating the vector p' with the vector p such that the corresponding matrices satisfies

$$p' = S\, p\, S^\dagger, \quad \det S = 1. \tag{5.108}$$

We will consider p as a contravariant 4-vector

$$(p_0, p_x, p_y, p_z) \sim (ct, x, y, z)\,. \tag{5.109}$$

If q is a covariant 4-vector

$$(q_0, q_x, q_y, q_z) \sim (ct, -x, -y, -z)\,, \tag{5.110}$$

we can set

$$q_0 = p_0, \quad q_x = -p_x, \quad q_y = -p_y, \quad q_z = -p_z, \tag{5.111}$$

and for the corresponding matrices:

$$q = q_0 + q_x\,\sigma_x + q_y\,\sigma_y + q_z\,\sigma_z = p^{-1}/\det p \sim p^{-1}, \tag{5.112}$$

since $\det p$ is an invariant. On performing a Lorentz transformation, from Eq. (5.108), we have

$$p'^{-1} = S^{-1\dagger}\,p^{-1}\,S^{-1}, \tag{5.113}$$

and, from Eq. (5.112):

$$q' = S^{-1\dagger}\,q\,S^{-1}, \quad \det S = 1. \tag{5.114}$$

The $S^{-1\dagger}$ matrices form the $\mathcal{D}_{1/2}$ representation, while the S matrices form the $\mathcal{D}'_{1/2}$ representation. If ψ is a quantity of the kind $\mathcal{D}_{1/2}$ and ϕ a quantity of the kind $\mathcal{D}'_{1/2}$ then

$$\psi' = S^{\dagger-1}\,\psi, \quad \phi' = S\,\phi, \tag{5.115}$$

and we have (see Sec. 5.1)

$$\sigma_y\,\psi^* \sim \phi, \quad \sigma_y\,\phi^* \sim \psi, \tag{5.116}$$

that is,

$$\begin{aligned} \phi_1, \quad \phi_2 &\sim \psi_2^*, \quad -\psi_1^*, \\ \psi_1, \quad \psi_2 &\sim \phi_2^*, \quad -\phi_1^*. \end{aligned} \tag{5.117}$$

Let $a = (a_1, a_2)$ and $b = (b_1, b_2)$ be two-component quantities. We have

$$a\,b^* = \frac{1}{2}\,(b^*a + b^*\sigma_x a\,\sigma_x + b^*\sigma_y a\,\sigma_y + b^*\sigma_z a\,\sigma_z)\,, \tag{5.118}$$

so that the following 4-vector is associated with the matrix ab^*:

$$\frac{1}{2}\,(b^*a,\; b^*\sigma_x a,\; b^*\sigma_y a,\; b^*\sigma_z a)\,. \tag{5.119}$$

Let us denote by $\psi, \Psi, \ldots$ some quantities of the kind $\mathcal{D}_{1/2}$ and with $\phi, \Phi, \ldots$ some quantities of the kind $\mathcal{D}'_{1/2}$. We then get

$$\begin{aligned} \psi' \Psi'^{\dagger} &= S^{-1\dagger}\, \psi\, \Psi^{\dagger}\, S^{-1}, \\ \phi' \Phi'^{\dagger} &= S\, \phi\, \Phi^{\dagger}\, S^{\dagger}, \end{aligned} \tag{5.120}$$

so that, from Eqs. (5.108), (5.114), (5.118), and (5.116):

$$\begin{array}{ccccc} & \Psi^{\dagger}\psi, & -\Psi^{\dagger}\sigma_x\psi, & -\Psi^{\dagger}\sigma_y\psi, & -\Psi^{\dagger}\sigma_z\psi; \\ \sim & \Phi^{\dagger}\phi, & \Phi^{\dagger}\sigma_x\phi, & \Phi^{\dagger}\sigma_y\phi, & \Phi^{\dagger}\sigma_z\phi; \\ \sim & ct, & x, & y, & z; \\ \sim & i\,\psi^*\sigma_y\phi, & \psi^*\sigma_z\phi, & i\psi^*\phi, & -\psi^*\sigma_x\phi. \end{array} \tag{5.121}$$

It follows that the 4-vectors transform according to $\mathcal{D}_{1/2}\times\mathcal{D}'_{1/2} = \mathcal{D}_{1/2,1/2}$ (apart from a change in the coordinates). There are also some special cases in which higher-order tensors have particular symmetry properties, and then their components do not transform according to irreducible representations. Thus, one of the ten components of the rank-2 symmetric tensor is an invariant, while the other nine components transform as $\mathcal{D}_{1,1}$. Similarly, three of the six components of the emisymmetric rank-2 tensor transform as $\mathcal{D}_{1,0} \equiv \mathcal{D}_1$, and the other three components as $\mathcal{D}_{0,1} \equiv \mathcal{D}'_1$.

The formulae in Eqs. (5.121) show how the product of the components of a quantity ψ times the components of a quantity ϕ transforms in some typical cases. This produces the irreducible representation $\mathcal{D}_{1/2,1/2}$. We will now consider the transformation law for products of quantities of the same kind ψ or ϕ. In this case, we will have non-irreducible representations, which are equivalent to

$$\mathcal{D}_{1/2}\times\mathcal{D}_{1/2} = \mathcal{D}_0 + \mathcal{D}_1 \quad \text{or} \quad \mathcal{D}'_{1/2}\times\mathcal{D}'_{1/2} = \mathcal{D}_0 + \mathcal{D}'_1,$$

so that, by using combinations of such products, we can construct an invariant and three quantities transforming as some three components (or the other three components) of the rank-2 emisymmetric tensor. The invariants in the typical cases discussed above are immediately found. In fact, from

$$\psi' = S^{-1\dagger}\, \psi \quad \text{and} \quad \phi' = S\, \phi,$$

it follows that

$$\psi'^{\dagger}\phi' = \psi^{\dagger}\, S^{-1} S\, \phi = \psi^{\dagger}\phi.$$

On using, as usual, Eq. (5.116), we find that the following quantities:

$$\begin{aligned} \psi^\dagger \phi = \left(\phi^\dagger \psi\right)^\dagger &= \psi_1^\dagger \phi_1 + \psi_2^\dagger \phi_2, \\ i\,\Psi^* \sigma_y \psi &= \Psi_1 \psi_2 - \Psi_2 \psi_1, \\ i\,\Phi^* \sigma_y \phi &= \Phi_1 \phi_2 - \Phi_2 \phi_1. \end{aligned} \tag{5.122}$$

are invariants. From this we have

$$\psi_i \Psi_k \sim \psi_i \Psi_k \left(\phi_1 \Phi_2 - \phi_2 \Phi_1\right). \tag{5.123}$$

Noting that the last of Eqs. (5.121) can be written as

$$\begin{array}{ccccc} & \psi_1\phi_1, & \psi_1\phi_2, & \psi_2\phi_1, & \psi_2\phi_2, \\ \sim & x - iy, & ct - z, & -ct - z, & -x - iy, \end{array} \tag{5.124}$$

substituting into the l.h.s. of Eq. (5.123), where we find products of the kind $(\psi_i \phi_\ell)(\Psi_k \Phi_m)$, and eliminating the invariant quantity $\psi_1 \Psi_2 - \psi_2 \Psi_1$ (this quantity is, in fact, invariant since it transforms as $c^2 t t_1 - x x_1 - y y_1 - z z_1$, which is an invariant), we find

$$\begin{aligned} \psi_1 \Psi_1 &\sim -c(t x_1 - x t_1) + i(y z_1 - z y_1) + ic(t y_1 - y t_1) \\ &\quad + (z x_1 - x z_1), \\ \frac{1}{2}\left(\psi_1 \Psi_2 + \psi_2 \Psi_1\right) &\sim c(t z_1 - z t_1) - i(x y_1 - y x_1), \\ \psi_2 \Psi_2 &\sim c(t x_1 - x t_1) - i(y z_1 - z y_1) + ic(t y_1 - y t_1) \\ &\quad + (z x_1 - x z_1). \end{aligned} \tag{5.125}$$

The terms on the left transform as the components of the electromagnetic field; more precisely,

$$\begin{aligned} E_x,\ E_y,\ E_z &\sim c(t x_1 - x t_1),\ c(t y_1 - y t_1),\ c(t z_1 - z t_1); \\ H_x,\ H_y,\ H_z &\sim y z_1 - z y_1,\ z x_1 - x z_1,\ x y_1 - y x_1, \end{aligned} \tag{5.126}$$

so that Eqs. (5.125) can be written as

$$\begin{aligned} \psi_1 \Psi_1 &\sim -(E_x - iH_x) + i(E_y - iH_y), \\ \frac{1}{2}\left(\psi_1 \Psi_2 + \psi_2 \Psi_1\right) &\sim E_z - iH_z, \\ \psi_2 \Psi_2 &\sim (E_x - iH_x) + i(E_y - iH_y). \end{aligned} \tag{5.127}$$

Using Eq. (5.116), we can derive the analogous formulae

$$\phi_1 \Phi_1 \sim -(E_x + iH_x) + i(E_y + iH_y),$$

$$\begin{aligned}\frac{1}{2}\left(\phi_1\Phi_2+\phi_2\Phi_1\right) &\sim E_z+iH_z,\\ \phi_2\Phi_2 &\sim (E_x+iH_x)+i(E_y+iH_y);\end{aligned}\tag{5.128}$$

$$\begin{aligned}\phi^\dagger\sigma_x\psi &\sim E_x-iH_x,\\ \phi^\dagger\sigma_y\psi &\sim E_y-iH_y,\\ \phi^\dagger\sigma_z\psi &\sim E_z-iH_z;\end{aligned}\tag{5.129}$$

$$\begin{aligned}\psi^\dagger\sigma_x\phi &\sim E_x+iH_x,\\ \psi^\dagger\sigma_y\phi &\sim E_y+iH_y,\\ \psi^\dagger\sigma_z\phi &\sim E_z+iH_z.\end{aligned}\tag{5.130}$$

If we now set $\Psi=(\psi,\phi)$, from Eqs. (5.129) and (5.130) it follows:

$$\begin{array}{lclcl}\Psi^\dagger\rho_1\sigma_x\Psi &\sim& E_x &\sim& -H_x,\\ \Psi^\dagger\rho_1\sigma_y\Psi &\sim& E_y &\sim& -H_y,\\ \Psi^\dagger\rho_1\sigma_z\Psi &\sim& E_z &\sim& -H_z,\\ \Psi^\dagger\rho_2\sigma_x\Psi &\sim& H_x &\sim& E_x,\\ \Psi^\dagger\rho_2\sigma_y\Psi &\sim& H_y &\sim& E_y,\\ \Psi^\dagger\rho_2\sigma_z\Psi &\sim& H_z &\sim& E_z.\end{array}\tag{5.131}$$

In our representation we have

$$\boldsymbol{\alpha} = \rho_3\,\boldsymbol{\sigma}, \quad \beta = \rho_1.\tag{5.132}$$

In order to obtain a generic representation corresponding to the equations

$$\left(\frac{W}{c}+\boldsymbol{\alpha}\cdot\mathbf{p}+\beta\,mc\right)\Psi = 0,\tag{5.133}$$

we need to make the following substitutions in the previous formulae:

$$\begin{array}{lcllcl}\rho_1 &\to& \beta, & \sigma_x &\to& -i\,\alpha_y\,\alpha_z,\\[1ex] \rho_2 &\to& \beta\,\alpha_x\,\alpha_y\,\alpha_z, & \sigma_y &\to& -i\,\alpha_z\,\alpha_x,\\[1ex] \rho_3 &\to& -i\,\alpha_x\,\alpha_y\,\alpha_z, & \sigma_z &\to& -i\,\alpha_x\,\alpha_y.\end{array}\tag{5.134}$$

We thus obtain the following transformation rules for all the possible combinations of the products $\Psi_r^* \Psi_s$:

$$\begin{aligned} \Psi^\dagger\Psi &\sim -i\Psi^\dagger\alpha_x\alpha_y\alpha_z\Psi \sim ct, \\ -\Psi^\dagger\alpha_x\Psi &\sim i\Psi^\dagger\alpha_y\alpha_z\Psi \sim x, \\ -\Psi^\dagger\alpha_y\Psi &\sim i\Psi^\dagger\alpha_z\alpha_x\Psi \sim y, \\ -\Psi^\dagger\alpha_z\Psi &\sim i\Psi^\dagger\alpha_x\alpha_y\Psi \sim z. \end{aligned} \tag{5.135}$$

$$\begin{array}{lll} i\Psi^\dagger\beta\alpha_x\Psi \sim E_x, & i\Psi^\dagger\beta\alpha_y\Psi \sim E_y, & i\Psi^\dagger\beta\alpha_z\Psi \sim E_z, \\ i\Psi^\dagger\beta\alpha_y\alpha_z\Psi \sim H_x, & i\Psi^\dagger\beta\alpha_z\alpha_x\Psi \sim H_y, & i\Psi^\dagger\beta\alpha_x\alpha_y\Psi \sim H_z, \end{array} \tag{5.136}$$

$$\Psi^\dagger\beta\Psi \ \sim \ \Psi^\dagger\beta\alpha_x\alpha_y\alpha_z\Psi \ \sim \ 1. \tag{5.137}$$

Let us set

$$\begin{array}{rclrcl} F^1 &=& \Psi^\dagger\Psi, & F^5 &=& -i\Psi^\dagger\alpha_x\alpha_y\alpha_z\Psi, \\ F^2 &=& -\Psi^\dagger\alpha_x\Psi, & F^6 &=& -i\Psi^\dagger\alpha_y\alpha_z\Psi, \\ F^3 &=& -\Psi^\dagger\alpha_y\Psi, & F^7 &=& -i\Psi^\dagger\alpha_z\alpha_x\Psi, \\ F^4 &=& -\Psi^\dagger\alpha_z\Psi, & F^8 &=& -i\Psi^\dagger\alpha_x\alpha_y\Psi, \\ F^9 &=& i\Psi^\dagger\beta\alpha_x\Psi, & F^{12} &=& +i\Psi^\dagger\beta\alpha_y\alpha_z\Psi, \\ F^{10} &=& i\Psi^\dagger\beta\alpha_y\Psi, & F^{13} &=& +i\Psi^\dagger\beta\alpha_z\alpha_x\Psi, \\ F^{11} &=& i\Psi^\dagger\beta\alpha_z\Psi, & F^{14} &=& +i\Psi^\dagger\beta\alpha_x\alpha_y\Psi, \\ F^{15} &=& \Psi^\dagger\beta\Psi, & F^{16} &=& \Psi^\dagger\beta\alpha_x\alpha_y\alpha_z\Psi. \end{array}$$

If we set, in general,

$$F^p = \sum_{r,s} F^p_{rs}\, \Psi_r^*\, \Psi_s, \tag{5.138}$$

the Hermitian matrices F_{rs} will be unitary. Moreover, from the properties of the group of the 32 different matrices of the form

$$\pm\, \beta^{n_0}\, \alpha_x^{n_1}\, \alpha_y^{n_2}\, \alpha_z^{n^3}, \tag{5.139}$$

the orthogonality relations

$$\sum_{p=1}^{16} F^{p*}_{rs} F^{p}_{r's'} = 4\,\delta_{rr'}\,\delta_{ss'} \tag{5.140}$$

follow, from which we get

$$\Psi^*_r \Psi_s = \frac{1}{4} \sum_{p=1}^{16} F^{p}_{rs} F^{p}. \tag{5.141}$$

7. SPHERICAL FUNCTIONS WITH SPIN 1/2

Let φ_ℓ^m be the ordinary spherical functions normalized in such a way that the phase constants yield the usual representation of the angular momentum with respect to the x, y, z axes, e.g.,

$$\varphi_\ell^m = (-1)^m \sqrt{\frac{2\ell+1}{4\pi}\frac{(\ell-m)!}{(\ell+m)!}}\, P_\ell^m(\cos\theta)\, \mathrm{e}^{im\phi}, \tag{5.142}$$

with P_ℓ^m being the Legendre polynomials

$$P_\ell^m(t) = \frac{1}{2^\ell\,\ell!}\left(1-t^2\right)^{m/2} \frac{\mathrm{d}^{\ell+m}(t^2-1)^\ell}{\mathrm{d}t^{\ell+m}}. \tag{5.143}$$

We have

$$\varphi_\ell^m = (-1)^m \varphi_\ell^{-m*}, \tag{5.144}$$

and, instead of Eq. (5.142), we can write

$$\varphi_\ell^m = \sqrt{\frac{2\ell+1}{4\pi}\frac{(\ell+m)!}{(\ell-m)!}}\, P_\ell^{-m}(\cos\theta)\, \mathrm{e}^{im\phi}. \tag{5.145}$$

The two-valued spherical functions with spin $s = 1/2$ that transform according to $\mathcal{D}_j$ ($j = 1/2, 3/2, 5/2, \ldots$) of given signature (thus belonging to given values of j and $\ell = j \mp 1/2$) are defined by

$$S_k^m = \left(\sqrt{\frac{k+m-1/2}{2k-1}}\,\varphi_\ell^{m-1/2},\quad (-1)^{k+\ell+1}\sqrt{\frac{k-m-1/2}{2k-1}}\,\varphi_\ell^{m+1/2}\right). \tag{5.146}$$

The integer number k ($k = \pm 1, \pm 2, \pm 3, \ldots$) defines both j and ℓ through the relation

$$k = j(j+1) - \ell(\ell+1) + \frac{1}{4} = \begin{cases} \ell + 1, & \text{for } j = \ell + 1/2, \\ -\ell, & \text{for } j = \ell - 1/2. \end{cases} \tag{5.147}$$

(a) Relations between ℓ, j, k:

$$\begin{aligned} \boldsymbol{\sigma}\cdot\mathbf{L} &= k - 1, \\ \ell(\ell+1) &= (k-1)k, \\ (j+m)(j-m+1) &= (k+m-1/2)\,(k-m+1/2)\,, \\ (\ell+m+1/2)\,(\ell-m+1/2) &= (k+m-1/2)\,(k-m-1/2)\,, \\ (\ell-a)(\ell+1+a) &= (k-1-a)(k+a). \end{aligned} \tag{5.148}$$

(b) Angular momentum matrices:

$$\begin{aligned} (J_x - i\,J_y)\,S_k^m &= \sqrt{(k+m-1/2)\,(k-m+1/2)}\,S_k^{m-1}, \\ (J_x + i\,J_y)\,S_k^m &= \sqrt{(k+m+1/2)\,(k-m-1/2)}\,S_k^{m+1}, \\ J_z\,S_k^m &= m\,S_k^m. \end{aligned} \tag{5.149}$$

(c) Relations between S_k^m, S_{-k}^m, and S_k^{-m}:

$$S_{-k}^m = \sigma_z\,S_k^m, \tag{5.150}$$

$$S_k^{-m} = i\,\sigma_y\,(-1)^{k+\ell+m-1/2}\,S_k^{m\,*}. \tag{5.151}$$

(d) Properties of the operator $\boldsymbol{\sigma}\cdot\mathbf{p}$ [6]:

$$\begin{aligned} \boldsymbol{\sigma}\cdot\mathbf{p}\,f(r)\,S_k^m &= \frac{\hbar}{i}\left(\frac{d}{dr} - \frac{k-1}{r}\right) f(r)\,S_{-k}^m, \\ \boldsymbol{\sigma}\cdot\mathbf{p}\,f(r)\,S_{-k}^m &= \frac{\hbar}{i}\left(\frac{d}{dr} + \frac{k+1}{r}\right) f(r)\,S_k^m. \end{aligned} \tag{5.152}$$

(e) Lowest order spherical functions with spin:

$$k = 1; \quad j = \frac{1}{2}, \quad \ell = 0.$$

[6]@ In the original manuscript, the old notation $h/2\pi$ is used instead of $\hbar$.

$$
\begin{aligned}
S_1^{1/2} &= \left(\varphi_0^0,\ 0\right) = \left(\frac{1}{\sqrt{4\pi}},\ 0\right),\\
S_1^{-1/2} &= \left(0,\ \varphi_0^0\right) = \left(0,\ \frac{1}{\sqrt{4\pi}}\right).
\end{aligned}
$$

$$
k=-1;\quad j=\frac{1}{2},\quad \ell=1.
$$

$$
\begin{aligned}
S_{-1}^{1/2} &= \left(\sqrt{\frac{1}{3}}\,\varphi_1^0,\ -\sqrt{\frac{2}{3}}\,\varphi_1^1\right) = \left(\frac{1}{\sqrt{4\pi}}\cos\theta,\ \frac{1}{\sqrt{4\pi}}\sin\theta\,\mathrm{e}^{i\phi}\right),\\
S_{-1}^{-1/2} &= \left(\sqrt{\frac{2}{3}}\,\varphi_1^{-1},\ -\sqrt{\frac{1}{3}}\,\varphi_1^0\right) = \left(\frac{1}{\sqrt{4\pi}}\sin\theta\,\mathrm{e}^{-i\phi},\ -\frac{1}{\sqrt{4\pi}}\cos\theta\right).
\end{aligned}
$$

$$
k=2;\quad j=\frac{3}{2},\quad \ell=1.
$$

$$
\begin{aligned}
S_2^{3/2} &= \left(\varphi_1^1,\ 0\right) = \left(-\sqrt{\frac{3}{8\pi}}\sin\theta\,\mathrm{e}^{i\phi},\ 0\right),\\
S_2^{1/2} &= \left(\sqrt{\frac{2}{3}}\,\varphi_1^0,\ \sqrt{\frac{1}{3}}\,\varphi_1^1\right) = \left(\frac{1}{\sqrt{2\pi}}\cos\theta,\ -\frac{1}{\sqrt{8\pi}}\sin\theta\,\mathrm{e}^{i\phi}\right),\\
S_2^{-1/2} &= \left(\sqrt{\frac{1}{3}}\,\varphi_1^{-1},\ \sqrt{\frac{2}{3}}\,\varphi_1^0\right) = \left(\frac{1}{\sqrt{8\pi}}\sin\theta\,\mathrm{e}^{-i\phi},\ \frac{1}{\sqrt{2\pi}}\cos\theta\right),\\
S_2^{-3/2} &= \left(0,\ \varphi_1^{-1}\right) = \left(0,\ \sqrt{\frac{3}{8\pi}}\sin\theta\,\mathrm{e}^{-i\phi}\right).
\end{aligned}
$$

$$
k=-2;\quad j=\frac{3}{2},\quad \ell=2.
$$

$$
\begin{aligned}
S_{-2}^{3/2} &= \left(\sqrt{\frac{1}{5}}\,\varphi_2^1,\ -\sqrt{\frac{4}{5}}\,\varphi_2^2\right)\\
&= \left(-\sqrt{\frac{3}{8\pi}}\sin\theta\,\cos\theta\,\mathrm{e}^{i\phi},\ -\sqrt{\frac{3}{8\pi}}\sin^2\theta\,\mathrm{e}^{2i\phi}\right),\\
S_{-2}^{1/2} &= \left(\sqrt{\frac{2}{5}}\,\varphi_2^0,\ -\sqrt{\frac{3}{5}}\,\varphi_2^1\right)
\end{aligned}
$$

$$
\begin{aligned}
&= \left(\sqrt{\frac{1}{2\pi}}\left(\frac{3}{2}\cos^2\theta - \frac{1}{2}\right),\ -\sqrt{\frac{9}{8\pi}}\sin\theta\,\cos\theta\,\mathrm{e}^{i\phi}\right), \\
S_{-2}^{-1/2} &= \left(\sqrt{\frac{3}{5}}\,\varphi_2^{-1},\ -\sqrt{\frac{2}{5}}\,\varphi_2^0\right) \\
&= \left(\sqrt{\frac{9}{8\pi}}\sin\theta\,\cos\theta\,\mathrm{e}^{-i\phi},\ -\sqrt{\frac{1}{2\pi}}\left(\frac{3}{2}\cos^2\theta - \frac{1}{2}\right)\right), \\
S_{-2}^{-3/2} &= \left(\sqrt{\frac{4}{5}}\,\varphi_2^{-2},\ -\sqrt{\frac{1}{5}}\,\varphi_2^{-1}\right) \\
&= \left(\sqrt{\frac{3}{8\pi}}\sin^2\theta\,\mathrm{e}^{-2i\phi},\ -\sqrt{\frac{3}{8\pi}}\sin\theta\,\cos\theta\,\mathrm{e}^{-i\phi}\right).
\end{aligned}
$$

(f) Matrices of the operators x/r, y/r, z/r in the basis of the ordinary spherical functions:

$$
\begin{aligned}
\frac{x-iy}{r}\,\varphi_\ell^m &= \sin\theta\,\mathrm{e}^{-i\phi}\,\varphi_\ell^m = -\sqrt{\frac{(\ell+m)(\ell+m-1)}{(2\ell+1)(2\ell-1)}}\,\varphi_{\ell-1}^{m-1} \\
&\quad + \sqrt{\frac{(\ell-m+1)(\ell-m+2)}{(2\ell+1)(2\ell+3)}}\,\varphi_{\ell+1}^{m-1}, \\
\frac{x+iy}{r}\,\varphi_\ell^m &= \sin\theta\,\mathrm{e}^{i\phi}\,\varphi_\ell^m = \sqrt{\frac{(\ell-m)(\ell-m-1)}{(2\ell-1)(2\ell+1)}}\,\varphi_{\ell-1}^{m+1} \\
&\quad - \sqrt{\frac{(\ell+m+1)(\ell+m+2)}{(2\ell+1)(2\ell+3)}}\,\varphi_{\ell+1}^{m+1}, \\
\frac{z}{r}\,\varphi_\ell^m &= \cos\theta\,\varphi_\ell^m = \sqrt{\frac{(\ell-m)(\ell+m)}{(2\ell-1)(2\ell+1)}}\,\varphi_{\ell-1}^m \\
&\quad + \sqrt{\frac{(\ell-m+1)(\ell+m+1)}{(2\ell+1)(2\ell+3)}}\,\varphi_{\ell+1}^m.
\end{aligned}
$$

(g) Matrices of the operators x/r, y/r, z/r in the basis of the spherical functions with spin:

$$
\begin{aligned}
\frac{x-iy}{r}\,S_k^m &= -\frac{1}{2k-1}\sqrt{\left(k+m-\frac{1}{2}\right)\left(k+m-\frac{3}{2}\right)}\,S_{k-1}^{m-1} \\
&\quad + \frac{1}{2k+1}\sqrt{\left(k-m+\frac{1}{2}\right)\left(k-m+\frac{3}{2}\right)}\,S_{k+1}^{m-1} \\
&\quad + \frac{2}{(2k-1)(2k+1)}\sqrt{\left(k-m+\frac{1}{2}\right)\left(k+m-\frac{1}{2}\right)}\,S_{-k}^{m-1},
\end{aligned}
$$

$$
\begin{aligned}
\frac{x+iy}{r}\, S_k^m &= \frac{1}{2k-1}\sqrt{\left(k-m-\frac{1}{2}\right)\left(k-m-\frac{3}{2}\right)}\, S_{k-1}^{m+1} \\
&\quad - \frac{1}{2k+1}\sqrt{\left(k+m+\frac{1}{2}\right)\left(k+m+\frac{3}{2}\right)}\, S_{k+1}^{m+1} \\
&\quad + \frac{2}{(2k-1)(2k+1)}\sqrt{\left(k-m-\frac{1}{2}\right)\left(k+m+\frac{1}{2}\right)}\, S_{-k}^{m+1}, \\
\frac{z}{r}\, S_k^m &= \frac{(-1)^{k+\ell+1}}{2k-1}\sqrt{\left(k+m-\frac{1}{2}\right)\left(k-m-\frac{1}{2}\right)}\, S_{k-1}^{m} \\
&\quad + \frac{(-1)^{k+\ell+1}}{2k+1}\sqrt{\left(k+m+\frac{1}{2}\right)\left(k-m+\frac{1}{2}\right)}\, S_{k+1}^{m} \\
&\quad + \frac{2}{(2k-1)(2k+1)}\, m\, S_{-k}^{m}.
\end{aligned}
$$

(h) Matrices of the operators L_x, L_y, L_z (note that $|2k-1| = 2\ell+1$):

$$
\begin{aligned}
(L_x - iL_y)\, S_k^m &= \frac{2k-2}{2k-1}\sqrt{\left(k+m-\frac{1}{2}\right)\left(k-m+\frac{1}{2}\right)}\, S_k^{m-1} \\
&\quad + \frac{1}{|2k-1|}\sqrt{\left(k+m-\frac{1}{2}\right)\left(k+m-\frac{3}{2}\right)}\, S_{-k+1}^{m-1}, \\
(L_x + iL_y)\, S_k^m &= \frac{2k-2}{2k-1}\sqrt{\left(k+m+\frac{1}{2}\right)\left(k-m-\frac{1}{2}\right)}\, S_k^{m+1} \\
&\quad - \frac{1}{|2k-1|}\sqrt{\left(k-m-\frac{1}{2}\right)\left(k-m-\frac{3}{2}\right)}\, S_{-k+1}^{m+1}, \\
L_z\, S_k^m &= \frac{2k-2}{2k-1}\, m\, S_k^m \\
&\quad - \frac{1}{|2k-1|}\sqrt{\left(k+m-\frac{1}{2}\right)\left(k-m-\frac{1}{2}\right)}\, S_{-k+1}^{m}.
\end{aligned}
$$

(i) Matrices of the operators σ_x, σ_y, σ_z:

$$
\begin{aligned}
(\sigma_x - i\sigma_y)\, S_k^m &= \frac{2}{2k-1}\sqrt{\left(k+m-\frac{1}{2}\right)\left(k-m+\frac{1}{2}\right)}\, S_k^{m-1} \\
&\quad - \frac{2}{|2k-1|}\sqrt{\left(k+m-\frac{1}{2}\right)\left(k+m-\frac{3}{2}\right)}\, S_{-k+1}^{m-1}, \\
(\sigma_x + i\sigma_y)\, S_k^m &= \frac{2}{2k-1}\sqrt{\left(k+m+\frac{1}{2}\right)\left(k-m-\frac{1}{2}\right)}\, S_k^{m+1}
\end{aligned}
$$

$$
\begin{aligned}
& + \frac{1}{|2k-1|} \sqrt{\left(k-m-\frac{1}{2}\right)\left(k-m-\frac{3}{2}\right)}\, S_{-k+1}^{m+1}, \\
\sigma_z\, S_k^m \;=\; & \frac{2}{2k-1}\, m\, S_k^m \\
& - \frac{1}{|2k-1|} \sqrt{\left(k+m-\frac{1}{2}\right)\left(k-m-\frac{1}{2}\right)}\, S_{-k+1}^{m}.
\end{aligned}
$$

8. INFINITE-DIMENSIONAL UNITARY REPRESENTATIONS OF THE LORENTZ GROUP

The representations of the Lorentz group considered in Sec. 5.1 are, except for the identity representation, essentially not unitary, i.e., they cannot be converted into unitary representations by some transformation. The reason for this is that the Lorentz group is an open group. However, in contrast to what happens for closed groups, open groups may have irreducible representations (even unitary) in infinite dimensions. In what follows, we shall give two classes of such representations for the Lorentz group, each of them composed of a continuous infinity of unitary representations.

A given representation may be defined in terms of the infinitesimal transformations that satisfy the commutation relations in (5.1). Instead of S_x, S_y, S_z, T_x, T_y, T_z we can introduce the matrices

$$
a_x = i\, S_x, \quad b_x = -i\, T_x, \quad \ldots . \tag{5.153}
$$

These will be Hermitian matrices in a unitary representation, and viceversa, and also obey the commutation relations [7]

$$
\begin{aligned}
[a_x,\, a_y] &= i\, a_z, \\
[b_x,\, b_y] &= -i\, a_z, \\
[a_x,\, b_x] &= 0, \\
[a_x,\, b_y] &= i\, b_z,
\end{aligned} \tag{5.154}
$$

[7]@ In the original manuscript the commutator is denoted with round brackets: (a, b). However we prefer to use the modern notation $[a, b]$. In the margin of the paper, the transformation properties of the matrices a, b (related to the transformation properties of the electromagnetic field) are also given: $(a_x, a_y, a_z, b_x, b_y, b_z) \sim (E_x, E_y, E_z, H_x, H_y, H_z) \sim (-H_x, -H_y, -H_z, E_x, E_y, E_z)$.

$$
\begin{aligned}
[b_x,\, a_y] &= i\, b_z, \\
&\text{etc.}
\end{aligned}
$$

Every representation considered here acts on an infinite-dimensional space whose unitary vectors are identified by two numbers, j and m (for the representations of the first class we have $j = 1/2, 3/2, 5/2, \ldots$, $m = j, j-1, \ldots, -j$, whereas for the second class $j = 0, 1, 2, \ldots$, $m = j, j-1, \ldots, -j$). Moreover, every representation is also labeled by a real number Z_0 that can take any value, both positive and negative, whose meaning will be clear in what follows. For the simplest representation $Z_0 = a_x b_x + a_y b_y + a_z b_z = 0$; the non-zero elements of $a_x - i a_y$, $a_x + i a_y$, a_z, $b_x - i b_y$, $b_x + i b_y$, b_z can be deduced as follows[8]:

$$
\begin{aligned}
< j, m \,|\, a_x - i a_y \,|\, j, m+1 > &= \sqrt{(j+m+1)(j-m)}, \\
< j, m \,|\, a_x + i a_y \,|\, j, m-1 > &= \sqrt{(j+m)(j-m+1)}, \\
< j, m \,|\, a_z \,|\, j, m > &= m, \\
< j, m \,|\, b_x - i b_y \,|\, j+1, m+1 > &= -\frac{1}{2}\sqrt{(j+m+1)(j+m+2)}, \\
< j, m \,|\, b_x - i b_y \,|\, j-1, m+1 > &= \frac{1}{2}\sqrt{(j-m)(j-m-1)}, \qquad (5.155) \\
< j, m \,|\, b_x + i b_y \,|\, j+1, m-1 > &= \frac{1}{2}\sqrt{(j-m+1)(j-m+2)}, \\
< j, m \,|\, b_x + i b_y \,|\, j-1, m-1 > &= -\frac{1}{2}\sqrt{(j+m)(j+m-1)}, \\
< j, m \,|\, b_z \,|\, j+1, m > &= \frac{1}{2}\sqrt{(j+m+1)(j-m+1)}, \\
< j, m \,|\, b_z \,|\, j-1, m > &= \frac{1}{2}\sqrt{(j+m)(j-m)}.
\end{aligned}
$$

Note the relations

$$
\begin{aligned}
a_x^2 + a_y^2 + a_z^2 &= j\,(j+1), \\
b_x^2 + b_y^2 + b_z^2 &= j\,(j+1) + 3/4;
\end{aligned}
\qquad (5.156)
$$

$$
\begin{aligned}
a_x\, b_x + a_y\, b_y + a_z\, b_z &= 0, \\
b_x^2 + b_y^2 + b_z^2 - a_x^2 - a_y^2 - a_z^2 &= 3/4.
\end{aligned}
\qquad (5.157)
$$

[8]@ In the original manuscript, the following scalar products are denoted by round brackets: $(\ldots|\ldots|\ldots)$. However, we prefer to use here the Dirac notation $< \ldots|\ldots|\ldots >$.

We now want to determine the matrices α_0, α_x, α_y, α_z in such a way that the equations

$$\left[\alpha_0\left(\frac{W_0}{c}+\frac{e}{c}\phi\right)+\boldsymbol{\alpha}\cdot\left(\mathbf{p}+\frac{e}{c}\mathbf{C}\right)-m\right]\psi = 0 \qquad (5.158)$$

are invariant. To this end, it is thus necessary that the operators α_0, α_x, α_y, α_z or the corresponding Hermitian forms (we're talking about unitary transformations!), transform as the components of a covariant vector (α_0, α_x, α_y, $\alpha_z \sim ct, -x, -y, -z$). In order to observe this requirement, it is necessary and sufficient that the following commutation relations be satisfied:

$$\begin{aligned}
[\alpha_0,\, a_x] &= 0,\\
[\alpha_0,\, b_x] &= i\,\alpha_x,\\
[\alpha_x,\, a_x] &= 0,\\
[\alpha_x,\, a_y] &= i\,\alpha_z,\\
[\alpha_x,\, a_z] &= -i\,\alpha_y,\\
[\alpha_x,\, b_x] &= i\,\alpha_0,\\
[\alpha_x,\, b_y] &= 0,\\
[\alpha_x,\, b_z] &= 0,
\end{aligned} \qquad (5.159)$$

etc.

From the first of these it follows that α_0 depends on j:

$$\alpha_0 = c_j, \qquad (5.160)$$

while, from the second and the sixth,

$$[[\alpha_0,\, b_x],\, b_x] = -\alpha_0, \quad \text{etc.}, \qquad (5.161)$$

that is,

$$-\alpha_0 = b_x^2\,\alpha_0 - 2\,b_x\,\alpha_0\,b_x + \alpha_0\,b_x^2. \qquad (5.162)$$

Therefore, by considering b_z, for example, it follows that

$$c_j - 2c_{j+1} + c_{j+2} = 0,$$

$$c_j = \frac{1}{2}\left(j^2 - m^2 + 2j + 1\right)\left(c_{j+1} - c_j\right) - \frac{1}{2}\left(j^2 - m^2\right)\left(c_j - c_{j-1}\right),$$

from which, apart from a constant factor, we get

$$c_j = j + 1/2. \tag{5.163}$$

Thus from Eq. (5.160), the second of Eqs. (5.159), and (5.155) we can univocally derive the matrices α_0, α_x, α_y, α_z, apart from a constant factor:

$$\begin{aligned}
\alpha_0 &= j + \frac{1}{2}, \\
< j, m \,|\, \alpha_x - i\alpha_y \,|\, j+1, m+1 > &= -\frac{i}{2}\sqrt{(j+m+1)(j+m+2)}, \\
< j, m \,|\, \alpha_x - i\alpha_y \,|\, j-1, m+1 > &= -\frac{i}{2}\sqrt{(j-m)(j-m-1)}, \\
< j, m \,|\, \alpha_x + i\alpha_y \,|\, j+1, m-1 > &= \frac{i}{2}\sqrt{(j-m+1)(j-m+2)}, \\
< j, m \,|\, \alpha_x + i\alpha_y \,|\, j-1, m-1 > &= \frac{i}{2}\sqrt{(j+m)(j+m-1)}, \\
< j, m \,|\, \alpha_z \,|\, j+1, m > &= \frac{i}{2}\sqrt{(j+m+1)(j-m+1)}, \\
< j, m \,|\, \alpha_z \,|\, j-1, m > &= -\frac{i}{2}\sqrt{(j+m)(j-m)},
\end{aligned}$$

where the elements that do not explicitly appear are zero.

For the representations with arbitrary real $Z_0 = a_x b_x + a_y b_y + a_z b_z$ the matrices a_x, a_y, a_z still are expressed by Eq. (5.155) as in the $Z_0 = 0$ case. In the general case, the non-zero elements of b_x, b_y, b_z are given by:

$$\begin{aligned}
< j, m \,|\, b_x - ib_y \,|\, j+1, m+1 > &= -\frac{\sqrt{4Z_0^2 + (j+1)^2}}{2(j+1)} \\
&\quad \times \sqrt{(j+m+1)(j+m+2)}, \\
< j, m \,|\, b_x - ib_y \,|\, j, m+1 > &= \frac{Z_0}{j(j+1)}\sqrt{(j+m+1)(j-m)}, \\
< j, m \,|\, b_x - ib_y \,|\, j-1, m+1 > &= \frac{\sqrt{4Z_0^2 + j^2}}{2j}\sqrt{(j-m)(j-m-1)}, \\
< j, m \,|\, b_x + ib_y \,|\, j+1, m-1 > &= \frac{\sqrt{4Z_0^2 + (j+1)^2}}{2(j+1)} \\
&\quad \times \sqrt{(j-m+1)(j-m+2)}, \\
< j, m \,|\, b_x + ib_y \,|\, j, m-1 > &= \frac{Z_0}{j(j+1)}\sqrt{(j+m)(j-m+1)},
\end{aligned}$$

$$
\begin{aligned}
< j,m\,|\,b_x+ib_y\,|\,j-1,m-1 > &= -\frac{\sqrt{4Z_0^2+j^2}}{2j}\sqrt{(j+m)(j+m-1)},\\
< j,m\,|\,b_z\,|\,j+1,m > &= \frac{\sqrt{4Z_0^2+(j+1)^2}}{2(j+1)}\\
&\quad\times\sqrt{(j+m+1)(j-m+1)},\\
< j,m\,|\,b_z\,|\,j,m > &= \frac{Z_0}{j(j+1)}\,m,\\
< j,m\,|\,b_z\,|\,j-1,m > &= \frac{\sqrt{4Z_0^2+j^2}}{2j}\sqrt{(j+m)(j-m)}.
\end{aligned}
$$

9. THE EQUATION $(\Box H + \lambda)A = p$

Let us define the symbol $\Box$ as

$$
\Box \equiv \frac{1}{c^2}\frac{\partial^2}{\partial t^2} - \frac{\partial^2}{\partial x^2} - \frac{\partial^2}{\partial y^2} - \frac{\partial^2}{\partial z^2} \qquad (5.164)
$$

and assume that λ is a positive constant (with dimensions $[L]^{-2}$), while $p = p(x,y,z,t)$ is an arbitrary function of space. The general solution of the equation

$$
(\Box\, H \,+\, \lambda)\; A \;=\; p(x,y,z,t) \qquad (5.165)
$$

is obtained by adding a particular solution to the general solution of the associated homogenous equation that results when one sets $p = 0$. A particular solution can be cast in the form

$$
A(q,t) \;=\; \int F(q,t;q',t')\,p(q',t')\,\mathrm{d}q'\,\mathrm{d}t', \qquad (5.166)
$$

and one can require that, for symmetry,

$$
F(q,t;q',t') \;=\; F(R,T), \qquad (5.167)
$$

where $R = \sqrt{X^2+Y^2+Z^2}$ and $X = x - x'$, $Y = y - y'$, $Z = z - z'$, $T = t - t'$. Moreover, we can also assume that $F(R,T)$ is different from zero only for $T \geq R/c$.

The function F must satisfy the homogeneous equation associated with Eq. (5.165) (in respect to the variables q,t or q',t') except that for $T = 0$ and thus for $R = 0$, where it should have an appropriate singular behavior. The function that satisfies the desired conditions also has a singular behavior at the boundary of the integration field, i.e, for $T = R/c$, so that the integral in Eq. (5.166) has to be evaluated as the sum of an integral over the negative optical cone and a four-dimensional integral on the interior of the same optical semi-cone. We find the following formula, which we shall verify later on:

$$A(q,t) = \frac{1}{4\pi}\int \frac{1}{R}\, p\left(q', t - \frac{R}{c}\right) \mathrm{d}q' - \frac{c\lambda}{4\pi}\int_{T>R/c} \frac{\mathcal{I}_1(\omega)}{\omega}\, p(q',t')\,\mathrm{d}q'\,\mathrm{d}t', \tag{5.168}$$

with $\mathcal{I}_1$ denoting the Bessel function of order 1 and

$$\omega = \sqrt{\lambda\,(c^2T^2 - R^2)}. \tag{5.169}$$

For $\lambda = 0$ only the first integral in Eq. (5.168) is non-vanishing, and this gives rise to the usual expression for the retarded potentials.

In order to verify Eq. (5.168), let us set, for fixed q and t,

$$A(q,t) = \int_{-\infty}^{t} u(t')\,\mathrm{d}t', \tag{5.170}$$

with $u(t')\mathrm{d}t'$ being the contribution to the two integrals on the r.h.s. of Eq. (5.168) coming from all points in the t' integration region lying between t' and $t' + \mathrm{d}t'$. We now want to show that $u(t')$ can be cast in the form

$$u(t') = \frac{\mathrm{d}v(t')}{\mathrm{d}t'}, \tag{5.171}$$

the function $v(t')$ being such that it can be expressed as the sum of two integrals computed in the region $t' =$ constant, one over the spherical surface $|q - q'| = R = cT = c(t - t')$ and the other over the region inside the same sphere. More precisely, we can set

$$\begin{aligned} v(t') = {} & \frac{1}{4\pi}\int_0^{4\pi c^2T^2} \left[\frac{1}{cR}\frac{\partial A(q',t')}{\partial t'} + \frac{1}{R}\frac{\partial A(q',t')}{\partial R} \right. \\ & \left. + \left(\frac{1}{R^2} - \frac{\lambda}{2}\right) A(q',t')\right] \mathrm{d}\sigma - \frac{\lambda}{4\pi}\int_0^{\frac{4}{3}\pi c^3T^3} \left[\frac{1}{c}\frac{\partial A(q',t')}{\partial t'}\frac{\mathcal{I}_1(\omega)}{\omega} \right. \\ & \left. - \lambda cT\, A(q',t')\,\frac{\mathcal{I}_1(\omega) - \omega\mathcal{I}_1'(\omega)}{\omega^3}\right] \mathrm{d}q'; \end{aligned} \tag{5.172}$$

note that $\partial/\partial R$ is the derivative performed along the outward normal. In order to prove this formula it is necessary to prove that $u(t')$ obtained from the derivative of $v(t')$ from Eq. (5.171) is the same as that computed using its definition (5.170). For the direct computation of $u(t')$

we need to replace the quantity $p(q',t')$ in Eq. (5.168) by its expression $(\Box+\lambda)A(q',t')$ derived from the differential equation (5.165).

On substituting Eq. (5.172) into Eq. (5.171), we find that even in this case $u(t')$ is expressed as a sum of an integral over the sphere $|q-q'|=R=cT$ and an integral over the region inside the sphere:

$$\begin{aligned}
u(t') &= \frac{1}{4\pi}\int_0^{4\pi c^2T^2}\left[\frac{1}{cR}\frac{\partial^2 A(q',t')}{\partial t'^2} - \frac{c}{R}\frac{\partial^2 A(q',t')}{\partial R^2}\right. \\
&\left. - \frac{2c}{R^2}\frac{\partial A(q',t')}{\partial R} + \left(\frac{1}{R}-\frac{\lambda R}{8}\right) c\lambda A(q',t') + \frac{c\lambda}{2}\frac{\partial A(q',t')}{\partial R}\right] \mathrm{d}\sigma \\
&- \frac{c\lambda}{4\pi}\int_0^{\frac{4}{3}\pi c^3T^3}\left[\frac{\mathcal{I}_1(\omega)}{\omega}\,(\Box+\lambda)\,A(q',t')\right. \\
&\left. + \sum_{(x,y,z)}\frac{\partial}{\partial x'}\left(\frac{\mathcal{I}_1(\omega)}{\omega}\frac{\partial A(q',t')}{\partial x'} + \lambda\times A(q',t')\,\frac{\mathcal{I}_1(\omega)-\omega\mathcal{I}_1'(\omega)}{\omega^3}\right)\right]\mathrm{d}q' \\
&= \frac{1}{4\pi T}\int_0^{4\pi c^2T^2}(\Box+\lambda)\,A(q',t')\,\mathrm{d}\sigma \\
&\qquad - \frac{c\lambda}{4\pi}\int_0^{\frac{4}{3}\pi c^3T^3}\frac{\mathcal{I}_1(\omega)}{\omega}\,(\Box+\lambda)\,A(q',t')\,\mathrm{d}q'.
\end{aligned} \tag{5.173}$$

In deriving this relation we have used the differential equation for $\mathcal{I}_1(\omega)$,

$$\mathcal{I}_1''(\omega) + \frac{1}{\omega}\mathcal{I}_1'(\omega) + \left(1-\frac{1}{\omega^2}\right)\mathcal{I}_1(\omega) = 0, \tag{5.174}$$

as well as Eq. (5.169) and the relations

$$\lim_{\omega\to 0}\frac{\mathcal{I}_1(\omega)}{\omega} = \frac{1}{2}, \qquad \lim_{\omega\to 0}\frac{\mathcal{I}_1(\omega)-\omega\mathcal{I}_1'(\omega)}{\omega^3} = \frac{1}{8}. \tag{5.175}$$

We immediately verify that $u(t')$ is precisely the function that we have introduced above in order to deduce Eq. (5.170) from Eq. (5.168). In fact, it is enough to replace p into Eq. (5.168) with $(\Box+\lambda)A$, in accordance with the differential equation (5.165). We have thus proven that the r.h.s. of Eq. (5.168) is

$$A'(q,t) = \lim_{t'\to t} v(t') - \lim_{t'\to-\infty} v(t'). \tag{5.176}$$

From Eq. (5.172) it follows that

$$\lim_{t'\to t} v(t') = A(q,t), \tag{5.177}$$

while, assuming that $A(q,t)$ goes to zero rapidly when $t\to-\infty$,

$$\lim_{t'\to-\infty} v(t') = 0. \tag{5.178}$$

It follows that

$$A'(q,t) = A(q,t), \tag{5.179}$$

so that we have proven Eq. (5.168), since *a posteriori* we can verify that the assumption above is realized when $A(q,t)$ is defined by Eq. (5.168) and $p(q,t)$ vanishes for sufficiently small values of t. However, we note that, even if the latter were not the case, $A(q,t)$ in Eq. (5.168) would again have the same form as Eq. (5.170), provided that we encounter no convergence problems.

Instead of Eq. (5.168) another particular solution of Eq. (5.165) can also be used. This is easily obtained by inverting the time axis:

$$B(q,t) = \frac{1}{4\pi}\int \frac{1}{R}\,p\left(q',t+\frac{R}{c}\right)\mathrm{d}q' - \frac{c\lambda}{4\pi}\int_{T<R/c}\frac{\mathcal{I}_1(\omega)}{\omega}\,p(q',t')\,\mathrm{d}q'\,\mathrm{d}t'. \tag{5.180}$$

In general, we obviously have that $B \neq A$, and the difference $B - A$ obeys the homogeneous differential equation (5.165) with $p = 0$.

The solutions (5.168) and (5.180) can also be used to determine the general integral of the homogeneous equation

$$(\Box + \lambda)\,A = 0. \tag{5.181}$$

The most general solution of Eq. (5.181) is determined by arbitrarily fixing at $t = 0$ the values of the function and of its time derivative:

$$A(q,0), \quad \dot{A}(q,0). \tag{5.182}$$

Let us consider a singular function $A_1(q,t)$:

$$A_1(q,t) = \begin{cases} A(q,t), & \text{for } t > 0, \\ 0, & \text{for } t < 0, \end{cases} \tag{5.183}$$

such that, knowing A_1, one can determine A for $t > 0$. If we now set

$$(\Box + \lambda)\,A_1 = p(q,t), \tag{5.184}$$

p is going to be singular for $t = 0$ and will vanish for $t > 0$ and $t < 0$. The function A_1 is precisely the particular solution of Eq. (5.184) that can be cast in the form (5.168). As far as the singular function $p(q,t)$ is concerned, it consists of a single layer at $t = 0$ having the density

$$s_0 = \frac{1}{c^2}\,\dot{A}(q,0) \tag{5.185}$$

and a double layer in the same space at $t = 0$ with density

$$s_1 = -\frac{1}{c^2}\,A(q,0). \tag{5.186}$$

On substituting these densities into Eq. (5.168) and using Eq. (5.183), for $t > 0$, we get

$$A(q,t) = \frac{1}{4\pi t}\int_0^{4\pi c^2 t^2} s_0(q')\,d\sigma - \frac{c\lambda}{4\pi}\int_0^{\frac{4}{3}\pi c^3 t^3} \frac{\mathcal{I}_1(\epsilon)}{\epsilon}\, s_0(q')\,dq'$$
$$-\frac{\partial}{\partial t}\left[\frac{1}{4\pi t}\int_0^{4\pi c^2 t^2} s_1(q')\,d\sigma - \frac{c\lambda}{4\pi}\int_0^{\frac{4}{3}\pi c^3 t^3} \frac{\mathcal{I}_1(\epsilon)}{\epsilon}\, s_1(q')\,dq'\right], \quad (5.187)$$

where

$$\epsilon = \sqrt{\lambda(c^2t^2 - R^2)}, \quad (5.188)$$

while the integrals are evaluated over the spherical surface or in the region inside the sphere of radius ct and center at q.

On substituting the expressions (5.185) and (5.186) for $s_0(q)$ and $s_1(q)$, after some further transformation, we find

$$A(q,t) = \frac{1}{4\pi c^2 t^2}\int_0^{4\pi c^2 t^2}\left[t\,\dot{A}(q',0) + \left(1 - \frac{\lambda R^2}{2}\right) A(q',0)\right.$$
$$\left. + R\,\frac{\partial A(q',0)}{\partial R}\right] d\sigma + \frac{\lambda^2 ct}{4\pi}\int_0^{\frac{4}{3}\pi c^3 t^3}\left[\frac{[\mathcal{I}]_\infty(\epsilon) - \epsilon \mathcal{I}'_\infty(\epsilon)}{\epsilon^3}\, A(q',0)\right.$$
$$\left. - \frac{1}{\lambda c^2 t}\frac{\mathcal{I}_1(\epsilon)}{\epsilon}\,\dot{A}(q',0)\right] dq' \quad (t > 0). \quad (5.189)$$

In this expression $\partial/\partial R$ denotes the derivative along the outward normal to the sphere of radius ct.

In a similar way, we can obtain A for $t < 0$ by making use of the particular solutions (5.180) of Eq. (5.165). The result can be easily foreseen from the invariance of Eq. (5.181) with respect to time reversal. We shall obtain the same expression (5.189) but the integrals will now be evaluated over the surface of the sphere of radius ct and center q or inside it; moreover, we have to change the sign of the integrand terms that contain the factor $\dot{A}(q',0)$.

10. RELEVANT FORMULAS FOR THE ATOMIC EIGENFUNCTIONS

(1) Dirac equations for a particle in a central field:

$$\left[\frac{W - V}{c} + \rho_1\,\boldsymbol{\sigma}\cdot\mathbf{p} + \rho_3\,mc\right]\psi = 0; \quad (5.190)$$

$$k = (2j+1)(j-\ell) = \begin{cases} \ell+1, & \text{for } j = \ell+1/2, \\ -\ell, & \text{for } j = \ell-1/2; \end{cases} \tag{5.191}$$

$$(\psi_3,\ \psi_4) = \frac{u(r)}{r}\, S_k^m, \tag{5.192}$$

$$(\psi_1,\ \psi_2) = i\,\frac{v(r)}{r}\, S_{-k}^m, \tag{5.193}$$

(see Sec. 5.7); we have [9]

$$\int \left(u^2 + v^2\right) \mathrm{d}r = 1, \tag{5.194}$$

$$\left(\frac{\mathrm{d}}{\mathrm{d}r} - \frac{k}{r}\right) u = \frac{1}{\hbar c}\left(W - V + mc^2\right) v, \tag{5.195}$$

$$\left(\frac{\mathrm{d}}{\mathrm{d}r} + \frac{k}{r}\right) u = -\frac{1}{\hbar c}\left(W - V - mc^2\right) u. \tag{5.196}$$

(2) The solution of

$$y'' + \left(\frac{2Z}{x} - \frac{\ell(\ell+1)}{x^2}\right) y = 0. \tag{5.197}$$

under the conditions

$$y(0) = 0, \quad \lim_{x\to 0} \frac{y}{x^{\ell+1}} = 1, \tag{5.198}$$

is

$$y = \frac{(2\ell+1)!}{(2Z)^{\ell+1}} \sqrt{2Zx}\, \mathcal{I}_{2\ell+1}\left(2\sqrt{2Zx}\right). \tag{5.199}$$

(3) Fine-structure formula:

$$E = mc^2 \left(1 + \frac{Z^2\alpha^2}{(S + \sqrt{k^2 - Z^2\alpha^2})^2}\right)^{-1/2} - mc^2, \tag{5.200}$$

with

$$\begin{aligned} S &= 0, 1, 2, \ldots, \quad \text{for } k > 0, \\ S &= 1, 2, 3, \ldots, \quad \text{for } k < 0. \end{aligned} \tag{5.201}$$

[9]@ In the original manuscript, the old notation $h/2\pi$ instead of $\hbar$ is used.

First approximation:

$$E = -\frac{Z^2}{n^2} Rh - \frac{Z^4}{n^3} \left(\frac{1}{|k|} - \frac{3}{4n} \right) \alpha^2 Rh \qquad (5.202)$$

$$(\alpha^2 Rh = 5.82 cm^{-1}).$$

Doublet shift:

$$\Delta E = \frac{Z^4}{n^3 \ell(\ell+1)} \alpha^2 Rh = Z a_0^3 \left(\ell + \frac{1}{2} \right) \overline{r^{-3}} \alpha^2 Rh, \qquad (5.203)$$

with

$$\overline{r^{-3}} = \frac{Z^3}{a_0^3} \frac{1}{n^3 \ell(\ell+1/2)(\ell+1)}. \qquad (5.204)$$

11. CLASSICAL THEORY OF MULTIPOLE RADIATION

Let us consider an oscillating electrical system of frequency ν, i.e., the charge and current densities can be expressed in the form

$$\begin{aligned} \rho &= \rho_0 \, e^{-2\pi\nu i t} + \rho_0^* \, e^{2\pi\nu i t}, \\ \mathbf{I} &= \mathbf{I}_0 \, e^{-2\pi\nu i t} + \mathbf{I}_0^* \, e^{2\pi\nu i t}. \end{aligned} \qquad (5.205)$$

We use the convention of expressing all quantities in electromagnetic units. From the continuity equation it follows that

$$\rho_0 = \frac{c}{2\pi\nu i} \nabla \cdot \mathbf{I}_0, \qquad (5.206)$$

and thus the system is completely determined by the arbitrary vector function $\mathbf{I}_0$. The radiation emitted by this system can be computed by finding a solution of the equations

$$\begin{aligned} \Box \phi &= 4\pi \rho, \\ \Box \mathbf{A} &= 4\pi \mathbf{I}, \end{aligned} \qquad (5.207)$$

with the "continuity condition" [10]

$$\frac{1}{c} \frac{\partial \phi}{\partial t} + \nabla \cdot \mathbf{A} = 0, \qquad (5.208)$$

[10]@ In modern terminology one speaks of a gauge condition; in particular, the author is considering the Lorenz gauge.

such that it can be cast in a form similar to Eq. (5.205) and satisfies the constraint that at large distances it describes a diverging wave (method of stationary waves). Alternatively, we can also compute the radiation emitted by the system by supposing that at the initial time the region of interest is free of any radiation and then determining the frequency distribution of the radiation after a time t (method of variation of parameters).

In either case, knowing the currents is sufficient to define the radiating system, while in order to evaluate how much energy has been radiated one must know the vector potential. We will then consider only the relations between $\mathbf{I}$ and $\mathbf{A}$ which, being vectorial quantities, transform according to a representation equivalent to $\mathcal{D}_1$. By choosing suitable linear combinations of the ordinary vector components, it is convenient to introduce the quantities $\mathbf{I} = (I_1, I_2, I_3)$ and $\mathbf{A} = (A_1, A_2, A_3)$, which transform exactly according to $\mathcal{D}_1$:

$$\begin{aligned} I_1 &= \left(1/\sqrt{2}\right)(-I_x + iI_y), & A_1 &= \left(1/\sqrt{2}\right)(-A_x + iA_y), \\ I_2 &= I_z, & A_2 &= A_z, \\ I_3 &= \left(1/\sqrt{2}\right)(I_x + iI_y), & A_3 &= \left(1/\sqrt{2}\right)(A_x + iA_y). \end{aligned} \tag{5.209}$$

Let us now define a suitable complete set of orthogonal functions in terms of which it is possible to expand any vector function $\mathbf{V} = (V_1, V_2, V_3)$. We choose the regular solutions of

$$\nabla^2 \mathbf{V} + k^2 \mathbf{V} = 0, \quad k > 0, \tag{5.210}$$

and label them with the continuous index k and with the discrete indices j and m having the usual meaning. It is easy to see that, for every value of k and for fixed (integer) j and m, there exist three regular independent solutions of Eq. (5.210) except when $j = 0$, in which case there is only one solution. In fact, if we introduce the "orbital" momentum ℓ (in units of $\hbar$), for every value of k we clearly find $3\cdot(2\ell+1)$ independent solutions of Eq. (5.210), which are regular at the origin. These are obtained by setting one of the components of $\mathbf{V}$ equal to

$$V_i = \frac{1}{\sqrt{r}} \mathcal{I}_{\ell+1/2}(kr)\, \varphi_\ell^{m_\ell}, \quad i = 1, 2, 3; \quad m_\ell = \ell, \ell-1, \ldots, -\ell, \tag{5.211}$$

and the other two equal to zero:

$$V_{i'} = 0, \quad i' \neq i. \tag{5.212}$$

These $3{\cdot}(2\ell+1)$ vector functions transform according to $\mathcal{D}_\ell \times \mathcal{D}_1$ and can therefore be expressed as combinations of three systems of independent functions transforming according to

$$\mathcal{D}_{\ell-1},\ \mathcal{D}_\ell,\ \mathcal{D}_{\ell+1} \tag{5.213}$$

except for the case $\ell = 0$ when only the $\mathcal{D}_{\ell+1}$ system survives. Every irreducible representation $\mathcal{D}_j$ appears, then, three times and yields regular solutions of Eq. (5.210) for a given value of k. Such a representation can, in fact, be obtained from $\ell = j+1, j, j-1$, except for the solution with $j = 0$, in which case it appears only once and derives from $\ell = 1$. Thus, in the general case, for any k, j, and m we have three solutions of Eq. (5.210), but we now need to establish a criterion to discriminate among them, which is necessary in order to list all the possible independent solutions of Eq. (5.210). The simplest way to achieve this stems from the previous considerations and consists in introducing an "orbital" momentum ℓ that can only take the values $j+1, j, j-1$ (only $\ell = 1$ for $j = 0$). Actually, this is not really the most convenient choice, since it does not make any distinction between longitudinal and transverse waves, which, in contrast, is a rather significant feature especially in practical applications. In fact, as is well known, the regular solutions of Eq. (5.210) may be expressed as combinations of two particular sets of solutions. The first set consists of longitudinal waves satisfying the additional condition

$$\nabla \times \mathbf{V} \;=\; 0, \tag{5.214}$$

from which it follows that

$$\mathbf{V} \;=\; \nabla\, v, \tag{5.215}$$

while the second set consists of transverse waves satisfying the condition

$$\nabla \cdot \mathbf{V} \;=\; 0. \tag{5.216}$$

Waves belonging to the two sets are orthogonal. Now, it is easy to see that for every value of k, j, and m (even for $j = 0$) we have only one longitudinal wave, which is obtained from Eq. (5.215) by setting, modulo a constant factor,

$$v \;=\; \frac{1}{\sqrt{kr}}\, \mathcal{I}_{j+1/2}(kr)\, \varphi_j^m. \tag{5.217}$$

Because of its symmetry properties under parity transformations, it follows that the longitudinal wave is a combination of the solutions of Eq. (5.210) corresponding to $\ell = j+1$ and $\ell = j-1$. The other combination of the same solutions that is orthogonal to the longitudinal wave (it exists for $j > 0$) is a transverse wave which, for a reason that will be explained later, we will call an "electric multipole wave of order j". Finally, the solution for $\ell = j$ (which also exists only for $j > 0$) is again a transverse wave that will be called a "magnetic multipole wave of order j". By performing explicitly the computations underlying the previous proofs, we find the following expressions for waves of the three different kinds:

(a) longitudinal waves:

$$\mathbf{V}^L_{k,j,m} = \sqrt{\frac{k}{r}}\left[\sqrt{\frac{j}{2j+1}}\,\mathcal{I}_{j-1/2}(kr)\,\varphi^m_{j,j-1} + \sqrt{\frac{j+1}{2j+1}}\,\mathcal{I}_{j+3/2}(kr)\,\varphi^m_{j,j+1}\right]; \tag{5.218}$$

(b) electric multipole wave:

$$\mathbf{V}^E_{k,j,m} = \sqrt{\frac{k}{r}}\left[\sqrt{\frac{j+1}{2j+1}}\,\mathcal{I}_{j-1/2}(kr)\,\varphi^m_{j,j-1} - \sqrt{\frac{j}{2j+1}}\,\mathcal{I}_{j+3/2}(kr)\,\varphi^m_{j,j+1}\right]; \tag{5.219}$$

(c) magnetic multipole wave:

$$\mathbf{V}^M_{k,j,m} = \sqrt{\frac{k}{r}}\,\mathcal{I}_{j+1/2}(kr)\,\varphi^m_{j,j}. \tag{5.220}$$

As stressed before and as it also clearly results from the expressions above, the transverse waves can exist only for $j > 0$. The set of orthogonal vector functions defined by Eqs. (5.218), (5.219), and (5.220) is complete and normalized with respect to dk. The latter property comes from the asymptotic form of the Bessel functions.

Let us expand $\mathbf{I}_0$ in Eq. (5.205) according to such set of orthogonal functions:

$$\mathbf{I}_0 = \sum_{j,m}\int_0^\infty \left[I^L_{k,j,m}\,\mathbf{V}^L_{k,j,m} + I^E_{k,j,m}\,\mathbf{V}^E_{k,j,m} + I^M_{k,j,m}\,\mathbf{V}^M_{k,j,m}\right] \mathrm{d}k, \tag{5.221}$$

with $I_{k,j,m}$ being constants. In a similar fashion, we set, at any time instant,

$$\mathbf{A} = \sum_{j,m} \int_0^\infty \sum_{\chi=L,E,M} \left(A^\chi_{k,j,m}\, \mathrm{e}^{-ikct} + A'^\chi_{k,j,m}\, \mathrm{e}^{ikct} \right) \mathbf{V}^\chi_{k,j,m}\, \mathrm{d}k, \tag{5.222}$$

$$\dot{\mathbf{A}} = \sum_{j,m} \int_0^\infty \sum_{\chi=L,E,M} ikc \left(A'^\chi_{k,j,m}\, \mathrm{e}^{ikct} - A^\chi_{k,j,m}\, \mathrm{e}^{-ikct} \right) \mathbf{V}^\chi_{k,j,m}\, \mathrm{d}k. \tag{5.223}$$

The constraint requiring the (Cartesian) components of $\mathbf{A}$ to be real gives

$$A'^\chi_{k,j,m} = \pm (-1)^m A^{\chi *}_{k,j,-m}, \qquad + \text{ for } \chi = L, E, \quad - \text{ for } \chi = M. \tag{5.224}$$

With the same prescription, we can obtain the expansion for I_0^* from Eq. (5.221).

Let us set

$$\mathbf{V}^L_{k,j,m} = \boldsymbol{\nabla}\, v_{k,j,m}. \tag{5.225}$$

The scalar function v is given in Eq. (5.217). However, the functions $v_{k,j,m}$ are not normalized with respect to $\mathrm{d}k$, whereas the same functions multiplied by k, $p_{k,j,m} = k v_{k,j,m}$, are. Let us expand the scalar potential and its time derivative in terms of $v_{k,j,m}$ and put

$$\phi = \sum_{j,m} \int_0^\infty \left(\phi_{k,j,m}\, \mathrm{e}^{-ikct} + \phi'_{k,j,m}\, \mathrm{e}^{ikct} \right) v^\chi_{k,j,m}\, \mathrm{d}k, \tag{5.226}$$

$$\dot{\phi} = \sum_{j,m} \int_0^\infty ikc \left(\phi'_{k,j,m}\, \mathrm{e}^{ikct} - \phi_{k,j,m}\, \mathrm{e}^{-ikct} \right) v^\chi_{k,j,m}\, \mathrm{d}k. \tag{5.227}$$

The reality condition gives

$$\phi'_{k,j,m} = (-1)^m \phi^*_{k,j,m}. \tag{5.228}$$

From the gauge condition [11] $\dot{\phi} = -c\, \boldsymbol{\nabla} \cdot A$, and the fact that the transverse waves have zero divergence and noting that $\boldsymbol{\nabla} \cdot \boldsymbol{\nabla}\, v_{k,j,m} = \boldsymbol{\nabla}^2\, v_{k,j,m} = -k^2 v_{k,j,m}$, we get

$$\phi_{k,j,m}\, \mathrm{e}^{-ikct} - \phi'_{k,j,m} \mathrm{e}^{ikct} = ik \left(A^L_{k,j,m}\, \mathrm{e}^{-ikct} + A'^L_{k,j,m}\, \mathrm{e}^{ikct} \right). \tag{5.229}$$

[11]@ For the sake of clarity, here and in the following we shall call Eq. (5.208) the gauge condition rather than the continuity equation (for the potentials) as it was labeled in the original manuscript.

We also set

$$\rho = \sum_{j,m} \int_0^\infty \rho_{k,j,m}\, v_{k,j,m}\, \mathrm{d}k. \tag{5.230}$$

Since, due to the gauge condition, the first of Eqs. (5.207) can be cast in the form

$$\Box\phi \equiv -\frac{1}{c}\boldsymbol{\nabla}\cdot\mathbf{A} - \boldsymbol{\nabla}^2\phi = 4\pi\rho, \tag{5.231}$$

on expanding in terms of the scalar solutions of (5.210), we find

$$\begin{aligned} &\phi_{k,j,m}\,\mathrm{e}^{-ikct} + \phi'_{k,j,m}\,\mathrm{e}^{ikct} \\ &= ik\left(A^L_{k,j,m}\,\mathrm{e}^{-ikct} - A'^L_{k,j,m}\,\mathrm{e}^{ikct}\right) + \frac{4\pi}{k^2}\rho_{k,j,m}. \end{aligned} \tag{5.232}$$

By combining Eqs. (5.229) and (5.232), we infer

$$\phi_{k,j,m} = ik\,A^L_{k,j,m} + \frac{2\pi}{k^2}\,\mathrm{e}^{ikct}\,\rho_{k,j,m}. \tag{5.233}$$

We now want to find an expression for the total energy of the electromagnetic field when both the expansion (5.222) for the vector potential and the expansion for the scalar potential through Eq. (5.233) is known. We then set

$$\mathcal{E} = \sum_\chi \sum_{j,m} \int_0^\infty E^\chi_{k,j,m}\,\mathbf{V}^\chi_{k,j,m}\,\mathrm{d}k \tag{5.234}$$

$$\mathcal{H} = \sum_\chi \sum_{j,m} \int_0^\infty H^\chi_{k,j,m}\,\mathbf{V}^\chi_{k,j,m}\,\mathrm{d}k. \tag{5.235}$$

From Eq. (5.232) and using

$$\begin{aligned} \boldsymbol{\nabla}\times\mathbf{V}^L_{k,j,m} &= 0, \\ \boldsymbol{\nabla}\times\mathbf{V}^E_{k,j,m} &= +ik\,\mathbf{V}^M_{k,j,m}, \\ \boldsymbol{\nabla}\times\mathbf{V}^M_{k,j,m} &= -ik\,\mathbf{V}^E_{k,j,m}, \end{aligned} \tag{5.236}$$

we easily find

$$\begin{aligned} E^L_{k,j,m} &= -\frac{4\pi}{k^2}\rho_{k,j,m}, \\ E^E_{k,j,m} &= ik\left(A^E_{k,j,m}\,\mathrm{e}^{-ikct} - A'^E_{k,j,m}\,\mathrm{e}^{ikct}\right), \\ E^M_{k,j,m} &= ik\left(A^M_{k,j,m}\,\mathrm{e}^{-ikct} - A'^M_{k,j,m}\,\mathrm{e}^{ikct}\right); \end{aligned} \tag{5.237}$$

$$\begin{aligned} H^L_{k,j,m} &= 0, \\ H^E_{k,j,m} &= -ik\left(A^M_{k,j,m}\,\mathrm{e}^{-ikct} + A'^M_{k,j,m}\,\mathrm{e}^{ikct}\right), \\ H^M_{k,j,m} &= ik\left(A^E_{k,j,m}\,\mathrm{e}^{-ikct} + A'^E_{k,j,m}\,\mathrm{e}^{ikct}\right). \end{aligned} \tag{5.238}$$

The total energy can thus be split into two parts:

$$W = W_{els} + W_R, \tag{5.239}$$

where the electrostatic energy is given by

$$\begin{aligned} W_{els} &= \sum_{j,m} \int_0^\infty \frac{2\pi}{k^4} \left|\rho_{k,h,m}\right|^2 \mathrm{d}k \\ &= \frac{1}{2} \int \frac{1}{|q-q'|} \rho(q)\, \rho(q')\, \mathrm{d}q\, \mathrm{d}q', \end{aligned} \tag{5.240}$$

and the radiated energy by

$$W_R = \frac{1}{2\pi} \sum_{j,m} \int_0^\infty k^2 \left(\left|A^E_{k,j,m}\right|^2 + \left|A^M_{k,j,m}\right|^2 \right) \mathrm{d}k. \tag{5.241}$$

Let us now go back to our oscillating system (5.205) in order to evaluate the radiated energy by using the method of variation of parameters. The electrostatic energy periodically oscillates between finite values with frequency ν, and we thus neglect it. Considering the radiated energy, let us assume that it is zero at the initial time, so that initially we have $A^E_{k,j,m} = A^M_{k,j,m}$ for all values of k, j, m. From the second of Eqs. (5.207) and from Eqs. (5.222) and (5.223), it follows that

$$\begin{aligned} \dot{A}^Y_{k,j,m}\, \mathrm{e}^{-ikct} &- \dot{A}'^Y_{k,j,m}\, \mathrm{e}^{ikct} \\ &= (4\pi\, ic/k) \left(I^Y_{k,j,m}\, \mathrm{e}^{-2\pi\nu it} + I'^Y_{k,j,m}\, \mathrm{e}^{2\pi\nu it} \right), \\ \dot{A}^Y_{k,j,m}\, \mathrm{e}^{-ikct} &+ \dot{A}'^Y_{k,j,m}\, \mathrm{e}^{ikct} = 0, \quad \text{Y=E,M}, \end{aligned} \tag{5.242}$$

where, in analogy to Eq. (5.224),

$$I'^Y_{k,j,m} = \pm (-1)^m\, I^{Y*}_{k,j,-m}; \qquad + \text{ for } Y = E, \quad - \text{ for } Y = M. \tag{5.243}$$

We thus derive

$$\dot{A}^Y_{k,j,m} = \frac{2\pi\, ic}{k} \left(I^Y_{k,j,m}\, \mathrm{e}^{i(kc-2\pi\nu)t} + I'^Y_{k,j,m}\, \mathrm{e}^{i(kc+2\pi\nu)t} \right), \tag{5.244}$$

$$A^Y_{k,j,m} = \frac{2\pi\, c}{k} \left(I^Y_{k,j,m} \frac{\mathrm{e}^{i(kc-2\pi\nu)t} - 1}{kc - 2\pi\nu} + I'^Y_{k,j,m} \frac{\mathrm{e}^{i(kc+2\pi\nu)t} - 1}{kc + 2\pi\nu} \right) \tag{5.245}$$

For $t \to \infty$ the main contributions to the integral in Eq. (5.241) come from values of k close to $k_0 = 2\pi v/c$, apart from quantities that do not exceed finite constant values. For k close to k_0 it follows from Eq. (5.245) that

$$\left|A^Y_{k,j,m}\right| = \frac{4\pi}{k_0} \frac{\sin(k-k_0)ct/2}{k-k_0} \left|I^Y_{k_0,j,m}\right|, \tag{5.246}$$

from which, by substituting into Eq. (5.241) and integrating between $-\infty$ and $+\infty$, instead of between 0 and ∞, as we can do in the considered limit:

$$W_R = 4\pi^2\, c\, t \sum_{j,m} \left(\left| I^E_{k_0,j,m} \right| + \left| I^M_{k_0,j,m} \right| \right). \tag{5.247}$$

It follows that the energy radiated in the unit time is

$$w_R = \frac{W_R}{t} = 4\pi^2\, c \sum_{j,m} \left(\left| I^E_{k_0,j,m} \right| + \left| I^M_{k_0,j,m} \right| \right). \tag{5.248}$$

The radiated energy can thus be computed by decomposing the oscillating system into transverse electric and magnetic multipoles of different orders and assuming that they radiate without interfering with each other. Naturally, the longitudinal multipoles do not radiate any energy. Every multipole corresponds to a given spherical wave with definite intensity distribution and polarization state along certain directions. In the quantum interpretation the numbers j and m represent the total angular momentum and its component along the z direction of the emitted quantum (in units of $\hbar$). The knowledge of these numbers does not determine completely the emitted wave, since this can be an electric multipole wave or a magnetic multipole wave. This ambiguity is due to the fact that different kinds of coupling between orbital momenta and intrinsic momenta of the emitted quantum are possible. As is well known, the intrinsic momentum is $\pm\hbar$ along the direction of motion and cannot be zero.

From Eq. (5.248), the intensity of a definite multipole is given by

$$w^Y_{j,m} = 4\pi^2\, c \left| I^Y_{k_0,j,m} \right|^2, \quad Y = E, M, \tag{5.249}$$

or, computing the coefficient $I^Y_{k_0,j,m}$ with the usual rule for the coefficients of expansion in terms of a set of orthogonal functions

$$w^Y_{j,m} = 4\pi^2\, c \left| \int \mathbf{V}^{Y\dagger}_{k_0,j,m} \cdot \mathbf{I}_0 \, \mathrm{d}q \right|^2. \tag{5.250}$$

The case in which the system has atomic dimensions very small compared with the wavelength of the emitted radiation $(2\pi/k_0)$ is of particular practical relevance. It is then possible to evaluate $w^Y_{j,m}$ in first approximation by substituting (into the integral appearing in the expression for $w^Y_{j,m}$) every Bessel function entering in the expressions of $V^Y_{k_0,j,m}$ by the first term of its series expansion. Of course, it is assumed that the radiating system is placed near the origin of the coordinate system. In the case of electric multipoles we have Bessel functions of order

$j+3/2$ as well as $j-1/2$; we can then neglect, inside the integral, the former functions and retain only the first term in the expansion of the latter ones. We are only interested in Bessel functions of order $n+1/2$ with integer n; and for these, in first approximation, we have

$$\begin{aligned} \mathcal{I}_{n+1/2} &= \sqrt{\frac{2}{\pi}}\,\frac{2^n \cdot n!}{(2n+1)!}\, x^{n+1/2} + \dots \\ &= \sqrt{\frac{2}{\pi}} \left(1\cdot\frac{1}{3}\cdot\frac{1}{5}\cdots\frac{1}{2n+1}\right) x^{n+1/2} + \dots . \end{aligned} \qquad (5.251)$$

We thus derive, in first approximation,

$$\begin{aligned} w_{j,m}^E &= 1\cdot\frac{1}{3^2}\cdot\frac{1}{5^2}\cdots\frac{1}{(2j-1)^2}\,\frac{j+1}{2j+1}\cdot 8\pi\, c \\ &\times \left(\frac{2\pi\nu}{c}\right)^{2j} \left|\int r^{j-1}\,\varphi_{j,j-1}^{m\dagger}\cdot \mathbf{I}_0\, \mathrm{d}q\right|^2, \end{aligned} \qquad (5.252)$$

$$\begin{aligned} w_{j,m}^M &= 1\cdot\frac{1}{3^2}\cdot\frac{1}{5^2}\cdots\frac{1}{(2j+1)^2}\cdot 8\pi\, c \\ &\times \left(\frac{2\pi\nu}{c}\right)^{2j+2} \left|\int r^{j}\,\varphi_{j,j}^{m\dagger}\cdot \mathbf{I}_0\, \mathrm{d}q\right|^2 . \end{aligned} \qquad (5.253)$$

Equation (5.252) can be cast in a different form that is more convenient for calculations and contains only the charge density ρ_0 instead of the current density. In fact, by using Cartesian coordinates, the integrand function in Eq. (5.252) can be written as

$$r^{j-1}\,\varphi_{j,j-1}^{m\dagger}\cdot\mathbf{I}_0 \;=\; r^{j-1}\,\varphi_{j,j-1}^{m*}\cdot\mathbf{I}_0. \qquad (5.254)$$

Since, quite generally,

$$\boldsymbol{\nabla}\, r^j\,\varphi_j^m \;=\; \sqrt{j(2j+1)}\, r^{j-1}\,\varphi_{j,j-1}^m, \qquad (5.255)$$

on taking Eq. (5.206) into account, we get

$$\begin{aligned} \int r^{j-1}\,\varphi_{j,j-1}^{m\dagger}\cdot\mathbf{I}_0\,\mathrm{d}q &= \int r^{j-1}\,\varphi_{j,j-1}^{m*}\cdot\mathbf{I}_0\,\mathrm{d}q \\ &= \frac{1}{\sqrt{j(2j+1)}}\int \boldsymbol{\nabla}\, r^j\,\varphi_j^{m*}\cdot\mathbf{I}_0\,\mathrm{d}q \\ &= -\frac{1}{\sqrt{j(2j+1)}}\int r^j\,\varphi_j^{m*}\,\boldsymbol{\nabla}\cdot\mathbf{I}_0\,\mathrm{d}q \\ &= -\frac{1}{\sqrt{j(2j+1)}}\,\frac{2\pi\,\nu\, i}{c}\int r^j\,\varphi_j^{m*}\,\rho_0\,\mathrm{d}q. \end{aligned} \qquad (5.256)$$

On substituting this in Eq. (5.252), we finally find

$$w^E_{j,m} = 1\cdot\frac{1}{3^2}\cdot\frac{1}{5^2}\cdots\frac{1}{(2j+1)^2}\,\frac{j+1}{j}\cdot 8\pi\, c \times\left(\frac{2\pi\nu}{c}\right)^{2j+2}\left|\int r^j\,\varphi_j^{m*}\,\rho_0\,\mathrm{d}q\right|^2. \tag{5.257}$$

We now want to study the radiation emitted by our oscillating system, using the method of stationary waves or, better, the method of periodic solutions. We thus look for a solution of Eq. (5.207) that, in analogy with Eqs. (5.205), can be cast in the form

$$\begin{aligned}\phi &= \phi_0\,\mathrm{e}^{-2\pi\nu i t} + \phi_0^*\,\mathrm{e}^{2\pi\nu i t},\\ \mathbf{A} &= \mathbf{A}_0\,\mathrm{e}^{-2\pi\nu i t} + \mathbf{A}_0^*\,\mathrm{e}^{2\pi\nu i t},\end{aligned} \tag{5.258}$$

with the additional constraint that the scalar and vector potential describe a diverging wave at infinity. We set

$$\mathbf{A}_0 = \sum_\chi\sum_{j,m} A^\chi_{j,m}(r)\,\mathbf{U}^\chi_{k_0,j,m}, \tag{5.259}$$

$$\phi_0 = \sum_{j,m}\phi_{j,m}(r)\,u_{k_0,j,m}, \qquad \chi = L, E, M. \tag{5.260}$$

The functions $\mathbf{U}^\chi_{k_0,j,m}$ and $u_{k_0,j,m}$ are obtained from $\mathbf{V}^\chi_{k_0,j,m}$ (given in Eqs.(5.218), (5.219), and (5.220)) and $v_{k_0,j,m}$ (given in Eq. (5.225)) by replacing the Bessel functions with the Hankel functions of the first kind. The condition that at infinity there exists only a diverging wave implies that the following limits exist:

$$A^\chi_{j,m}(\infty) = B^\chi_{j,m}, \tag{5.261}$$

$$\phi_{j,m}(\infty) = \Phi_{j,m}. \tag{5.262}$$

From the gauge condition (5.208), we derive, in analogy with Eq. (5.206),

$$\phi_0 = \frac{c}{2\pi\,\nu\, i}\,\boldsymbol{\nabla}\cdot\mathbf{A}_0. \tag{5.263}$$

12. HYDROGEN EIGENFUNCTIONS

In electronic units we have

$$\boldsymbol{\nabla}^2\,\psi + \left(2E + \frac{2}{r}\right)\psi = 0, \tag{5.264}$$

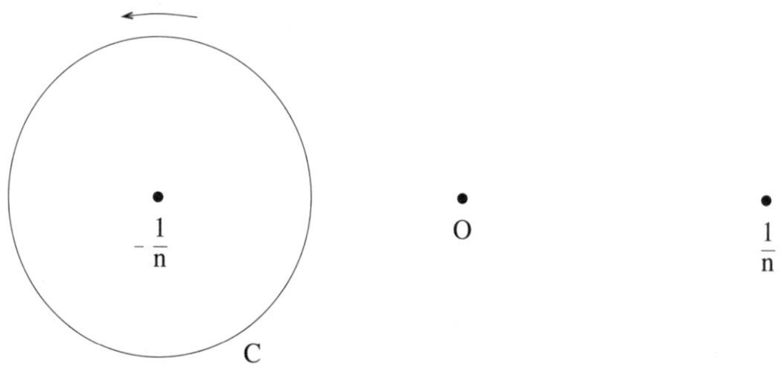

Fig. 5.1. The contour line for the integral defining the hydrogen eigenfunctions.

and, by setting $\psi = (y/r)\,\varphi_\ell^m$, we get

$$y'' + \left(2E + \frac{2}{r} - \frac{\ell(\ell+1)}{r^2}\right) y = 0. \tag{5.265}$$

For the discrete energy spectrum we have

$$E = -\frac{1}{2}\frac{1}{n^2}, \quad n = \ell+1, \ell+2, \ldots, \tag{5.266}$$

$$N\, y = A\, r^{\ell+1} \int_C \left(t + \frac{1}{n}\right)^{\ell-n} \left(t - \frac{1}{n}\right)^{\ell+n} \mathrm{e}^{tr}\, \mathrm{d}t, \tag{5.267}$$

where

$$A = -\left(\frac{n}{2}\right)^{2\ell+1} \Big/ 2\pi i \left(\begin{array}{c} n+\ell \\ 2\ell+1 \end{array}\right). \tag{5.268}$$

Integration using the method of residues leads to

$$N\, y = \sum_{p=0}^{n-\ell-1} (-1)^p \frac{(n-\ell-1)(n-\ell-2)\cdots(n-\ell-p)}{(2\ell+2)(2\ell+3)\cdots(2\ell+1+p)} \left(\frac{2}{n}\right)^p \times \frac{r^{\ell+1+p}}{p!}\, \mathrm{e}^{-r/n}. \tag{5.269}$$

The normalization constant is given by

$$N^2 = \left(\frac{(2\ell+1)!}{2^{\ell+1}}\right) \frac{(n-\ell-1)!\, n^{2\ell+4}}{(n+\ell)!}. \tag{5.270}$$

For example, for $\ell = 0, 1, 2$ we have, respectively,

$$N^2 = \frac{1}{4} n^3, \tag{5.271}$$

$$N^2 = \frac{9}{4}\frac{n^5}{n^2-1}, \tag{5.272}$$

$$N^2 = 225 \frac{n^7}{(n^2-1)(n^2-4)}. \tag{5.273}$$

Below we list the first eigenfunctions

$$1s: \quad N\,y = r\,\mathrm{e}^{-r} \qquad \left(N = \frac{1}{2}\right).$$

$$2s: \quad N\,y = \left(r - \frac{1}{2}r^2\right)\mathrm{e}^{-r/2} \qquad \left(N = \sqrt{2}\right).$$

$$3s: \quad N\,y = \left(r - \frac{2}{3}r^2 + \frac{2}{27}r^3\right)\mathrm{e}^{-r/3} \qquad \left(N = \frac{\sqrt{27}}{2}\right).$$

$$2p: \quad N\,y = r^2\,\mathrm{e}^{-r/2} \qquad \left(N = \sqrt{24}\right).$$

$$3p: \quad N\,y = \left(r^2 - \frac{1}{6}r^3\right)\mathrm{e}^{-r/3} \qquad \left(N = \sqrt{\frac{2187}{32}}\right).$$

$$3d: \quad N\,y = r^3\,\mathrm{e}^{-r/3} \qquad \left(N = 81\sqrt{\frac{15}{8}}\right).$$

The asymptotic expression for $r \to \infty$ is

$$y \sim (-1)^{n-\ell-1} \frac{2^n}{n^{n+1}\sqrt{(n+\ell)!(n-\ell-1)!}}\, r^n\,\mathrm{e}^{-r/n}. \tag{5.274}$$

SELECTED FACSIMILE PAGES

quando si badi che

$$u - \frac{1}{6} r^2 \Delta_2 u = \frac{1}{6}\left(u \Delta_2 r^2 - r^2 \Delta_2 u\right)$$

Riprendiamo la (22) e facciamo delle approssimazioni. Supponiamo in primo luogo r grande rispetto alla lunghezza d'onda, così che si può trascurare l'unità di fronte a iKr; supponiamo inoltre che σ sia una superficie d'onda di un'onda progressiva, con raggio minimo di curvatura grande anch'esso rispetto alla lunghezza d'onda; si potrà allora considerare l'onda come piana per un tratto breve e sarà approssimativamente:

$$\frac{\partial u}{\partial n} = \pm\, i K u \qquad (29)$$

secondo che l'onda si avvicina a P_0 o se ne allontana: la (22) si riduce allora con le fatte approssimazioni a:

$$u(P_0) = \frac{Ki}{4\pi}\int_\sigma \frac{u(\cos\alpha \pm 1)e^{-iKr}}{r}\, d\sigma \qquad (30)$$

ovvero introducendo la lunghezza d'onda in base alla relazione:

$$K = \frac{2\pi}{\lambda} \qquad (31)$$

$$u(P_0) = \frac{i}{\lambda}\int_\sigma \frac{\cos\alpha \pm 1}{2}\,\frac{u\, e^{-\frac{2\pi i}{\lambda} r}}{r}\, d\sigma \qquad (32)$$

Se α è piccolo e l'onda si avvicina:

$$u(P_0) = \frac{i}{\lambda}\int_\sigma \frac{e^{-\frac{2\pi i}{\lambda} r}}{r}\, u\, d\sigma \qquad (33)$$

3 - Equilibrio di una massa liquida eterogenea in rotazione

(Problema di Clairaut)

Si suppone che la massa rotante risulti dalla sovrapposizione di strati liquidi incompressibili di densità differenti. La velocità angolare di rotazione ω si suppone piccola; le deformazioni che la massa subisce per effetto della rotazione sono allora dell'ordine di ω^2. Si riguarderà perciò ω^2 come infinitesimo principale.

Se particelle liquide si attirano secondo la legge di Newton, in cui si supporrà di ridurre ~~all'unità~~ il coefficiente d'attrazione, mediante una conveniente scelta di unità.

Finchè la massa è in riposo sarà la densità una funzione mai crescente della distanza dal centro:

$$\rho = \rho(r) \qquad \rho' \leq 0 \qquad (1)$$

Analogamente il potenziale newtoniano (potenziale delle forze) dipenderà da r:

$$V_0 = V_0(r) \qquad (2)$$

Indicheremo con D la densità media della massa che si trova a distanza minore di r:

$$D = \frac{\int \rho r^2 dr}{\frac{1}{3} r^3} \qquad (3)$$

Segue:

$$r^3 D = 3\int_0^r \rho r^2 dr$$

$$3r^2 D + r^3 D' = 3\rho r^2$$

cioè:

$$3\rho = 3D + rD' \qquad (4)$$

da cui derivando:

$$3\rho' = 4D' + rD'' \qquad (5)$$

che servirà in seguito.
La forza che si esercita a distanza r su una massa unitaria sarà

$$\frac{\int_0^r 4\pi r^2 \rho \, dr}{r^2} = \frac{4}{3}\pi r D$$

cosicché:

$$V_0' = -\frac{4}{3}\pi r D \qquad (6)$$

Si ponga ora la massa in rotazione; una particella che si trovava in P si porterà in P' nella nuova configurazione di equilibrio. Poniamo:

$$\eta = \overline{PP'} \cos(r, \overline{PP'}) \qquad (7)$$

Lo spostamento normale η si potrà sviluppare secondo le funzioni sferiche Y:

$$\eta = \Sigma H Y \qquad (8)$$

essendo le H funzioni del raggio.
Se la rotazione ha luogo intorno all'asse z compariranno nello sviluppo (8) solo le funzioni sferiche simmetriche intorno all'asse z, le quali saranno esprimibili mediante i polinomi di Legendre:

$$P_n(\cos\theta)$$

Inoltre scambiando $-z$ in z, η deve rimanere inalterato. Dovremo quindi limitarci alle funzioni sferiche d'ordine pari. Inoltre sulla superficie sferica di raggio r si avrà:

$$\int \eta \, d\sigma = 0$$

Deve quindi mancare la funzione

sferica d'ordine zero. La prima ad apparire sarà la funzione di secondo ordine che prenderemo solo in considerazione:

$$Y = \frac{x^2 + y^2 - 2z^2}{r^2} \qquad (9)$$

Noi vogliamo supporre che per tutte le arti sia $n = 0$. Questo equivale a supporre che le superficie di egual densità sono in prima approssimazione degli ellissoidi. Tale ipotesi è evidentemente verificata per la superficie libera.
La (8) si riduce allora a:

$$\eta = H Y \qquad (10)$$

con Y definita dalla (9).
Lo schiacciamento della superficie di egual densità e di raggio medio r è evidentemente:

$$s = \frac{3H}{r} \qquad (11)$$

Analogamente supporremo che il potenziale newtoniano sia in prima approssimazione:

$$V = V_0 + LY \qquad (12)$$

Aggiungendo il potenziale delle forze centrifuga si ottiene il potenziale totale che deve essere considerato per l'equilibrio relativo:

$$U = V + \frac{1}{2}\omega^2(x^2 + y^2) =$$

$$= V_0 + \frac{1}{3}\omega^2 r^2 + \left(L + \frac{1}{6}\omega^2 r^2\right) Y \qquad (13)$$

La densità ρ_1 del fluido in rotazione sarà in prima approssimazione, a causa di (7) e (10):

$$\rho_1 = \rho - \eta\rho' = \rho - H\rho' Y \qquad (14)$$

Per determinare H e L che sono attualmente le incognite del nostro problema dobbiamo servirci dell'equazione di Poisson e della condizione che le superficie di egual densità coincidano con le superficie equipotenziali.
L'equazione di Poisson ci dà:

$$\Delta_2 V = -4\pi\rho_1$$

cioè essendo

$$\Delta_2 V_0 = 4\pi\rho$$

$$V - V_0 = LY$$

$$\rho_1 - \rho = -H\rho' Y$$

semplicemente:

$$\Delta_2 LY = 4\pi\rho' Y H$$

Ovvero, dividendo per Y:

$$4\pi H\rho' = L'' + \frac{2}{r}L' - \frac{6}{r^2}L \qquad (15)$$

Le superficie equipotenziali (U = cost.) sono in prima approssimazione degli ellissoidi di rivoluzione intorno a z. Lo schiacciamento della sezione meridiana sarà in prima approssimazione:

$$s_u = -3\,\frac{L + \frac{1}{8}\omega^2 r^2}{r\,V_0'} = +3\,\frac{L + \frac{1}{8}\omega^2 r^2}{4\pi r^2 D/3}$$

Le superficie di egual densità sono, come si è visto, anche esse ellissoidi di rivoluzione il cui schiacciamento è dato dalla (11). Perchè le due famiglie di superficie coincidano dovrà essere:

$$s = s_u$$

cioè:

$$H = \frac{L + \frac{1}{8}\omega^2 r^2}{\frac{4}{3}\pi r D} \qquad (16)$$

Risolvendo la (16) rispetto a L si ha:

$$L = \frac{4}{3}\pi r D H - \frac{1}{8}\omega^2 r^2$$

$$L' = \frac{4}{3}\pi D H + \frac{4}{3}\pi r D' H + \frac{4}{3}\pi r D H' - \frac{1}{4}\omega^2 r$$

$$L'' = \frac{8}{3}\pi D' H + \frac{8}{3}\pi D H' + \frac{8}{3}\pi r D' H' +$$
$$+ \frac{4}{3}\pi r D'' H + \frac{4}{3}\pi r D H'' - \frac{1}{4}\omega^2$$

Sostituendo in (15) si elimina L:

$$3H\rho' = -\frac{4DH}{r} + 4D'H + 4DH' + 2rD'H' + rD''H + rDH''$$

ma per la (5):

$$3H\rho' = 4D'H + rD''H$$

così che rimane:

$$0 = -\frac{4DH}{r} + 4DH' + 2rD'H' + rDH''$$

o anche:

$$D\left(-4 + 4r\frac{H'}{H} + r^2\frac{H''}{H}\right) + 2rD'\frac{rH'}{H} = 0 \qquad (17)$$

Poniamo:

$$q = \frac{r s'}{s} \qquad (18)$$

ricordando che $s = \frac{3H}{r}$, sarà:

$$\frac{s'}{s} = \frac{H'}{H} - \frac{1}{r}$$

$$q = r\frac{H'}{H} - 1 \qquad (19)$$

da cui:

$$r\frac{H'}{H} = 1 + q \qquad (20)$$

$$\frac{H'}{H} + r\frac{H''}{H} - r\frac{H'^2}{H^2} = q'$$

$$r\frac{H'}{H} + r^2\frac{H''}{H} - r^2\frac{H'^2}{H^2} = rq'$$

$$1+q+r^2\frac{H''}{H}-(1+q)^2=rq'$$

$$r^2\frac{H''}{H}=rq'+q+q^2 \qquad (21)$$

Sostituendo mediante (20) e (21) in (17) si trova:

$$D(rq'+5q+q^2)+2rD'(1+q)=0 \qquad (22)$$

che è l'equazione di Clairaut.
Se $\frac{rD'}{D}$ tende a zero per r che tende a zero dovrà essere per $r=0$

$$5q+q^2=0$$

cioè:

$$q=\begin{cases}0\\-5\end{cases}$$

Ora nel centro della massa rotante si avrà sviluppando V:

$$V=V(0)+A(x^2+y^2)+Bz^2+\dots$$

Se si suppone $\varrho'(0)$ finito (in particolare nullo) V sarà sviluppabile secondo x, y, z e per ragioni di simmetria mancheranno i termini di grado dispari. Indicando con ε una funzione infinitesima con r del quarto ordine, avremo:

$$U=V(0)+A(x^2+y^2)+\tfrac{1}{2}\omega^2(x^2+y^2)+Bz^2+\varepsilon$$

e sarà naturalmente:

$$4A+2B=-4\pi\varrho(0)$$

Poniamo:

$$U=\text{cost.}$$

sarà:

$$A_1=-A \qquad B_1=-B$$

$$(A_1+\tfrac{1}{2}\omega^2)(x^2+y^2)+B_1z^2+\varepsilon=\text{cost}$$

ε è a meno di infinitesimi del secondo ordine

sarà

$$s=\frac{\frac{1}{\sqrt{A_1-\frac{1}{2}\omega^2}}-\frac{1}{\sqrt{B_1}}}{\frac{1}{\sqrt{A_1-\frac{1}{2}\omega^2}}}=1-\sqrt{\frac{A_1-\frac{1}{2}\omega^2}{B_1}}$$

Sarà quindi:

$$s'(0)=0$$

e "a fortiori":

$$q(0)=\frac{rs'(0)}{s(0)}=0$$

La curva integrale di (22) che risolve il problema passa dunque per il punto:

$$r=0\ ,\ q=0$$

Supponiamo che D possa svilupparsi secondo le potenze pari di r:

$$D=D(0)+ar^2+br^4+cr^6+\dots \qquad (23)$$

e analogamente:

$$q=q_0+\alpha r^2+\beta r^4+\gamma r^6+\dots \qquad (24)$$

ovvero, ponendo:

$$a_0=D_0 \quad a_2=a \quad a_4=b \quad a_6=c \quad \dots$$

$$\alpha_0=q_0=0 \quad \alpha_2=\alpha \quad \alpha_4=\beta \quad \alpha_6=\gamma$$

$$D=\sum_0 a_{2n}r^{2n} \qquad (23)$$

$$q=\sum_0 \alpha_{2n}r^{2n} \qquad (24)$$

e sostituendo in (22):

$$\left(\sum a_{2n}r^{2n}\right)\left[\sum(2n\alpha_{2n}r^{2n}+5\alpha_{2n}r^{2n})+\sum{}^2\alpha_{2n}r^{2n}\right]$$
$$+2\sum 2n\,a_{2n}r^{2n}\left(1+\sum\alpha_{2n}r^{2n}\right)=0$$

Calcoliamo i primi coefficienti dello sviluppo di q

Avremo:

$$D = D_0 + a r^2 + b r^4 + c r^6 + \dots$$

$$r D' = 2a r^2 + 4b r^4 + 6c r^6 + \dots$$

$$q = \alpha r^2 + \beta r^4 + \gamma r^6 + \dots$$

$$r q' = 2\alpha r^2 + 4\beta r^4 + 6\gamma r^6 + \dots$$

$$q^2 = \alpha^2 r^4 + 2\alpha\beta r^6 + \dots$$

$$r q' + 5q + q^2 = 7\alpha r^2 + (9\beta + \alpha^2) r^4 + (11\gamma + 2\alpha\beta) r^6 + \dots$$

$$1 + q = 1 + \alpha r^2 + \beta r^4 + \gamma r^6 + \dots$$

$$D(rq' + q^2 + 5q) = 7\alpha D_0 r^2 + \left[(9\beta + \alpha^2) D_0 + 7\alpha a\right] r^4 + \left[(11\gamma + 2\alpha\beta) D_0 + (9\beta + \alpha^2) a + 7\alpha b\right] r^6 + \dots$$

$$2rD'(1+q) = 4a r^2 + (4\alpha a + 8b) r^4 + (4\beta a + 8\alpha b + 12c) r^6 + \dots$$

Segue:

$$7\alpha D_0 + 4a = 0$$

$$(9\beta + \alpha^2) D_0 + 7\alpha a + 4\alpha a + 8b = 0$$

$$(11\gamma + 2\alpha\beta) D_0 + (9\beta + \alpha^2) a + 7\alpha b + 4\beta a + 8\alpha b + 12c = 0$$

. . . . . .

Se M è la massa del pianeta, l'attrazione in punti esterni, in particolare sulla superficie libera, ammette per potenziale in prima approssimazione:

$$V = \frac{M}{r} + \frac{J_0 - \frac{3}{2} J}{r^3} \qquad (23)$$

in J è il momento di inerzia rispetto alla retta OP e J_0 è il momento di inerzia (non assiale, ma polare) rispetto al baricentro.

Sulla superficie libera sarà:

$$u = \frac{M}{r} + \frac{J_0 - \frac{3}{2} J}{r^3} + \frac{1}{2}\omega^2 (x^2 + y^2) = \text{cost.} \qquad (24)$$

Indicando con R_p il raggio polare e con R_e il raggio equatoriale, sarà:

$$\frac{M}{R_p} + \frac{C - A}{R_p^3} = \frac{M}{R_e} + \frac{1}{2}\frac{C - A}{R_e^3} + \frac{1}{2}\omega^2 R_e^2 \qquad (25)$$

essendosi indicato con C il momento di inerzia rispetto all'asse polare e con A il momento di inerzia rispetto a un asse equatoriale; di modo che

$$J_0 = A + \frac{1}{2} C$$

Indichiamo con f il rapporto tra forza centrifuga e gravità all'equatore e con r_1 il raggio medio del pianeta; sarà in prima approssimazione

$$f = \frac{\omega^2 r_1^3}{M} \qquad (26)$$

Segue da (25) in prima approssimazione:

$$M\left(\frac{1}{R_p} - \frac{1}{R_e}\right) = \frac{3}{2}\frac{C - A}{r_1^3} + \frac{1}{2}\frac{f}{r_1} M \qquad (27)$$

e ponendo al solito:

$$s_1 = \frac{R_e - R_p}{R_e}$$

sarà, sempre in prima approssimazione,

$$s_1 = \frac{3}{2}\frac{C - A}{M r_1^2} + \frac{1}{2} f \qquad (28)$$

O anche indicando con D_1 la densità media dell'intero pianeta:

$$s_1 - \frac{1}{2} f = \frac{9(C - A)}{8\pi r_1^5 D_1} \qquad (29)$$

Il momento di inerzia medio della Terra sarà:

$$J = \frac{8\pi}{3} \int_0^{r_1} \rho r^4 dr \qquad (30)$$

Ora:

$$\int_0^{r_1} \rho r^4 dr = \int_0^{r_1} r^2 \rho r^2 dr = \int_0^{r_1} r^2 d\left(\tfrac{1}{3} r^3 D\right)$$

$$= \frac{1}{3} r_1^5 D_1 - \frac{2}{3} \int_0^{r_1} r^4 D\, dr$$

segue:

$$J = \frac{8\pi}{9}\left(r_1^5 D_1 - 2\int_0^{r_1} r^4 D\, dr\right) \qquad (31)$$

In prima approssimazione sarà

$$\bar{C} \cong J$$

e sostituendo in (29) si trova:

$$\varepsilon_1 - \frac{1}{2} f = \frac{C-A}{C}\left(1 - \frac{2\int r^4 D\, dr}{r_1^5 D_1}\right) \qquad (32)$$

Riprendiamo l'equazione (22) di Clairaut

$$D(r^2 q' + 5q + q^2) + 2r D'(1+q) = 0 \qquad (22)$$

e calcoliamo l'espressione:

$$\frac{d}{dr}\left(r^5 D \sqrt{1+q}\right) = 5r^4 D\sqrt{1+q} + r^5 D'\sqrt{1+q} + r^5 D \frac{q'}{2\sqrt{1+q}}$$

$$= \frac{5r^4 D}{\sqrt{1+q}}\left(1 + q + \frac{rD'}{5D}(1+q) + \frac{rq'}{10}\right) \qquad (33)$$

Poiché da (22) si trae:

$$\frac{rD'}{5D}(1+q) = -\frac{rq'}{10} - \frac{q}{2} - \frac{q^2}{10}$$

sostituendo in (33) si ricava:

$$\frac{d}{dr}\left(r^5 D\sqrt{1+q}\right) = \frac{5r^4 D}{\sqrt{1+q}}\left(1 + \frac{1}{2} q - \frac{1}{10} q^2\right)$$

da cui:

$$r_1^5 D_1 \sqrt{1+q_1} = \int_0^{r_1} \frac{5r^4 D}{\sqrt{1+q}}\left(1 + \frac{q}{2} - \frac{q^2}{10}\right) dr \qquad (34)$$

Poniamo:

$$K = \frac{1 + \frac{q}{2} - \frac{q^2}{10}}{\sqrt{1+q}} \qquad (35)$$

e la (34) diventa:

$$r_1^5 D_1 \sqrt{1+q_1} = \int_0^{r_1} 5r^4 D K\, dr \qquad (36)$$

Se q è ancora piccolo, K è molto prossimo all'unità:

q		K	
0		1	
0,1			
0,2			
0,3			
3	massimo	1,00074	$\frac{26\sqrt{3}}{45}$
0,4			
0,5		1,00021	
0,6			
0,7			
0,8			
0,9			
1		0,98995	
2			
3			

Se [illegible] il valore minimo al centro: $r=0$, $q=0$.

Calcoliamo q_1 [illegible] equipotenziali all'esterno del pianeta [illegible]: [28]

$$\varepsilon = \frac{1}{2} f + \frac{3}{2}\frac{C-A}{M r^2} \qquad (37)$$

Il primo termine del secondo termine [illegible] in prima approssimazione come r^3. Si ha quindi derivando:

$$r s' = \frac{3}{2} f - 3 \frac{C-A}{M r^2} \qquad (38)$$

confrontiamo con (37):

$$r s' + 2 s = \frac{5}{2} f$$

$$r s' = \frac{5}{2} f - 2 s \qquad (39)$$

$$q = \frac{r s'}{s} = \frac{5}{2} \frac{f}{s} - 2$$

In particolare, riferendo f alla superficie libera del pianeta:

$$q_1 = \frac{5}{2} \frac{f}{s_1} - 2 \qquad (40)$$

Nel caso della terra si ha $q_1 = 0.57$. Sarà allora K sempre molto vicino all'unità. Supponendo $K=1$ (e tale ipotesi è lecita tutte le volte che la densità del pianeta non sia eccessivamente disuniforme) la (36) diventa:

$$r_1^5 D_1 \sqrt{\frac{5}{2} \frac{f}{s_1} - 1} \cong \int 5 r^4 D \, dr \qquad (41)$$

ovvero confrontiamo con (32):

$$\sqrt{\frac{5}{2} \frac{f}{s_1} - 1} \cong \frac{5}{2} - \frac{5}{2} \frac{C}{C-A} (s_1 - \frac{1}{2} f), \qquad (42)$$

Nel caso della terra si ha:

$$f = \frac{1}{288} \qquad (42)$$

$$\frac{C}{C-A} = 305 \qquad (43)$$

si trova allora:

$$s_1 = \frac{1}{297} \qquad (44)$$

in perfetto accordo con l'esperienza. Sostituendo in (38) si ha:

$$\frac{1}{297} = \frac{1}{2} \frac{1}{288} + \frac{3}{2} \frac{C-A}{M r_1^2}$$

da cui:

$$C - A = \frac{1}{920} M r_1^2 \qquad (45)$$

e per la (43):

$$C = 0,332 \, M r_1^2 \qquad (46)$$

mentre se la densità fosse costante si avrebbe $\mathcal{J} = 0,4 \, M r_1^2$. Valori stabiliti dalla conferenza di Madrid:

$$\left.\begin{aligned} r_e &\cong 6378 \\ r_p &= 6357 \\ s &= \frac{1}{297} \\ D_1 &= 5,515 \end{aligned}\right\} \qquad (47)$$

Supponiamo che la densità all'interno della terra sia esprimibile sotto la forma

$$\rho = a + b r^2 + c r^4 \qquad (48)$$

Vogliamo determinare i coefficienti con le condizioni:

$$\left.\begin{aligned} D_1 &= 5.515 \\ \rho_1 &= 2,5 \\ \mathcal{J} &= 0,332 \, M r_1^2 \end{aligned}\right\} \qquad (49)$$

Si avrà:

$$\rho_1 = a + b r_1^2 + c r_1^4 \qquad (50)$$

$$\frac{4}{3} r_1^3 D_1 = \int_0^{r_1} (a r^2 + b r^4 + c r^6) \, dr =$$

$$= \frac{1}{3} a r_1^3 + \frac{1}{5} b r_1^5 + \frac{1}{7} c r_1^7$$

cioè:

$$D_1 = a + \frac{3}{5} b r_1^2 + \frac{3}{7} c r_1^4 \qquad (51)$$

Inoltre:

$$\mathcal{J} = \frac{8\pi}{3} \int_0^{r_1} (a r_1^4 + b r_1^6 + c r_1^8) \, dr =$$

$$= \frac{8\pi}{3} \left(\frac{1}{5} a r_1^5 + \frac{1}{7} b r_1^7 + \frac{1}{9} c r_1^9 \right),$$

cioè:

$$\frac{y}{r_1^5} = \frac{8\pi}{3}\left(\frac{1}{5}a + \frac{1}{7}b r_1^2 + \frac{1}{9}c r_1^4\right) \qquad (52)$$

D'altra parte

$$\frac{M}{r_1^3} = \frac{4}{3}\pi D_1 = \frac{4}{3}\pi\left(a + \frac{3}{5}b r_1^2 + \frac{3}{7}c r_1^4\right) \qquad (53)$$

da cui segue:

$$\frac{y}{M r_1^2} = \frac{2}{5}\,\frac{a + \frac{5}{7}b r_1^2 + \frac{5}{9}c r_1^4}{a + \frac{3}{5}b r_1^2 + \frac{3}{7}c r_1^4} \qquad (54)$$

I primi membri delle (50) (51) e (54) riguardandoli come noti, abbiamo il sistema di equazioni lineari ~~nei coefficienti~~ nelle incognite a, br_1^2, cr_1^4:

$$\left.\begin{aligned} & a + b r_1^2 + c r_1^4 = e_1 \\ & a(1-\delta) + b r_1^2\left(\tfrac{5}{7} - \delta\tfrac{3}{5}\right) + c r_1^4\left(\tfrac{5}{9} - \delta\tfrac{3}{7}\right) = 0 \\ & a + \tfrac{3}{5} b r_1^2 + \tfrac{3}{7} c r_1^4 = D_1 \end{aligned}\right\} \qquad (55)$$

nella seconda delle quali si è posto:

$$\delta = \frac{5}{2}\,\frac{y}{M r_1^2} \qquad (56)$$

Poniamo inoltre:

$$\varepsilon = \frac{e_1}{D_1} \qquad (57)$$

Segue dalle (55):

$$a(1-\delta) + b r_1^2\left(\frac{5}{7} - \delta\frac{3}{5}\right) + c r_1^4\left(\frac{5}{9} - \delta\frac{3}{7}\right) = 0$$

$$a(1-\varepsilon) + b r_1^2\left(1 - \frac{3}{5}\varepsilon\right) + c r_1^4\left(1 - \frac{3}{7}\varepsilon\right) = 0$$

da cui:

$$r_1^2 b = -\frac{\frac{[illegible]}{9} - \frac{4}{7}\delta + \frac{8}{63}\varepsilon}{\frac{10}{63} - \frac{6}{35}\delta + \frac{4}{147}\varepsilon}\, a$$

$$c r_1^4 = \frac{\frac{2}{7} - \frac{2}{5}\delta + \frac{4}{35}\varepsilon}{\frac{10}{63} - \frac{6}{35}\delta + \frac{4}{147}\varepsilon}\, a$$

Segue:

$$\left.\begin{aligned} a &= l\,(175 - 189\,\delta + 30\,\varepsilon) \\ r_1^2 b &= -\,l\,(490 - 630\,\delta + 140\,\varepsilon) \\ r_1^4 c &= l\,(315 - 441\,\delta + 126\,\varepsilon) \end{aligned}\right\} \qquad (59)$$

Sostituendo nelle (55) si ha:

$$\left.\begin{aligned} 16\,\varepsilon\, l &= e_1 \\ 0 &= 0 \\ 16\, l &= D_1 \end{aligned}\right\} \qquad (60)$$

Da cui, ricordando la (57):

$$l = \frac{D_1}{16}$$

Risulta infine:

$$e = \frac{(175 - 189\delta + 30\varepsilon) D_1}{16} + \\ - \frac{(490 - 630\delta + 140\varepsilon) D_1}{16}\,\frac{r^2}{r_1^2} + \qquad (61) \\ + \frac{(315 - 441\delta + 126\varepsilon) D_1}{16}\,\frac{r^4}{r_1^4}$$

con δ e ε espressi mediante (56) e (57). Nel caso della Terra segue dalle (59):

$$\left.\begin{aligned} \delta &= 0{,}83 \\ \varepsilon &= 0{,}45 \end{aligned}\right\} \qquad (62)$$

Sostituendo questi valori nelle (61) si ha:

$$e = D_1\left(1{,}977 - 1{,}881\,\frac{r^2}{r_1^2} + 0{,}354\,\frac{r^4}{r_1^4}\right) \qquad (63)$$

La densità massima (al centro della

verrà) risulterebbe:

(64) $\quad \rho_0 = 1{,}977 D_1 = 1{,}977 \cdot 5{,}515 = 10{,}90$

$1{,}977 - 1{,}881 + 0{,}354 = 0{,}45$

$1{,}977 - \frac{3}{5} 1{,}881 + \frac{3}{7} 0{,}354 = 1{.}000$

$\frac{2}{5} 1{,}977 - \frac{2}{7} 1{,}881 + \frac{2}{9} 0{,}354 = 0{,}35$

cioè se si pone

$$\rho = D_1 \left(\alpha + \beta \frac{r^2}{r_1^2} + \gamma \frac{r^4}{r_1^4} \right) \qquad (64)$$

i coefficienti α, β e γ soddisfanno alle equazioni:

$$\left.\begin{aligned} \alpha + \beta + \gamma &= \frac{\rho_1}{D_1} \\ \alpha + \frac{3}{5}\beta + \frac{3}{7}\gamma &= 1 \\ \frac{2}{5}\alpha + \frac{2}{7}\beta + \frac{2}{9}\gamma &= \frac{3}{M r_1^2} \end{aligned}\right\} \qquad (65)$$

più semplici delle (55).

$\frac{r}{r_1}$	$\frac{\rho}{D_1}$
0	1,977
0,2	1,902
0,4	1,685
0,6	1,346
0,8	0,918
1	0,450

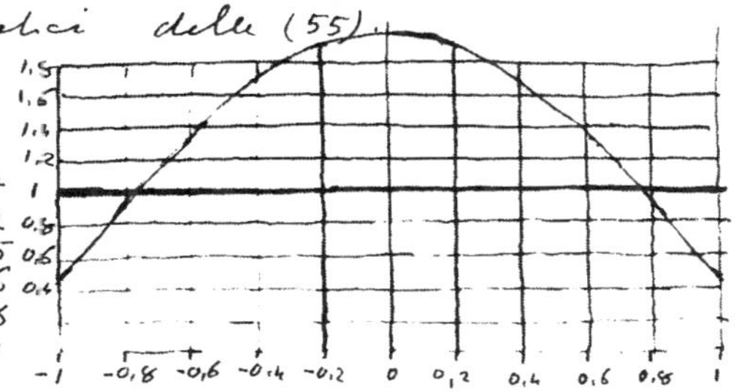

4 - Determinazione di una funzione quando sono noti i momenti di tutti gli ordini

Sia y una funzione di x:

$$y = y(x)$$

e supponiamo che per $x^2 > a^2$ si abbia $y = 0$; supponiamo inoltre che sia finito l'integrale

$$\int_{-\infty}^{\infty} |y| \, dx$$

Definiamo i momenti μ_0 ... [illegible] rispettivamente d'ordine [illegible]:

$$\left.\begin{aligned} \mu_0 &= \int y \, dx \\ \mu_1 &= \int x y \, dx \\ &\cdots \\ \mu_n &= \int x^n y \, dx \end{aligned}\right\} \qquad (1)$$

Poniamo:

$$z(t) = \int y e^{ixt} dx \qquad (2)$$

e sarà:

$$y = \frac{1}{2\pi} \int_{-\infty}^{\infty} e^{-ixt} z \, dt \qquad (3)$$

Segue da (2):

$$\frac{dz}{dt} = i \int x y e^{ixt} dx$$

$$\frac{d^n z}{dt^n} = i^n \int x^n y e^{ixt} dx \qquad (4)$$

INDEX

α particles
  scattering with a nucleus, 177
Π network method, 18
Action
  4-dimensional, 97
  quantum of, 50
Admittance, 17
Ammonia
  oscillation frequencies, 382
Angular momentum, 441, 463
Atomic eigenfunctions, 454
Atomic weight, 433
Atoms
  many-electron, 190
  two-electron, 267
Azimuthal quantum number, 245, 425
Balmer
  formula, 127
  term, 294
Bernoulli
  numbers, 242
  polynomials, 242
Bessel equation, 452
Bessel functions, 339, 354, 389, 459
  half order, 427
    zeros, 427
  integral representation, 352
Binomial coefficients, 55, 150, 195–197
Blackbody, 120
Bohr
  magneton, 249
  radius, 249
Boltzmann
  constant, 50, 249
  law, 57, 144
Born method, 345
  relativistic limit, 396
Bose-Einstein statistics, 429
Capacitance, 15, 31
  gravitational, 61, 64
Capacitor
  oscillating discharge, 31
Center of mass, 12
Central field
  selection rules, 301
Centrifugal force, 128
Charge
  circular distribution, 28
    energy of, 33
  density, 19, 33, 456
Circular integrals, 381
Clairaut
  equation, 216
  problem, 212
Clapeyron equation, 49
Classical mechanics, 51
Coil, 32
  electromagnetic energy, 32
  finite length, 36, 41
  self-induction, 32
Commutation relations, 419
Commutators, 243, 265
Conductor
  electric, 6, 18
    energy of, 11
    self-induction, 43
Conformal transformations, 74
Conjugate momenta
  for the relativistic electron, 99
Continuity equation, 456
Coulomb's law, 155
Crickets' simile, 53
Curl theorem, 14
Current density, 6, 456
  linear, 18
  surface, 18
Current intensity, 7, 15, 18, 29
Current line, 7, 30
Current
  alternate, 6, 15, 18
  circular distribution, 29
  elementary, 19
D'Alembert operator, 450
De Broglie waves, 154
Definite integrals, 69
Density
  of mass, 18, 55
Depopulation rate
  for atoms in an e.m. field, 132
Differential equations, 156
  complete set of, 311
Differential operators
  formulae involving, 14
Dirac equation, 287, 439
  central field, 454
  plane wave solution, 359
Dirac equations, 424
Dirac matrices
  and the Lorentz group, 279
Disintegration of light nuclei resonances, 414
Displacement current, 20
Divergence theorem, 14, 211
Earth
  angular velocity, 129
  flattening factor, 128
  moment of inertia, 127, 130, 220
  rotation of its axis, 130
Eccentricity, 65, 67
Electric charge, 155
Electric dipole moment
  for an atom in its ground state, 319
Electric energy of a radiation, 122

Electric field, 19
Electric lines, 15
Electric moment, 430
Electric resistivity, 7
Electromagnetic field
  energy of, 461
  radiated energy, 462
  spin of, 463
Electromagnetic oscillators, 122
Electromotive force, 7, 10, 30
Electron distribution in heavy atoms, 105
Electron gas, 252
Electron scattering by radiation, 150
Electron
  charge, 249
  electromagnetic mass, 78
  mass, 249
  origin of its mass, 78
  relativistic equations of motion, 98
  self-energy, 79
  spinning, 287
Electrostatic energy, 11, 462
Electrostatic stress, 25
Ellipsoid
  fluid, 66
  gravitational attraction by, 58
  gravitational attraction, 63
  gravitational capacitance, 61, 64
  prolate, 63
Energy conservation, 10
Entropy, 10
  for a system in equilibrium, 428
Equatorial quantum number, 245
Equilibrium
  of a rotating fluid, 66
Equipotential surfaces, 13, 58, 214, 252
Error function
  series expansion, 150
Euler constant, 333
Factorial, 47
Faraday constant, 249
Fermi-Dirac statistics, 429
Fine structure formula, 455
Fine-structure formula, 294
Formulae
  involving differential operators, 14
Fourier integral expansion, 379
Free energy
  of a diatomic gas, 434
Frequency cutoff, 125
Function determination
  from its moments, 226
Gas
  diatomic, 430
  monoatomic, 430
  perfect, 429
Gauge
  condition, 456
  Lorenz, 160, 456
  transformations, 160
Gravitating masses, 12
Gravitational potential, 127
Green function, 330
Green theorems, 14
Group velocity, 154
Group
  $O(3)$, 271, 298
  $SU(2)$, 298, 421
  continuous, 265
  infinitesimal transformations, 265
  Lorentz, 276, 419, 446
  of rotations, 271, 355, 421
  permutations, 335, 355
Hamiltonian
  for the electromagnetic field, 122
  for the relativistic motion of the electron, 97, 123
Hankel functions, 353, 389
  half order, 427
Harmonic function, 76
  polynomial expansion, 85
Harmonic functions, 76
Harmonic group, 28
Harmonic oscillator, 402
  energy levels, 90
  quantization, 87
  selection rules, 88
  zero point energy E_0, 88, 90
Heat well, 53
Heat, 8, 10
  displacement, 55, 71
  one-dimensional propagation, 71
  propagation along a bar, 53
  propagation along a finite bar, 72
  propagation in an isotropic and homogeneous medium, 71
  propagation, 20, 71
  specific, 71
  transmission coefficient, 55, 71
Homography, 14
Homothetic, 62, 66
Homothety, 2, 33
Huygens principle, 210
Hydrogen atom
  eigenfunctions, 372
  in an electric field, 322
  polarization forces, 350
  spontaneous ionization, 161
Hydrogen eigenfunctions, 465
Hysteresis
  in the limit skin effect, 26
Impedance, 17
Inertia
  moment of, 219

Infinite products, 241
Infinitesimal transformations, 422
Integral logarithm
  series expansion, 333
Integrals, 146
  definite, 69
Interaction energy, 5
Interatomic potential, 105
Joule effect, 7, 21, 27
Keplerian motion
  kinematics, 137
  perturbed (adiabatic approximation), 141
  perturbed, 137
  revolution period, 138
Laplace transformation, 348, 378
Laplace's equation, 60, 348, 353, 372
Laplacian
  in cylindrical coordinates, 85
  in parabolic coordinates, 322
  in spherical coordinates, 83
Larmor
  frequency, 249, 309
  wave number, 249
Legendre polynomials, 82, 329, 339, 435, 441
  multiplication rules, 329
Lines of force, 60
Longitude, 129
Longitudinal multipoles, 463
Lorentz
  contraction, 78
  group, 276, 279
  transformations, 419, 435
Lorenz
  gauge, 160, 456
Magnetic energy of a radiation, 122
Magnetic field, 18
  influence on the melting point, 48
Magnetic permeability, 7, 26, 30, 49
Magnetization, 18
  magnitude of, 19
Matrices, 76
Matrix
  associative product, 92
  diagonalization of a, 90
Mean distances
  of line elements, 38
  of surface elements, 38
  of volume elements, 38
Melting temperature, 49
Method
  of periodic solutions, 465
  of residues, 466
  of sources, 71
  of stationary waves, 457, 465
  of variation of parameters, 457, 462
Molecular weight, 433
Molecule
  potential in a, 252
Moment of inertia, 13, 56, 127, 219
Moments of a function, 226
Multipole radiation
  classical theory, 456
Neumann functions, 354
  half order, 427
Neutron, 425
Newton's
  gravitational constant, 127
  law, 18, 59, 212
  potential, 12, 18, 213
Observables and matrices, 133
Occupation number, 434
Operators
  improper, 369
Orthogonal functions, 377
Orthogonal transformations, 276
Oscillator
  specific heat, 50
Parabolic coordinates, 322
Paschen-Back effect, 311
Pauli's principle, 105
Perfect gas, 429
  constant R, 433
Permutations, 335
Perpetual motion, 49
Phase velocity, 154
Photon spin, 463
Planck law, 308
Planck relation, 121
Plane wave, 154
  expansion in spherical harmonics, 339
  in the Dirac theory, 359
Poisson brackets, 243, 314
Poisson equation, 214
  for the local potential in an atom, 116
Polarizability
  atomic, 117
Potential scattering, 375
Potential
  advanced, 5
  electric, 6, 11, 15
    mean, 35
  in a molecule, 252
  Newtonian, 12, 18, 213
  retarded, 4, 188, 451
Poynting theorem, 27
Pressure
  hydrostatic, 49
Probability, 55
  curves, 235
  mean error, 45
Proton, 425
Quantization rules, 87
Quasi-stationary states, 403
Quaternions, 256, 272

Radiation theory, 121, 131, 135, 144, 150
  Einstein coefficients, 144
Random variables, 193
Resistance, 9, 15, 21, 31, 44
Resistivity, 20–21
Retarded potential, 4, 188, 451
Ritz method, 351
Rotating fluid
  equilibrium, 66
Rotator, 56
  specific heat, 57
Rutherford formula
  and Born Method, 345
  and Classical Mechanics, 342, 374
Rydberg energy, 249
Rydberg frequency, 249
Rydberg wave number, 249
Scattering
  between protons and neutrons, 425
Schrödinger equation, 77, 289, 425
Schwartz theorem, 75
Schwarz formula, 192
Selection rules, 301
Self-inductance, 15
Self-induction
  coefficient of, 31
  in a coil with finite length, 36, 41
  of a circular coil, 32
  variation due to the skin effect, 43
Series expansions, 39, 148, 205
Similitude, 76
Skin effect, 6, 43
  limit, 18, 23
    hysteresis, 26
    in conductors with elliptic cross sections, 23
    in conductors with irregular cross sections, 24
  weak - in conductors with elliptic cross sections, 30
Sommerfeld
  conditions, 94
  constraints, 56
Sources
  method of, 71
Specific heat, 55, 71
  of an oscillator, 50
Speed of light, 249
Spherical functions, 328
  with spin, 385, 417, 441
Spherical harmonics, 339
Spheroid, 66
Spin functions, 369
Spin, 287
  nuclear, 430
  photon, 463
Spinors
  transformation of, 435
Stark effect, 320, 326
Static equations for a perfect charged fluid, 109
Statistical weight, 431
Stirling's formula, 147
String
  tension, 80
Strings, 80
Surface density, 58
Susceptibility (electric)
  for an atom in its ground state, 319
T–network method, 18
Thermodynamics
  statistical theory, 428
Thermoelectric cell, 10
Thomas-Fermi function, 101
  application to heavy atoms, 108
  Majorana approach, 105
  second approximation for the potential inside an atom, 116
Thomson effect, 10
Three-phase system
  unbalance, 46
Transformations
  conformal, 74
Transverse multipoles, 463
Uncertainty principle, 176
Unitary representations
  and the Lorentz group, 446
Unitary transformations
  in two variables, 255
Vector potential, 121, 123
Wallis formula, 241
Wave mechanics
  in a conservative field, 77
  variational approach, 77
Wavefront, 154
Wavepacket, 154
Waves
  damped, 20
  longitudinal, 458
  transverse, 458
Zeeman effect, 311
  anomalous, 307

Fundamental Theories of Physics

94. V. Dietrich, K. Habetha and G. Jank (eds.): *Clifford Algebras and Their Application in Mathematical Physics.* Aachen 1996. 1998 ISBN 0-7923-5037-5
95. J.P. Blaizot, X. Campi and M. Ploszajczak (eds.): *Nuclear Matter in Different Phases and Transitions.* 1999 ISBN 0-7923-5660-8
96. V.P. Frolov and I.D. Novikov: *Black Hole Physics.* Basic Concepts and New Developments. 1998 ISBN 0-7923-5145-2; Pb 0-7923-5146
97. G. Hunter, S. Jeffers and J-P. Vigier (eds.): *Causality and Locality in Modern Physics.* 1998 ISBN 0-7923-5227-0
98. G.J. Erickson, J.T. Rychert and C.R. Smith (eds.): *Maximum Entropy and Bayesian Methods.* 1998 ISBN 0-7923-5047-2
99. D. Hestenes: *New Foundations for Classical Mechanics (Second Edition).* 1999 ISBN 0-7923-5302-1; Pb ISBN 0-7923-5514-8
100. B.R. Iyer and B. Bhawal (eds.): *Black Holes, Gravitational Radiation and the Universe.* Essays in Honor of C. V. Vishveshwara. 1999 ISBN 0-7923-5308-0
101. P.L. Antonelli and T.J. Zastawniak: *Fundamentals of Finslerian Diffusion with Applications.* 1998 ISBN 0-7923-5511-3
102. H. Atmanspacher, A. Amann and U. Müller-Herold: *On Quanta, Mind and Matter Hans Primas in Context.* 1999 ISBN 0-7923-5696-9
103. M.A. Trump and W.C. Schieve: *Classical Relativistic Many-Body Dynamics.* 1999 ISBN 0-7923-5737-X
104. A.I. Maimistov and A.M. Basharov: *Nonlinear Optical Waves.* 1999 ISBN 0-7923-5752-3
105. W. von der Linden, V. Dose, R. Fischer and R. Preuss (eds.): *Maximum Entropy and Bayesian Methods Garching, Germany 1998.* 1999 ISBN 0-7923-5766-3
106. M.W. Evans: *The Enigmatic Photon Volume 5: O(3) Electrodynamics.* 1999 ISBN 0-7923-5792-2
107. G.N. Afanasiev: *Topological Effects in Quantum Mecvhanics.* 1999 ISBN 0-7923-5800-7
108. V. Devanathan: *Angular Momentum Techniques in Quantum Mechanics.* 1999 ISBN 0-7923-5866-X
109. P.L. Antonelli (ed.): *Finslerian Geometries A Meeting of Minds.* 1999 ISBN 0-7923-6115-6
110. M.B. Mensky: *Quantum Measurements and Decoherence Models and Phenomenology.* 2000 ISBN 0-7923-6227-6
111. B. Coecke, D. Moore and A. Wilce (eds.): *Current Research in Operation Quantum Logic.* Algebras, Categories, Languages. 2000 ISBN 0-7923-6258-6
112. G. Jumarie: *Maximum Entropy, Information Without Probability and Complex Fractals.* Classical and Quantum Approach. 2000 ISBN 0-7923-6330-2
113. B. Fain: *Irreversibilities in Quantum Mechanics.* 2000 ISBN 0-7923-6581-X
114. T. Borne, G. Lochak and H. Stumpf: *Nonperturbative Quantum Field Theory and the Structure of Matter.* 2001 ISBN 0-7923-6803-7
115. J. Keller: *Theory of the Electron.* A Theory of Matter from START. 2001 ISBN 0-7923-6819-3
116. M. Rivas: *Kinematical Theory of Spinning Particles.* Classical and Quantum Mechanical Formalism of Elementary Particles. 2001 ISBN 0-7923-6824-X
117. A.A. Ungar: *Beyond the Einstein Addition Law and its Gyroscopic Thomas Precession.* The Theory of Gyrogroups and Gyrovector Spaces. 2001 ISBN 0-7923-6909-2
118. R. Miron, D. Hrimiuc, H. Shimada and S.V. Sabau: *The Geometry of Hamilton and Lagrange Spaces.* 2001 ISBN 0-7923-6926-2

Fundamental Theories of Physics

119. M. Pavšič: *The Landscape of Theoretical Physics: A Global View*. From Point Particles to the Brane World and Beyond in Search of a Unifying Principle. 2001 ISBN 0-7923-7006-6
120. R.M. Santilli: *Foundations of Hadronic Chemistry*. With Applications to New Clean Energies and Fuels. 2001 ISBN 1-4020-0087-1
121. S. Fujita and S. Godoy: *Theory of High Temperature Superconductivity*. 2001 ISBN 1-4020-0149-5
122. R. Luzzi, A.R. Vasconcellos and J. Galvão Ramos: *Predictive Statitical Mechanics*. A Nonequilibrium Ensemble Formalism. 2002 ISBN 1-4020-0482-6
123. V.V. Kulish: *Hierarchical Methods*. Hierarchy and Hierarchical Asymptotic Methods in Electrodynamics, Volume 1. 2002 ISBN 1-4020-0757-4; Set: 1-4020-0758-2
124. B.C. Eu: *Generalized Thermodynamics*. Thermodynamics of Irreversible Processes and Generalized Hydrodynamics. 2002 ISBN 1-4020-0788-4
125. A. Mourachkine: *High-Temperature Superconductivity in Cuprates*. The Nonlinear Mechanism and Tunneling Measurements. 2002 ISBN 1-4020-0810-4
126. R.L. Amoroso, G. Hunter, M. Kafatos and J.-P. Vigier (eds.): *Gravitation and Cosmology: From the Hubble Radius to the Planck Scale*. Proceedings of a Symposium in Honour of the 80th Birthday of Jean-Pierre Vigier. 2002 ISBN 1-4020-0885-6
127. W.M. de Muynck: *Foundations of Quantum Mechanics, an Empiricist Approach*. 2002 ISBN 1-4020-0932-1
128. V.V. Kulish: *Hierarchical Methods*. Undulative Electrodynamical Systems, Volume 2. 2002 ISBN 1-4020-0968-2; Set: 1-4020-0758-2
129. M. Mugur-Schächter and A. van der Merwe (eds.): *Quantum Mechanics, Mathematics, Cognition and Action*. Proposals for a Formalized Epistemology. 2002 ISBN 1-4020-1120-2
130. P. Bandyopadhyay: *Geometry, Topology and Quantum Field Theory*. 2003 ISBN 1-4020-1414-7
131. V. Garzó and A. Santos: *Kinetic Theory of Gases in Shear Flows*. Nonlinear Transport. 2003 ISBN 1-4020-1436-8
132. R. Miron: *The Geometry of Higher-Order Hamilton Spaces*. Applications to Hamiltonian Mechanics. 2003 ISBN 1-4020-1574-7
133. S. Esposito, E. Majorana Jr., A. van der Merwe and E. Recami. *Ettore Majorana: Notes on Theoretical Physics*. 2003 ISBN 1-4020-1649-2

KLUWER ACADEMIC PUBLISHERS – DORDRECHT / BOSTON / LONDON